THE SCIENCE OF ELECTRONICS
DC / AC

DAVID M. BUCHLA

THOMAS L. FLOYD

Upper Saddle River, New Jersey
Columbus, Ohio

TK
7816
.B788
2005

To the best publishing team any authors could ask for:

Dennie, Jane, Kate, Lois, and Rex

Library of Congress Cataloging-in-Publication Data

Buchla, David M.
  The science of electronics. DC/AC / David M. Buchla, Thomas L. Floyd.
    p. cm.
  ISBN 0–13–087565–1 (alk. paper)
  1. Electronics–Textbooks. 2. Electronic circuits–Textbooks. I. Floyd, Thomas L. II. Title.

TK7816.B788 2005
621.381–dc22   2003069033

Editor in Chief: Stephen Helba
Acquisitions Editor: Dennis Williams
Development Editor: Kate Linsner
Production Editor: Rex Davidson
Design Coordinator: Diane Ernsberger
Cover Designer: Linda Sorrells-Smith
Cover art: Getty Images
Production Manager: Pat Tonneman
Marketing Manager: Ben Leonard

This book was set in Times Roman and Kabel by Carlisle Communications, Ltd. It was printed and bound by Courier Kendallville, Inc. The cover was printed by Phoenix Color Corp.

Copyright © 2005 by Pearson Education, Inc., Upper Saddle River, New Jersey 07458. Pearson Prentice Hall. All rights reserved. Printed in the United States of America. This publication is protected by Copyright and permission should be obtained from the publisher prior to any prohibited reproduction, storage in a retrieval system, or transmission in any form or by any means, electronic, mechanical, photocopying, recording, or likewise. For information regarding permission(s), write to: Rights and Permissions Department.

Pearson Prentice Hall™ is a trademark of Pearson Education, Inc.
Pearson® is a registered trademark of Pearson plc
Prentice Hall® is a registered trademark of Pearson Education, Inc.

Pearson Education Ltd.   Pearson Education Australia Pty. Limited
Pearson Education Singapore Pte. Ltd.   Pearson Education North Asia Ltd.
Pearson Education Canada, Ltd.   Pearson Educación de Mexico, S.A. de C.V.
Pearson Education—Japan   Pearson Education Malaysia Pte. Ltd.

10 9 8 7 6 5 4 3 2 1
ISBN: 0-13-087565-1

PREFACE

Introduction to *The Science of Electronics* Series

The Science of Electronics: DC/AC is one of a series that also includes *The Science of Electronics: Digital* and *The Science of Electronics: Analog Devices*. This series presents basic electronics theory in a clear and simple, yet complete, format and shows the close relationship of electronics to other sciences. These texts are written at a level that makes them suitable as introductory texts for secondary schools as well as technical and community college programs. Pedagogical features are numerous, and they are designed to enhance the learning process and make the use of each text in the series an enjoyable experience. The same author team prepared all of the texts and lab manuals in the series, providing consistency of approach and format.

The *DC/AC* text begins the series with some of the basic physics behind electronics, such as fundamental and derived units, work, energy, and energy conservation laws. Important ideas in measurement science, such as accuracy, precision, significant digits, and measurement units, are covered. In addition, the text covers passive dc and ac circuits, magnetic circuits, motors, and generators and instruments.

The *Digital* text introduces traditional topics, including number systems, Boolean algebra, combinational logic, and sequential logic, plus topics not normally found in an introductory text. The trend in industry is toward programmable devices, computers, and digital signal processing. A chapter is devoted to each of these important topics. Despite their complexity, these topics are treated with the same basic approach. Inexpensive laboratory exercises for these topics are included in the accompanying lab manual.

The *Analog Devices* text includes five chapters that cover diodes, transistors, and discrete amplifiers, followed by six chapters of operational amplifier coverage. Measurements are particularly important in all of the sciences, so the final chapter covers measurement and control circuits, including transducers and thyristors.

A "Science Highlight" feature opens every chapter in all textbooks. This highlight looks at scientific advances in an area related to the coverage in that chapter. Science Highlights include important related ideas in the fields of physics, chemistry, biology, computer science, and more. Electronics is a dynamic field of science, and we have tried to bring the excitement of some of the latest discoveries and advancements to the forefront for readers even as they begin their studies.

Key Features of *The Science of Electronics* Series

- A Science Highlight in each chapter looks at scientific advances in an area that is related to the coverage in that chapter.
- Easy to read and well illustrated format
- Full-color format
- "To the Student" provides an overview of the field of electronics, including careers, important safety rules, and workplace information, as well as a brief history of electronics.
- Many types of exercises reinforce knowledge and check progress, including worked-out examples, example questions, section review and chapter questions, chapter checkups, basic and basic-plus problems, and Multisim circuit simulations.
- Two-page chapter openers include a chapter outline, key objectives, list of key terms, a computer simulations directory to the appropriate figure in the chapter, and a lab experiments directory with titles of relevant exercises from the accompanying lab manual.

PREFACE

- Computer simulations throughout all of the books allow the student to see how specific circuits actually work.
- Safety Notes in the margins throughout all of the books continually remind students of the importance of safety.
- Historical Notes appear in margins throughout all of the books and are linked to ideas or persons mentioned in the text.
- An "On the Job" feature appears on selected chapter openers and discusses important aspects of employment.
- Key terms are indicated in red throughout the text, in addition to being defined in a key term glossary at the end of each chapter.
- A comprehensive glossary is included at the end of the book with all key terms and boldface terms from the chapters included.
- Important facts and formulas are summarized at the end of the chapters.

Introduction to *The Science of Electronics: DC/AC*

This text begins with a background chapter on important physics concepts needed in the study of electronics. Math skills frequently need to be honed in DC/AC classes, so we have included a chapter on math specifically for electronics (Chapter 2). In addition, if you plan to teach phasor math, the necessary trigonometry concepts are introduced at the beginning of Chapter 12. For some programs, this may not be necessary, so it can be omitted. Phasors and phasor math are not discussed in earlier chapters.

One unique approach in this text is the way we have organized information for circuit problem solving into tabular form in many examples and problems. Years of teaching experience have shown us that students understand the tabular approach because of its inherent organization of information. We have used this approach in both the DC and AC examples and problems at the end of the chapters. In some cases, the tabular approach requires students to enter the same information several times, such as the current in a series circuit, but this reinforces important concepts better than additional explanations ever can.

Another important concept that has been mentioned is the close relationship between science and electronics. All electronics instructors are very aware of this link, but many texts seem to ignore it. We have highlighted this tie throughout the text and augmented the coverage with science highlights that illustrate how electronics is rooted in science.

This text includes a range of topics to accommodate a variety of program requirements and is designed to permit freedom in choosing many of the topics. For example, some instructors may choose to omit some or all of the topics in Chapters 7 and 8, which deal with magnetism, motors, and generators. This material can be omitted with no loss in continuity.

Accompanying Student Resources

- *The Science of Electronics: DC/AC Lab Manual* by David M. Buchla
- *Companion Website, www.prenhall.com/SOE*. This website, created for *The Science of Electronics* series, includes:
  Computer simulation circuits designed to accompany selected examples in the text.
  Chapter-by-chapter quizzes in true/false, fill-in-the-blank, and multiple choice formats that enable a student to check his or her understanding of the material presented in the text.
- *Prentice Hall Electronics Supersite*. This website offers math study aids, links to career opportunities in the industry, and other useful information.

PREFACE

Instructor Resources

- *PowerPoint® Slides* on CD-ROM.
- *Companion Website (www.prenhall.com/SOE)*. For the instructor, this website offers the ability to post your syllabus online with Syllabus Manager™. This is a great solution for classes taught online, self-paced, or in any computer-assisted manner.
- *Online Course Support*. If your program is offered in a distance-learning format, please contact your local Prentice Hall sales representative for a list of product solutions.
- *Instructor's Edition*. Includes answers to all chapter questions and worked-out solutions to all problems.
- *Lab Manual Solutions*. A separate solutions manual for all experiments in the series is available.
- *Test Item File*. A test bank of multiple choice, true/false, and fill-in-the-blank questions.
- *Prentice Hall TestGen*. This is an electronic version of the Test Item File, allowing an instructor to customize tests.
- *Prentice Hall Electronics Supersite*. Instructors can access various resources on this site, including password-protected files for the instructor supplements accompanying this text. Contact your local Prentice Hall sales representative for your user name and password.

Illustration of Chapter Features

Chapter Opener
Each chapter begins with a two-page spread, as shown in Figure P–1. The left page includes a list of the sections in the chapter and an introduction to the chapter. The right page has a list of key objectives for each section, a computer simulations directory, a laboratory experiments directory, and a list of key terms to look for in the chapter. Selected chapters contain a special feature called "On the Job."

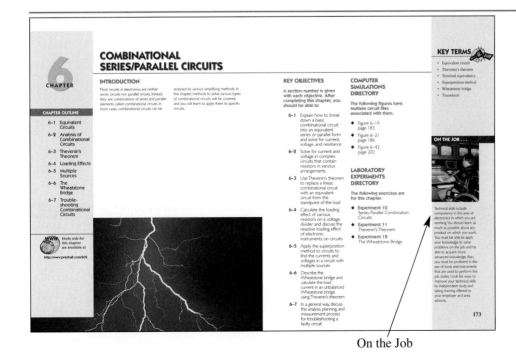

FIGURE P–1

Chapter opener.

PREFACE

Science Highlight
Immediately following the chapter opener is a *Sci Hi* feature that presents advanced concepts and science-related topics that tie in with the coverage of the text. A typical *Sci Hi* is shown in Figure P–2.

FIGURE P–2

Science highlight.

PREFACE

Section Opener

Each section in a chapter begins with a brief introduction that includes a general overview. An illustration is shown in Figure P–3. This particular page also shows an example of a Safety Note. Safety Notes are appropriately placed throughout the text.

Review Questions

Each section ends with a review consisting of five questions that emphasize the main concepts presented in the section. This feature is also shown in Figure P–3. Answers to the Section Reviews are at the end of the chapter.

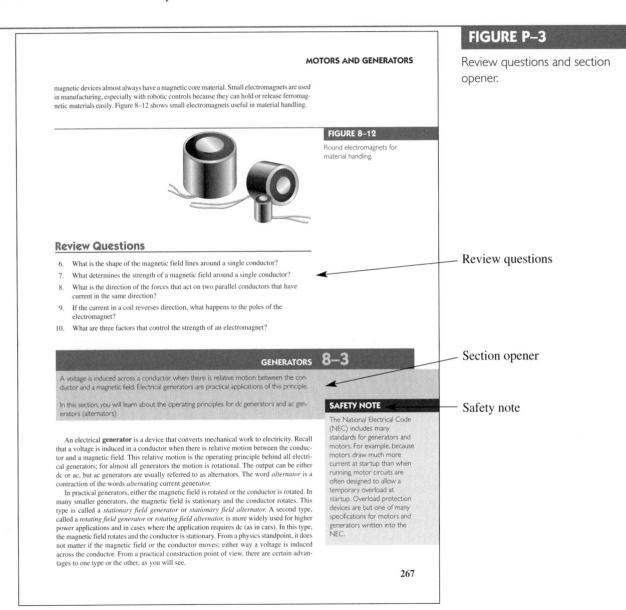

FIGURE P–3

Review questions and section opener.

PREFACE

FIGURE P–4 Worked examples and questions.

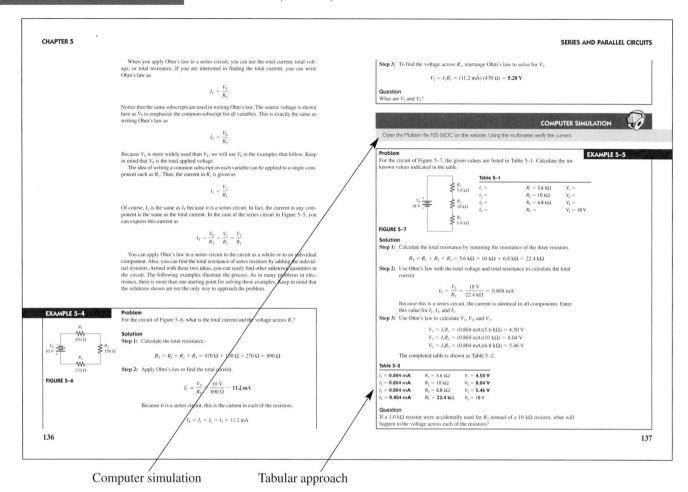

Computer simulation Tabular approach

Worked Examples and Questions
There is an abundance of worked-out examples that help to illustrate and clarify basic concepts or specific procedures. Each example ends with a question that is related to the example. Typical examples are shown in Figure P–4.

Computer Simulation
Numerous circuits are provided online in Multisim. Filenames are keyed to figures within the text with the format Fxx-yyDC and Fxx-yyAC. The xx-yy is the figure number and DC or AC represents a file for this text (DC/AC). The DC files are for Chapters 1–6; the AC files are for Chapters 10–13. (There are no simulation files for Chapters 7–9). These simulations can be used to verify the operation of selected circuits that are studied in the text. An example of a computer simulation feature is shown in Figure P–4. The Multisim circuits can be accessed on the website by going to *www.prenhall.com/SOE* and selecting this text. Choose the chapter you wish to study by clicking on that chapter, then click on the module entitled "Multisim." There you will see an introductory page with a link to the circuits for the chapter.

viii

FIGURE P–5

Troubleshooting.

Troubleshooting
Many chapters include troubleshooting techniques and the use of test instruments as they relate to the topics covered. Figure P–5 shows typical troubleshooting coverage.

Chapter Review
Each chapter ends in a special color section that is intended to highlight important chapter ideas. Several features are illustrated in Figure P–6. The chapter review includes:

- *Key Terms Glossary.* Terms that are in red within the chapter are defined here and in the glossary at the end of the book.
- *Important facts.* Major points from the chapter are summarized.
- *Formulas.* Numbered formulas in the chapter are summarized.
- *Chapter Checkup.* This is a set of multiple-choice questions with answers at the end of the chapter.
- *Questions.* This is a set of questions pertaining to the chapter. Answers to odd-numbered questions appear at the end of the book.

PREFACE

FIGURE P–6 Chapter review.

Problems
Pedagogical features continue with two levels of problems: Basic and Basic-Plus. In general, the Basic-Plus problems are more difficult than the Basic problems. Answers to all odd-numbered problems are at the end of the book.

Answers
Each chapter concludes with selected answers for questions within that chapter. These include:
- Answers to Example Questions
- Answers to Review Questions
- Answers to Chapter Checkups

End-of-Book Features

- Three Appendixes:
  Definitions of the Fundamental Units
  Table of Standard Resistor Values
  Selected figures from the lab manual in color
- Comprehensive glossary
- Answers to odd-numbered questions
- Answers to odd-numbered problems
- Index

CONTENTS

PART 1
DC

CHAPTER 1
Physics for Electronics 2
- 1–1 Electronics Today 4
- 1–2 Electrical Safety 5
- 1–3 Work and Energy 7
- 1–4 Static Electricity 10
- 1–5 The Atom 12

CHAPTER 2
Mathematics for Electronics 22
- 2–1 Scientific and Engineering Notation 24
- 2–2 Metric Prefixes 28
- 2–3 The SI System and Electronic Units 32
- 2–4 Measured Numbers 35
- 2–5 Basic Algebra Review 38
- 2–6 Graphing 44

CHAPTER 3
Electrical Quantities and Measurements . . . 56
- 3–1 Conductors, Insulators, and Semiconductors 58
- 3–2 Current and the Electric Circuit 62
- 3–3 Voltage Sources 67
- 3–4 Resistance and Resistors 72
- 3–5 Basic Electrical Measurements 81

CHAPTER 4
Ohm's Law and Watt's Law 94
- 4–1 Ohm's Law 96
- 4–2 Applications of Ohm's Law 103
- 4–3 Electrical Energy and Power 108
- 4–4 Watt's Law 111
- 4–5 Applications of Watt's Law 113
- 4–6 Nonlinear Resistance 117

CHAPTER 5
Series and Parallel Circuits 130
- 5–1 Resistors in Series 132
- 5–2 Applying Ohm's Law in Series Circuits 135
- 5–3 Kirchhoff's Voltage Law 139
- 5–4 Voltage Dividers 142
- 5–5 Resistors in Parallel 147
- 5–6 Applying Ohm's Law in Parallel Circuits 152
- 5–7 Kirchhoff's Current Law 156

CHAPTER 6
Combinational Series/Parallel Circuits 172
- 6–1 Equivalent Circuits 174
- 6–2 Analysis of Combinational Circuits 180
- 6–3 Thevenin's Theorem 188
- 6–4 Loading Effects 191
- 6–5 Multiple Sources 196
- 6–6 The Wheatstone Bridge 200
- 6–7 Troubleshooting Combinational Circuits 205

CHAPTER 7
Magnetism and Magnetic Circuits 220
- 7–1 Magnetic Quantities 222
- 7–2 Magnetic Materials 228
- 7–3 Magnetic Circuits 233
- 7–4 Transformers 237
- 7–5 Solenoids and Relays 240
- 7–6 Programmable Logic Controllers 243

CHAPTER 8
Motors and Generators 256
- 8–1 Voltage Induced Across a Moving Conductor 258
- 8–2 Force on a Current-Carrying Conductor 263

CONTENTS

8–3 Generators 267
8–4 DC Motors 275
8–5 AC Motors 279

PART 2
AC

CHAPTER 9
Alternating Current 290

9–1 The Sine Wave 292
9–2 Phasor Representation of a Sine Wave . . . 300
9–3 Nonsinusoidal Waveforms 308
9–4 Function Generators 312
9–5 Oscilloscopes 315

CHAPTER 10
Capacitors 334

10–1 Capacitance 336
10–2 Types of Capacitors 341
10–3 Series and Parallel Capacitors 345
10–4 Capacitors in DC Circuits 349
10–5 Capacitors in AC Circuits 356
10–6 Applications of Capacitors 360

CHAPTER 11
Inductors 374

11–1 Inductance 376
11–2 Types of Inductors 379
11–3 Series and Parallel Inductors 381
11–4 Inductors in DC Circuits 382
11–5 Inductors in AC Circuits 389
11–6 Applications of Inductors 392

CHAPTER 12
Series AC Circuits 402

12–1 Representing Phasor Quantities 404
12–2 Impedance 411
12–3 Series RC Circuits 415
12–4 Series RL Circuits 421
12–5 Series RLC Circuits 424
12–6 Series Resonance 428

CHAPTER 13
Parallel AC Circuits   442

13–1 Admittance 444
13–2 Parallel RC Circuits 451
13–3 Parallel RL Circuits 456
13–4 Parallel RLC Circuits 458
13–5 Parallel Resonance 462

APPENDIX A Definitions of the Fundamental Units 475

APPENDIX B Table of Standard Resistor Values 476

APPENDIX C Lab Manual Figures 477

Answers to Odd-Numbered Questions 479

Answers to Odd-Numbered Problems 486

Glossary 497

Index 503

TO THE STUDENT

Introduction to *The Science of Electronics: DC/AC*

We believe that you will find *The Science of Electronics: DC/AC* an effective tool in your preparation for a career and should find this text useful in further studies. When you have finished this course, this book should become a valuable reference for more advanced courses or even after you have entered into the job market. We hope it provides a foundation for your continued studies in electronics.

The most complicated system in electronics can be broken down into a collection of simpler circuits. These include passive circuits (resistors, capacitors, inductors) and active circuits (integrated circuits including digital and analog devices). With a solid foundation in these topics, understanding large systems is simplified. Electronics is not an easy subject, but we have endeavored to provide help along the way to make it interesting and informative and to provide you with the preparation you need for a career in this exciting field.

Many examples in the text are worked out in detail. You should follow the steps in the examples and check your understanding with the related question. Check your understanding of each section by answering the review questions and checking your answers. At the end of each chapter are summaries, glossary terms, formulas, questions, and problems as well as many answers. If you can answer all of the questions and work the problems at the end of a chapter, you are well under way toward mastering the material presented.

Careers in Electronics

The field of electronics is diverse, and career opportunities are available in many related areas. Because electronics is currently found in so many different applications and because new technology is being developed at a fast rate, the future appears limitless. There is hardly an area of our lives that is not enhanced to some degree by electronics technology. Those who acquire a sound, basic knowledge of electrical and electronic principles and are willing to continue learning will always be in demand.

The importance of obtaining a thorough understanding of the basic principles contained in this text cannot be overemphasized. Most employers prefer to hire people who have both a thorough grounding in the basics and the ability and eagerness to grasp new concepts and techniques. If you have a good training in the basics, an employer will train you in the specifics of the job to which you are assigned.

There are many types of job classifications for which a person with training in electronics technology may qualify. Common job functions are described in the Bureau of Labor Statistics (BLS) occupational outlook handbook, which can be found on the Internet at *http://www.bls.gov/oco*. Two engineering technician's job descriptions from the BLS are as follows:

- *Electrical and electronics engineering technicians* help design, develop, test, and manufacture electrical and electronic equipment such as communication equipment, radar, industrial and medical measuring or control devices, navigational equipment, and computers. They may work in product evaluation and testing, using measuring and diagnostic devices to adjust, test, and repair equipment.
- *Broadcast and sound engineering technicians* install, test, repair, set up, and operate the electronic equipment used to record and transmit radio and television programs, cable programs, and motion pictures.

(Fluke Corporation. Reproduced with permission.)

xiii

TO THE STUDENT

Many other technical jobs are available in the electronics field for the properly trained person:

- *Service technicians* are involved in the repair or adjustment of both commercial and consumer electronic equipment that is returned to the dealer or manufacturer for service.
- *Industrial manufacturing technicians* are involved in the testing of electronic products at the assembly-line level or in the maintenance and troubleshooting of electronic and electromechanical systems used in the testing and manufacturing of products.
- *Laboratory technicians* are involved in testing new or modified electronic systems in research and development laboratories.
- *Field-service technicians* repair electronic equipment at the customer's site; these systems include computers, radars, automatic banking equipment, and security systems.
- *User-support technicians* are the first people called when a computer or "high-tech" electronic equipment acts up. User-support technicians must know their product inside and out and be able to troubleshoot a product over the phone. An ability to communicate well is vital.

Related jobs in electronics include technical writers, technical sales people, x-ray technicians, auto mechanics, cable installers, and many others.

Getting a Job

Once you have successfully completed a course of study in the field of electronics, the next step is to find employment. You must consider several things in the process of finding and obtaining a job.

Resources and Considerations for Locating a Job

One consideration is the job location. You must determine if you are willing to move to another town or state or if you prefer to remain near home. Depending on the economy and the current job market, you may not have much choice in location. The important thing is to get a job and gain experience. This will allow you to move up to better or more desirable jobs later. You should also try to find out if a job is one for which you are reasonably suited in terms of personality, skills, and interest.

A good resource for locating a potential employer is the classified ads in your local newspaper or in newspapers from other cities. The Internet is also an important resource for finding employment. Many large employers have a "job-line", which is a phone service dedicated to describing current openings. Another way to find a job is through an employment agency that specializes in technical jobs, but fees may be substantial with private agencies. Often, especially at the college level, employers will come to the campus to interview prospective employees. If you know someone in a technical job, contact that person to find out about job openings at his or her company.

The Resume and Application

A resume is a record of your skills, education, and job experience. Many employers request a resume before you actually apply for a job. This allows the employer to sort through many prospective applicants and narrow the field to a few of the most qualified. For this reason, your resume is very important. Don't wait until you are about to look for a job to start working on it.

The resume is the initial way in which you present yourself to a potential employer, so it is important that you create a well-organized document. There are different types of resumes, but all resumes should have certain specific information. Here are some basic guidelines:

- Your resume should be one page long unless you have significant experience or education. A shorter resume will be more likely to be read, so shorter is generally better.

- Your identification information (name, address, phone number, e-mail address) should come first.
- List your educational achievements such as diplomas, degrees, special certifications, and awards. Include the year for each item.
- List the specific courses that you have completed that relate directly to the type of job for which you are applying. Generally, you should list all the math, science, and electronics courses when applying for a job in electronics technology.
- List all of your prior job experience, especially if it is related to the job you are seeking. Most people prefer to organize it in reverse chronological order (most recent first). Show the employer, dates of employment, and a short description of your duties.
- You may include brief personal data that you feel may be to your advantage such as hobbies and interests (especially if they relate to the job).
- Do not attach any letters of recommendation, certificates, or documents to the resume when you submit it. You may indicate something like, "Letters of recommendation available on request."

If a prospective employer likes your resume, he or she will usually ask you to come in and fill out an application. As with the resume, neatness and completeness are important when completing an employment application. Often, references are required when submitting an application. Make sure you ask the person whom you want to use as reference if it's okay with him or her. Usually a reference must be from outside your family. Previous employers, teachers, school administrators, and friends are excellent choices to use as references.

The Job Interview

The interview is the most critical part of getting a job. The resume and the application are important steps because they get you to the point of having an interview. Although you might look good on paper, it is the personal contact with a prospective employer that usually determines whether or not you get the job. Two main steps in the interview process are preparing for the interview and the interview itself.

An interview helps the employer choose the best person for the job. Your goal is to prove to the employer that you are that person. Here are several guidelines for preparing for a job interview:

- Learn as much about the company as you can. Use the Internet or other local resources such as other people who work for the company to obtain information.
- Practice answering some typical questions that an interviewer may ask you.
- Make sure you know how to get to the employer's location and try to get the name of the person who will be interviewing you, if possible.
- Dress appropriately and neatly. Get your clothes and anything you plan to carry ready the night before the interview so that you will not have to rush right before leaving. The employer is looking at you as a potential representative of the company, so make a good impression with your choice of clothes.
- Bring a copy of your resume and diplomas, certificates, or other documents you believe may be of interest to the interviewer.
- Be on time. You should arrive at the interview location at least 15 minutes before the scheduled time.

During the interview itself, there are a few things to keep in mind:

- Greet the interviewer by name and introduce yourself.
- Be polite.
- Maintain good posture.

(Getty Images)

TO THE STUDENT

- Keep eye contact.
- Do not interrupt the interviewer.
- Answer all questions as honestly as you can. If you do not know something, say so.
- Be prepared to tell the interviewer why you believe you are the best person for the job.
- Show interest and enthusiasm. Have some questions in mind and ask the interviewer at appropriate times about the company and the job, but avoid questions about salary, raises, etc.
- When the interview is over, thank the interviewer for his/her time.

After You Get the Job

The job itself is what it's all about. It is the reason that you went through the job search, preparation, and interview. Now you have to come through for the employer because you are being paid to do a job. You must retain interest and enthusiasm for the job and apply your technical skills to the best of your ability. In addition, you will need to work as a "team player," so you must bring basic social and communication skills to the job.

Safety in the Workplace

One of the most important skills a technician brings to the job is knowledge of safe operating practice and recognition of unsafe operation, especially for shock and burn hazards. Employers expect you to work safely. It is important to learn about specific safety issues for an employer when you start a new job, as there may be special requirements that depend on the job. Workplace safety is an important issue to virtually all employers. Most employers will cover issues in an initial job orientation and/or in an employee handbook. Electricity is never to be taken lightly, and safe operation in the laboratory is vital to your own well-being and that of other employees.

Companies have to comply with regulations prescribed by OSHA to ensure a safe and healthy workplace. OSHA is the Occupational Safety and Health Administration which is part of the U.S. Department of Labor. You can obtain information by accessing the OSHA website at *www.osha.gov*. In addition, you might have to be familiar with certain aspects of the National Electrical Code (NEC) and the standards issued by the American Society for Testing Materials (ASTA). Information on NEC can be obtained from the NFPA website at *www.nfpa.org* and ASTA can be accessed at *www.asta.org*. Of course, all standard electrical and electronic symbols as well as other related areas and standards are issued by IEEE (Institute of Electrical and Electronics Engineers). Their website is *www.ieee.com*.

(Fluke Corporation. Reproduced with Permission.)

Product Development, Marketing, and Servicing

The company you work for may be involved in developing, marketing, and servicing of products. Most electronic companies have a role in one or more of these activities. It is useful to understand the various steps in the process of bringing a new product to market.

1. Identifying the need
2. Designing
3. Prototyping and Evaluating
4. Producing
5. Marketing
6. Servicing

Identifying the Need
Before a new product is ever produced, someone (a person or group) must identify a problem or need, and suggest one or more ways a new product can solve the problem or meet the need. A company may be interested in the product but will want to do marketing analysis before expending money and time developing it. If the analysis shows a potential market for the product, design is started.

Designing
Most electronics projects can be implemented with more than one approach, so all ideas for the design of a new product are gathered together and evaluated. Usually this involves meetings with various specialists including electronic designers, test engineers, manufacturing engineers, as well as purchasing and marketing people. Selection considerations for a design must include input from these specialists. The best design is then chosen from the various ideas and a time line for completion is agreed upon. After a preliminary design is accomplished, components are selected based on cost, reliability, and availability.

Prototyping and Evaluating
Once the design of a product is complete, it must be proven that it will work properly by building and testing a prototype. Prototyping is often accomplished in two phases, preproduction and production. Usually technicians will construct a prototype of the design in a preliminary form. Test technicians will measure and evaluate the performance and report results to the design engineer. After approval and any necessary modification, the first prototype is converted to a production prototype by technicians. It is thoroughly tested and evaluated again before final approval for production.

Producing
A manufacturing engineer, familiar with the many processes that must be accomplished to produce a new product, is in charge of the production process. This person must verify the final layout and configuration and determine the sequence of all operations. The manufacturing engineer works with purchasing to determine cost of production and must ensure that the product will meet all necessary safety and reliability standards. Assemblers and technicians will produce the product and test the first production models to assure it works in accordance with the design specifications. Quality assurance technicians will test subsequent models to assure that products are ready to be shipped to customers.

Marketing
Once a product is produced, it is turned over to the marketing organization. The best product in the world is not much good if it can't be sold, so marketing is the key to success for a product. Marketing involves advertising, distribution, and followup. Technical marketing is a specialty requiring persons with both technical and marketing skills.

Servicing
Before a product is marketed, the cost of servicing it must be considered. A product may require servicing in order to assure maximum customer satisfaction and to create repeat business. Most electronics organizations maintain a service center to repair products returned for servicing. The type of servicing will depend on the cost of the product and replacement parts and the level of automation involved in the production process. Many products can be tested with automated test systems that can pinpoint a fault. Test engineers will set up the program to check the product and service technicians will correct problems discovered as a result of testing.

TO THE STUDENT

The Social and Cultural Impact of Electronics Technology

Electronics technology has had a tremendous impact on our lives. The rapid advancement of technology has had both positive and negative influences on our society but by far the positives outweigh the negatives. Three areas of technological advancement that have had the most impact are the computer, communications technology, and medical technology. These areas are all interrelated because computer technology has influenced most other areas.

The Computer and the Internet

In terms of the effect on society, the computer and the Internet have changed in a relatively short time the way we get information, communicate with friends and business associates, learn about new topics, compose letters, and pay bills. In electronics work, instruments can be connected to the computer and data sent anywhere in the world via the Internet. Computers have reduced personal contact to some degree because now we can e-mail someone instead of writing a letter or making a phone call. We tend to spend time on the computer instead of visiting with friends or neighbors, and it has made us, in a sense, more isolated. On the other hand, we can be in touch with just about anyplace in the world via the Internet and we can "talk" to people we have never met through chat rooms.

(Fluke Corporation. Reproduced with permission.)

Communications

Obviously, the computer is closely related to the way we communicate with each other because of the Internet. The world has "shrunk" as a result of modern communications and the instant availability of information. Television is another way in which our cultural and social values have been influenced for both good and bad. We can watch events as they unfold around the world because of television and satellite technology. Political candidates are often elected by how they present themselves on television. We learn about different cultures, people, and topics through the television media in addition to being entertained and enlightened. Unfortunately, we are also exposed to many factors that have a negative influence including significant levels of violence portrayed as "entertainment."

Another recent development that has had a major impact on our ability to communicate is the cellular telephone. Now, it is possible to contact anyone or be contacted no matter where you are. This, of course, facilitates doing business as well as providing personal benefits. Although it has added to our ability to stay in touch, it can be distracting; for example, when one is operating a motor vehicle.

(Seth Joel/Getty Images)

Medical Technology

Great advancements in medical technology have resulted in improvement in the quality and length of life for many people. New medical tools have provided health care professionals with the ability to diagnose illnesses, analyze test results, and determine the best course of treatment. Imaging technologies such as MRI, CATSCAN, XRAY, ultrasound, and others make effective diagnosis possible. Electronic monitoring equipment helps to supervise patients, keeping a constant watch on their condition. Operating rooms use electronic tools such as lasers and various video monitors to permit doctors to perform ever more complex surgical procedures and examinations.

Increased life span due to modern medical achievements has an impact on our society. People live longer, more productive lives and can contribute more to the improvement of the social and cultural aspects of life. On the other hand, medical advances have extended lives in some cases because of expensive life support, which may put a strain on the social and economic resources.

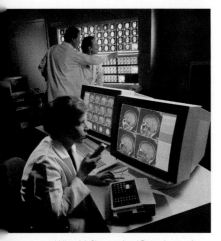
(Allan H. Shoemaker/Getty Images)

History of Electronics

Early experiments with electronics involved electric currents in vacuum tubes. Heinrich Geissler (1814–1879) removed most of the air from a glass tube and found that the tube glowed when there was current through it. Later, Sir William Crookes (1832–1919) found the current in vacuum tubes seemed to consist of particles. Thomas Edison (1847–1931) experimented with carbon filament bulbs with plates and discovered that there was a current from the hot filament to a positively charged plate. He patented the idea but never used it.

Other early experimenters measured the properties of the particles that flowed in vacuum tubes. Sir Joseph Thompson (1856–1940) measured properties of these particles, later called *electrons.*

Although wireless telegraphic communication dates back to 1844, electronics is basically a 20th century concept that began with the invention of the vacuum tube amplifier. An early vacuum tube that allowed current in only one direction was constructed by John A. Fleming in 1904. Called the Fleming valve, it was the forerunner of vacuum tube diodes. In 1907, Lee deForest added a grid to the vacuum tube. The new device, called the audiotron, could amplify a weak signal. By adding the control element, deForest ushered in the electronics revolution. It was with an improved version of his device that made transcontinental telephone service and radios possible. In 1912, a radio amateur in San Jose, California, was regularly broadcasting music!

In 1921, the secretary of commerce, Herbert Hoover, issued the first license to a broadcast radio station; within two years over 600 licenses were issued. By the end of the 1920s radios were in many homes. A new type of radio, the superheterodyne radio, invented by Edwin Armstrong, solved problems with high-frequency communication. In 1923, Vladimir Zworykin, an American researcher, invented the first television picture tube, and in 1927 Philo T. Farnsworth applied for a patent for a complete television system.

The 1930s saw many developments in radio, including metal tubes, automatic gain control, "midget" radios, and directional antennas. Also started in this decade was the development of the first electronic computers. Modern computers trace their origins to the work of John Atanasoff at Iowa State University. Beginning in 1937, he envisioned a binary machine that could do complex mathematical work. By 1939, he and graduate student Clifford Berry had constructed a binary machine called ABC, (for Atanasoff-Berry Computer) that used vacuum tubes for logic and condensers (capacitors) for memory. In 1939, the magnetron, a microwave oscillator, was invented in Britain by Henry Boot and John Randall. In the same year, the klystron microwave tube was invented in America by Russell and Sigurd Varian.

The decade of the 1940s opened with World War II. The war spurred rapid advancements in electronics. Radar and very high-frequency communication were made possible by the magnetron and klystron. Cathode ray tubes were improved for use in radar. Computer work continued during the war. By 1946, John von Neumann had developed the first stored program computer, the Eniac, at the University of Pennsylvania. One of the most significant inventions ever occurred in 1947 with the invention of the transistor. The inventors were Walter Brattain, John Bardeen, and William Shockley. All three won Nobel prizes for their invention. PC (printed circuit) boards were also introduced in 1947. Commercial manufacturing of transistors didn't begin until 1951 in Allentown, Pennsylvania.

The most important invention of the 1950s was the integrated circuit. On September 12, 1958, Jack Kilby, at Texas Instruments, made the first integrated circuit, for which he was awarded a Nobel prize in the fall of 2000. This invention literally created the modern computer age and brought about sweeping changes in medicine, communication, manufacturing, and the entertainment industry. Many billions of "chips"—as integrated circuits came to be called—have since been manufactured.

The 1960s saw the space race begin and spurred work on miniaturization and computers. The space race was the driving force behind the rapid changes in electronics that followed. The first successful "op-amp" was designed by Bob Widlar at Fairchild Semiconductor in 1965. Called the μA709, it was very successful but suffered from "latch-up" and other problems. Later, the most popular op-amp ever, the 741, was taking shape at Fairchild. This op-amp became the industry standard and influenced design of op-amps for years to come. Precursors to the Internet began in the 1960s with remote networked computers. Systems were in place within Lawrence Livermore National Laboratory that connected over 100 terminals to a computer system (colorfully called the "Octopus system" and used by one of this text's authors). In an experiment in 1969 with very remote computers, an exchange took place between researchers at UCLA and Stanford. The UCLA group hoped to connect to a Stanford computer and began by typing the word "login" on its terminal. A separate telephone connection was set up and the following conversation occurred.

> The UCLA group asked over the phone, "Do you see the letter L?"
> "Yes, we see the L."
> The UCLA group typed an O. "Do you see the letter O?"
> "Yes, we see the O."

The UCLA group typed a G. At this point the system crashed. Such was technology, but a revolution was in the making.

By 1971, a new company that had been formed by a group from Fairchild introduced the first microprocessor. The company was Intel and the product was the 4004 chip, which had the same processing power as the Eniac computer. Later in that same year, Intel announced the first 8-bit processor, the 8008. In 1975, the first personal computer was introduced by Altair, and *Popular Science* magazine featured it on the cover of the January, 1975, issue. The 1970s also saw the introduction of the pocket calculator and new developments in optical integrated circuits.

By the 1980s, half of all U.S. homes were using cable hookups instead of television antennas. The reliability, speed, and miniaturization of electronics continued throughout the 1980s, including automated testing and calibrating of PC boards. The computer became a part of instrumentation and the virtual instrument was created. Computers became a standard tool on the workbench.

The 1990s saw a widespread application of the Internet. In 1993, there were only 130 websites; by the start of the new century (in 2001) there were over 24 million. In the 1990s, companies scrambled to establish a home page and many of the early developments of radio broadcasting had parallels with the Internet. The exchange of information and e-commerce fueled the tremendous economic growth of the 1990s. The Internet became especially important to scientists and engineers, becoming one of the most important scientific communication tools ever.

In 1995, the FCC allocated spectrum space for a new service called Digital Audio Radio Service. Digital television standards were adopted in 1996 by the FCC for the nation's next generation of broadcast television. As the 20th century drew toward a close, historians could only breathe a sigh of relief. As one person put it, "I'm all for new technologies, but I wish they'd let the old ones wear out first."

TO THE STUDENT

The 21st century dawned on January 1, 2001 (although most people celebrated the new century the previous year, known as "Y2K"). The major story was the continuing explosive growth of the Internet; shortly thereafter, scientists were planning a new supercomputer system that would make massive amounts of information accessible in a computer network. The new international data grid will be an even greater resource than the World Wide Web, giving people the capability to access enormous amounts of information and the resources to run simulations on a supercomputer. Research in the 21st century continues along lines of faster and smaller circuits using new technologies. One promising area of research involves carbon nanotubes, which have been found to have properties of semiconductors in certain configurations.

ACKNOWLEDGMENTS

This text is the result of the work and the skills of many people. We think you will find this book and all the others in *The Science of Electronics* series to be valuable tools in teaching your students the basics of various areas of electronics.

Those at Prentice Hall who have, as always, contributed a great amount of time, talent, and effort to move this project through its many phases in order to produce the book as you see it, include but are not limited to Rex Davidson, Kate Linsner, and Dennis Williams. We are grateful that Lois Porter once again agreed to edit the manuscript on this project. She has done an outstanding job and we appreciate the incredible attention to detail that she has provided. Also, Jane Lopez has done a beautiful job with the graphics. Another individual who contributed significantly to this book is Doug Joksch, of Yuba College, who created all of the Multisim circuit files for the website and helped with checking the problems for accuracy. Our thanks and appreciation go to all of these people and others who were directly involved in this project.

We depend on expert input from many reviewers to create successful textbooks. We wish to express our sincere thanks to the following reviewers who submitted many valuable suggestions and provided lots of constructive criticism: Bruce Bush, Albuquerque Technical-Vocational Institute; Gary DiGiacomo, Broome Community College; Brent Donham, Richland College; J.D. Harrell, South Plains College; Benjamin Jun, Ivy Tech State College; David McKeen, Rogue Community College; Jerry Newman, Southwest Tennessee Community College; Philip W. Pursley, Amarillo College; Robert E. Magoon, Erie Institute of Technology; Dale Schaper, Lane Community College; and Arlyn L. Smith, Alfred State College.

David Buchla
Tom Floyd

THE SCIENCE OF ELECTRONICS
DC / AC

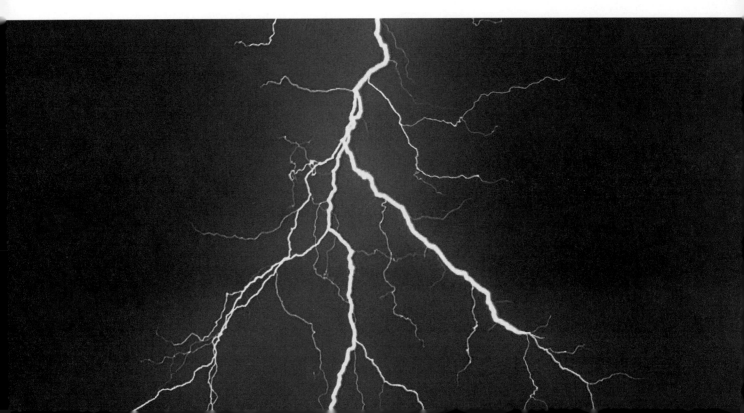

CHAPTER 1

CHAPTER OUTLINE

1-1 Electronics Today
1-2 Electrical Safety
1-3 Work and Energy
1-4 Static Electricity
1-5 The Atom

Study aids for this chapter are available at

http://www.prenhall.com/SOE

PHYSICS FOR ELECTRONICS

INTRODUCTION

Electronics began at the turn of the 20th century with the invention of the vacuum tube, a revolutionary device, which made possible radio, television, and computers. The inventions of the transistor in 1947 and the integrated circuit in 1958 began a second revolution in the field of electronics. Amazingly, as newer more complicated devices were developed, their size and cost went down. Today, electronic devices are faster, less expensive, more reliable, and far more capable than in the past.

Your study of this exciting field begins with some historical background and a discussion of physical laws that govern how all electrical circuits and devices work. Electronics is grounded in physics; throughout the text, the science behind electronics is never far away. The science highlights (called Sci Hi) are linked to the material in the chapter. Also, you will be introduced to important safety ideas in this chapter that you need to keep in mind as you work on electronics circuits.

DC/AC circuits is an important practical foundation course for the study of electronics. Concepts from DC/AC circuit studies are applied to all circuits.

KEY OBJECTIVES

A section number is given for each objective. After completing this chapter, you should be able to

1-1 Summarize milestones and career opportunities in the field of electronics

1-2 List basic safety rules when working around electricity

1-3 Explain the terms *work* and *energy* and describe the three forms of energy

1-4 Discuss Coulomb's law

1-5 Describe the structure of an atom and how different solids are chemically bonded

LABORATORY EXPERIMENTS DIRECTORY

The following exercise is for this chapter. The lab manual is entitled *The Science of Electronics: DC/AC Lab Manual* by David M. Buchla (ISBN 0-13-087566-X). © 2005 Prentice Hall.

◆ **Experiment 1**
Static Electricity

KEY TERMS

- Transistor
- Integrated circuit
- Work
- Newton-meter
- Joule
- Energy
- Potential energy
- Kinetic energy
- Rest energy
- Conductor
- Insulator
- Semiconductor

CHAPTER 1

One of the fundamental ideas of science is that energy is conserved in any process. Energy can be converted from one form to another but cannot be created or destroyed. The conservation of energy has been found to be true in every possible situation, from the atomic-size particles that interact in particle accelerators to the energy radiated from the sun. This idea of conservation is a useful concept for much of physics. For example, the energy stored in a battery as chemical energy is transformed into electrical energy and heat as the battery discharges.

Natural quantities, such as electric charge, are always conserved and are unchanged in any physical process. The total electric charge in any process is the same before and after the process. You might ask how scientists know that energy and charge are always conserved. The answer is based on careful observation and logical thinking. No experiment has ever contradicted the conservation law; however, in science nothing is taken for granted. New ideas can always replace older ideas if they are supported by observational evidence.

1-1 ELECTRONICS TODAY

Every day, your life is affected by electronics—from watching television, listening to the radio, using your computer, riding in your car, or buying a product. Because of the many facets of electronics, career opportunities are wide ranging.

In this section, you will learn about milestones and career opportunities in electronics.

Milestones

Early experimenters of electricity found that they could produce a current in an evacuated tube (a vacuum), and they conducted various experiments with vacuum tubes. Thomas Edison (1847–1931) experimented with carbon filament bulbs with plates and discovered that there was a current from the hot filament to the positively charged plate. John A. Fleming in 1904 noted that current would only go one way in a similar device that was called a Fleming valve. These devices were the forerunners of electronics. The first truly electronic device was the triode vacuum tube, invented by Lee DeForest, an American inventor, in 1907. DeForest added a third electrode to the Fleming valve and discovered that he could control the flow of electrons in the tube by applying a signal voltage to this electrode. As a result he was able to amplify the signal. The electrode was essentially a wire that he wrapped back and forth that reminded him of a football field, so he called it a "grid." During the decades that followed (until about 1950), electronics was primarily restricted to communication systems, and the vacuum tube was used as the principal active component.

In 1947, Walter Brattain, John Bardeen, and William Shockley made one of the most significant inventions of the 20th century at the Bell Telephone labs. The transistor with its small size and low-power dissipation revolutionized electronic devices. A **transistor*** is a three-terminal, solid-state device fabricated from silicon or germanium that has the ability to amplify a signal. Transistors are much more efficient and reliable than the vacuum tubes they replaced. Vacuum tubes became relegated to special applications (such as high-power broadcasting requirements).

In 1958, Jack Kilby of Texas Instruments made the first integrated circuit. Simultaneously, Robert Noyce at Fairchild Semiconductor perfected a "planar" method for interconnecting transistors on a single piece of silicon. An **integrated circuit** is a combination of circuit elements constructed together as a single unit on a small supporting material called the **substrate**. Because the integrated circuit can be mass-produced, electronic products have dropped signif-

---

* *Key terms are shown in red.*

icantly in price and have increased in reliability and functionality. The transistor is the key component of integrated circuits, and millions of miniature transistors can be put onto a single integrated circuit. Today, integrated circuits have become the mainstay of virtually all electronic products; literally billions of these circuits have been manufactured.

Electronic products are part of every aspect of our lives and are so reliable that we seldom give them a second thought. Consider just one example, your car. Not so many years ago, cars had only one electronic system, a radio. It had all of the requirements of any electronic system: an input (antenna), a process (convert radio waves to audio), and an output (sound). Today, cars have electronic systems for controlling the engine (ignition, fuel, emissions, exhaust), driveline control, vehicle motion control (cruise control), various readouts, safety (airbags, wipers, antilock brakes), entertainment, communication, and even navigation. In the future, you may see vehicles that are equipped with systems that will automatically adjust your speed as dictated by traffic. Maybe you will even ride in a computer controlled car; just tell the car where you want to go!

Careers in Electronics

Many electronics careers involve hardware in assembly, manufacturing, testing, repair, or product design. In addition to careers directly linked to products, electronics specialists work in broadcasting, industrial process control, telephone communication, computers and networking, information technology, technical writing, sales, and education.

The educational requirements depend on the specific job and the employer. Most employers require some formal training in electronics for all jobs in electronics. Assemblers and operators may obtain specialized "in-house" training from their employer. Technicians generally must have an associate's degree in electronics technology from a community college or technical school. Engineers are persons that are typically involved in design of products and normally will have a minimum of a bachelor's degree in electrical engineering from a four-year school. These requirements may be waived for persons with considerable experience.

Working conditions depend on the specific job. Most technicians work in a clean, well-lighted, and air-conditioned room; however, the specific conditions depend on the job requirements and may include outside work in some cases. The work of all technicians requires knowledge of safe working practice.

A large amount of information about electronic careers is available on the Internet. A good starting point is at the Bureau of Labor Statistics site at *http://www.bls.gov/emp/*.

Review Questions

Answers are at the end of the chapter.

1. Who was Lee DeForest? What did he invent?
2. What are two materials used in the construction of transistors?
3. What is an integrated circuit?
4. What education level is required by most employers for an electronics technician?
5. List some career opportunities for a person trained in electronics technology.

ELECTRICAL SAFETY 1-2

Electricity is a form of energy and can injure or kill if one does not follow safety procedures. The most important hazard is electrical shock. A current of a few milliamps can produce severe shock.

In this section, you will learn to recognize safety hazards and learn basic safety rules when working around electricity.

CHAPTER 1

> **SAFETY NOTE**
> **Eye Protection**
>
> You need to have eye protection with approved safety glass whenever you are doing electrical or electronic hardware repair or using tools that can produce flying cuttings such as drills, saws, hammers, air hoses, or grinders. Flying cuttings can also be produced whenever small wires are clipped from a circuit board. In addition to wearing eye protection, you should cut in a manner that directs the cuttings downward, not into the lab area.

Electricity should never be taken for granted. The most common injury associated with electrically powered devices is shock. The severity of an electrical shock depends on a number of factors, which include the voltage level, the specific path through the body, the body resistance (opposition), and the length of time of the shock. Voltage will be defined more precisely later, but for now, consider it the driving "force" of electricity. The voltage level and body resistance determine how much current is in the body. Current is the important factor in electric shock. Body resistance varies dramatically and drops significantly when the body is damp due to perspiration or the environment. One dangerous aspect of a high-voltage shock is that it can break down the skin resistance and cause an even more pronounced effect.

Voltages below 32 V are considered to be extra low voltages and generally do not present a shock hazard, but they should still be respected, particularly for high-energy sources (such as a car battery). Low voltages *can* shock under adverse circumstances (such as sweating profusely) and can generate high voltages in certain situations (such as inductive switching circuits). In the laboratory, you should avoid direct contact with any voltage source, even an extra low voltage source.

In addition to shock, electricity can cause burns and be the ignition source for fires. Thermal burns can occur even with low voltages. Jewelry can provide a low resistance path for electricity and become hot almost instantly, literally cooking the wearer. Even if the wearer is not burned, the sudden high current when jewelry contacts a voltage source can damage a sensitive circuit. For these reasons, you should never wear metallic jewelry while you work on circuits. Electricity can provide the heat necessary to ignite a fire. A fire hazard is created when a circuit is overloaded; the temptation to plug in too many electrical appliances is prevalent in electronics laboratories because of the amount of equipment. Multiple outlet boxes should never be connected to each other ("daisy-chained"); they should be connected only to a wall outlet. Fire hazards are also created when volatile chemicals, such as solvents, are used near live electrical equipment. The circuit should be deenergized before using solvents.

If an electrical fire should occur, you need to disconnect the power first. Pull the plug or trip an emergency shut off. Use only an approved extinguisher with category C for electrical fires; *never* use water. Aim the extinguisher at the base of the fire and sweep it back and forth. If an extinguisher is not available, try to smother the fire. If the room is filling with smoke, get everyone out; smoke can be deadly.

Persons who are trained in electronics must also be trained in basic safety and should be able to recognize safety hazards associated with their work. Electrical wiring in buildings must be in compliance with a set of rules and regulations known as the National Electrical Code (NEC), which is published by the National Fire Protection Association (NFPA)[*]. The NEC specifies many provisions required for safety. Electronics and electrical workers should be familiar with the basic provisions of the NEC. Two other important safety organizations are the Occupational Safety and Health Administration (OSHA)[**] and Underwriter's Laboratories (UL)[†].

You should bring to your work a constant awareness of safety and try to anticipate hazards by planning. One serious accident involved an employee in a plant that used a tape measure near an electrical distribution box. He received a serious shock when the tape measure made contact with a "hot" electrical contact. If the employee had thought about the possible consequences and planned his actions, the accident could have been avoided. In another case, a technician lost a finger when a wedding ring made contact with a high current source. A basic safety rule was forgotten!

---

[*] *http://www.nfpa.org*
[**] *http://www.osha.gov*
[†] *http://www.ul.com*

It is not possible to write a list of rules that will cover all potential hazards (like the example of the tape measure incident described previously). However, a few basic rules are important to begin with and are given as follows:

1. Never work on high-voltage electrical circuits alone.
2. If you are not sure of a procedure, ask your instructor or supervisor.
3. Know and follow the rules for the laboratory.
4. Report any hazard or unsafe condition (such as overloaded circuits, frayed cords, broken or missing ground leads).
5. Keep your work area neat to avoid hidden hazards.
6. Know the location of the emergency power cutoff switch.
7. Do not wear metallic jewelry when you work on circuits.

Review Questions

6. What are factors that determine the severity of an electrical shock?
7. What is the NEC?
8. Why should you avoid wearing metallic jewelry when you work on circuits?
9. Why isn't it okay to "daisy-chain" multiple outlet boxes?
10. Where is the electrical cutoff switch located in your laboratory?

WORK AND ENERGY 1-3

In science and engineering, it is important to use definitions that everyone agrees on for defining quantities. Terms such as *work*, *torque*, and *energy* are precise terms that have definitions agreed upon by all scientists. The measurement units are defined by international agreements.

In this section, you will learn to define the terms *work* and *energy* and learn about the three forms of energy.

Work

If you slide a heavy object across the floor or lift a box onto a counter, you have done work. These activities meet the general requirement for the definition of work. **Work** is done when a force is applied to an object and moves the object over a distance. The force applied must be measured in the same direction as the distance. In the metric system, the unit of work is the **newton-meter**. As a mathematical formula,

$$W = Fd$$

where W is work in newton-meters (N-m), F is force in newtons (N), and d is distance (or displacement) in meters (m).

You may not be familiar with the units for force and work. The metric unit for force is the newton, symbolized by N (in the English system it is the pound). A force of 1 N is fairly small; a newton is a little less than 1/4 lb (1 N = 0.225 lb). A meter (m) is a little longer than 3 ft (1 m = 3.28 ft). Thus, a newton-meter (N-m) is about the same amount of work that would be done by lifting a fourth of a pound three feet. In the English system, the unit of work is the foot-pound or ft-lb (1 N-m = 0.7376 ft-lb).

CHAPTER 1

EXAMPLE 1–1

Problem
To move a piano, three people push with a total steady force of 800 N (about 180 lb). The piano is moved 8 m. How much work is done?

Solution
$$W = Fd = (800 \text{ N})(8 \text{ m}) = \mathbf{6400 \text{ N-m}}$$

Notice how the units of newtons and meters are combined in the answer, giving the unit for work of newton-meter. In the English system, this is equivalent to 4721 foot-pounds.

Question*
How much work is done if the piano is moved with the same force a total of 12 meters?

EXAMPLE 1–2

Problem
A grocery clerk, showing his strength, holds a 22 N (approximately 5 lb) bag of potatoes at arm's length for 30 s. How much work has he done?

Solution
$$W = Fd = (22 \text{ N})(0 \text{ m}) = \mathbf{0 \text{ N-m}}$$

No work is done because there is no distance involved. Although the grocery clerk may be tired, he technically has done no work.

Question
What is the work done if the 22 N bag of potatoes is lifted 0.8 m?

HISTORICAL NOTE

The unit of energy, the joule, is named in honor of James Prescott Joule (1818–1889), a British physicist. Joule is known for his research in electricity and thermodynamics. He formulated the relationship that states that the amount of heat produced by an electrical current is proportional to the current squared times the resistance. Joule is also credited with inventing the arc welder and the displacement pump.

The newton-meter has been given the name **joule**, in honor of the British physicist James Prescott Joule, who helped develop theories in the field of thermodynamics (heat). Thus, one newton-meter (1 N-m) is equal to one joule (1 J). Either unit can be used when you express the unit for work.

Torque

Like work, torque is also the product of a force and a distance but with an important difference. To do work, the force and the distance are in the same direction. Torque tends to rotate an object either clockwise or counterclockwise. As such, the force and distance are at right angles to each other. Torque on a solid beam will cause it to flex, an important idea used on many scales.

Figure 1–1 shows an example of torque. The torque applied by the boy is clockwise and is given by the product of his weight times the perpendicular distance from the pivot point. In this case, the torque exerted by the boy is 62.5 lb × 4 ft = 250 ft-lb. The torque exerted by the girl is equal but is counterclockwise. She exerts 50 lb × 5 ft = 250 ft-lb. No motion will occur until the children shift their weight slightly to change the torque.

Energy

As you know, you do work whenever you move an object by applying a force over some distance. In order to do work, you must supply energy. **Energy** is the ability or capacity for doing work; it comes in three major forms: potential, kinetic, and rest. It is difficult to visualize energy because it is an abstract idea.

---

*Answers are at the end of the chapter.

PHYSICS FOR ELECTRONICS

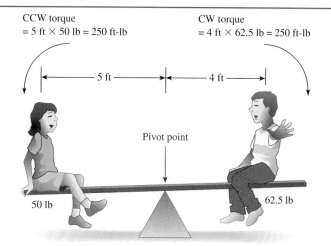

FIGURE 1–1

Torque is the product of a moment arm and a force, which act at right angles.

Under certain conditions, an object has the capacity to do work because of its position or configuration. This stored energy is called **potential energy** and is the result of work having been done to put the object in that position. In Figure 1–2, the weight has potential energy as a result of its position. Work will be done on the scientist if a force causes him to move upward. Thus, the stationary weight can do work if it is allowed to fall. In this case, the weight is said to have gravitational potential energy.

FIGURE 1–2

When the weight strikes the board, the potential energy of the weight is converted to kinetic energy of the scientist as he moves upward.

Matter can have potential energy because of its configuration. A compressed gas or a compressed spring has potential energy because either can do work. Similarly, the water stored behind a dam has potential energy because it can do work when it falls, turning a generator. Fuels like gasoline have potential energy because they can do work in an engine as a result of combustion. A battery stores chemical energy. The stored chemical energy in a battery is converted to electrical energy when it is connected to an external circuit. You can probably think of many other examples of potential energy.

Another major form of energy is called kinetic energy. **Kinetic energy** is the ability to do work because of motion. The moving matter can be a gas, a liquid, or a solid. An example of a gas in motion is simply the wind. The wind can turn a windmill to pump water. Water falling into a turbine is a liquid with kinetic energy. A moving car can also do work; therefore, it has kinetic energy because of its motion. Heat is still another basic form of kinetic energy. In this case, the motion is that of moving molecules that can do work. You can probably think of many examples of moving matter—each has kinetic energy.

CHAPTER 1

The third major form of energy is called rest energy. **Rest energy** is the equivalent energy of matter because it has mass. Einstein, in his famous equation $E = mc^2$, showed that mass and energy are equivalent. In this equation, E is energy (in joules), m is mass (in kilograms), and c is the speed of light (in meters per second). It is this equivalence between mass and energy that is accounted for when we speak of rest energy.

Unit for Energy

Because energy is the capacity to do work, energy and work are measured in the same units. Energy is usually expressed in joules (J). As you know, a joule is the work done when 1 N of force is applied over a distance of 1 m.

Review Questions

11. What two quantities are required for work to be done?
12. What is the metric unit for work?
13. What are three forms of energy?
14. What form is the energy of a battery?
15. What form is the energy of a bowling ball rolling along a flat surface?

1–4 STATIC ELECTRICITY

The electrical force between the charges in all matter is one of the fundamental forces in nature. It is considerably stronger than the gravitational force, also a fundamental force that you experience daily. Because of the nearly exact balance in positive and negative charges in most material, we never directly experience the true strength of the electrical force, which is enormous.

In this section, you will learn about Coulomb's law, which is a fundamental law that describes the electrical force between charges.

Electrical Charges

HISTORICAL NOTE

The unit of electric charge, the coulomb, is named for Charles Augustin de Coulomb (1736–1806), a Frenchman who spent many years as a military engineer. He is best known for his work on electricity and magnetism due to his development of the inverse square law for the force between two charges.

Static electricity is literally the study of stationary electrical charge. While most useful applications of electricity involve charges in motion, some applications such as electronic copiers rely on static electricity for their operation. Static electricity was historically studied much earlier than the study of moving charges.

If you rub a hard rubber rod with fur and suspend it with a silk thread, you can observe the effects of static charges. The rubber rod will acquire an excess of electrons, and become negatively charged. A second identical rubber rod also rubbed with fur will repel the first. If you now bring a glass rod that has been rubbed with a silk cloth near the rubber rod, the rubber rod is attracted to the glass rod. This is because rubbing the glass rod with the silk cloth left the glass rod with a deficiency of electrons, or a net positive charge, and the silk cloth with a negative charge. Repeating the experiment using two glass rods shows that they repel each other.

Interestingly, only a very tiny fraction of the electrons available are transferred by friction; most stay behind and do not affect the force between the rods. This experiment has its roots in the experiments of Thales of Greece (600 B.C.). The experiment shows that like charges repel and unlike charges attract. This idea is illustrated with the charged spheres shown in Figure 1–3.

Lightning is perhaps the most dramatic example of static electricity. While the exact mechanism of charge buildup is not understood, it is known that a huge number of electrons from a cloud can be transferred over time to another cloud or the ground. The spark that re-

PHYSICS FOR ELECTRONICS

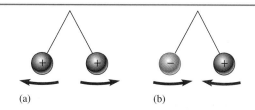

FIGURE 1-3

When the charges are the same, they repel each other. When they are different, they attract.

sults from the charge difference is known as lightning. A miniature version of lightning can be made in the laboratory with a Van de Graaff generator, which carries free electrons to a smooth conductor on top. A small Van de Graaff generator is often used to show effects of static electricity in the lab. Figure 1–4 shows a girl holding onto a Van de Graaff generator. The excess electrons in the girl's hair causes each hair to repel other hairs that also have an excess of electrons.

Coulomb's Law

Charles Augustus Coulomb (1736-1806) was the first to measure the electrical forces of attraction and repulsion of static electricity. The nature of electrical charge was not well understood at the time. Coulomb formulated the basic law that bears his name. His name was also given to the unit of charge, the coulomb, abbreviated C. Coulomb's law indicates the electrical force between two stationary point charges separated by a distance and is given as

$$F = k\left(\frac{Q_1 Q_2}{r^2}\right)$$

where F is force between the charges in newtons (N), k is the Coulomb constant, which is 8.99×10^9 N-m^2/C^2, Q_1 and Q_2 are the magnitudes of the charges in coulombs (C), and r is distance between the charges in meters (m).

Coulomb's law works for like charges or unlike charges. If the signs of both charges are the same, the force is repulsive; if the signs are different, the force is attractive. Coulomb's law is one of the most important in physics because it describes the electrical force, a fundamental force. Although Coulomb worked with large charges, his law applies to charges as small as those in an atom.

Long after Coulomb's work with static electricity, J. J. Thomson, an English physicist, discovered the electron and found that it carried a negative charge. The **electron** is the basic atomic particle that accounts for the flow of charge in solid conductors. The charge on the electron is very, very tiny, so literally trillions of electrons are involved in practical circuits. The charge was first measured by Robert Millikan, an American physicist, and found to be only 1.60×10^{-19} C.

FIGURE 1-4

A Van de Graaff generator. The girl's hair is negatively charged by an excess of electrons; therefore, each hair repels the other hairs. (Photo courtesy of Discovery Science Center, Santa Ana, CA.)

EXAMPLE 1-3

Problem
Assume that two equal positive charges of 0.002 C could be separated by 0.1 m. What is the force between the two charges? Is it an attractive or a repulsive force? Is this a practical problem?

Solution

$$F = k\left(\frac{Q_1 Q_2}{r^2}\right) = 8.99 \times 10^9 \text{ N-m}^2/\text{C}^2 \left(\frac{(0.002 \text{ C})(0.002 \text{ C})}{(0.1 \text{ m})^2}\right) = \mathbf{3.60 \times 10^6 \text{ N}}$$

CHAPTER 1

Because both charges have the same sign, the force is repulsive. This example isn't very practical but illustrates that the electrical force is huge! The force between the charges is greater than the weight of the largest steam locomotive ever built! Of course, practical lab experiments with static electricity involve very much smaller charges.

Question
What is the force if *both* charges are doubled and the distance is the same?

EXAMPLE 1–4

Problem
If the distance separating two charges is doubled, what happens to the force between them?

Solution
Coulomb's law is called a square law equation in physics because the distance term is squared in the denominator.

$$F = k\left(\frac{Q_1 Q_2}{r^2}\right)$$

When the distance, r, is doubled to $2r$, the force, F, is reduced by four times ($2^2 = 4$). Thus, the force will be 1/4 of its former value.

$$F = k\left(\frac{Q_1 Q_2}{(2r)^2}\right) = k\left(\frac{Q_1 Q_2}{4r^2}\right) = \left(\frac{1}{4}\right) k\left(\frac{Q_1 Q_2}{r^2}\right)$$

Question
What happens if the distance between two charges is tripled?

Review Questions

16. What is the rule for deciding if two charges attract each other or repel each other?
17. What is the unit for charge?
18. What is the polarity of the charge on an electron?
19. What is Coulomb's law?
20. What happens to the force between two charges if the distance separating them is halved?

1–5 THE ATOM

The ancient Greeks debated as to how far matter could be subdivided. Democritus, in the 5th century B.C., believed that all matter consisted of tiny, indivisible parts called atomos, meaning indivisible. We know today that there are only 92 naturally occurring elements in which the smallest uniquely identifiable part is the atom.

In this section, you will learn about the structure of an atom and how different solids are chemically bonded.

PHYSICS FOR ELECTRONICS

Atomic Structure

The electrical properties of materials are explained by their atomic structure. At the beginning of the 20th century, scientists conducted experiments to determine the nature of the atom. In 1913, Neils Bohr used experimental evidence to formulate a model of the atom that looked like a miniature "solar system," as shown in Figure 1–5. Unlike our solar system, the orbits of electrons are three dimensional, as pictured. At the center is a **nucleus**, which holds positively charged **protons** and uncharged **neutrons**. Around the nucleus are the negatively charged electrons that orbit only at certain discrete (separate and distinct) distances. The orbiting electrons are attracted to the positively charged nucleus with a force described by Coulomb's law. Although the Bohr model of the atom has been updated by a more complicated mathematical model, it still provides a useful mental picture of atomic structure and was the foundation of modern quantum mechanics.

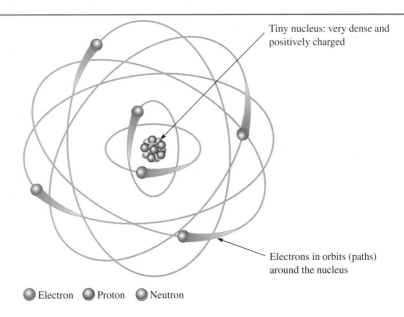

FIGURE 1–5

The Bohr "solar system" model of the atom. The nucleus is the center and the electrons move at high speeds in circular or elliptical orbits around the nucleus.

The simplest atom is the hydrogen atom. Hydrogen always has exactly one proton in its nucleus (but may have zero, one, or two neutrons). The number of protons in the nucleus is known as the **atomic number**, which is used by chemists to identify a specific element. The neutral hydrogen atom has one proton in the nucleus and one orbiting electron. Other types of atoms have more protons in the nucleus. Lithium, for example, has exactly three protons in the nucleus and usually has three neutrons (but may have four). Any atom with exactly three protons in the nucleus is lithium. This idea is illustrated in Figure 1–6.

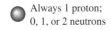

(a) Hydrogen nucleus — Always 1 proton; 0, 1, or 2 neutrons

(b) Lithium nucleus — Always 3 protons; 3 or 4 neutrons

FIGURE 1–6

The number of protons in the nucleus determines the element.

Electron Shells and Orbits

Electrons move around the positively charged nucleus. In a neutral atom, the number of electrons is the same as the number of protons. The distance from the nucleus determines the electron's energy. Electrons near the nucleus are harder to remove and, therefore, have less energy than those farther out because they are more tightly bound to the atom.

CHAPTER 1

Only certain energy levels are permitted for electrons within the atom. These energy levels are known as **shells**. The shells are designated 1, 2, 3, 4 and so forth, with 1 being closest to the nucleus. Each shell has a certain maximum permissible number of electrons and then is a **closed shell** (filled to capacity). The maximum number of electrons in a shell is given by the formula $2n^2$ where n is the shell number. The first shell ($n = 1$) can hold a maximum of only two electrons and is then closed. The second shell ($n = 2$) has room for eight electrons, which are at a higher energy level than the electrons in the first shell. The third shell ($n = 3$) can hold a maximum of 18 electrons and is at a higher energy level than the first and second shells. The fourth shell can hold up to 32 electrons. Larger atoms have even more shells that contain electrons. This concept is illustrated in Figure 1–7.

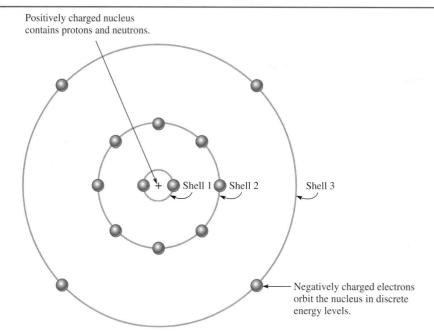

FIGURE 1–7

Energy levels increase as the distance from the nucleus of the atom increases. A neutral silicon atom (with 14 electrons in shells and 14 protons in the nucleus) is shown.

Valence Electrons, Conduction Electrons, and Ions

Electrons in orbits farther from the nucleus are less tightly bound to the atom than those closer to the nucleus. This is because the force of attraction between the positively charged nucleus and the negatively charged electron decreases with increasing distance as described by Coulomb's law.

Electrons in the outermost-filled shell are called **valence electrons**. All electrons are identical, but the outer-shell electrons are involved in chemical reactions and account for the electrical properties of an atom. The valence electrons have the highest energy and are relatively loosely bound to their parent atom (easiest to remove). The energy to remove an electron from an atom is called the **ionization potential**. For the silicon atom in Figure 1–7, the outer shell is the third one and it contains four valence electrons.

Because of the way atoms fill their shells, there are never more than eight electrons in the outer shell. Metals are those materials that have only one, two, or three outer-shell electrons that are loosely bound to the atom. Nonmetals are materials that have five or more outer-shell electrons; these electrons are held more tightly by the nucleus than in the case of metals. If an atom contains exactly four outer-shell electrons, it has properties between metals and nonmetals.

When a valence electron acquires enough thermal energy, it can break free of its parent atom. This free electron is called a **conduction electron** because it is no longer bound to any particular atom. When a negatively charged electron is freed from an atom, the remaining part of the atom has a net positive charge and is said to be a positive ion. In some chemical reactions, the freed electron attaches itself to a neutral atom (or group of atoms), forming a negative ion. Because **ions** are charged particles and can move in solutions, they account for the electrical properties of ionic solutions such as the acid found in automobile batteries.

In chemistry, **compounds** are substances in which two or more elements combine in a definite proportion. Ordinary table salt is a compound that is composed of sodium ions and chloride ions. The sodium ions are positively charged, and the chloride ions are negatively charged. As a solid, they form a lattice of alternating sodium and chloride ions, called an ionic crystal or ionic compound, as shown in Figure 1–8. The lattice has certain properties that give sodium chloride certain distinctive properties. Salt has characteristics in common with other ionic crystals; for example, it is a poor conductor of electricity because it has no free electrons. (Conductors and insulators are described further in Chapter 3).

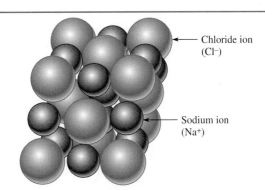

FIGURE 1–8

A crystal of sodium chloride (table salt). The crystal is held together by the electrostatic attractions between ions.

Interestingly, when salt is dissolved in water, the ions of the salt crystal are separated in the process. These charged ions are now free to move in the water solution. Although pure water and pure (dry) salt are both poor conductors, in combination they form a good conductor. The conduction properties of saltwater depend on the proportions of salt and water. Because it can be added in any proportion, the resulting solution is a called a **mixture** in chemistry.

Metallic Bonds

Except for mercury, all metals are solid at room temperature. The nucleus and inner-shell electrons of solid metals occupy fixed lattice positions. The outer-shell valence electrons are held loosely by all of the atoms of the crystal and are free to move about. This "sea" of negatively charged electrons holds the positive ions of the metal together in fixed positions, forming **metallic bonding**. With the large number of atoms in the metallic crystal, the discrete energy levels for the valence electrons are "blurred" into a band called the **valence band**. These electrons are not bound to a particular atom. It is these mobile electrons that account for the thermal and electrical conductivity of metals. A **conductor** is any material that allows the free movement of charge. Solid metals are conductors because of mobile valence electrons.

In addition to the valence energy band, the next (normally unoccupied) level in the atom is also blurred into a band of energies called the **conduction band**. Figure 1–9 compares the energy level diagrams for three types of solids. Notice that for conductors, shown in Figure 1–9(a), the conduction and valence bands are overlapping. Electrons can easily move between them by absorbing light. The movement, back and forth, of electrons between the valence and conduction bands account for the luster of metals.

CHAPTER 1

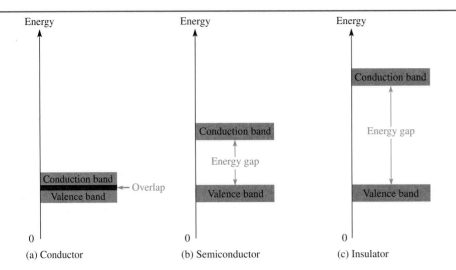

FIGURE 1–9

Energy diagrams for three categories of materials. The upper band is the conduction band; the lower band is the valence band.

Covalent Bonds

Some materials, such as diamond, also form crystals. In the case of diamond, the crystal is a three-dimensional structure held together by strong bonds between the atoms. Sharing of four valence electrons with adjacent atoms forms the bonds. This sharing of valence electrons produces the strong covalent bonds that hold the atoms together. The shared electrons in diamond are not mobile; each electron is associated with the shared bond between the atoms of the crystal. As in the case of salt (an ionic bond), there is a large gap between the valence band and the conduction band, resulting in almost no mobile electrons. As a consequence, diamonds and other covalently bonded substances are normally nonconductors of electricity. A material that prevents the free movement of charge is called an **insulator**. Figure 1–9(c) shows the energy bands for a solid insulator.

Semiconductors

Between the highly conductive metals and the insulators is a class of materials called semiconductors. A **semiconductor** is a crystalline material that has four electrons in its outer shell. The two elemental semiconductors used in electronics are silicon and germanium* although silicon is more commonly used. The valence and conduction bands are closer together than in insulators but do not overlap. In their very purest form semiconductors have few free electrons, which causes them to be poor conductors. However, their conduction characteristics can be readily changed by controlling how much of certain impurities are added to the crystal.

Review Questions

21. The atomic number of carbon is 6. How many protons are in the carbon atom?
22. What is a valence electron?
23. Why isn't dry table salt a good conductor?
24. Why does metallic bonding produce good conductors?
25. What is a semiconductor?

Interestingly, work on artificially produced diamonds holds promise that in the future very pure diamonds may be reclassified as a semiconductor in special circumstances.

CHAPTER REVIEW

Key Terms

Conductor A material that allows the free movement of charge; it can be a solid, liquid, or gas.

Energy The ability or capacity for doing work.

Insulator A material that prevents the free movement of charge; a nonconductor.

Integrated circuit A combination of circuit elements connected together as a single entity on a small piece of semiconductive material.

Joule The unit for energy or work; it is equal to one newton-meter.

Kinetic energy The energy due to motion of matter.

Newton-meter The metric unit for work; it is equal to one joule.

Potential energy The capacity to do work because of the position or configuration of a substance.

Rest energy The equivalent energy of matter because it has mass.

Semiconductor A material that has four electrons in the outer shell and whose conductive properties can be controlled by the addition of certain impurities.

Transistor A three-terminal, solid-state device fabricated from silicon or germanium that has the ability to amplify a signal or act as an electronic switch.

Work The product of a force and a distance; the force applied must be measured in the same direction as the distance.

Important Facts

- Three important milestones in the development of electronics include the inventions of the triode vacuum tube in 1907, the transistor in 1947, and the integrated circuit in 1958.
- Persons trained in electronics are trained in basic safety and must be able to recognize safety hazards associated with their work.
- Work is the product of the force that is applied to an object and the distance that an object moves ($W = Fd$). The force applied must be measured in the same direction as the distance.
- Energy is found in potential, kinetic, and rest forms and is defined as the ability or capacity for doing work. It is measured in units of newton-meters (N-m) or joules (J).
- A battery stores potential energy in the form of chemical energy.
- Coulomb's law describes the force between two stationary point charges and is given as

$$F = k\left(\frac{Q_1 Q_2}{r^2}\right)$$

- The electron is the subatomic electrical particle that accounts for the flow of charge in solid conductors.
- The nucleus of an atom holds positively charged protons and uncharged neutrons.
- Only certain energy levels are permitted for electrons within the atom. Electrons farther from the nucleus have the highest energy and are held relatively loosely compared to those closer to the nucleus.

- The energy to remove an electron from an atom is called ionization potential.
- Materials with free (mobile) electrons are conductors; those without free electrons are insulators.
- Semiconductors are crystalline materials that have four electrons in the outer shell.

Chapter Checkup

Answers are at the end of the chapter.

1. The first device that could amplify a signal was the
    (a) Fleming valve (b) triode vacuum tube
    (c) transistor (d) integrated circuit
2. Generally, the minimum educational requirement for an electronics technician is
    (a) high school (b) two-year college
    (c) four-year college
3. When you are working on "live" electrical circuits, it is okay to
    (a) wear metal jewelry (b) work alone
    (c) use solvents (d) neither (a), (b), nor (c)
4. The metric unit for work is the
    (a) newton (b) meter
    (c) newton-meter (d) foot-pound
5. A joule is the same as a
    (a) newton (b) meter
    (c) newton-meter (d) foot-pound
6. Energy can be measured using the same unit that is used for
    (a) force (b) distance
    (c) work (d) charge
7. Another word for "energy of motion" is
    (a) kinetic energy (b) potential energy
    (c) rest energy (d) work
8. The type of energy represented by a pendulum at the top of its arc is
    (a) kinetic energy (b) potential energy
    (c) rest energy
9. The potential energy stored in a charged battery is
    (a) electrical (b) mechanical
    (c) chemical (d) rest
10. The basic particles that account for the flow of charge in solid conductors are
    (a) electrons (b) protons
    (c) neutrons (d) ions
11. If the distance between two charges is increased by a factor of four, the force between the particles will be
    (a) smaller by a factor of 8 (b) smaller by a factor of 16
    (c) larger by a factor of 8 (d) larger by a factor of 16

12. If the distance between two charges is decreased by a factor of two, the force between the particles will be

    (a) smaller by a factor of 2   (b) smaller by a factor of 4

    (c) larger by a factor of 2   (d) larger by a factor of 4

13. A positive charge and a negative charge will

    (a) attract each other   (b) repel each other

    (c) neither attract nor repel each other

14. A valence electron is one that

    (a) has a positive charge   (b) is not bound to a particular atom

    (c) is shared between atoms   (d) is in the outer shell of an atom

15. The number of electrons in the outer shell of semiconductors is

    (a) one   (b) two

    (c) three   (d) four

16. The nucleus of the atom does not contain

    (a) electrons   (b) protons

    (c) neutrons

17. The maximum number of electrons that can be held in any given shell

    (a) is 2   (b) is 4

    (c) is 8   (d) depends on the particular shell

18. An example of a semiconductor is

    (a) silicon   (b) gold

    (c) saltwater   (d) iron

Questions

Answers to odd-numbered questions are at the end of the book.

1. What are the principal advantages of the transistor over the vacuum tube?
2. What is the key component used in integrated circuits?
3. Describe the typical job of electronic engineers.
4. What is the principal hazard for working with voltages below 32 V?
5. What organization publishes the National Electrical Code (NEC)?
6. Why is a neat work area important for electrical/electronics work?
7. How much work is done if 500 N of force is applied to a box, but the box does not budge?
8. What metric unit is equal to a newton-meter?
9. What is the difference between kinetic and potential energy?
10. What is rest energy?
11. What form is the energy of fuels such as oil?
12. Why does rubbing a glass rod with a silk cloth cause the rod to become positively charged?
13. Why do lab experiments with static electricity involve a very tiny amount of charge?
14. Describe Bohr's "solar system" model of the atom.

15. The hydrogen atom has a single proton and a single electron. What law describes the force between the proton and electron in the hydrogen atom?
16. What determines the energy of an electron within the atom?
17. What is meant by atomic number?
18. What is an ion?
19. What accounts for the fact that metals are good conductors but nonmetals are not?
20. How are the conductive properties of semiconductors controlled?

Basic Problems

Answers to odd-numbered problems are at the end of the book.

1. What work is done by a hoist that is used to lift a 22,000 N car by 3 m?
2. How much work is done in lifting a 150 N boy 1.5 m?
3. What is the force between two equal positive charges of 0.000 010 C if they are separated by 0.55 m?
4. (a) What is the force between two tiny charged spheres separated by 0.20 m, if each sphere is charged with 0.000 020 C?
   (b) What can you conclude about the charges on the spheres if the force is an attractive force?

Example Questions

1–1: 9600 N-m

1–2: 17.6 N-m

1–3: 1.44×10^7 N

1–4: The force is reduced by a factor of 9.

Review Questions

1. Lee DeForest was an American who invented the first vacuum tube amplifier.
2. Silicon and germanium
3. An integrated circuit is a combination of circuit elements constructed together as a unit on a small supporting material called the substrate.
4. A two-year associate's degree in electronics technology
5. Opportunities include assembly, manufacturing, testing, repair, or design of products. Others include broadcasting, industrial process control, telephone communication, computers and networking, information technology, technical writing, sales, and education.
6. The voltage level, the specific path through the body, body resistance, and the length of time of the shock
7. The National Electrical Code
8. Metallic jewelry is a conductor of electricity and heat. It can provide a ready path for a shock or can be rapidly heated in high-energy circuits. In addition, it can damage a circuit.
9. Daisy-chaining multiple outlet boxes tends to overload a circuit.

10. Answer depends on the lab. You should be aware of its location.
11. Force has to operate over a distance.
12. The newton-meter (N-m) or joule (J) is the metric unit for work.
13. Potential, kinetic, and rest are three forms of energy.
14. Potential (chemical) energy
15. Kinetic
16. Unlike charges attract; like charges repel.
17. The coulomb is the unit of charge.
18. Negative
19. Coulomb's law is a mathematical statement that indicates the force between two stationary point charges separated by a distance.
20. The force between the charges is four times larger.
21. 6
22. A valence electron is an electron in the outer shell of an atom.
23. Dry table salt does not have any mobile charges (electrons).
24. The metallic bond has a "sea" of valence and conduction band electrons that are free to move in the metallic crystal.
25. A semiconductor is a crystalline material with four electrons in the outer shell. Their conduction characteristics can be controlled with the addition of impurities to the crystal.

Chapter Checkup

1. (b)	2. (b)	3. (d)	4. (c)	5. (c)
6. (c)	7. (a)	8. (b)	9. (c)	10. (a)
11. (b)	12. (d)	13. (a)	14. (d)	15. (d)
16. (a)	17. (d)	18. (a)		

CHAPTER 2

MATHEMATICS FOR ELECTRONICS

CHAPTER OUTLINE

2–1 Scientific and Engineering Notation
2–2 Metric Prefixes
2–3 The SI System and Electronic Units
2–4 Measured Numbers
2–5 Basic Algebra Review
2–6 Graphing

INTRODUCTION

Electronics technicians and engineers must be able to apply certain fundamental scientific and mathematical principles to their work. This includes working with very large and very small numbers, reading and interpreting metric units, manipulating and solving basic equations, and graphing data. This chapter includes an overview of these important skills. This overview cannot replace rigorous mathematical studies, but it is intended to review and strengthen important skills needed for electronics work.

Although a number of equations are introduced in this chapter, they are intended to reinforce and explain mathematical skills used in electronics work. These skills include expressing quantities with metric prefixes and scientific notation, manipulating equations, and graphing.

Study aids for this chapter are available at

http://www.prenhall.com/SOE

KEY OBJECTIVES

A section number is given for each objective. After completing this chapter, you should be able to

2–1 Perform mathematical operations using scientific and engineering notation

2–2 List the key engineering metric prefixes and convert between them

2–3 Describe the basic SI system and the SI units for voltage, current, resistance, and power

2–4 Explain how to report measured data with the proper number of significant digits

2–5 Manipulate linear algebraic equations to solve for an unknown variable

2–6 Describe the steps for preparing a linear graph

LABORATORY EXPERIMENTS DIRECTORY

The following exercises are for this chapter. The lab manual is entitled *The Science of Electronics: DC/AC Lab Manual* by David M. Buchla (ISBN 0-13-087566-X). © 2005 Prentice Hall.

◆ **Experiment 2**
Constructing and Observing a Circuit

◆ **Experiment 3**
Graphing Data from an Experiment

KEY TERMS

- Scientific notation
- Engineering notation
- Ampere
- Volt
- Ohm
- Watt
- Accuracy
- Error
- Precision
- Significant digits
- Variable
- Constant
- Coefficient
- Independent variable
- Dependent variable

CHAPTER 2

Coulomb's law, introduced in Chapter 1, is similar in form to other laws in physics such as Newton's universal law of gravity. Both are said to be *inverse square laws*. In both cases, the force between two objects changes by the square of the distance of separation. The distance of separation is given as r, so the equation has an r^2 term in the denominator, which is why it is called an inverse square law. Thus, if the distance between two bodies such as the earth and the moon were doubled, the force would be one-fourth as much. Another law in physics that follows this same idea is the intensity of light from a point source such as a star. If two stars with equal intrinsic brightness are different distances from us, we can apply the inverse square law to estimate the ratio of distances.

2–1 SCIENTIFIC AND ENGINEERING NOTATION

Very large and very small numbers are often needed in electronics. These numbers are generally written with either scientific notation or engineering notation. In both of these methods, a number is written as an ordinary number multiplied by a power of ten.

In this section, you will learn to use scientific and engineering notation to perform mathematical operations on numbers.

A common requirement in electronics work is to express very large or very small numbers. For example, the number of cycles that occur each second from a k-band radar is approximately 20,000,000,000. The time between cycles is only 0.000 000 000 050 s (seconds). Such numbers that are expressed with a decimal point and a number of zeros as needed are called **fixed point** (or fixed format). They adhere to rules for place value associated with basic number systems. The decimal point cannot be moved without changing the value. (Some calculators refer to this format as NORM). Fixed-point numbers are not convenient for very large and very small quantities. Numbers with more than ten digits cannot even be entered in standard form into most calculators. Scientific notation and engineering notation were developed to simplify writing very large and very small numbers.

Scientific Notation

Scientific notation is a means of expressing very large or very small numbers in concise form using power of ten notation. Ten is the base for numbers expressed in scientific notation because it is the base of our counting system and simplifies converting to or from fixed-point form.

To express a number using scientific notation, write the number as an ordinary number from 1 to 10 times a power of ten. The power of ten shows how much and which way to move the decimal point. When you convert a fixed number to scientific notation by moving the decimal point to the left, the power of ten is positive; if you move the decimal point to the right, the power of ten is negative. The following examples illustrate the conversion between fixed format and scientific notation.

$$57{,}000{,}000. = 5.7 \times 10{,}000{,}000. = 5.7 \times 10^7$$

Notice that the decimal point has been moved seven places to the left to make the first part of the number between 1 and 10. To retain the same value, the number 5.7 is multiplied by 10,000,000; this is the same as 10^7. A calculator will generally show this number as 5.7 [07] if it is in scientific (SCI) mode.

For a number less than 1, the decimal point is moved to the right to convert from the fixed value to scientific notation.

$$0.000\,042 = 4.2 \times 0.000\,01 = 4.2 \times 10^{-5}$$

This time, the exponent has a negative sign to indicate that the number is less than 1. If 4.2 is multiplied by 0.000 01, the result is equal to the original fixed-point number. In scientific notation, the exponent of 10 is -5, which indicates that the decimal has been moved five places to the right. In SCI (scientific) mode, a calculator will show this number as 4.5^{-05} where the number -05 is understood to be a power of ten. On some calculators, the display will show an E (for Exponent), for example, 4.2 E-5.

Don't confuse a negative sign in the exponent with a negative number. Both numbers in the examples given previously are positive numbers. The number 4.2×10^{-5} is simply a tiny positive number whereas 5.7×10^{7} is a large positive number. To show a negative number, the minus sign is written in front of the first part. Thus, -5.1×10^4 and -9.2×10^{-5} are *both* negative numbers.

A special case arises when the power of ten is zero. Sometimes, a power of zero occurs in a problem because of a mathematical operation with exponents. Any number raised to the zero power is 1. Strictly speaking, the conversion of a fixed-format number that is between 1 and 10 to scientific notation requires that it be multiplied by 10^0. Thus, a number such as 3.2 is represented in scientific notation as 3.2×10^0. A calculator that is in SCI or ENG (engineering) mode will show this as $3.2^{\,00}$ (or 3.2 E 00). Keep in mind that the superscripted ^{00} is the power of 10. Because $10^0 = 1$, it is unnecessary (and rarely done) to show the power of ten in this special case if you are writing the number.

EXAMPLE 2–1

Problem
Convert each of the fixed-format numbers to scientific notation:

(a) 0.000 33
(b) $-290,000$
(c) $-0.000\,005\,2$
(d) 11,480
(e) 5.1

Solution
The arrow indicates the number of places the decimal point is moved.

(a) $0.000\,33 = 0.000\,33 = \mathbf{3.3 \times 10^{-4}}$
(b) $-290,000. = -290,000. = \mathbf{-2.9 \times 10^{5}}$
(c) $-0.000\,005\,2 = -0.000\,005\,2 = \mathbf{-5.2 \times 10^{-6}}$
(d) $11,480 = 11,480. = \mathbf{1.148 \times 10^{4}}$
(e) $5.1 = 5.1 \times 10^0 = \mathbf{5.1}$

Question*
In what order would the five numbers in the problem be if they were arranged from the largest positive to the largest negative number?

Answers are at the end of the chapter.

CHAPTER 2

Sometimes you need to convert a number written in scientific notation to a fixed number. In this case, the process is simply reversed. The sign of the exponent shows the direction to move the decimal point; the value shows how many places.

EXAMPLE 2–2

Problem
Convert each of the scientific notation numbers to fixed format:

(a) 6.3×10^{-4} (b) -9.8×10^{5}
(c) -6.6×10^{-2} (d) 1.6×10^{7}
(e) 4.8×10^{0}

Solution
(a) $6.3 \times 10^{-4} = 0.000\,63 = \mathbf{0.000\,63}$
(b) $-9.8 \times 10^{5} = -980{,}000. = \mathbf{-980{,}000}$
(c) $-6.6 \times 10^{-2} = -0.066 = \mathbf{-0.066}$
(d) $1.6 \times 10^{7} = 16{,}000{,}000. = \mathbf{16{,}000{,}000}$
(e) $4.8 \times 10^{0} = 4.8 \times 1 = \mathbf{4.8}$

Question
Express the number of electrons in a coulomb (6.25×10^{18}) as a fixed number.

Adding and Subtracting in Scientific Notation

If you want to add or subtract numbers written in scientific notation, both numbers must have the same exponent, which will also be the exponent in the answer. If the numbers do not already have the same exponent, then one of the numbers must be converted to the same exponent as the other before the arithmetic is performed. Then both numbers and the answer will have that exponent. For example, assume you want to add 1.50×10^{4} and 3.0×10^{5}.

$$\begin{array}{ll} \text{Convert } 1.5 \times 10^{4} \text{ to} & 0.15 \times 10^{5} \\ \text{Then add} & \underline{3.00 \times 10^{5}} \\ & 3.15 \times 10^{5} \end{array}$$

Notice that all exponents, including the answer, are the same.
Alternatively, you can convert 3.00×10^{5} to 30.0×10^{4}.

$$\begin{array}{ll} \text{Convert } 3.00 \times 10^{5} \text{ to } 30.0 \times 10^{4} \\ \text{Then add} & \underline{1.5 \times 10^{4}} \\ & 31.5 \times 10^{4} \end{array}$$

The answers are equivalent for both methods, but only the first answer is in scientific notation.

A calculator handles addition of numbers written in scientific notation automatically. Although there are a number of different calculators on the market, most work with an algebraic operating system (AOS) that allows numbers to be entered and manipulated in a manner described in this text. AOS calculators work with precedence rules that are equivalent to the way equations are evaluated. Various calculators have slightly different steps for certain calculator operations. To simplify explanations, key strokes given in this book are based on the Texas Instruments TI-36X calculator shown in Figure 2–1. Most other AOS calculators will be similar; however, you should consult your owner's manual if you aren't sure.

MATHEMATICS FOR ELECTRONICS

Like all scientific calculators, the TI-36X can show numbers in more than one format. If you select floating-decimal display (press [3rd] [5] FLO), the display will show up to ten digits, a sign, and a decimal point. The decimal point can be in any location on the display, hence the term **floating-point**. If the number is too large or too small for this format, it is automatically displayed in scientific notation. You can also select scientific notation directly (press [3rd] [6] SCI).

For the TI-36X calculator, the keystrokes for the addition problem given previously are [1] [.] [5] [EE] [4] [+] [3] [.] [0] [EE] [5] [=]. The answer will be shown in the display as 315000 if floating-point is selected or as 3.15 ^{05} if scientific notation is selected. In scientific notation, the 05 is shown to the right as a superscript, but it is understood to represent the exponent of ten (10^5). Some calculators will show an E before the exponent (3.15 E05).

FIGURE 2–1

The TI-36X calculator. (Photo Courtesy of Texas Instruments.)

Engineering Notation

Engineering notation is used much more often in electronics work than scientific notation because the exponents can easily be converted into the standard metric prefixes used in engineering. Engineering notation is like scientific notation; it also shows a number multiplied by a power of ten. In **engineering notation**, the exponent can only take on values that are evenly divisible by three (and zero). Thus, exponents of $-6, -3, 0, 3,$ and 6 are all valid in engineering notation, but exponents of $-5, -4, 1, 2, 4$ are not valid. Because the exponent can only be a number that is a multiple of three, the restriction for scientific notation that the first number is from 1 to 10 does not apply. By convention, the first part of a number written in engineering notation is generally from 1 to 1000. For example, the number 295,000,000 in engineering notation is 295×10^6. You may find in a given calculation that it is more convenient to write this as 0.295×10^9. This form is equivalent but not standard.

EXAMPLE 2–3

Problem
Convert each of the following fixed-point numbers to engineering notation. Show the first part as a number between 1 and 1000.

(a) 0.000 33
(b) −290,000
(c) −0.000 005 2
(d) 11,480
(e) 5.1

Solution

(a) $0.000\ 33 = 0.000\ 330. = \mathbf{330 \times 10^{-6}}$

(b) $-290,000 = -290,000. = \mathbf{-290 \times 10^3}$

(c) $-0.000\ 005\ 2 = -0.000\ 005\ 2 = \mathbf{-5.2 \times 10^{-6}}$

(d) $11,480 = 11,480. = \mathbf{11.48 \times 10^3}$

(e) $5.1 = 5.1 \times 10^0 = \mathbf{5.1}$ (not necessary to show exponent)

On the TI-36X calculator, select the ENG format by pressing [3rd] [+] ENG. Then simply key in the number and press [=].

Question
Which of the answers are correct for both scientific notation and engineering notation?

CHAPTER 2

Review Questions

Answers are at the end of the chapter.

1. What is meant by *floating-point*?
2. What is the advantage of power-of-ten notation over ordinary numbers?
3. What is the value of 10^0?
4. How does engineering notation differ from scientific notation?
5. Why is engineering notation preferred for electronics work?

2–2 METRIC PREFIXES

In order to express very large and very small quantities, a prefix can be used in conjunction with the measurement unit. Metric prefixes are based on the decimal system and stand for powers of ten. They are widely used in electronics because many quantities are either very large or very small.

In this section, you will learn the key engineering metric prefixes and learn to convert between various metric prefixes.

The metric system is based on powers of ten. The names and abbreviations of common metric prefixes, together with the power of ten they stand for, are shown in Table 2–1. These prefixes are used in conjunction with all measurement units to indicate a large or small quantity. For electrical measurements, engineering prefixes that represent an exponent that is a multiple of three are preferred.

TABLE 2–1

Names and abbreviations of metric prefixes. Engineering prefixes and their symbols are shown in red.

Number	Metric prefix	Symbol
$1,000,000,000,000,000 = 10^{15}$	peta	P
$1,000,000,000,000 = 10^{12}$	tera	T
$1,000,000,000 = 10^9$	giga	G
$1,000,000 = 10^6$	mega	M
$1,000 = 10^3$	kilo	k
$100 = 10^2$	hecto	h
$10 = 10^1$	deca	da
$1 = 10^0$	—	—
$0.1 = 10^{-1}$	deci	d
$0.01 = 10^{-2}$	centi	c
$0.001 = 10^{-3}$	milli	m
$0.000001 = 10^{-6}$	micro	μ
$0.000000001 = 10^{-9}$	nano	n
$0.000000000001 = 10^{-12}$	pico	p
$0.000000000000001 = 10^{-15}$	femto	f

Use of Metric Prefixes

Metric prefixes are widely used to express large or small quantities. For example, a large power station can produce 1,600,000,000 W. This can be expressed as 1.6×10^9 W. More commonly, a large number will be expressed with a metric prefix affixed to the measurement

MATHEMATICS FOR ELECTRONICS

unit. Since 10^9 is equal to the metric prefix *giga* (G), the metric prefix can be used instead of 10^9. Thus, the power generated can be expressed as 1.6 GW (1.6 gigawatts). Alternatively, the number could be expressed as 1600 MW (1600 megawatts). As you can see, conversion between units simply involves moving the decimal point.

As another example, a small electrical current of 0.000 010 A (amperes) can be expressed as 10×10^{-6} A. The metric prefix *micro* (μ) is used for 10^{-6}, so 0.000 010 A becomes 10 μA (10 microamps) when the decimal point is moved six places to the right.

EXAMPLE 2–4

Problem
Express each quantity using a metric prefix.

(a) 0.000 027 A (b) 81,000 Ω
(c) 0.000 000 010 s (d) 0.002 3 m

Solution
(a) 0.000 027 A = 27×10^{-6} A = **27 μA**
(b) 81,000 Ω = 81×10^3 Ω = **81 kΩ**
(c) 0.000 000 010 s = 10×10^{-9} s = **10 ns**
(d) 0.002 3 m = **2.3 mm.** Notice that in this case, the first *m* stands for the metric prefix *milli* and the second *m* stands for the base unit *meter*.

Question
Express 0.005 s with a metric prefix.

Entering Numbers on the Calculator

To enter a number expressed in engineering notation, use the [EE] key on your calculator (it is labeled [EXP] on some calculators). The exponent can be entered as a negative value by pressing the change sign [+/−] key on most calculators, including the TI-36X.

EXAMPLE 2–5

Problem

(a) Enter 47 kΩ (47×10^3 Ω) on your calculator.
(b) Enter 260 μA (260×10^{-6} A) on your calculator.

Solution
For the TI-36X calculator,

(a) Press the key sequence: [4] [7] [EE] [3]
(b) Press the key sequence: [2] [6] [0] [EE] [+/−] [6]

Question
How is a negative number entered in the calculator?

Conversion Between Metric Prefixes

Sometimes it is useful to convert a quantity with a given metric prefix to one with a different prefix. For example, a 470 kΩ (kilohm) resistor is in series with a 2.2 MΩ (megohm) resistor and you want to find the total resistance by adding the values. Before adding, the

CHAPTER 2

prefixes must be the same. In this case, the 470 kΩ resistor can be converted to a 0.47 MΩ resistor and then added to the 2.2 MΩ value for a total of 2.67 MΩ.

The process for conversion between prefixed units involves moving the decimal point.

1. When converting from a larger unit to a smaller unit, move the decimal point to the right. Remember, a smaller unit means the number must be larger.
2. When converting from a smaller unit to a larger unit, move the decimal point to the left. Remember, a larger unit means the number must be smaller.
3. Determine the number of places that the decimal point is moved by finding the difference in the powers of ten of the units being converted.

For example, when converting from milliamperes (mA) to microamperes (μA), move the decimal point three places to the right because there is a three-place difference between the two units (mA is 10^{-3} A and μA is 10^{-6} A). A microampere is a smaller unit than a milliampere, so the number of microamperes will be larger than the equivalent number of milliamperes.

A way to visualize the conversion process is shown in Table 2–2. Any number can be converted from one engineering prefix to another (or no prefix) using the table. As an example, a unit that commonly has a very small value is the farad, abbreviated F. If you want to convert the number 0.000 000 010 F to one with any other metric prefix, you could write it on the line with the decimal point lined up under the 10^0 column as shown in the first line of the example. The decimal point is then placed under any desired metric prefix and that prefix is read with the converted unit. In the second line of the example, the number is 0.010 μF and in the third line it is 10,000 pF. All of these are ways to express the same value; that is, 0.000 000 010 F = 0.010 μF = 10,000 pF.

TABLE 2–2

Metric conversions. Apply the prefix located above the decimal point to the number.

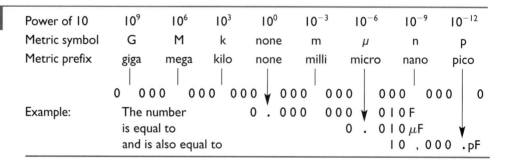

Power of 10	10^9	10^6	10^3	10^0	10^{-3}	10^{-6}	10^{-9}	10^{-12}
Metric symbol	G	M	k	none	m	μ	n	p
Metric prefix	giga	mega	kilo	none	milli	micro	nano	pico

Example: The number 0 000 000 000 0.000 000 000 0 F
is equal to 0.000 000 010 F → 0.010 μF
and is also equal to 10,000 pF

EXAMPLE 2–6

Problem
Convert each of the values in the Change column to the ones shown in the To column of Table 2–3.

Table 2-3

Change	To	Change	To
2.2 ms	s	50,000 W	kW
100 pF	μF	910 Ω	kΩ
0.0015 s	ms	200 mV	V
470 kΩ	MΩ	0.043 m	mm
10 MHz	GHz	86 ns	μs

MATHEMATICS FOR ELECTRONICS

Solution
See Table 2–4.

Table 2–4

Change	To	Change	To
2.2 ms	0.0022 s	50,000 W	50 kW
100 pF	0.000 100 µF	910 Ω	0.910 kΩ
0.0015 s	1.5 ms	200 mV	0.200 V
470 kΩ	0.47 MΩ	0.043 m	43 mm
10 MHz	0.010 GHz	86 ns	0.086 µs

Question
How can you express 220 kΩ using the prefix *mega*?

Adding and Subtracting Numbers with Metric Prefixes

To add or subtract numbers with metric-prefixed units, you must make the prefixes be the same; the answer will also have this same prefix. For example, if you wanted to subtract 330 kΩ from 1.5 MΩ, you need to convert one of the numbers to the prefix of the other. If you convert 330 kΩ to 0.33 MΩ, then the problem is expressed as

$$1.5 \text{ M}\Omega - 0.33 \text{ M}\Omega = 1.17 \text{ M}\Omega$$

If, instead, you convert 1.5 MΩ to 1500 kΩ, then the problem is written as

$$1500 \text{ k}\Omega - 330 \text{ k}\Omega = 1170 \text{ k}\Omega$$

The answers are equivalent and either is correct.

On a calculator, you can enter each metric prefix as a number in engineering notation without having to manually convert both numbers to the same power of ten; the calculator will take care of the decimal points automatically. The sequence for the TI-36X calculator is

To see the result in engineering notation, press [3rd] [ENG/+]. The display will show 1.17 ⁰⁶, which is equivalent to 1.17 MΩ.

Review Questions

6. What are the engineering prefixes that mean 10^3, 10^6, 10^9, and 10^{12}?
7. What are the engineering prefixes that mean 10^{-3}, 10^{-6}, 10^{-9}, and 10^{-12}?
8. When converting from a unit with a larger metric prefix to one with a smaller metric prefix, which way do you move the decimal point?
9. Add 0.022 µF and 1500 pF. Express the answer in nF.
10. Add 270 kΩ and 1.5 MΩ. Express the answer in kΩ.

CHAPTER 2

2-3 THE SI SYSTEM AND ELECTRONIC UNITS

The need for an international standard became apparent in the 19th century. In 1875, at a conference in Paris called by the French, 18 countries signed a treaty to set up and maintain a standard body of weights and measures. Today, the International System of Units (Le Système International d'Unites, abbreviated SI) is an outgrowth of that treaty.

In this section, you will learn about the SI system and learn the SI units for voltage, current, resistance, and power.

Measurement Units

Any given quantity is measured with certain measurement units. Measurement units are the means of specifying the magnitude of a quantity. For example, speed can be measured in kilometers per hour. To ensure consistent usage, measurement units must mean the same thing to everyone using them. The definition for various units are modified and improved from time to time as new standards and refined measuring techniques become available.

The dominant system of measurement units in use today is called the International System of Units (*Le Système International d'Unites,* abbreviated SI). The SI system is based on seven **fundamental units** and two **supplementary units**, which are the "building blocks" of all measurements. Table 2–5 lists the fundamental SI units and Table 2–6 lists the supplementary SI units. Definitions of the fundamental units are given in Appendix A.

TABLE 2–5

Fundamental SI units.

Quantity	Base unit	Abbreviation
length	meter	m
mass	kilogram	kg
time	second	s
electric current	ampere	A
temperature	kelvin	K
luminous intensity	candela	cd
amount of substance	mole	mol

TABLE 2–6

Supplementary SI units.

Quantity	Base unit	Abbreviation
plane angle	radian	r
solid angle	steradian	sr

In addition to the fundamental and supplementary units, 27 derived units are part of the SI system. A derived unit is one that uses two or more fundamental or supplementary units in its definition. For example, the SI unit for energy is the joule, introduced in Section 1–3. A joule uses kilograms, meters, and seconds in its definition. All of these are fundamental units (for mass, length, and time). Another example of a derived unit is that of velocity, which is measured as meters per second, both fundamental units. Any measure of rate must involve a time unit.

Basic Electronic Quantities and Units

Current (*I*)

Current is the fundamental electrical unit defined in the SI standards. **Current** is defined as a rate of charge flow equal to one coulomb per second, as illustrated in Figure 2–2. (This is equivalent to 6.25×10^{18} electrons per second passing a point in one second, an enormous number of electrons.)

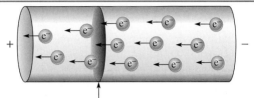

FIGURE 2–2

Current is defined as the charge per time passing a point.

If one coulomb of electrons pass through the cross-sectional area in 1 second, the current is 1 ampere.

The unit of current is the **ampere (A)**. Keep in mind that the word *rate* always involves *time*. The defining equation for current is

$$I = \frac{Q}{t} \qquad (2\text{–}1)$$

where *I* is current in amperes, *Q* is charge in coulombs, and *t* is time in seconds. This equation is rarely applied to actual circuit work because the quantity of charge is generally not as important as the current itself. For practical reasons, the *rate* charge moves past a point is defined rather than the quantity of charge. Thus, the SI fundamental electrical unit (current) has another fundamental unit (time) in its definition.

It is a simple matter to measure current directly. Current is measured by placing a meter "in line" in a circuit. Thus, current is said to be through the meter, never across the meter. In typical electronic circuits, currents are much smaller than an ampere, although in power applications, a few amperes is not uncommon.

HISTORICAL NOTE

The unit of electrical current, the ampere, is named in honor of Andre Marie Ampere (1775–1836). In 1820 Ampere, a Frenchman, developed a theory of electricity and magnetism that was fundamental for 19th century developments in the field. He was the first to build an instrument to measure charge flow (current).

Voltage (*V*)

Another important electrical quantity is voltage, named in honor of Alessandro Volta, who experimented with batteries. **Voltage** is the work required (or energy expended) to move a unit of positive charge from a negative point to a more positive point. In this definition, work is done to move the charge just as energy is required to move a car up a hill.

The measurement unit for voltage is the **volt (V)**. Like all electrical units except the ampere, the volt is a derived unit, which can be derived from fundamental units of length, mass, time, and current. The defining equation for voltage includes both work (or energy) and charge.

$$V = \frac{W}{Q} \qquad (2\text{–}2)$$

where *V* is voltage in volts, *W* is work (or energy) in joules, and *Q* is charge in coulombs. Alternatively, voltage can be thought of as the work obtained (or energy released) when a unit of positive charge moves from the more positive point to a more negative point (equivalent to allowing a car to coast down a hill). Keep in mind that the work done (or obtained) is done to a charge; hence, the definition for a volt has both the units of joules and coulombs in it.

We always measure voltage between two points in a circuit. Frequently, you may hear of a voltage at a point. In this case, the voltage is measured with respect to a second point called *ground*. The most common symbol for ground is shown in Figure 2–3(a). This particular symbol is actually an earth ground—the potential of the earth. Other ground symbols used in electronics are shown in Figure 2–3(b).

HISTORICAL NOTE

The unit for electrical potential, the volt, is named in honor of Alessandro Volta (1745–1827). Volta, an Italian, invented a device to generate static electricity and he discovered methane gas. He also investigated reactions between dissimilar metals and developed the first battery in 1800.

FIGURE 2–3

Ground symbols.

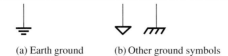

(a) Earth ground   (b) Other ground symbols

Resistance (R)

Resistance is the opposition to current. Resistors, the most common component in electronics, place a specific amount of "opposition" in a circuit path and act to impede current. In addition to the amount of opposition, resistors are specified by the amount of power they can dissipate. We'll look much more at resistors in Chapter 3. For now, remember that resistance is opposition to current.

The unit of resistance is the **ohm (Ω)**, in honor of Georg Simon Ohm, a German mathematics and science teacher who studied electricity in the science lab. The ohm can be derived from fundamental units for length, mass, and time.

Ohm wrote the famous law that bears his name. Ohm's law shows the relationship between voltage, current, and resistance. Based on Ohm's law, the defining equation for resistance is

$$R = \frac{V}{I} \tag{2-3}$$

where R is resistance in ohms, V is voltage in volts, and I is current in amperes. Ohm's law will be discussed in more detail in Chapter 4.

Power (P)

Power is the rate that energy is transferred or converted to heat or to some other form such as sound or light. It is a quantity that is widely used in all of the physical sciences.

The unit for power is the **watt (W)**, named after James Watt, a Scottish scientist and inventor. One watt is equal to one joule per second. The power rating on a component, such as a resistor, tells how fast that component can dissipate heat energy. The defining equation for power is

$$P = \frac{W}{t} \tag{2-4}$$

where P is power in watts, W is energy in joules, and t is time in seconds.

Standards

Whenever you measure current, voltage, or resistance, it is important that you know that the instrument you are using is giving accurate results. Accurate measurements require that the instrument is calibrated and that the calibration information is maintained and up-to-date.

Most organizations working with electronics have highly accurate standards for current, voltage, or resistance that can be used to verify the accuracy of instruments. **Accuracy** is the difference between the measured value and accepted (or true) value of a measurement.

A **working standard** is an instrument or device that is used for routine calibration and certification of test equipment or a component. An example of a working standard is a standard resistor that is highly accurate and used to compare resistance measurements in the laboratory. Working standards require periodic calibration and certification against higher standards to assure that they maintain their accuracy.

Important attributes of all standards include long-term stability, accuracy, and insensitivity to environmental conditions. Eventually, an accurate measurement can be traced through proper documentation to a national standards laboratory. In the United States, this is the National Institute for Standards and Technology (NIST).

Review Questions

11. What is a fundamental unit?
12. How many fundamental units are defined?
13. What is a derived unit?
14. What are the units for current, voltage, resistance, and power?
15. What is a working standard?

MEASURED NUMBERS 2–4

Whenever a quantity is measured, there is uncertainty in the result due to limitations of the instruments used. When a measured quantity contains approximate numbers, the digits known to be correct are called significant digits. When reporting measured quantities, the number of digits that should be retained are the significant digits and no more than one uncertain digit.

In this section, you will learn how to report measured data with the proper number of significant digits.

Error, Accuracy, and Precision

Data taken in experiments are not perfect because the accuracy of the data depends on the accuracy of the test equipment and the conditions under which the measurement was made. In order to properly report measured data, the error associated with the measurement should be taken into account. Experimental error should not be thought of as a mistake. All measurements that do not involve counting are approximations of the true value. **Error** is the difference between the true or best-accepted value of some quantity and the measured value. A measurement is said to be accurate if the error is small. Accuracy refers to the difference between the measured and accepted or "true" value. For example, if you measure thickness of a 10.00 mm gauge block with a micrometer and find that it is 10.8 mm, the reading is not accurate because a gauge block is considered to be a working standard. If you measure 10.02 mm, the reading is accurate because it is in reasonable agreement with the standard.

Another term associated with the quality of a measurement is *precision*. **Precision** is a measure of the repeatability (or consistency) of a measurement of some quantity. It is possible to have a precise measurement in which a series of readings are not scattered, but each measurement is inaccurate because of an instrument error. For example, a meter may be out of calibration and produce inaccurate but consistent (precise) results. However, it is not possible to have an accurate instrument unless it is also precise.

Significant Digits

The digits in a measured number that are known to be correct are called **significant digits**. Most measuring instruments show the proper number of significant digits, but some instruments can show digits that are not significant, leaving it to the user to determine what should be reported. This may occur because of an effect called *loading* (discussed in Section 6–4). A meter can change the actual reading in a circuit by its very presence. It is important to recognize when a reading may be inaccurate; you should not report digits that are known to be inaccurate.

CHAPTER 2

Another problem with significant digits occurs when you perform mathematical operations with numbers. The number of significant digits should never exceed the number in the original measurement. For example, if 1.0 V is divided by 3.0 Ω, a calculator will show 0.33333333. Since the original numbers each contain 2 significant digits, the answer should be reported as 0.33 A, the same number of significant digits.

The rules for determining if a reported digit is significant are

Rule 1: Nonzero digits are always considered to be significant.
Rule 2: Zeros to the left of the first nonzero digit are never significant.
Rule 3: Zeros between nonzero digits are always significant.
Rule 4: Zeros to the right of the decimal point for a decimal number are significant.
Rule 5: Zeros to the left of the decimal point with a whole number may or may not be significant depending on the measurement. For example, the number 12,100 Ω has 3, 4, or 5 significant figures. To clarify the significant digits, scientific notation (or a metric prefix) should be used. For example, 12.10 kΩ has 4 significant figures.

When a measured value is reported, one uncertain digit may be retained but other uncertain digits should be discarded. To find the number of significant digits in a number, ignore the decimal point, and count the number of digits from left to right starting with the first nonzero digit and ending with the last digit to the right. All of the digits counted are significant except zeros to the right end of the number, which may or may not be significant. In the absence of other information, the significance of the right-hand zeros is uncertain. Generally, zeros that are placeholders, and not part of a measurement, are considered to be not significant. To avoid confusion, numbers should be shown using scientific or engineering notation if it is necessary to show the significant zeros.

EXAMPLE 2–7

Problem
Show how to express the measured number 4300 with 2, 3, and 4 significant digits.

Solution
Zeros to the right of the decimal point in a decimal number are significant. Therefore, to show two significant digits, write 4.3×10^3 (no zeros to the right). To show three significant digits, write 4.30×10^3. To show four significant digits, write 4.300×10^3.

Question
How would you show the number 10,000 showing three significant digits?

EXAMPLE 2–8

Problem
Underline the significant digits in each of the following measurements:

(a) 40.0
(b) 0.3040
(c) 1.20×10^5
(d) 120,000
(e) 0.00502

Solution
(a) 40.0 has three significant digits; see rule 4.
(b) 0.3040 has four significant digits; see rules 2 and 3.
(c) 1.20×10^5 has three significant digits; see rule 4.

(d) 120,000 has at least two significant digits. Although the number has the same value as in (c), zeros in this example are uncertain; see rule 5. This is *not* a recommended method for reporting a measured quantity; use scientific notation or a metric prefix in this case. See Example 2–7.

(e) 0.00502 has three significant digits; see rules 2 and 3.

Question
What is the difference between a measured quantity of 10 and 10.0?

Rounding Off Numbers

Since they always contain approximate numbers, measurements should be shown only with those digits that are significant plus no more than one uncertain digit. The number of digits shown is indicative of the precision of the measurement. For this reason, you should **round off** a number by dropping one or more digits to the right of the last significant digit. Use only the most significant dropped digit to decide how to round off. The rules for rounding off are

Rule 1: If the most significant digit dropped is greater than 5, increase the last retained digit by 1.

Rule 2: If the digit dropped is less than 5, do not change the last retained digit.

Rule 3: If the digit dropped is 5, increase the last retained digit *if* it makes it an even number, otherwise do not. This is called the "round-to-even" rule.

EXAMPLE 2–9

Problem
Round each of the following numbers to three significant digits:

(a) 10.071   (b) 29.961   (c) 6.3948
(d) 123.52   (e) 122.52

Solution
(a) 10.1   See rule 1.
(b) 30.0   See rule 1.
(c) 6.39   See rule 2.
(d) 124   See rule 3.
(e) 122   See rule 3.

Question
Why is the round-to-even rule used?

In most electronics work, components have tolerances greater than 1% (5% and 10% are common). Most measuring instruments have accuracy specifications better than this, but it is unusual for measurements to be made with higher accuracy than 1 part in 1000. For this reason, three significant digits are appropriate for numbers that represent measured quantities in all but the most exacting work. If you are working with a problem with several intermediate results, keep all digits in your calculator, but round the answers to three when reporting a result. This is the method used in example problems in this text.

CHAPTER 2

Review Questions

16. What is the rule for showing zeros to the right of the decimal point?
17. What is the round-to-even rule?
18. On schematics, you will frequently see a 1000 Ω resistor listed as 1.0 kΩ. What does this imply about the value of the resistor?
19. If a power supply is required to be set to 10.00 V, what does this imply about the accuracy needed for the measuring instrument?
20. How can scientific or engineering notation be used to show the correct number of significant digits in a measurement?

2–5 BASIC ALGEBRA REVIEW

In scientific and engineering work, algebraic equations are used to express the relationship between various quantities. Algebraic equations are necessary in electronics to solve real-world problems. Think of an equation as a shorthand method of showing a relationship between quantities.

In this section, you will learn to manipulate linear algebraic equations to solve for an unknown variable.

Constants and Variables

A **variable** is a quantity that changes. It is usually indicated by an italic letter that represents the quantity. There are many quantities in electronics that can be variables. Some examples include voltage, power, and time. For a commonly used variable, the letter that is associated with it is often the first letter in the name of the quantity; thus, V is used for voltage, P for power, and t for time. Variables are frequently written with subscripts to represent different variables. For example, R_1, R_2, and R_3 are three different variables, each representing a different resistor in a circuit.

A **constant** is a quantity that never changes. Some familiar constants are pi (3.14) and the speed of light (3.00×10^8 m/s). Familiar constants are often indicated with a letter. For example, pi is indicated with the Greek letter π and the speed of light is indicated with the letter c.

Terms and Expressions

A **term** is a group of variables and/or constants that are not separated with plus or minus signs. A term may have a number written in front of other quantities that is referred to as a **coefficient**. An example of a term is $3ax^2$; the coefficient is 3 because it is the number in front of other quantities. Other terms are $2\pi fL$ and πr^2. A term may simply be a numeric value such as the number 12.

Like terms are terms with the same variables and exponents. The coefficients of the terms may be different. The expression $R_1 + 2R_1 + 4R_1$ contains the same variable, R_1, with different coefficients.

An **expression** consists of one or more terms. Numeric expressions consist of only numbers and signs whereas literal expressions contain one or more variables. Mathematicians also classify expressions by the number of terms they contain. An expression that contains only a single term is called a **monomial**; two terms form a **binomial**, and more than two terms are classified as a **polynomial**. The expression $R_1 + R_2 + R_3$ is a polynomial because there are three terms.

EXAMPLE 2–10

Problem
Identify the number of terms in each of the following expressions:

(a) I^2R
(b) $V_1 + V_2 + V_3$
(c) $3I_1 - 1.2$
(d) $\dfrac{1}{2\pi fC}$
(e) $22 - 5$

Solution
(a) one term
(b) three terms
(c) two terms
(d) one term
(e) two terms

Question
Which of the expressions are literal expressions?

Equations

An **equation** is a mathematical statement that sets one expression or quantity equal to another. For example, you may be familiar with an important law from physics called Newton's second law. Newton's second law, which describes the relationship between acceleration, force, and mass, is a good example to illustrate basic equations. In sentence form Newton's second law can be stated, "The amount of acceleration of a certain object is determined by the net force applied to the object divided by the mass of the object." In shorthand equation form, this sentence is translated to the algebraic equation: $a = F/m$. Although this equation shows three variables (a, F, and m), in practice one of these quantities may be held constant—usually the mass in this case.

Linear Equations
An equation that contains variables with no exponents is called a linear equation. Newton's second law is just one example of a linear equation. A linear equation can be plotted as a straight line (discussed in the next section). Linear equations can be represented by any of several basic mathematical forms. The most popular form for a linear equation is called the **slope-intercept form** and is written as

$$y = mx + b$$

This is the general equation for a straight line. The letters x and y stand for variables. The letter m stands for the **slope** (constant for a straight line) and is a measure of how steep the graph of x and y is. The letter b is a constant called the intercept. It tells where the line crosses the y-axis.

A lot of the equations you will work with in electronics are linear equations (including the most important equation in electronics, Ohm's law). These equations will make more sense if you keep in mind the general form for linear equations.

CHAPTER 2

Solving Linear Equations

If a linear equation (or any other for that matter) contains only one unknown value, it is possible to rearrange the equation to solve for the unknown. The methods for rearranging linear equations are straightforward but require you to "maintain the balance" on either side of the equal sign. This means that any mathematical operation performed on one side of an equation must be performed on the other side as well.

The steps selected for solving an equation are not always "cut and dried" but depend on the particular equation. Of course, you must follow certain basic rules. Typically, it is useful to start solving an equation by eliminating any parentheses by using the distribution rule of algebra. This important rule is written symbolically as

$$a(b + c) = ab + ac$$

As a statement, the distribution rule can be given as "multiplication distributes over addition." The reverse operation of distribution is called *factoring*. In symbolic form, factoring, as expressed for three variables, is written as

$$ab + ac = a(b + c)$$

A common next step in solving equations is to move terms that contain the unknown to one side of the equation and to move terms that do not contain the unknown to the other side of the equation. To do this, you can add or subtract the same term from both sides of the equation. Then, you can factor the unknown from the terms containing the unknown and perform a mathematical operation (addition, subtraction, multiplication, or division) to finally isolate the unknown to a single variable. In solving an equation, additions and subtractions are done first, then multiplication and division. Sometimes a different sequence may be necessary, depending on the particular equation. In all cases, the same operation is performed on both sides of the equation to maintain the equality.

As an example of solving for an unknown, assume you read the temperature on a thermometer as 10°C (centigrade). You wonder what is the temperature in °F (Fahrenheit), and you remember that the equation for converting between the two scales is

$$°C = \frac{5}{9}(°F - 32°)$$

Because °F is the unknown, you need to rearrange the equation to solve for °F. It is customary to write equations with the unknown appearing on the left side. Therefore, your first objective is to rewrite the equation with °F on the left and everything else on the right. Then you can substitute the known value for °C (10°) and solve for the unknown.

You could begin by applying the distribution rule, but in this case the solution will be easier if you multiply both sides by 9/5 in order to cancel out the fraction 5/9 on the right.

$$\frac{9}{5} \times °C = \frac{\cancel{9}}{\cancel{5}} \times \frac{\cancel{5}}{\cancel{9}}(°F - 32°)$$

Because the fraction on the right cancels, the effect of this operation is to "move" the fraction to the left side. The parentheses can now be cleared, and 32 is added to both sides of the equation.

$$\frac{9}{5} \times °C + 32° = °F - \cancel{32°} + \cancel{32°}$$

This has the effect of isolating the unknown, but it is on the right side. Because $a = b$ implies that $b = a$, the equation can be written in conventional format as

MATHEMATICS FOR ELECTRONICS

$$°F = \frac{9}{5} \times °C + 32°$$

The first objective has been met. Now, by substituting the known Celsius temperature of 10°, you can find the equivalent Fahrenheit temperature.

$$°F = \frac{9}{5}(10°) + 32° = 18° + 32° = 50°$$

EXAMPLE 2–11

Problem
Isolate x on the left side for the following equation. Then solve for x if the value of y is 14.

$$y = 4\left(\frac{3x}{2} + 5\right)$$

Solution
Apply the distribution rule to remove parentheses.

$$y = \frac{12x}{2} + 20$$

Reduce the fraction and subtract 20 from both sides.

$$y - 20 = 6x$$

Divide both sides of the equation by 6 and reverse the left and right sides.

$$x = \frac{y - 20}{6}$$

Substitute the known value of y (14) into the expression on the right.

$$x = \frac{14 - 20}{6} = \frac{-6}{6} = -1$$

As a check, you can substitute this value into the original equation and verify that $y = 14$.

Question
What is the value of x if $y = 20$?

Example 2–12 introduces a common electronics formula called the product-over-sum rule for parallel resistors. You will study this rule in Section 5–5 on parallel circuits. Although starting with the product-over-sum rule isn't the only way to arrive at an equation for R_2, the example illustrates the process of solving an equation for an unknown.

EXAMPLE 2–12

Problem
The total resistance (R_T) of two parallel resistors (R_1 and R_2) is given by the product-over-sum rule written in equation form as

$$R_T = \frac{R_1 R_2}{R_1 + R_2}$$

CHAPTER 2

Assume you want to obtain a total resistance of 300 Ω using two resistors. R_1 is 820 Ω. What should be the value of R_2? Start with the product-over-sum rule to find the equation for R_2.

Solution
Multiply both sides of the equation by $R_1 + R_2$ to clear the fraction on the right side. Notice that parentheses must be shown when multiplying because these two terms are grouped under the fraction bar. For clarity, the terms under the fraction bar are also placed in parentheses.

$$R_T(R_1 + R_2) = \frac{R_1 R_2}{\cancel{(R_1 + R_2)}} \cancel{(R_1 + R_2)}$$

Apply the distribution law to the left side to remove the parentheses.

$$R_T R_1 + R_T R_2 = R_1 R_2$$

Subtract $R_T R_2$ from both sides in order to place the two terms with the unknown (R_2) together on the right side.

$$R_T R_1 + \cancel{R_T R_2} - \cancel{R_T R_2} = R_1 R_2 - R_T R_2$$

Factor the unknown from the terms on the right side.

$$R_T R_1 = R_2(R_1 - R_T)$$

Divide both sides of the equation by $(R_1 - R_T)$, eliminate the parentheses, and switch the left and right sides.

$$R_2 = \frac{R_T R_1}{R_1 - R_T}$$

The equation is now in the form with the unknown on the left side. Substitute the values given in the problem to find the value of R_2.

$$R_2 = \frac{R_T R_1}{R_1 - R_T} = \frac{300 \,\Omega \times 820 \,\Omega}{820 \,\Omega - 300 \,\Omega} = \mathbf{473 \,\Omega}$$

Question
Assume the required total resistance is 350 Ω and R_1 is still 820 Ω. What is the new value of R_2 that will produce this total?

Nonlinear Equations
Frequently, equations used in electronics will have an exponent on one or more variables. When an exponent other than 1 is in an equation, the equation is a nonlinear equation. We will restrict the discussion to equations that are commonly used in electrical circuits. For instance, a common problem is to find the power in a resistor (P) when the voltage (V) and resistance (R) are known. The power in a resistor is found from the equation

$$P = \frac{V^2}{R}$$

Sometimes you will need to rearrange an equation like this to solve in terms of another variable. If you know the power and resistance but you need to solve in terms of voltage, the basic rules for equation manipulation are much the same as for linear equations. The requirement that you do the same thing to both sides is the key. In this case, start by multiplying both sides by R.

$$PR = V^2$$

Now, rearrange so that the unknown (V) appears on the left side.

$$V^2 = PR$$

Finally, take the square root of each side of the equation.

$$V = \sqrt{PR}$$

Notice on the left side that taking the square root of a squared term leaves the variable without an exponent.

EXAMPLE 2-13

Problem
In a certain type of circuit called a tuned circuit, f_r is the resonant frequency, L is the inductance, and C is the capacitance. The equation for resonant frequency is

$$f_r = \frac{1}{2\pi\sqrt{LC}}$$

Solve this equation in terms of L by rearranging the equation so that L is on the left side and the other variables are on the right.

Solution
Multiply both sides by $\sqrt{LC}$. This effectively "moves" $\sqrt{LC}$ to the left side.

$$\sqrt{LC}\,f_r = \frac{1}{2\pi}$$

Divide both sides by f_r.

$$\sqrt{LC} = \frac{1}{2\pi f_r}$$

Square both sides.

$$LC = \frac{1^2}{(2\pi f_r)^2} = \frac{1}{4\pi^2 f_r^2}$$

Divide both sides by C.

$$L = \frac{1}{4\pi^2 f_r^2 C}$$

Question
What is the result of the same equation if you solve for C?

CHAPTER 2

Review Questions

21. What is a term?
22. What is a coefficient?
23. Explain the difference between *distribution* and *factoring*.
24. The formula for finding the resistance of a wire is $R = \rho L/A$. Solve this equation for L.
25. The total time for a 555 timer (a type of integrated circuit) is given by the equation $T = 0.693(R_A + 2R_B)$. Solve this equation for R_B.

2-6 GRAPHING

An important means of showing data is the linear graph. A graph is a pictorial representation of data that enables you to see the effect of one variable on another. Graphs are widely used in electronics work to present information because they show variations in magnitude, slope, and direction between two quantities.

In this section, you will learn how to prepare a linear graph.

The relationship between two variables can be visualized by plotting the ordered, paired data points on a graph. The data may be taken in an experiment or may come from an equation. In any case, the graph is an excellent way to visualize and understand the data. For example, the relationship between the temperature in °F and the temperature in °C was defined earlier by the equation:

$$°C = \frac{5}{9}(°F - 32°)$$

Table 2–7 illustrates some ordered, paired data points for this equation.

TABLE 2–7

Temperature (°C)	Temperature (°F)
−17.8°	0°
0°	32°
15.6°	60°
23.9°	75°
37.8°	100°

The Cartesian Coordinate System

Linear plots are set up with a standard plotting method that uses a horizontal and a vertical numbered line. The horizontal line is called the *x*-axis or **abscissa**; the vertical axis is called the *y*-axis or **ordinate**. The intersection of the two lines is called the **origin**. This system is called the Cartesian coordinate system, after its inventor, René Descartes.

The two axis lines divide the space into four quadrants, which are numbered counterclockwise as shown in Figure 2–4. This system allows both positive and negative values of *x* or *y* to be plotted. The *x*-axis is positive to the right of the origin, and the *y*-axis is positive above the origin. Quadrant 1 is used when both *x* and *y* values are positive. Quadrant 2 is negative for *x* and positive for *y*. Quadrant 3 is negative for both *x* and *y*. Quadrant 4 is positive for *x* and negative for *y*. Sometimes graphs only involve positive numbers; in these cases, only Quadrant 1 is shown on the plot.

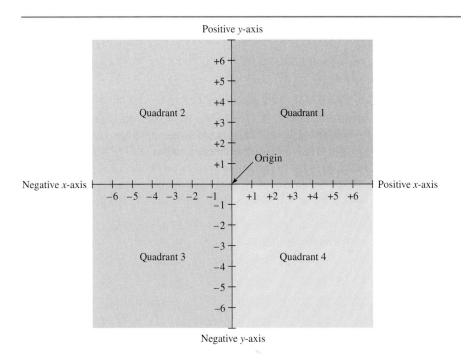

FIGURE 2-4

The Cartesian coordinate system.

Independent and Dependent Variables

Data from an experiment or other source is plotted on a graph in a specific way. Consider the equation for Newton's second law, given in Section 2-5 as $a = F/m$ where a stands for acceleration, F stands for force, and m stands for mass. Assume that a student takes experimental data using a 5.5 kg mass, varying the force on it, and measuring the resulting acceleration. Mass is not changing, so it is a constant. Force is controlled (or varied) by the experimenter, so it is called the **independent variable**. In mathematics, an independent variable is one that is controlled or changed to observe the effect on another variable, which is called the **dependent variable**. In this case, the dependent variable is *acceleration*, which responds to the force. The dependent variable is influenced by changes in the independent variable.

The experimental data for Newton's second law will look like Figure 2–5 but may vary a little because of experimental error. The plot is a straight line as you might expect. As with

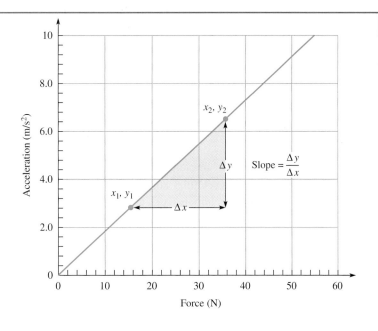

FIGURE 2-5

Plot of acceleration as a function of force in an experiment. The mass is constant and is equal to 5.5 kg.

any straight line, it can be written in the general slope-intercept form given in the last section. Recall that the general form for writing a straight line with the slope-intercept method is

$$y = mx + b$$

Thus, when Newton's second law is written in slope-intercept form, y represents the acceleration (the dependent variable), and x represents the force (the independent variable). The y intercept (b) is zero because if there is no force, there is no acceleration so the y intercept is at the origin. The slope of the line is a measure of how steep it is. To find the slope, divide a change in the y variable (indicated with the mathematical notation Δy) by a corresponding change in the x variable (indicated with the mathematical notation Δx), as illustrated in Figure 2–5. If the two points are (x_1, y_1) and (x_2, y_2), the slope can be written as

$$m = \frac{y_2 - y_1}{x_2 - x_1} = \frac{\Delta y}{\Delta x}$$

The yellow triangle shows where the slope was measured. Notice that the slope has units of acceleration over force.

Many equations in science and in electronics are linear equations; some are not. Keep in mind that linear equations can always be arranged to plot a straight line and can be written in the slope-intercept format given here.

Choosing a Scale for a Graph

The **scale** (or scale factor) is the value of each division along the x-axis or the y-axis. With a linear graph, each division has equal weight. When you assign numbers to either axis, you must increase the value assigned to each division by the same amount. For most graphs, the scale does not need to be the same for both the x-axis and the y-axis. To determine a reasonable scale to use, divide the largest number you will plot by the number of divisions on the appropriate axis and choose a scale that is equal or larger than the result. Usually, numbers that are multiples of 1, 2, and 5 units per division are preferred. For example, if the largest x number is 370, and there are 10 divisions, the scale should be larger than 37. You could choose 50 as a reasonable scale.

The following steps will guide you in preparing a graph:

Step 1: Determine the type of plot that you will draw. A linear plot is the most frequently used and will be discussed here.

Step 2: Choose a scale that enables all of the data to be plotted on the graph without being cramped. Start both axes from 0 unless the data covers less than half of the length of the coordinate.

Step 3: Number the *major* divisions along each axis. Do not number each small division as it will make the graph appear cluttered. Each division must have equal weight. (*Note:* The experimental data itself is not used to number the divisions.)

Step 4: Label each axis to indicate the quantity being measured and the measurement units. Usually, the measurement units are given in parentheses.

Step 5: Plot the data points with a small dot. If additional sets of data are plotted, use other distinctive symbols (such as triangles) to identify each set.

Step 6: Draw a smooth line that represents the data trend. It is normal practice to consider data points but to ignore minor variations due to experimental errors. (*Exception:* Calibration curves, and other discontinuous data, are connected "dot-to-dot".)

Step 7: Title the graph, indicating with the title what the graph represents. The completed graph should be self-explanatory.

EXAMPLE 2–14

Problem
A student closes a switch on an electrical circuit and reads the voltage across a component at certain times after closing the switch. The data is recorded in Table 2–8. Plot the data according to the guidelines given in this section.

Table 2–8

Time (s)	Voltage (V)
2.0	8.0
5.0	5.9
8.0	4.1
14	2.1
20	1.2
25	0.65

Solution

Step 1: Select a linear plot with 10 divisions along each axis. Because all of the data is positive, you only need quadrant 1. The voltage is dependent on the time; therefore, plot the time on the x-axis, and plot the voltage along the y-axis.

Step 2: Choose a scale for the x-axis that is equal or larger than 25 s/10 div = 2.5 s/div. This is a reasonable scale if you number every other division by 5. It also enables you to plot all of the data on the graph without being cramped. For the y-axis, an appropriate scale is larger than 8 V/10 div = 0.8 V/div. Choose 1 V/div as a reasonable scale.

Step 3: Number the *major* divisions along each axis. The progress to this point is shown in Figure 2–6.

Step 4: Label the x-axis as Time (s) and the y-axis as Volts (V) to indicate the quantity measured and the units of measurement. The measurement units are given in parentheses.

Step 5: Plot each data point with a dot.

Step 6: Draw a smooth line that represents the data trend.

Step 7: Title the graph. The completed graph is shown in Figure 2–7.

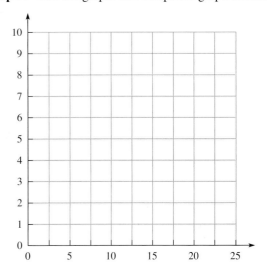

FIGURE 2–6
Preparing the graph.

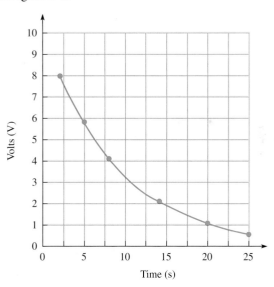

Voltage versus time for an electrical component after a switch is closed

FIGURE 2–7
Completed graph.

Question

(a) From the graph, predict the voltage that would be on the component at a time of 10 s.

(b) From the graph, at what time do you expect the voltage to be 0.1 V?

CHAPTER 2

Review Questions

26. Which axis (*x* or *y*) is the abscissa?
27. Which axis (*x* or *y*) is the ordinate?
28. What is meant by the independent variable?
29. What is meant by the dependent variable?
30. What labels should be on a graph?

CHAPTER REVIEW

Key Terms

Accuracy The difference between the measured value and accepted or true value of a measurement.

Ampere The unit for current; the fundamental electrical unit. One ampere is one coulomb per second.

Coefficient A number written in front of a term.

Constant A quantity that never changes.

Dependent variable A variable that is influenced by changes in another variable.

Engineering notation A method of writing a large or small number as an ordinary number (between 1 and 1000) times a power of ten. The exponent can only take on values that are evenly divisible by three.

Error The difference between the true or best-accepted value of some quantity and the measured value.

Independent variable A variable that is controlled or changed to observe the effect on another variable.

Ohm The unit of resistance. One ohm is one volt per ampere.

Precision A measure of the repeatability (or consistency) of a series of data points.

Scientific notation A method of writing a large or small number as an ordinary number between 1 and 10 times a power of ten.

Significant digits The digits known to be correct in a measured number.

Variable A quantity that changes.

Volt The unit of voltage. One volt is one joule per coulomb.

Watt The unit of power. One watt is equal to one joule per second.

Important Facts

❏ Scientific and engineering notations use powers of ten to simplify writing very large and very small numbers.

❏ Metric prefixes, especially those referred to as engineering prefixes, are widely used to express large and small units.

- The SI system is based on seven fundamental units and two supplementary units that form the "building blocks" of all measurements in the metric system.
- When a measured value is reported, the least significant uncertain digit may be retained but other uncertain digits should be discarded.
- A linear equation with one unknown can be rearranged by performing identical operations on both sides of the equal sign.
- A graph is a pictorial representation of data that enables you to see the effect of one variable on another.

Formulas

Definition of current:

$$I = \frac{Q}{t} \qquad (2\text{–}1)$$

Definition of voltage:

$$V = \frac{W}{Q} \qquad (2\text{–}2)$$

Definition of resistance:

$$R = \frac{V}{I} \qquad (2\text{–}3)$$

Definition of power:

$$P = \frac{W}{t} \qquad (2\text{–}4)$$

Chapter Checkup

Answers are at the end of the chapter.

1. The metric prefix that stands for 10^{-12} is
   - (a) milli
   - (b) micro
   - (c) nano
   - (d) pico

2. The number of nanoseconds in 1 microsecond is
   - (a) 10
   - (b) 100
   - (c) 1,000
   - (d) 1,000,000

3. The two SI units that are for measuring angles are called
   - (a) supplementary units
   - (b) degree units
   - (c) fundamental units
   - (d) derived units

4. The fundamental electrical unit is
   - (a) coulomb
   - (b) volt
   - (c) ampere
   - (d) second

5. How many significant digits are in the number 0.1050?
   - (a) two
   - (b) three
   - (c) four
   - (d) five

6. The coefficient of the term $2\pi fL$ is
   (a) 2
   (b) π
   (c) f
   (d) L

7. If the equation, $X_L = 2\pi fL$ is solved for L, the result is
   (a) $L = \dfrac{1}{2\pi fX_L}$
   (b) $L = \dfrac{2\pi f}{X_L}$
   (c) $L = 2\pi fX_L$
   (d) $L = \dfrac{X_L}{2\pi f}$

8. Quadrant 1 of the Cartesian coordinate system is in which corner?
   (a) upper left
   (b) upper right
   (c) lower left
   (d) lower right

9. A variable that is controlled or changed to observe the effect on another variable is called the
   (a) independent variable
   (b) controlled variable
   (c) significant variable
   (d) dependent variable

10. The place where the x-axis and the y-axis cross is called the
    (a) crossing point
    (b) zero point
    (c) starting point
    (d) origin

Questions

Answers to odd-numbered questions are at the end of the book.

1. How is 0.000 050 written in engineering notation?
2. How many picoseconds are in 1 microsecond?
3. What power of ten does the prefix *mega* represent?
4. What power of ten does the prefix *nano* represent?
5. What power of ten does the prefix *giga* represent?
6. What power of ten does the prefix *pico* represent?
7. What is meant by a derived unit in the SI system?
8. Is voltage a fundamental or a derived unit?
9. What is a working standard?
10. What is the NIST responsible for?
11. What are three desired attributes of a measurement standard?
12. Is it possible to have a precise but inaccurate measurement? Explain your answer.
13. Is it possible to have an accurate but imprecise measurement? Explain your answer.
14. What is the "round-to-even" rule?
15. What is a binomial?
16. What is meant by the distribution rule in algebra?
17. How are the quadrants numbered in the Cartesian coordinate system?
18. Which axis should normally be used for plotting the independent variable?
19. Is it necessary for a line to pass through every data point in a linear plot? Explain.

Basic Problems

Answers to odd-numbered problems are at the end of the book.

1. Express each of the fixed numbers in scientific notation:
   (a) 73,000
   (b) 0.000 22
   (c) −92,000,000
   (d) −0.005 1
   (e) 470,000

2. Express each number in Problem 1 in engineering notation.

3. Express each of the numbers written in scientific notation as a fixed number:
   (a) 1.6×10^5
   (b) 5.6×10^{-3}
   (c) -8.0×10^5
   (d) -7.1×10^{-2}
   (e) 4.2×10^{-1}

4. Convert each of the numbers in Problem 3 to engineering notation.

5. Write each of the following numbers with a metric prefix:
   (a) 220,000 Ω
   (b) 0.000 047 F
   (c) 5000 W
   (d) 0.000 000 055 s
   (e) 1.5×10^6 Hz
   (f) 10×10^{-3} H

6. Express each number written with a metric prefix as a number in engineering notation:
   (a) 56 kΩ
   (b) 50 ps
   (c) 100 kW
   (d) 37 ms
   (e) 5.0 GW
   (f) 22 μs

7. Convert the following:
   (a) 2.2 MΩ to kΩ
   (b) 100 ns to μs
   (c) 10,000 kW to MW
   (d) 220 Ω to kΩ
   (e) 4.0 mV to μV
   (f) 100 ps to μs

8. How many significant digits are in each of the following numbers?
   (a) 1.00×10^3
   (b) 0.0057
   (c) 1502.0
   (d) 0.000 036
   (e) 0.105
   (f) 2.6×10^2

9. Round each of the following numbers to three significant digits. Use the "round-to-even" rule.
   (a) 50,505
   (b) 220.45
   (c) 4646
   (d) 10.99
   (e) 1.005

10. Isolate *x* on the left side for each of the following equations. Then solve for *x*, assuming the value of *y* is 5 in all cases.
    (a) $y = 1.5\left(\dfrac{3x - 2}{6}\right)$
    (b) $y = 2x - 9$
    (c) $y = \dfrac{x - 3}{4}$
    (d) $y = \dfrac{4}{x + 3}$

11. The equation for finding the resistance of a length of wire is $R = \rho l/A$. Solve this equation for l (that is, isolate l on the left side).

12. The equation for finding the resistance of a meter shunt is $R_s = R_m/(N - 1)$. Solve this equation for N (that is, isolate N on the left side).

13. The equation for characteristic impedance is: $Z_0 = \sqrt{L/C}$. Solve this equation for C (that is, isolate C on the left side).

14. An equation for power dissipated in a resistor is $P = I^2 R$. Solve this equation for I (that is, isolate I on the left side).

15. Current (I_D) for a light-emitting diode is recorded as voltage (V_D) is increased. Plot the data given in Table 2–9, according to the guidelines given in Section 2–6. The voltage is the independent variable.

Table 2–9

Voltage, V_D (V)	Current, I_D (mA)
1.00	0.0
1.50	0.0
1.75	0.065
2.00	2.9
2.25	7.21
2.50	11.8

16. The drain current (I_D) for a field-effect transistor is measured as the gate-source voltage (V_{GS}) is increased. Notice that the gate-source voltage is negative. You should plot the data in Quadrant 2. Plot the data given in Table 2–10, according to the guidelines given in Section 2–6. The gate-source voltage is the independent variable.

Table 2–10

Voltage, V_{GS} (V)	Current, I_D (mA)
0.00	1.26
−0.20	0.92
−0.40	0.68
−0.60	0.42
−0.80	0.27
−1.00	0.13
−1.20	0.03
−1.40	0.00

Basic-Plus Problems

17. Add. Express your answer in engineering notation.
    (a) $6.5 \times 10^2 + 1.60 \times 10^3$
    (b) $910 \times 10^3 + 1.5 \times 10^6$
    (c) $12{,}000 + 4.5 \times 10^3$
    (d) $0.0050 + 3.5 \times 10^{-4}$

18. Subtract. Express your answer in engineering notation.
    (a) $3.2 \times 10^4 - 1.60 \times 10^4$
    (b) $1.50 \times 10^5 - 12 \times 10^4$
    (c) $9.10 \times 10^{-3} - 8.0 \times 10^{-4}$
    (d) $6.45 \times 10^{-9} - 4.5 \times 10^{-10}$

19. Add:
    (a) 470 kΩ + 2.2 MΩ
    (b) 10,000 pF + 0.01 μF
    (c) 120 MHz + 0.550 GHz
    (d) 650 pF + 10 nF
20. Subtract:
    (a) 5.10 kΩ − 330 Ω
    (b) 500 nF − 10,000 pF
    (c) 1.0 μF − 200 nF
    (d) 3.3 s − 300 ms
21. Pythagorean's theorem states $c = \sqrt{a^2 + b^2}$. Solve this equation for a (that is, isolate a on the left side).
22. The formula for the volume, V, of a cone is given by $V = \pi r^2 h/3$. Solve this equation for r (that is, isolate r on the left side).
23. The inductance of a coil is given by the formula $L = N^2 \mu A/l$. Solve this equation for N (that is, isolate N on the left side).
24. The energy stored in the magnetic field of a coil is given by the equation $W = LI^2/2$. Solve this equation for I (that is, isolate I on the left side).
25. The equation $C_T = 1/(1/C_1 + 1/C_2)$ is for the total capacitance, C_T, of series capacitors. Solve this equation for C_1 (that is, isolate C_1 on the left side).
26. A plot of input current for an LM318 op-amp is taken as the differential input voltage is varied. Plot the data given in the Table 2–11 according to the guidelines given in Section 2–6.

    Table 2–11

Differential voltage (V)	Input current (nA)
−0.8	575
−0.7	80
−0.6	−24
−0.4	−100
−0.2	−105
0	−109
+0.4	−90
+0.6	−175
+0.7	−390

27. Plot the data given in Table 2–7 according to the guidelines given in Section 2–6.

Example Questions

ANSWERS

2–1: (1) 11,480   (2) 5.1   (3) 0.000 33   (4) −0.000 005 2   (5) −290,000

2–2: 6,250,000,000,000,000,000

2–3: Answers (c) and (e)

2–4: 5 ms

2–5: The ▭ key is pressed before entering the digits.

2–6: 0.22 MΩ

2–7: 10.0×10^3

2–8: The number 10 has two significant digits; the number 10.0 has three.

2–9: There are four numbers larger than 5 and four numbers smaller than 5 (not counting zero). Ideally, 5 should be retained half the time to avoid weighting rounded numbers too high or too low; hence the round-to-even rule.

2–10: All are literal expressions (contain variables) except (e).

2–11: 0

2–12: 611 Ω

2–13: $C = 1/4\pi^2 f_r^2 L$

2–14: (a) 3.3 V (b) about 35 s

Review Questions

1. *Floating-point* means the decimal is not in a fixed location.
2. Power of ten notation simplifies writing large and small numbers.
3. 1
4. Engineering notation uses a fixed number multiplied by a power of ten that is a multiple of 3. Scientific notation uses a fixed number between 1 and 10 and a power of ten.
5. Engineering exponents can easily be converted into standard metric prefixes.
6. 10^3 = kilo, 10^6 = mega, 10^9 = giga, and 10^{12} = tera
7. 10^{-3} = milli, 10^{-6} = micro, 10^{-9} = nano, and 10^{-12} = pico
8. To the right
9. 23.5 nF
10. 1770 kΩ
11. A fundamental unit is one of the seven units defined in the SI system that, along with two supplementary units, can define all other measurement units.
12. Seven
13. A derived unit is one that is formed from a combination of two or more fundamental and supplementary units.
14. The unit for current is the ampere, the unit for voltage is the volt, the unit for resistance is the ohm, and the unit for power is the watt.
15. A working standard is an instrument or device that is used for routine calibration and certification of test equipment.
16. Zeros should be retained only if they are significant because if they are shown, they are considered significant.
17. If the digit dropped is 5, increase the last retained digit *if* it makes it even, otherwise do not.
18. A zero to the right of the decimal point implies that the resistor is accurate to the nearest 100 Ω (0.1 kΩ).
19. The instrument must be accurate to four significant digits.
20. Scientific and engineering notation can show any number of digits to the right of a decimal. Numbers to the right of the decimal are always considered significant.
21. A term is a group of variables and/or constants, which are not separated with plus or minus signs.
22. A coefficient is a number written in front of a term.
23. Distribution is the process of expanding an expression to eliminate parentheses. The general form of the distribution law is written as

$$a(b + c) = ab + ac$$

Factoring is the reverse operation. A common variable is removed from two or more terms; the general form of factoring is

$$ab + ac = a(b + c)$$

24. $L = \dfrac{RA}{\rho}$

25. $R_B = \dfrac{1}{2}\left(\dfrac{T}{0.693} - R_A\right)$

26. The x-axis

27. The y-axis

28. The independent variable is a variable that is controlled or changed to observe the effect on another variable.

29. The dependent variable is a variable that is influenced by changes in another variable.

30. Each axis should be labeled to indicate the quantity being measured and the measurement units. Usually, the measurement units are given in parentheses. In addition, a title should be included for the overall plot.

Chapter Checkup

1. (d)   2. (c)   3. (a)   4. (c)   5. (c)
6. (a)   7. (d)   8. (b)   9. (a)   10. (d)

ELECTRICAL QUANTITIES AND MEASUREMENTS

CHAPTER OUTLINE

3–1 Conductors, Insulators, and Semiconductors

3–2 Current and the Electric Circuit

3–3 Voltage Sources

3–4 Resistance and Resistors

3–5 Basic Electrical Measurements

INTRODUCTION

The units for current, voltage, and resistance were introduced in the last chapter. In the first part of this chapter, these units are considered further in a discussion of different types of voltage sources such as batteries, generators, and electronic power supplies. In the last part of the chapter, the resistor color code is introduced. In addition, two types of meters are discussed, VOMs (volts-ohms-milliammeter) and DMMs (digital multimeter). Because meters are fundamental measurement tools in electronics, you should be familiar with their proper use. An ammeter, which measures current, can damage a circuit under test or be damaged by the circuit if it is improperly connected.

Study aids for this chapter are available at

http://www.prenhall.com/SOE

KEY OBJECTIVES

A section number is given for each objective. After completing this chapter, you should be able to

3–1 Compare the atomic structure of conductors, insulators, and semiconductors

3–2 Compare direct current (dc) and alternating current (ac) and explain terms used with ac

3–3 Explain how a battery changes chemical energy into electrical energy and describe generators and power supplies in general

3–4 Read four-band and five-band resistor color codes

3–5 Explain how to use a digital multimeter (DMM) and an analog volt-ohm-milliammeter (VOM) to measure voltage, current, or resistance

LABORATORY EXPERIMENTS DIRECTORY

The following exercises are for this chapter. The lab manual is entitled *The Science of Electronics: DC/AC Lab Manual* by David M. Buchla (ISBN 0-13-087566-X). © 2005 Prentice Hall.

◆ **Experiment 4**
Resistance

◆ **Experiment 5**
DC Voltage Measurement

KEY TERMS

- Electrolyte
- Direct current
- Load
- Schematic
- Alternating current
- Sinusoidal wave
- Frequency
- Hertz (Hz)
- Period
- RMS current
- RMS voltage
- Power supply
- Efficiency
- Resistivity
- DMM
- VOM

CHAPTER 3

Although most people have heard of three states of matter (solids, liquids, and gases), most would be hard-pressed to name a fourth state. Yet more than 99% of the observable universe is in this fourth state. Can you name it? Although it is common in man-made lights, it is not normally found on earth in nature except in the aurora, high above the earth's surface. It is so hot that electrons have been stripped from all of the atoms. Scientists call this state a plasma. Plasmas are fully ionized gases but behave so differently from ordinary gases that they have their own classification. Because charges are free to move, plasmas are excellent conductors. The sun, the solar wind, and the trapped radiation belts surrounding the earth are all plasmas. The most common place to find a plasma on earth is in a "neon" sign.

A recent discovery announced by particle physicists is a fifth state of matter that only has been made in a particle accelerator for a few microseconds. This state consists of a plasma of quarks and gluons, elementary particles that are the basic constituents of protons and neutrons as well as more exotic particles. A quark-gluon plasma is thought to have existed for a few microseconds after the "big bang".

3–1 CONDUCTORS, INSULATORS, AND SEMICONDUCTORS

Electricity consists of moving charge. Most of the time when we think of the path for electricity, we think of wires. In metallic solids, electrons make up the charge that moves; however, charge can also move through certain liquids and gases as well as space itself.

In this section, you will learn more about the atomic structure of conductors, insulators, and semiconductors.

Conductors

Recall from Section 1–5 that conductors are materials that allow the free movement of charge and can be composed of solids, liquids, or gases. Nearly all circuits use metallic solids such as copper or aluminum to provide a path for electricity. Wires and traces on pc boards are the most common form of solid conductor for electrical signals. Wires are either solid or stranded. Stranded wire is used for meter leads or in lamp cords. For its size, stranded wire has the same current capacity as solid wire but provides more flexibility than solid wire.

For electronic circuits, copper is the most widely used material because it is an excellent conductor, can easily be drawn into wires, is corrosion resistant, and is cost effective. Copper is such a good conductor because of its atomic structure. Recall that a positive ion is an atom that has "lost" one (or more) of its outer-shell electrons. Solid metals such as copper have a regular arrangement of positive ions, which form the metallic crystal that was described in Section 1–5. These outer-shell electrons are free to move about but tend to distribute themselves so as to maintain overall electrical neutrality, as shown in Figure 3–1.

FIGURE 3–1

Metallic bonding in copper.

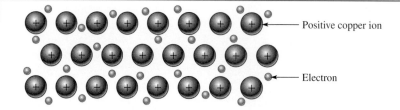

ELECTRICAL QUANTITIES AND MEASUREMENTS

The negatively charged electrons hold the positive ions of the metal together, forming the metallic bond. The "sea" of free electrons accounts for copper's excellent electrical properties and is responsible for its metallic luster and other properties.

Other solid metals have a similar structure to copper. Most are excellent conductors and have other properties that make them suited to specific electrical and electronic applications. Gold, for example, can be drawn into extremely fine wires; it is noncorrosive, so it is used in switch contacts and as a plating on plugs. Alloys (mixtures of two or more metals) also have many electronic applications. Most solder used in electrical work is an alloy of lead and tin and is widely used for making solid connections on circuit boards, cables, and connectors.

In liquids, the moving charge is composed of positive and negative ions, never electrons. Materials known as **electrolytes** form ions in water solution and are good conductors. The three general categories of substances that form ions in water solution are acids, bases, and salts. An example of a common acid that is used in lead-acid batteries is sulfuric acid; an example of a common base is sodium hydroxide (lye); and an example of a salt is ordinary table salt. In addition to water solutions of salts, molten salts can conduct electricity also. Interestingly, pure water is not a good conductor but can conduct slightly due to the slight natural ionization of pure water. Ordinary table salt is composed of a crystalline structure of sodium (Na^+) ions and chlorine ions (Cl^-). When table salt is added to water, the crystal breaks apart and separate ions enter the solution. In solution, the sodium and chlorine ions are the charge carriers. The solution can conduct electricity due to the movement of these ions, as illustrated in Figure 3–2.

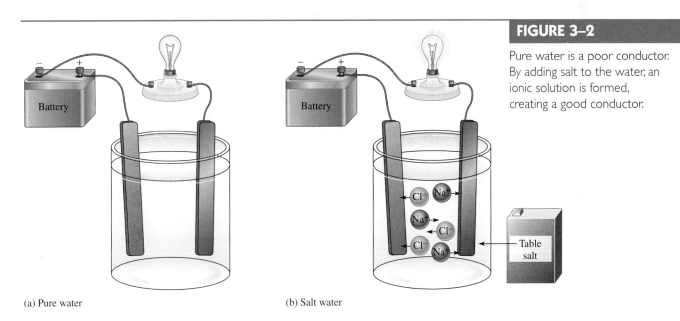

FIGURE 3–2

Pure water is a poor conductor. By adding salt to the water, an ionic solution is formed, creating a good conductor.

(a) Pure water (b) Salt water

Gases can also conduct electricity when they are broken down into ions and electrons. Unlike liquid solutions, the charge carriers in a gas include electrons. A good example of a gas conductor is the fluorescent lamp, diagrammed in Figure 3–3. The tube contains low-pressure argon gas (not shown) and a small quantity of mercury. The instant the lamp is turned on, there are no free electrons or ions to allow current. An alternating voltage is applied to the electrodes, through the ballast, which controls the starting and regulates the current. The voltage on the electrodes rapidly heats the electrodes, causing electrons to boil off and start the process. These electrons rapidly move back and forth between the electrodes because of the voltage difference between them. The electrons ionize the mercury atoms in the tube, causing them to emit high-energy photons. Finally, the high-energy photons from the mercury interact with the phosphor coating on the inside of the bulb, emitting light composed of low-energy photons. All of this happens very rapidly in modern lamps, so the light appears to come on almost instantaneously.

FIGURE 3–3

In a fluorescent lamp, mercury ions emit high-energy photons, which are converted by the phosphor coating to visible light.

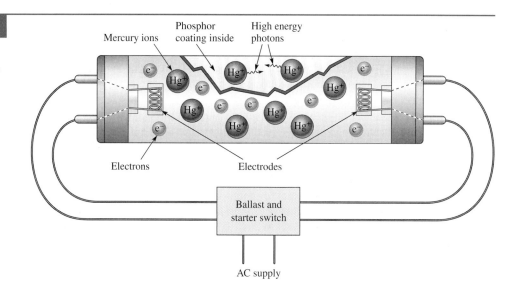

Insulators

Recall from Section 1–5 that insulators are materials that prevent the free movement of charge; hence, they are considered nonconductors. While no material is a perfect insulator, many materials are such poor conductors that we classify them as nonconductors. The atoms in these materials tend to form bonds with other atoms that leave no "extra" electrons in the structure. The most common class of electronic insulating material is plastics. Other common solid materials used as insulators include ceramics, paper, and glass.

Plastics are widely used as the insulation layer on wiring. The insulating coating provides isolation from other conductors and also is important for protection against shock hazard. Sometimes a second coating is added to provide moisture- or flame-resistance to wiring. Two or more wires are commonly bundled together within a jacket to form a wire cable, such as the one shown in Figure 3–4. The individual wires are color-coded for identification.

FIGURE 3–4

Small wire cable with a connector.

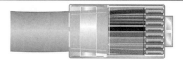

As in the case of conductors, insulators can be solids, liquids, or gases. Although solids are the most common in electronics, all three forms of matter are used in specific applications. Glass, rubber, and mica are solids commonly used in electronics. Special oils are liquids widely used in large utility system transformers because of their insulating capability and ability to carry heat. The most common gas is simply air, which is an insulator under normal conditions but can be a conductor when a sufficient voltage difference is applied across a layer of air. Lightning is a dramatic example of air breaking down and becoming a conductor. Solid conductors (with relatively low voltage differences) are sometimes separated by air to provide the required isolation from other conductors.

Semiconductors

Recall from Section 1–5 that a semiconductor is a crystalline material that has four electrons in its outer shell and with properties between those of conductors (metals) and insulators (nonmetals). Chemically, silicon acts like a nonmetal, but it has metallic luster and

ELECTRICAL QUANTITIES AND MEASUREMENTS

electrical behavior closer to a metal. It is the most widely used semiconductor for electronics. Although pure silicon is not a good conductor, silicon can become a good conductor when certain impurities are added to it.

There are two different types of impurities that can be added to the crystal structure of the semiconductor. One common impurity is phosphorous, which has five electrons in its outer shell. It is called an **n-type impurity** (*n* stands for negative) because four of the outer-shell electrons bond with adjacent silicon atoms in the crystal, leaving an "extra" one available for conduction. A different type of impurity is represented by boron, which has only three electrons in its outer shell. This is called a **p-type impurity** (*p* stands for positive). When it is introduced into the silicon crystal, one bonding position for the silicon is missing a bonding electron. This missing position is called a *hole*.

When the *p*-materials and *n*-materials are manufactured together as one solid, some electrons on the *n* side move to fill holes on the *p* side, creating a region in the middle called the *depletion region*. This is illustrated in Figure 3–5. As a result the material has an interesting electrical property. Because of this depletion region, the resulting device is a good conductor if the voltage is applied in one direction and a good insulator if the voltage is applied in the other direction. Thus, current can pass in only one direction. This type of device is called a **diode**, a very important electronic device. Because diodes allow current in only one direction, they are used in power supplies as a key component for converting ac to dc. One type of diode is the **light-emitting diode (LED)**, so called because it emits light when current passes through it.

FIGURE 3–5

A diode is composed of two types of semiconductor materials joined together as shown in (a). At the time of formation, a depletion region is formed as shown in (b).

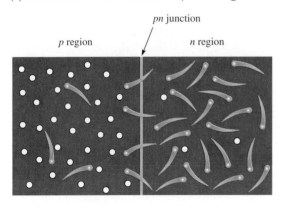

(a) At the instant of junction formation, free electrons in the *n* region near the *pn* junction begin to diffuse across the junction and fall into holes near the junction in the *p* region. Free electrons are shown with "tails" to indicate mobility.

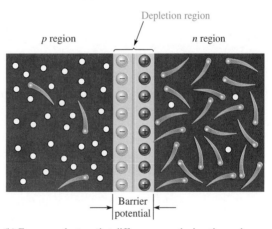

(b) For every electron that diffuses across the junction and combines with a hole, a positive charge is left in the *n* region and a negative charge is created in the *p* region, forming a barrier potential. This action continues until the voltage of the barrier repels further diffusion.

Various electronic devices are constructed from semiconductors, but the most important device is the **transistor**. The first successful transistor at Bell labs (see Section 1–1) was a semiconductor that used three layers of *p*-type and *n*-type materials. The device could amplify (make larger) a small signal such as the electrical pattern that is made when you speak into a telephone. Figure 3–6 shows this first transistor. It was an active device, meaning it

CHAPTER 3

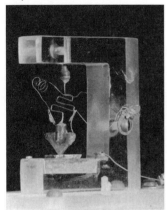

FIGURE 3–6

The first transistor. Reprinted with the permission of Lucent Technologies, Inc./Bell Labs—http://www.bell-labs.com.

could take a small amount of power from a dc source and convert it to signal power. (A passive device, like a resistor, cannot increase signal power.)

The transistor formed the backbone of small amplifiers that were important in the improvement of long distance telephone service. Today, a large variety of transistors are available. Transistors have many applications; for example, they are the primary components in integrated circuits. Typically, transistors are formed into a three-layer sandwich of either *npn* or *pnp* materials. These and other semiconductor components are studied in detail in courses dealing with electronic devices.

Review Questions

Answers are at the end of the chapter.

1. What type of particle forms the structure of metallic solids?
2. Why are metals good conductors?
3. What is the name for liquids that are good conductors?
4. What are two purposes for the insulated coating on wiring?
5. What is a semiconductor?

3–2 CURRENT AND THE ELECTRIC CIRCUIT

Current was defined in Section 2–3 as the rate that charge moves past a point. Recall that the definition of one ampere is one coulomb of charge moving past a point in one second.

In this section, you will learn the difference between direct and alternating current and terms associated with ac.

Direction of Current

Current is the flow of charge past a point. The word *current* implies a *flow,* so it is redundant to say "current flow." The original definition of current was based on Benjamin Franklin's belief that electricity was an unseen substance that moved from positive to negative. Today, we define **conventional current** based on this original assumption of positive to negative. Engineers still use this definition, and many schools and textbooks show current with arrows drawn with this viewpoint.

Today, it is known that in metallic conductors, the moving charge is actually negatively charged electrons. Electrons move from the negative to the positive point, just *opposite* to the defined direction of conventional current. The movement of electrons in a conductor is called **electron flow**. A large number of schools and textbooks show electron flow with arrows drawn in the opposite direction of conventional current.

Unfortunately, the controversy between whether it is better to show conventional current or electron flow in representing circuit behavior has continued for many years and does not appear to be subsiding. It is immaterial which direction you use to form a mental picture of current. In practice, there is only *one* correct way to connect a dc ammeter to make current measurements. Since this text is intended as a practical guide, we show the proper polarity for dc meters where appropriate rather than specifying a direction for current with arrows. All current paths are indicated with the meter symbols shown in Figure 3–7. These are similar to the symbols used in Multisim. The symbols can be used to show the relative current for comparison purposes or a specific amount of current. It is important to notice the polarity of the meters when you see them in circuits.

ELECTRICAL QUANTITIES AND MEASUREMENTS

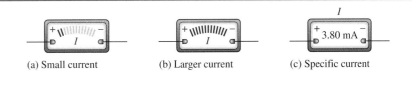

(a) Small current   (b) Larger current   (c) Specific current

FIGURE 3–7

Special meter symbols for showing current. The number of bars indicates the relative current.

Many of the circuit drawings in this text show the polarity of the voltage source with the positive terminal up; others show it down. Although it is of no particular significance to draw it one way or the other, you must observe the polarity when you connect meters.

Direct and Alternating Current

Direct Current (dc)

Current that is uniform in one direction is called **direct current**, or just dc. A constant applied voltage causes direct current from a source, such as a battery, in a circuit. To have current in a circuit, the circuit must be complete, which means it has a source, a load, and a closed path. The **load** is a component that uses the power provided by the rest of the circuit.

A simple complete dc circuit is shown in Figure 3–8(a) with a battery as the source and a light bulb for the load. The current in this circuit is steady, as illustrated in Figure 3–8(b). Because the conductors are wires, electrons comprise the moving charge in the circuit itself; but ions move in the battery. The flow of electrons is from the negative terminal of the battery, through the load, and back to the positive terminal of the battery where they complete a chemical reaction by combining with the ions (described in Section 3–3).

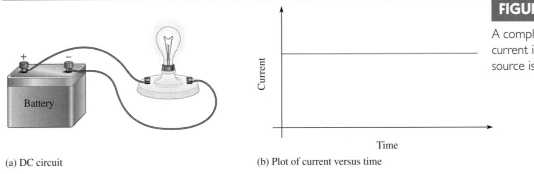

(a) DC circuit   (b) Plot of current versus time

FIGURE 3–8

A complete dc circuit. The current is dc because the source is a battery.

A closed water system is analogous to a basic circuit. The path is a closed loop with a pump and a turbine, as shown in Figure 3–9. The pump increases the pressure from the inlet side to the outlet side and is analogous to the voltage source in an electric circuit. The pump pushes water (current) through a pipe, which represents the conductive path in

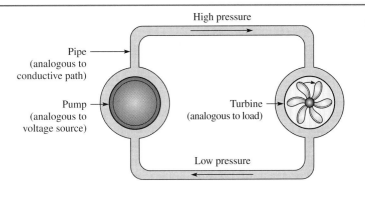

FIGURE 3–9

A water system analogy to an electrical circuit.

a circuit. As in an electric circuit, the final outcome is useful work; in this case, the turbine is the load. Of course, voltage is not a true "pressure" as in the case of a water pump, but the analogy is a useful mental picture of the processes occurring in a circuit.

A familiar example of a dc circuit is a two-cell flashlight, as shown in Figure 3–10. The voltage applied to the bulb is the sum of the two battery voltages. The way the two batteries are connected to increase the voltage is referred to as **series aiding**. If one battery were reversed, the voltage would be nearly zero—this is often called **series opposing**. Electrons leave the negative terminal of the right battery, travel through a spring, a metal strip, a switch, and a metal reflector to the bulb. The other end of the bulb is connected to the positive side of the left battery. As in all circuits, there must be a continuous path through conducting materials. The switch is used to open or close this path, which turns the flashlight on and off. The term *open* means the path is broken and current cannot pass; the term *closed* means the path is continuous and current can pass.

FIGURE 3–10

A flashlight circuit.

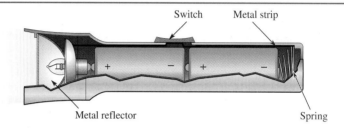

For most circuits, it is easier to show the electrical connections with a schematic. A **schematic** is a drawing that uses standard symbols to show the logical connection of the various components. A schematic of the flashlight is shown in Figure 3–11. Notice the symbol for the two cells is shown as one battery symbol, which is the symbol for more than one cell with the longer line representing the positive side. The battery symbol is also used as a generalized symbol for a dc source. The switch is a mechanical control for turning on and off the flashlight. The switch symbol shown is a single-pole–single-throw (SPST) type in the closed position. The term *single-pole* refers to the fact that there is only one movable arm on the switch. The term *single-throw* refers to the fact that there is only one contact on the switch. Other details of the flashlight, such as the spring and metal reflector, are not indicated on a schematic.

FIGURE 3–11

Schematic of flashlight.

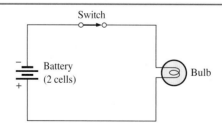

A variation on the basic flashlight is a light-emitting diode (LED) flashlight. The LED flashlight is useful for astronomers working at night because the pure red light emitted by the diode helps prevent loss of night vision. Figure 3–12 shows the schematic of this flashlight. It has the basic requirements of any circuit: a voltage source, a complete path, and a load. In this case, the load is the resistor R and the LED. The schematic shows the resistor and the LED. Notice that arrows point away from the LED to remind you that it emits light. This circuit is an interesting one to construct in the lab and test. LEDs are useful as indicators in many different devices; the lab manual that accompanies this text uses them in selected experiments as a "visual tool" to show that current is present.

ELECTRICAL QUANTITIES AND MEASUREMENTS

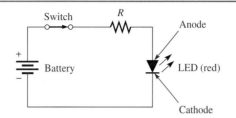

FIGURE 3–12

Schematic of an LED flashlight. R is a resistor that limits current.

Alternating Current (ac)

Electric utility companies distribute current that changes direction or alternates back and forth. AC generators (called alternators) and many function generators used in electronic testing produce **alternating current** (ac). In one cycle, the current goes from zero, to a positive peak, back to zero, to a negative peak, and back to zero. One cycle consists of a positive and a negative **alternation** (half-cycle). The cyclic pattern of ac is called a **sinusoidal wave** (or sine wave for short) because it has the same shape as the trigonometric sine function in mathematics. The sine wave will be described in more detail in Section 9–1.

In North America, ac alternates one complete cycle 60 times per second; in other parts of the world it is 50 times per second. The number of complete cycles that occur in one second is known as the **frequency**, f. Frequency is measured in units of **hertz (Hz)**, named for Heinrich Hertz, a German physicist. Figure 3–13 illustrates the definition of frequency.

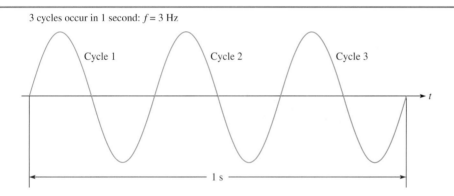

FIGURE 3–13

Example of a frequency of 3 Hz.

The **period** (T) of a sine wave is the time required for 1 cycle. For example, if there are 3 cycles in one second, each cycle takes 1/3 of a second. This is illustrated in Figure 3–14. From this definition, you can see that there is a simple relationship between frequency and period.

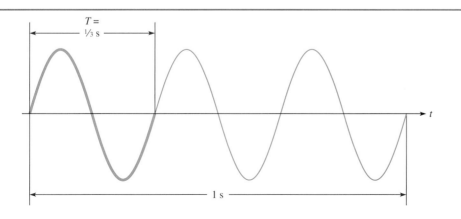

FIGURE 3–14

Example of the definition of period. The period shown is 1/3 s because it is the time for one cycle.

CHAPTER 3

$$f = \frac{1}{T} \qquad (3\text{--}1)$$

and

$$T = \frac{1}{f} \qquad (3\text{--}2)$$

where f is frequency in Hz and T is period in s.

An ac circuit is illustrated Figure 3–15(a). The difference in this drawing and the flashlight shown in Figure 3-10 is that the source is not shown explicitly and may be located miles away at the power company's generating plant. All circuits require a source, a complete path, and a load. Notice that both the dc flashlight circuit and the ac light circuit have these essential elements.

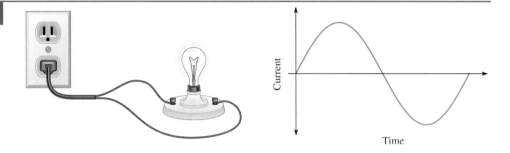

FIGURE 3–15

A complete ac circuit. AC is supplied by the electric utility company.

Comparison of AC and DC

You might wonder how a steady current (dc) and a cyclic current (ac) can be compared. In ac, the charge reverses direction many times in a second, so current is constantly changing. The fact that it moves in one direction and then the other averages to zero, yet energy is still delivered to the load. We call the current at any instant in time **instantaneous current** to describe a specific point on the cycle. Instantaneous current can be any value from the negative peak value to the positive peak value (including zero current).

In order to make sense of ac and compare it to dc, engineers devised a measurement of current and voltage based on the equivalent electrical energy that is delivered. As you know, the product of current and voltage is power. If the ac is specified as **rms current** (rms is root-mean-square) and **rms voltage**, the product is exactly the same power as a steady direct current and direct voltage of the same values. This is the way all ac is specified (unless stated otherwise). Thus, 1 A of ac (rms) is equivalent to 1 A of dc and 1 Vrms is equivalent to 1 Vdc. Section 9–1 has more on rms current and voltage. For now, keep in mind that ac and dc can be compared directly when the ac is stated as an rms value (the usual case).

EXAMPLE 3–1

Problem

Determine the direct current (dc) for each of the following cases:

(a) 50 C passes a point in 4 s

(b) 0.5 mC passes a point in 0.1 s

(c) 0.06 C passes a point in 20 ms

Solution

(a) $I = \dfrac{Q}{t} = \dfrac{50 \text{ C}}{4 \text{ s}} = \mathbf{12.5 \text{ A}}$

(b) $I = \dfrac{Q}{t} = \dfrac{0.5 \text{ mC}}{0.1 \text{ s}} = \mathbf{5 \text{ mA}}$

(c) $I = \dfrac{Q}{t} = \dfrac{0.06 \text{ C}}{20 \text{ ms}} = \mathbf{3 \text{ A}}$

Question*

How long does it take a current of 50 mA (dc) to move 1 C of charge past a point?

Review Questions

6. What is a load?
7. What are the two general types of current?
8. What type of current is supplied by electric utilities?
9. How many times does 60 Hz change directions in 1 s?
10. What is the difference between instantaneous current and rms current?

VOLTAGE SOURCES 3–3

In general, voltage sources have in common the ability to convert energy from one form to another or from one voltage to another. Three methods to produce a voltage in electronic systems are batteries, generators, and electronic power supplies.

In this section, you will learn how a battery changes chemical energy into electrical energy and about generators and power supplies.

Batteries

A **battery** is a dc voltage source that converts chemical energy to electrical energy. Recall that voltage is the energy (or work) per charge. A 1.5 V battery supplies 1.5 joules of energy to each coulomb of charge that moves between its terminals. If a resistor is connected across the terminals of this battery, then each coulomb of charge that moves through the resistor will give up its 1.5 joules of energy to heat in the resistor. The battery can be thought of as a source of energy for the charge that flows in the external circuit. This charge gives up its energy to useful work, such as turning a motor, or to heat in the external circuit.

When the battery is first constructed, chemical energy (through a chemical reaction) converts the materials in the battery (reactants) to electrostatic potential energy. In a very short time, the voltage builds on the positive terminal, and the work the battery must do through the chemical reaction increases until the reaction stops. The battery is then in a state of equilibrium.

Electrons on the negative terminal are attracted to the positive terminal of the battery but must have an external conductive path to follow. When an external path is provided, chemical reactions can occur at each terminal of the battery, allowing work to be done on additional

---

*Answers are at the end of the chapter.

CHAPTER 3

charges provided by the materials in the battery. As long as the external path is present, the chemical reactions can continue, and the battery continues to do work. Of course, the battery eventually runs out of reactants.

The variety of types of batteries and the various chemical reactions that occur within those batteries suggest a close relationship between chemistry and electricity. All batteries exchange chemical energy for electrical energy. As you know, energy (or work) per charge defines voltage (measured in volts), and the battery provides energy to each unit of charge. It is something of a misnomer to talk about "charging a battery" because a battery does not store charge but rather stores chemical potential energy.

The type of chemical reaction that occurs in batteries is known as an **oxidation-reduction** reaction. This is a class of chemical reaction that occurs and always involves electron transfer from one substance to another. In a battery, the electron transfer can only occur in the external circuit, not in the battery itself because the reactants themselves are never in physical contact with each other. In effect, the connection of the external circuit is the mechanism that permits the oxidation-reduction reaction to occur.

Figure 3–16 illustrates how a single-cell copper-zinc battery could be constructed in a chemistry lab. A copper and a zinc electrode are immersed in solutions of zinc sulfate ($ZnSO_4$) and copper sulfate ($CuSO_4$), which are separated by a porous barrier. The barrier allows ions to pass through it, keeping the charge balance of the cell. There are no free electrons in the solution, so an external path for electrons is provided to allow the reactions to proceed. On the anode side, electrons are given up by the zinc; on the cathode side, they combine with copper ions from the solution to form copper metal. The chemical reactions (shown in the figure) occur at the electrodes. Different types of batteries have different reactions, but all involve transfer of electrons in the external circuit.

FIGURE 3–16

A copper-zinc battery. The reaction can only occur if an external path is provided for the electrons.

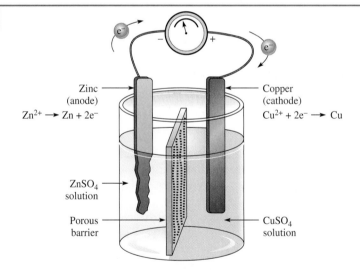

A **cell**, such as the one described in Figure 3–16, is the basic unit of a battery. A single cell will have a certain fixed voltage. In the zinc-copper cell, the voltage is 1.1 V. Other combinations of materials will have different voltages.

The Lead-acid Battery

A more useful cell than the zinc-copper cell uses lead and lead-oxide electrodes that are immersed in sulfuric acid. These lead-acid batteries are widely used in cars because they can provide very high current needed to start the car. They are particularly useful because the

ELECTRICAL QUANTITIES AND MEASUREMENTS

chemical reaction can be reversed (referred to as "recharging"). The cell voltage is approximately 2.1 V. For this reason, a standard 12 V car battery requires six such cells and has a voltage of approximately 12.6 V when fully charged.

In the lead-acid battery, the sulfuric acid forms ions in solution (H^+ and SO_4^{2-}). These ions carry charge through the solution to the lead and lead-oxide strips, where a reaction takes place that converts the acid ions and the lead metal to lead sulfate ($PbSO_4$) and water. On the negative terminal, electrons are released as the oxidation part of the chemical reaction, and they travel through the load to the positive terminal. At the positive terminal, the electrons enable the reduction part of the chemical reaction to proceed. The lead sulfate forms on both strips. Eventually, the lead sulfate prevents further oxidation-reduction reactions from occurring and the battery is discharged.

When the cell is discharged, much of the original acid has been converted to water. This causes the density (mass per volume) of the battery acid to be lower. One method for testing batteries is to measure the specific gravity of the acid; if it is low, the battery is discharged. **Specific gravity** is a dimensionless number that compares the density of a substance to that of water. Lead-acid batteries can be recharged by applying an external voltage. This causes current to proceed in the reverse direction, restoring the strips and the acid to their original condition.

Ampere-hour Ratings of Batteries

Because of their limited source of chemical energy, batteries have a certain capacity that limits the amount of time over which they can produce a given amount of current. This capacity is measured in **ampere-hours (Ah)**. The ampere-hour rating (*Ah*) indicates the length of time that a battery can deliver a certain amount of current to a load at the rated voltage. When a battery is unable to furnish current, the voltage drops below the rated level. While the ampere-hour rating is certainly not the only important specification for batteries, it is a good indicator of a battery's life in a given application. Table 3–1 shows the ampere-hour ratings of some common batteries.

> **SAFETY NOTE**
> **Battery Safety**
>
> Lead-acid batteries can be dangerous because sulfuric acid is highly corrosive and battery gases are explosive (primarily hydrogen). The acid in the battery can cause serious eye damage if it contacts the eye and can cause skin burns or destroy clothing. You should always wear eye protection when you work on or around lead-acid batteries and wash well after handling batteries.
>
> When removing a battery from service, make sure the switch is off. When a cable is removed, a spark may be created and ignite the explosive battery gases.

Battery type	Voltage (V)	Typical battery rating (Ah)
1235 (9 V)	9.0	0.4
AA Carbon-zinc	1.5	0.95
C Carbon-zinc	1.5	3.0
D Carbon-zinc	1.5	5.9
AA alkaline	1.5	2.85
C Alkaline	1.5	8.35
D Alkaline	1.5	18.0
Automobile - Lead acid	12.6	125

TABLE 3–1

Some representative ampere-hour ratings of batteries. Specifications vary significantly, particularly for automobile batteries. AA, C, and D are single-cell batteries.

A rating of 1 Ah means that a battery can deliver 1 A of current to a load for 1 h at the rated voltage output. This same battery can deliver 2 A for one-half hour. The more current the battery is required to deliver, the shorter is the life of the battery. In practice, a battery usually is rated for a specified current level and output voltage. For example, a 12 V lead-acid battery may be rated for 70 Ah at 3.5 A. This means that it can produce 3.5 A for 20 h. As an equation,

$$Ah = It \tag{3-3}$$

where *Ah* is the ampere hour rating of the battery in ampere-hours, *I* is current in amperes, and *t* is time in hours.

CHAPTER 3

EXAMPLE 3–2

Problem
For how many hours can a battery deliver 10 A if it is rated at 120 Ah?

Solution
The ampere-hour rating of the battery is the current times the number of hours.

$$Ah = It$$

Solving for t,

$$t = \frac{Ah}{I} = \frac{120 \text{ Ah}}{10 \text{ A}} = \mathbf{12\ h}$$

Notice that the ampere units cancel in the above equation, leaving time in hours.

Question
How long will a fully charged battery rated for 100 Ah last if the current is 20 A?

Generators

Generators are another important type of voltage source. Generators convert mechanical energy to electrical energy by the action of rotating a coil of wire in a magnetic field. As the coil rotates, a voltage is induced across it. The mechanical energy to turn the generator can be supplied by any of a number of sources. In a hydroelectric plant, falling water is used to turn the generator. In most power plants, generators are typically turned by a steam turbine. Burning a fuel such as coal or producing heat from a nuclear reaction creates steam for the turbine.

Generators can be designed to produce either ac or dc. Nearly all large power generators are ac types. DC generators are used in many manufacturing applications such as the semiconductor and glass industries. A dc generator is shown in Figure 3–17. Incidentally, the term ac or dc was derived to describe alternating *current* or direct *current* but is sometimes applied to voltage. You may hear of "ac voltage" which, taken literally, means "alternating current voltage." It is generally understood to mean alternating voltage. Both ac and dc generators are discussed in detail in Chapter 8.

FIGURE 3–17

A cutaway view of a dc generator.

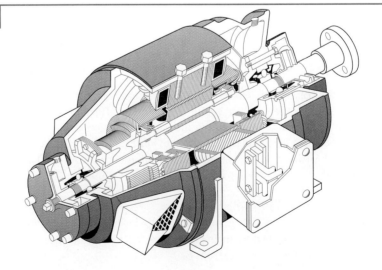

Electronic Power Supplies

All circuits require a source of energy to operate. In most cases, dc is required for electronic circuits and systems. Batteries are particularly useful for small portable systems such as cell phones or laptop computers because they provide a constant voltage level, but they are too expensive and have too short a useful life for larger systems. For larger, nonportable systems, a power supply is required to provide the dc.

Power supplies could be more properly called *voltage converters* because typically they change the alternating utility voltage into a constant voltage. DC power supplies convert ac from the electric utility to precisely regulated dc. This means that the output voltage is automatically maintained by internal regulation circuitry, even if the input varies or the load changes.

Occasionally, an application requires regulated ac instead of dc. AC power supplies usually provide the user with the means to vary the output amplitude and frequency and maintain the set quantities precisely. One application of ac power supplies is for testing electronic systems to determine the response to changes in the utility voltage.

Testing circuits is an essential part of laboratory work, so a dc power supply is a fundamental instrument in electronics laboratories. Figure 3–18 shows a typical laboratory dc regulated power supply. This power supply has three outputs in one unit; all are precisely regulated. One of the supplies is a fixed 5.0 V supply, a typical voltage for operating logic circuits; the other two can be varied, either independently or together. The power supply meters allow the user to monitor voltage and current or set the supplies prior to connecting them.

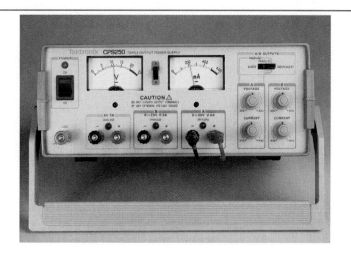

FIGURE 3–18

A laboratory power supply. (Used with permission from Tektronix, Inc.)

Power Supply Efficiency

An important characteristic of electronic power supplies is efficiency. **Efficiency** is the ratio of the output power, P_{OUT}, to the input power, P_{IN}. Recall that power is measured in watts (W). Efficiency is usually expressed as a percentage.

$$\text{Efficiency} = \frac{P_{OUT}}{P_{IN}} \times 100\% \qquad (3\text{--}4)$$

For example, if the input power is 100 W and the output power is 60 W, the efficiency is $(60 \text{ W}/100 \text{ W}) \times 100\% = 60\%$.

All power supplies require that power be supplied to them. Most laboratory power supplies use the ac power from a wall outlet as the input. The output of the supply is typically regulated dc. The output power is always less than the input power because some of the total power must be used internally to operate the power supply circuitry. This amount is called the *power loss*.

CHAPTER 3

The output power is the input power minus the amount of internal power lost. High efficiency means that very little power is dissipated within the power supply (as heat) and there is a higher proportion of output power for a given input power. Figure 3–19 illustrates this idea. The efficiency of power supplies depends on the type of internal circuitry. Highly efficient supplies are desirable in products such as computers to avoid heat buildup due to the confined space. Power supply efficiency can range from about 40% to as high as 85%.

FIGURE 3–19

The input power is divided between the output power and the power lost. The power lost is dissipated as heat.

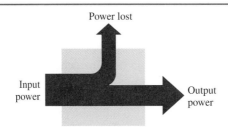

EXAMPLE 3–3

Problem
A power supply delivers 40 W of power and has an efficiency of 70%. What is the input power?

Solution

$$\text{Efficiency} = \frac{P_{\text{OUT}}}{P_{\text{IN}}} \times 100\%$$

$$P_{\text{IN}} = \frac{P_{\text{OUT}}}{\text{Efficiency}} \times 100\% = \frac{40 \text{ W}}{70\%} \times 100\% = \mathbf{57.1 \text{ W}}$$

Question
What is the output power if the input power to a power supply is 500 W and the efficiency is 85%?

Review Questions

11. What type of chemical reaction takes place in all batteries?
12. Why is it important to wear eye protection when working with lead-acid batteries?
13. What is meant by the ampere-hour rating of a battery?
14. What form of energy is the input to a generator?
15. If a power supply has an input power of 200 W and 60 W is lost as heat, what is the efficiency?

3–4 RESISTANCE

Resistance is the opposition to current and is a basic property of materials. Components that are designed to have a specific amount of resistance are called resistors.

In this section, you will learn more about resistance and how to read resistor color codes.

ELECTRICAL QUANTITIES AND MEASUREMENTS

Resistance

As defined in Section 2–3, resistance is the opposition to current. Resistance is measured in units of ohms symbolized by the Greek letter Ω. One ohm (1 Ω) of resistance exists when there is one ampere (1 A) of current in a material with one volt (1 V) applied across the material.

With the exception of a special class of materials called superconductors, all materials have resistance. Conductors are metallic materials with low resistance whereas insulators are materials with high resistance. **Superconductors** are capable of carrying current with no resistance but in general are specialized materials that must be kept extremely cold to function (the best ones are ceramic materials composed of metallic oxides). They are not found in ordinary circuits or systems.

When there is current through a resistive material, heat is produced by the collisions of electrons and atoms. Even wire, which has a very small resistance, can become warm when there is sufficient current through it.

Resistors

Resistors are components that are designed to have a certain amount of resistance between their leads or terminals. The principal applications of resistors are to limit current, divide voltage, and, in certain cases, generate heat. Although there is a variety of different types of resistors that come in many shapes and sizes, they can all be placed in one of two main categories: fixed or variable. The symbols for a fixed resistor and two variable resistors are shown in Figure 3–20. Variable resistors are subdivided into **potentiometers** ("pots") with three terminals and **rheostats** with two terminals. Notice, however, that a potentiometer can be connected as a rheostat, as shown, by connecting the center terminal to one side. Potentiometers are used to control voltage; rheostats are used to control current.

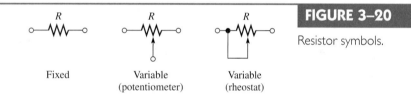

FIGURE 3–20

Resistor symbols.

Fixed Resistors

Fixed resistors are available with a large selection of resistance values and power ratings that are set by the manufacturer. Fixed resistors are constructed using various methods and materials. A common type of fixed resistor is the carbon-composition type, which is made with a mixture of finely ground carbon, insulating filler, and a resin binder. Figure 3–21(a) shows the construction of a typical carbon-composition resistor.

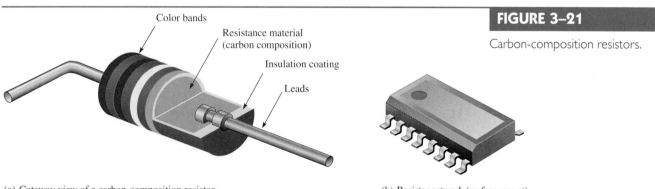

FIGURE 3–21

Carbon-composition resistors.

(a) Cutaway view of a carbon-composition resistor

(b) Resistor network (surface mount)

The ratio of carbon to insulating filler sets the resistance value. The mixture is formed into rods, and wire leads are added. The entire resistor is then encapsulated in an insulated coating for protection.

Figure 3–21(b) shows a typical surface-mounted network of resistors. Surface-mounted resistors are designed to be entirely on one surface of a mounting plane; holes are not required for the leads. Surface-mounted resistors are also available in a large number of types.

Precision resistors are available for critical applications (such as circuits requiring long-term temperature stability or low noise). One popular precision resistor is the metal-film type. These types have higher accuracy and greater temperature stability than carbon-composition resistors but are more expensive. Metal-film resistors are manufactured by depositing a resistive film material evenly onto a high-grade ceramic rod. The resistive film may be carbon (carbon film) or nickel chromium (metal film).

Another type of resistor for low-resistance precision applications is the wirewound resistor. Wirewound resistors are available either as small precision resistors or as large power resistors. They have excellent low-frequency characteristics but are not suitable for use at high frequencies. They are constructed by wrapping a resistive wire around an insulating rod and then sealed. Figure 3–22 shows two different types of wirewound power resistors.

> **SAFETY NOTE**
>
> Resistors and other components may become very hot, particularly if a circuit malfunctions. Do not touch any component while power is applied to a circuit. After power has been removed, allow time for components to cool down.

FIGURE 3–22

Typical wirewound resistors.

Variable Resistors

Variable resistors are also available in different types including carbon-composition, film, and wirewound. Potentiometers are constructed with a rotating shaft connected to a wiper, similar to the one shown in Figure 3–23. The center terminal is connected to the wiper arm, which is controlled by rotating the shaft. For applications requiring more precise control, potentiometers are available with multiple turns and in a multiturn linear arrangement such as shown in Figure 3–24.

FIGURE 3–23

Potentiometer.

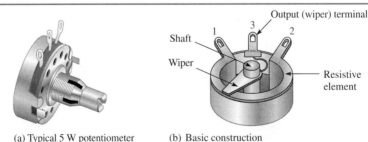

(a) Typical 5 W potentiometer   (b) Basic construction

FIGURE 3–24

Typical multiturn potentiometer.

 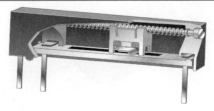

(a) Outside view   (b) Cutaway view

Variable resistors are used as volume controls, speed controls, or in other applications that require voltage or current control. General-purpose variable resistors are usually the carbon-composition type. Wirewound resistors are also commonly used in general-purpose applications.

A resistive sensor is another type of variable resistor that "automatically" changes its resistance as some physical parameter (such as temperature or light intensity) changes. One example of a resistive sensor is a thermistor ("thermal resistor"), a device that looks like a small bead but has a large resistance change with temperature. Thermistors are widely used as temperature sensors in thermostats because of their small size, low cost, and sensitivity. Their resistance can change by a factor of 1000 over the useful range of temperatures.

Resistor Specifications

The most important specifications for resistors are the amount of resistance and the power rating. The physical size of a resistor is related to the power rating; a larger resistor can dissipate more power. Other specifications include the operating temperature range and the stability. Carbon-composition resistors are available from about 1 Ω to 100 MΩ, a huge range of values. They are available with wattage ratings from 1/8 W to 2 W. Power resistors are available with power ratings up to 1000 W.

Variable resistors are also specified for the amount of resistance and the power rating. The amount of resistance is the maximum value, as measured between the outside terminals. Power ratings are again determined by the size; variable resistors have power ratings that range from 50 mW for small trim pots to about 2 W for larger potentiometers.

Another specification for potentiometers is the **taper**; potentiometers are available with either a linear or nonlinear taper. A linear taper means that the resistance is proportional to the angular position of the control; a nonlinear taper means that the resistance is not proportional to the angular position.

Resistor Color Codes

Many small resistors have color-coded bands to indicate the resistance value and the tolerance. Resistors with tolerances of 5% or 10% are coded with four bands, as indicated in Figure 3–25.* A 5% tolerance means that the actual resistance value should be within ±5% of the color-coded value. Table 3–2 indicates how the colors are interpreted. By looking for the gold or silver color band on a four-band resistor, you know how to orient the resistor.

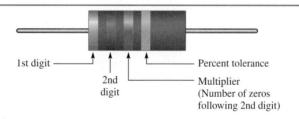

FIGURE 3–25

Four-band resistor color code.

Read the color code on a 4-band resistor as follows:

1. Beginning at the banded end, the first band is the first digit of the resistance value. If it is not clear which is the banded end, start from the end that does not begin with a gold or silver band.
2. The second band is the second digit.
3. The third band, called the multiplier band, indicates a power of ten as shown in Table 3–2. This value is multiplied by the previous digits to arrive at the total resistance.

---

*An obsolete method was to use three bands for 20% resistors.

CHAPTER 3

TABLE 3–2
Resistor color codes.

		Color	Digit	Multiplier	Tolerance
Resistance value, first three bands: First band—1st digit Second band—2nd digit *Third band—multiplier (number of zeros following the 2nd digit)		Black	0	10^0	
		Brown	1	10^1	1% (five band)
		Red	2	10^2	2% (five band)
		Orange	3	10^3	
		Yellow	4	10^4	
		Green	5	10^5	
		Blue	6	10^6	
		Violet	7	10^7	
		Gray	8	10^8	
		White	9	10^9	
Fourth band—tolerance		Gold	±5%	10^{-1}	5% (four band)
		Silver	±10%	10^{-2}	10% (four band)
		No band	±20%		

*For resistance values less than 10 Ω, the third band is either gold or silver. Gold is for a multiplier of 0.1 and silver is for a multiplier of 0.01.

4. The fourth band indicates the tolerance and is gold for 5% tolerance or silver for 10%. The tolerance band indicates the maximum variation in the actual resistance from the color-coded markings.

For resistance values less than 10 Ω, the *third* band is either gold or silver. Gold represents a multiplier of $10^{-1} = 0.1$, and silver represents $10^{-2} = 0.01$. For example, a color code of red, violet, gold, and gold represents 2.7 Ω with a tolerance of ±5%. The gold in the third position is read as a multiplier, whereas the gold in the fourth position is read as a tolerance. Gold and silver are not used in the digit bands.

EXAMPLE 3–4

Problem
Find the resistance values in ohms and the percent tolerance for each of the color-coded resistors shown in Figure 3–26.

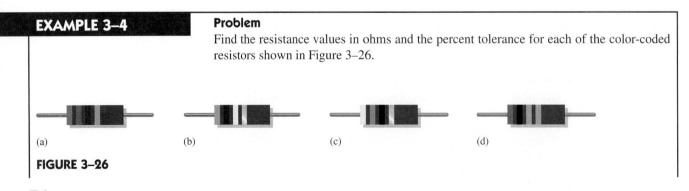

(a)　　　　　　(b)　　　　　　(c)　　　　　　(d)

FIGURE 3–26

ELECTRICAL QUANTITIES AND MEASUREMENTS

Solution
(a) Digits: first band is green = 5, second band is brown = 1
   Multiplier: third band is red = $\times 10^2$ = $\times 100$
   Tolerance: fourth band is gold = 5% tolerance
   $R = 5{,}100\ \Omega \pm 5\%$

(b) Digits: first band is gray = 8, second band is red = 2
   Multiplier: third band is yellow = $\times 10^4$ = $\times 10{,}000$
   Tolerance: fourth band is silver = 10% tolerance
   $R = 820{,}000\ \Omega \pm 10\%$

(c) Digits: first band is yellow = 4, second band is violet = 7
   Multiplier: third band is black = $\times 10^0$ = $\times 1$
   Tolerance: fourth band is silver = 10% tolerance
   $R = 47\ \Omega \pm 10\%$

(d) Digits: first band is brown = 1, second band is black = 0
   Multiplier: third band is gold = $\times 10^{-1}$ = $\times 0.1$
   Tolerance: fourth band is gold = 5% tolerance
   $R = 1.0\ \Omega \pm 5\%$

Question
A certain resistor has a brown first band, a black second band, a brown third band, and a gold fourth band. What are the value in ohms and the tolerance?

EXAMPLE 3–5

Problem
Determine the color codes for each of the following resistors. Assume all are four-band resistors with 5% tolerance.

(a) 330 Ω
(b) 15 Ω
(c) 680 kΩ
(d) 5.1 MΩ

Solution
(a) 330 Ω ± 5% = orange, orange, brown, gold
(b) 15 Ω ± 5% = brown, green, black, gold
(c) 680 kΩ ± 5% = blue, gray, yellow, gold
(d) 5.1 MΩ ± 5% = green, brown, green, gold

Question
What are the colors of a four-band 910 Ω ± 10% resistor?

Resistors with 5% tolerance values are commonly used in most applications because they can be made inexpensively. Long ago, both 5% and 10% were standardized for certain values to provide overlap between values. In this text, standard values are used throughout to help familiarize you with those values. Appendix B lists the standard 5% and 10% values.

When more precise values are required in a circuit, precision resistors with tolerances of 1% or 2% are available. These are color-coded with five bands instead of four. Beginning at the banded end, the first band is the first digit of the resistance value, the second band is the second digit, the third band is the third digit, the fourth band is the multiplier, and the fifth band indicates the tolerance. Table 3–2 digit and multiplier colors have the same meaning.

CHAPTER 3

In a five-band resistor, the fourth band is the multiplier and is normally either black, brown, red, orange, green, gold, or silver. Gold in the fourth band indicates a multiplier of $10^{-1} = 0.1$ and silver indicates a multiplier of $10^{-2} = 0.01$. Only brown or red appear in the fifth (tolerance) band. Brown in the fifth band indicates 1% tolerance, and red indicates 2% tolerance.

EXAMPLE 3–6

Problem
Find the resistance values in ohms and the percent tolerance for each of the color-coded resistors shown in Figure 3–27.

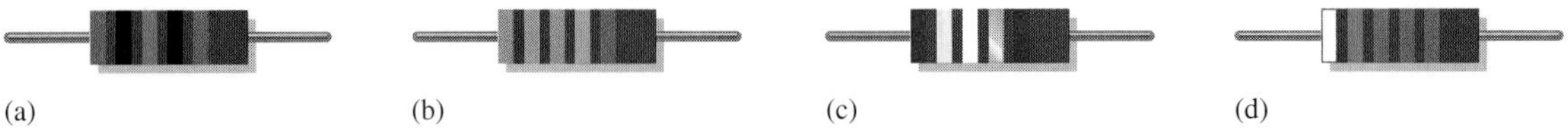

FIGURE 3–27

Solution
These are all five-band resistors; therefore, the first three bands are for the digits.

(a) Digits: first band is brown = 1, second band is black = 0, third band is green = 5
Multiplier: fourth band is black = $\times 10^0$ = $\times 1$
Tolerance: fifth band is brown = 1% tolerance
$R = 105\ \Omega \pm 1\%$

(b) Digits: first band is violet = 7, second band is gray = 8, third band is violet = 7
Multiplier: fourth band is gold = $\times 10^{-1}$ = $\times 0.1$
Tolerance: fifth band is brown = 1% tolerance
$R = 78.7\ \Omega \pm 1\%$

(c) Digits: first band is red = 2, second band is yellow = 4, third band is white = 9
Multiplier: fourth band is silver = $\times 10^{-2}$ = $\times 0.01$
Tolerance: fifth band is red = 2% tolerance
$R = 2.49\ \Omega \pm 1\%$

(d) Digits: first band is white = 9, second band is green = 5, third band is orange = 3
Multiplier: fourth band is brown = $\times 10^1$ = $\times 10$
Tolerance: fifth band is brown = 1% tolerance
$R = 9.53\ \text{k}\Omega \pm 1\%$

Question
What are the colors of the bands for a five-band resistor that is 5.62 kΩ $\pm$ 1%?

Wire Resistance

Although copper is an excellent conductor and is widely used in electronic circuits, it has some resistance. For many applications, the resistance of wire can be ignored; but in some cases, it can affect a circuit and needs to be considered. You may have seen low-voltage yard lights that grow dimmer the farther they are from the source. This is a case where the wire resistance has affected the circuit. The problem is reduced if a larger-diameter wire (greater cross-sectional area) is used or if the lights are moved closer to the source because a shorter wire has less resistance.

The size of wire that is appropriate depends on the application. The diameter of wire is arranged according to standard gauge numbers called **American Wire Gauge (AWG)** sizes. The larger the gauge number, the smaller the wire. For example, 12 gauge is common in household wiring and 24 gauge is useful for telephone circuits. Wire as small as a human hair is number 40 gauge; even smaller sizes are used in certain applications. The AWG is related to the cross-sectional area of the wire; the common English unit for cross-sectional area of wires is the circular mil, abbreviated CM. One circular mil is the area of a wire that is one thousandth of an inch (one mil) in diameter. A circular mil is illustrated in Figure 3–28.

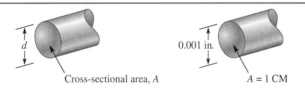

FIGURE 3–28

A circular mil.

Another common unit for cross-sectional area of wires is the square millimeter. There are 1000 millimeters in one meter, so a square millimeter has one millionth ($10^{-3} \times 10^{-3}$) as much area as a square meter. For reference, the cross-sectional area and resistance per 1000 feet of some common AWG wires are given in Table 3–3.

AWG Number	Cross-sectional area		Ω/1000 ft for copper wire*
	Circular mils (CM)	Square millimeters (mm²)	
10	10,380	5.261	0.999
12	6530	3.310	1.588
14	4110	2.083	2.525
16	2580	1.308	4.016
18	1,600	0.811	6.385
20	1022	0.518	10.15
22	642	0.325	16.14
24	404	0.204	25.67
26	253	0.128	40.81

TABLE 3–3

Cross-sectional area and resistance per 1000 feet of selected sizes of copper wire.

*Resistance is specified for 20° C.

Two wires of exactly the same size and AWG number but made from different materials will also have different resistances. This is caused by a characteristic called the **resistivity**, ρ (the Greek letter rho), which is a constant for each type of material that is related to its ability to conduct electricity (for example, all copper has the same resistivity). Resistivity can be determined by measuring the resistance across opposite faces of a cube that is 1 cm on each side. Resistivity is specified with various units so care needs to be taken in solving problems to assure that the units are consistent. In the metric system, the unit for resistivity is the ohm-meter (Ω-m) or the ohm-millimeter squared/meter (Ω-mm²/m). The latter unit is more convenient for working with wire calculations. The equivalent English-system unit for wires is the ohm-circular mil per foot (Ω-CM/ft). All of these units are essentially the product of a resistance and a length unit. Resistivity of several common conductors is given in Table 3–4.

The resistance of wire depends principally on the three factors previously mentioned: the cross-sectional area, A; the length, L; and the resistivity, ρ. These three factors are related with the following equation for wire resistance:

$$R = \frac{\rho l}{A} \qquad (3-5)$$

CHAPTER 3

TABLE 3–4
Resistivity of some common conductors.

Material	Resistivity, ρ (Ω-m)	(Ω-mm^2/m)	(Ω-CM/ft)
aluminum	2.67×10^{-8}	0.0267	16.4
copper	1.72×10^{-8}	0.0172	10.4
gold	2.2×10^{-8}	0.022	13.5
silver	1.63×10^{-8}	0.0163	10.0

EXAMPLE 3–7

Problem
Find the resistance of 700 feet of AWG #24 copper wire.

Solution
This problem can be solved by proportion. From Table 3–3, the resistance of AWG #24 copper wire is 25.67 Ω/1000 ft. This means that 700 ft will have a resistance that is

$$R = 700 \text{ ft} \times \frac{25.67 \text{ }\Omega}{1000 \text{ ft}} = 18.0 \text{ }\Omega$$

The problem can also be solved by using Equation 3–5.

$$R = \frac{\rho l}{A}$$

Using English units, the resistivity of copper is 10.4 Ω-CM/ft (see Table 3–4), and the cross-sectional area (from Table 3–3) is 404 CM. Substituting,

$$R = \frac{\rho l}{A} = \frac{10.4 \text{ }\Omega\text{-CM/ft} \times 700 \text{ ft}}{404 \text{ CM}} = 18.0 \text{ }\Omega$$

The resistance can also be found using metric units. There are 0.3048 feet in a meter, therefore 700 ft is 213.4 m. The resistivity of copper is 0.0172 Ω-mm^2/m. The area of #24 copper wire (from Table 3–3) is 0.204 mm^2. Substituting,

$$R = \frac{\rho l}{A} = \frac{0.0172 \text{ }\Omega\text{-mm}^2/\text{m} \times 213.4 \text{ m}}{0.204 \text{ mm}^2} = 18.0 \text{ }\Omega$$

Notice how the units canceled in each application of Equation 3–5.

Question
What is the resistance of 500 feet of #10 aluminum wire?

Review Questions

16. What is a superconductor?
17. What is the difference between a rheostat and a potentiometer?
18. With respect to a potentiometer, what is meant by a linear taper?
19. What does it mean if the third band of a four-band resistor is gold?
20. What is the color code for a four-band 1.0 Ω ± 5% resistor?

ELECTRICAL QUANTITIES AND MEASUREMENTS

BASIC ELECTRICAL MEASUREMENTS 3–5

In order to test and troubleshoot circuits, it is necessary to understand how to use basic electronic measuring instruments. The DMM is the most widely used electronic-measuring instrument. The older analog meter called the volt-ohm-milliammeter (VOM) is still used in some applications.

In this section, you will learn how to use the digital multimeter (DMM) and the VOM.

Digital Multimeter

A **digital multimeter (DMM)** is an instrument that can measure several basic electrical quantities and shows the measurement with a number in a display. All multimeters measure ac and dc voltage, ac and dc current, and resistance. Some can also measure other electrical quantities such as frequency or even temperature. A typical digital multimeter is shown in Figure 3–29.

DC Voltage Measurements

To measure voltage, connect the meter leads across the component to be measured (this is a parallel connection as you will see later). With digital meters, the meter normally sets the polarity automatically, and the sign is indicated on the display. It is convenient to use a red lead in the red voltage jack and a black one on ground. Then positive or negative readings are easy to interpret. Many DMMs have **autoranging**, a feature in which the DMM selects the optimum range automatically for displaying the reading. The range is the maximum voltage that can be displayed with a particular setting. Lower-priced meters have **manual ranging**, which requires the user to select an appropriate range for the measurement.

To make a voltage measurement, first move the selector switch to DC VOLTS; and, if the meter is a manual-ranging type, choose a range larger than the expected voltage (this is not necessary with an autoranging meter). Sometimes the dc volts position is indicated by the letter *V* and a straight line over a dashed line (to remind you of dc). Only after selecting the proper function (DC VOLTS) and range should you connect the meter to the circuit. Most meters will indicate the voltage no matter which way the leads are connected to the

HISTORICAL NOTE

The forerunner to the modern digital multimeter (DMM) was a digital voltmeter (DVM) with a three-digit display (999 full scale). Accuracy was quoted as 0.25% of full scale with little knowledge of long-term stability. Today, digital meters are available with 8 1/2 digits in the display and accuracy specifications of a few parts per million. In the same year that the DVM was introduced, Hewlett-Packard announced a new type of ammeter. It was a clip-on type that sensed the strength of the magnetic field produced by the current being measured instead of using standard moving-coil methods.

FIGURE 3–29

A DMM.

CHAPTER 3

circuit. If the positive meter lead is connected to the more positive voltage, then the voltage reading is displayed as a positive value; otherwise, it is indicated as a negative value. An example of measuring the voltage in a basic circuit is shown in Figure 3–30.

FIGURE 3–30

DC voltage measurement. The meter is connected directly across the component to be measured; in this case, the reading is the drop across the resistor.

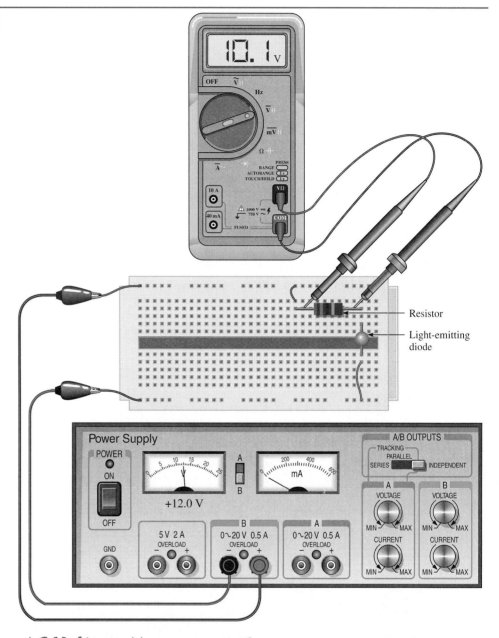

AC Voltage Measurements

AC voltage measurements are done in the same manner as dc measurements except the selector switch must be in the AC VOLTS position. Some meters indicate ac with a small sine wave. As in the case with dc measurements, select the function and range (if necessary) before you hook the meter to the circuit. Because ac goes between positive and negative values, the polarity is not important and will be displayed as a positive value in any case. It is important to realize that the reading will be displayed as an *rms value* for ac.

Another important consideration in ac voltage measurements is the frequency you are measuring. DMMs vary widely in their ability to measure higher frequencies; many are accurate only for a very small range of frequencies. Typically, the range may be from 45 Hz to 1 kHz, but some meters can be used up to 1 MHz. You should check on the accuracy specification of a meter if you aren't sure.

ELECTRICAL QUANTITIES AND MEASUREMENTS

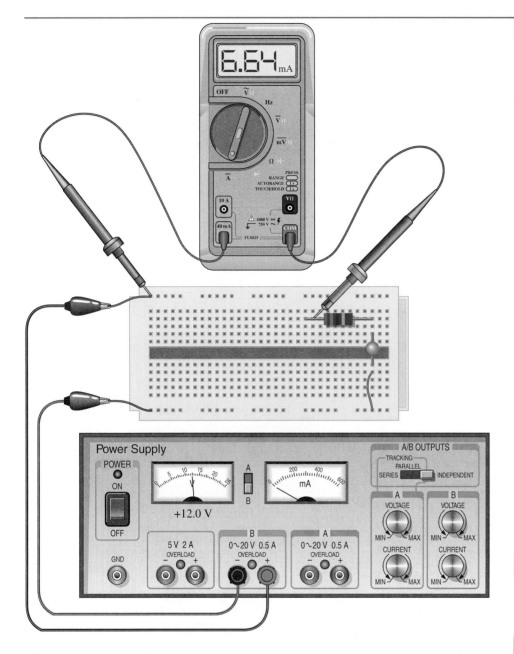

FIGURE 3–31

Current measurement. Notice that the meter is part of the path. The probes are moved to the low-current jack and the common jack for the meter shown.

FIGURE 3–32

A technician is measuring current with a clamp-on ammeter. (Courtesy of Fluke)

Current Measurements

For DMMs, select either the AC or DC function, and it is necessary to move the probes to separate current jacks. As in any current measurement, connect the meter in series ("in-line") with the circuit under test. Note that if you are measuring ac, it will be shown on the display as the rms value. A current measurement is illustrated in Figure 3–31. Caution! If the meter is inadvertently connected across a voltage source, very high current will flow through the meter, causing either a fuse to blow or damage to the meter.

A useful meter for measuring alternating current is the clamp-on ammeter, which does not require the circuit to be broken to insert the meter. The meter may be a stand-alone unit or part of a traditional DMM. Figure 3–32 shows a technician using a clamp-on ammeter. The sensing element is a set of jaws that can be opened and closed around a single conductor. The meter senses the changing magnetic field from the current and converts it directly to an rms current reading.

CHAPTER 3

To measure an unknown current with any ammeter, you should always select the scale with the highest amperage range to start. If the reading is within the limits of the next lower range, select it for a more accurate reading. If you are trying to measure a very low current in a flexible wire, you can wrap several turns inside the jaws to increase the sensitivity. Divide the reading on the meter by the number of turns in the jaws.

Resistance Measurements

DMMs measure resistance by using an internal current source to provide an accurate, fixed current through the unknown resistance. This develops a voltage across the unknown resistance that is proportional to the resistance. The voltage developed across the unknown is converted internally by the meter to a resistance value and displayed.

Resistance is always measured with one (or both) ends of the resistor under test disconnected from the circuit. This is to ensure that another path in the circuit or another voltage is not present that can cause the meter to produce an erroneous reading or (worse!) damage the meter. Figure 3–33 illustrates a resistance reading.

FIGURE 3–33

Resistance measurement. Notice that the path is broken by removing a jumper wire. Merely turning off the power supply is *not* a way to break the path.

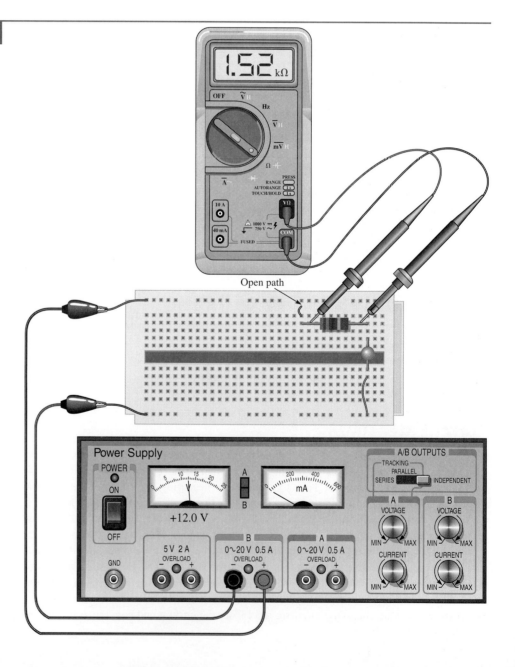

ELECTRICAL QUANTITIES AND MEASUREMENTS

DMMs frequently have continuity testing as a part of the ohmmeter function. An audible "beep" indicates a conducting path between the probes, making the test a quick check of wiring by not having to look at the display. The continuity tester is a handy feature for tracing wiring and testing for open or shorted paths on circuit boards. Some meters use two tones to differentiate between a reading that is less than 1% of the selected range and a reading that is less than 10% of the selected range.

Resolution and Accuracy of DMMs

Two important specifications for DMMs are resolution and accuracy.

Resolution

The number of digits shown for a digital multimeter is generally expressed as a whole number and a fraction such as 3 1/2 or 5 1/2. The whole number indicates the number of nines the display is able to show. The fraction indicates that the display is unable to show all possible decimal numbers in the most significant position. The fraction 1/2 indicates that the most significant digit (MSD) can be either a 0 (normally blanked off) or a 1. A 3 1/2 digit display can show numbers from 0000 to 1999; this represents a resolution of 1 part in 2000 or 0.05%.

The resolution can also be expressed as the smallest difference in voltage, current, or resistance that the meter can display. For example, a 3 1/2 digit meter on the 20 V range has 10 mV resolution. On the same meter, the 2 V range gives a resolution of 1 mV. Manufacturers' specifications often list the available ranges and resolution on this basis.

Accuracy

The accuracy of digital multimeters is specified in various ways. Typically, it is specified as a percentage of the reading plus or minus a number of counts, to reflect the behavior of the circuitry. Occasionally, it is specified in parts per million (ppm) or other way. However it is stated, accuracy is only valid if the instrument is given the specified warm-up time, is operated within the stated temperature range, and the time since calibration of the instrument is valid. Manufacturers may specify an operating temperature or range for which accuracy specifications are valid (normally 25° C).

The VOM

The **volt-ohm-milliammeter** (VOM for short) is a portable analog multimeter that can measure voltage, current, or resistance. Today, VOMs are largely replaced by DMMs, but in some applications technicians prefer them, so a short introduction is warranted. Instead of a number shown on the display, the VOM shows the reading by a needle, which is read by noting the proper scale and value on that scale.

A VOM is shown in Figure 3–34. It has built-in ammeter, voltmeter, and ohmmeter functions. VOMs like the one shown have a single multipurpose switch to select the function (volts, ohms, milliamps) and the range (which of several scales to read). The meter has multiple scales that correspond to the various switch positions. The correct scale to read is determined by the function selected as well as the range. Often the scales are color-coded in some way to make it easier to identify the proper one to read. Sometimes the scale that is read is a factor of ten larger or smaller than the range switch. In those cases, the reading must be scaled for the proper factor of ten.

The resistance scale on most VOMs is different than the other scales. Usually, it is the top scale and is read from right to left instead of left to right. It also is a nonlinear scale, meaning that values are crowded together at the top end of the scale. To read resistance, the meter leads are first shorted together and a zero adjust potentiometer is adjusted for zero resistance. Then the probes are placed across the resistor to be measured and the reading noted.

The measurement methods described for DMMs for measuring voltage and current apply to VOMs as well. Note that certain functions may require moving the probes to a separate set of jacks on the front of the VOM, the same as in a DMM.

SAFETY NOTE

Multimeter Safety

All multimeters have a limited voltage that can be applied safely between any one terminal and ground (typically 1000 Vdc or 750 Vac). The dc limit will be greater than the ac limit because the meter must be rated for the ac peak voltage, not the rms voltage. You should check the manufacturer's rating before using the meter. It is dangerous to use a meter for measurements that exceed its rating.

Before using a meter, you need to be certain that test leads are in the proper sockets and the correct function is selected. A meter with the function switch in the volts position but with the test leads in the amps socket will blow a fuse (or worse!) when connected across a voltage source.

For more information on meter safety, go to www.fluke.com and download the application note "ABCs of Multimeter Safety."

CHAPTER 3

FIGURE 3–34

A VOM. (Photo courtesy of Triplett Corporation)

EXAMPLE 3–8

Problem

Determine the reading on the meter in Figure 3–35 for each of the following settings:

(a) The function switch is on the Rx100 position.

(b) The function switch is on ac voltage 6 V position.

(c) The function switch is on ac amps 3 A position.

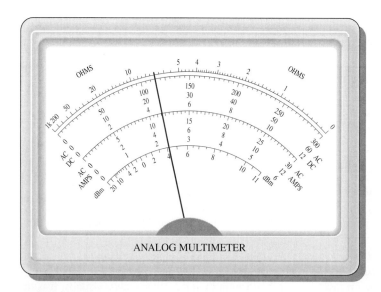

FIGURE 3–35

Solution

(a) The resistance scale is the top scale and points to the 7. Because the Rx100 position is selected, the reading is $7 \times 100 = 700\ \Omega$.

ELECTRICAL QUANTITIES AND MEASUREMENTS

(b) When the 6 V range is selected, the 60 V scale is read because there is no 6 V scale. You will need to move the decimal after making the reading. The reading is 23 but must be divided by 10. Therefore, the voltage is 2.3 V.

(c) When the 3 A position is selected, the reading is made on the 30 A scale because there is no 3 A scale. The reading is 11.6 but must be divided by 10. Therefore, the current is 1.16 A.

Question
What is the reading in Figure 3–35 if the function switch is on the Rx1000 position?

Review Questions

21. What measurements can be made with a basic DMM or VOM?
22. What are two procedures that must be done when connecting a DMM for a current measurement?
23. Why is it necessary to disconnect a voltage source when using a DMM to measure resistance?
24. What range of digits can be displayed on a 3 1/2 digit DMM?
25. What is different about the resistance scale on a typical VOM than the other scales?

CHAPTER REVIEW

Key Terms

Alternating current (ac) Current that changes direction or "alternates" back and forth between positive and negative values.

Digital multimeter (DMM) An instrument that can measure ac or dc voltage, ac and dc current, and resistance and shows the result as a number in a display.

Direct current (dc) Current that is uniform in one direction.

Efficiency The ratio of the output power, P_{OUT}, to the input power, P_{IN}, of a device.

Electrolyte A material that forms ions in water solution, thus becoming a good conductor.

Frequency The number of complete cycles that occur in one second; it is measured in hertz.

Hertz The measurement unit for frequency; one hertz is equal to one cycle per second.

Load A component that uses the power provided by the rest of the circuit.

Period The time required for one complete cycle or repetition of a periodic waveform.

Power supply Generally refers to a dc device that changes the alternating utility voltage into a constant voltage. AC power supplies convert the utility voltage to an ac output.

Resistivity A constant for each type of material that is related to its ability to conduct electricity. Metric units are the ohm-meter (Ω-m) or the ohm-millimeter squared/meter (Ω-mm^2/m). The English system unit for wires is the ohm-circular mil per foot (Ω-CM/ft).

rms current A measure of ac; it is the amount of current that delivers the same power to a load as dc.

rms voltage A measure of ac; it is the amount of voltage that delivers the same power to a given load as a constant (dc) voltage.

Schematic A drawing that uses standard symbols to show the logical connection of the various components.

Sinusoidal wave The cyclic pattern of alternating current, which has the same shape as the trigonometric sine function in mathematics.

Volt-Ohm-Milliammeter (VOM) A portable analog multimeter that can measure voltage, current, or resistance. It uses a needle against a scale to show the reading.

Important Facts

- Conductors can be solids, liquids, or gases.
- Three general categories of ion-forming liquids (electrolytes) are acids, bases, and salts.
- Three important electronic devices made from semiconductors are diodes, transistors, and integrated circuits.
- Current that is uniform in one direction is called direct current, or dc.
- Current that changes direction or "alternates" back and forth is called alternating current, or ac.
- The number of complete cycles of ac that occur in one second is known as the frequency, which is measured in units of hertz (Hz).
- AC is generally specified as rms current and voltage that is based on the equivalent power when compared to dc of the same current and voltage.
- Batteries exchange stored chemical energy for electrical energy using a type of chemical reaction known as an oxidation-reduction reaction.
- The capacity of a battery is measured in ampere-hours (Ah), which determines the length of time that a battery can deliver a certain amount of current to a load at the rated voltage.
- A dc power supply converts the alternating utility voltage into the constant voltage required by most electronic devices.
- Efficiency is the ratio of the output power, P_{OUT}, to the input power, P_{IN}.
- Resistors are components that are designed to have a certain amount of resistance between their leads or terminals. Small resistors have color-coded bands to indicate the resistance value and the tolerance.
- Two types of multimeters are the digital multimeter (DMM) and an analog type called a volt-ohm-milliammeter (VOM).

Formulas

Frequency of a repetitive wave:

$$f = \frac{1}{T} \tag{3-1}$$

Period of a repetitive wave:

$$T = \frac{1}{f} \qquad (3\text{-}2)$$

Ampere-hour rating of a battery:

$$Ah = It \qquad (3\text{-}3)$$

Efficiency of a power supply:

$$\text{Efficiency} = \frac{P_{OUT}}{P_{IN}} \times 100\% \qquad (3\text{-}4)$$

Wire resistance:

$$R = \frac{\rho l}{A} \qquad (3\text{-}5)$$

Chapter Checkup

Answers are at the end of the chapter.

1. An example of an electrolyte is
   (a) salt water (b) pure water
   (c) mercury vapor (d) glass

2. A device that allows current in only one direction is a
   (a) conductor (b) semiconductor
   (c) diode (d) integrated circuit

3. Conventional current is from
   (a) negative to positive (b) positive to negative
   (c) either of the above

4. AC that has the same heating value as dc (at the same voltage) is called
   (a) average current (b) peak current
   (c) instantaneous current (d) rms current

5. The form of energy stored in a battery is
   (a) thermal (b) electrical
   (c) chemical (d) mechanical

6. When an automobile battery is discharged, the density of the acid is
   (a) lower (b) unchanged
   (c) higher

7. If a battery rated for 80 Ah supplies 0.5 A, you could expect it to last about
   (a) 40 h (b) 80 h
   (c) 120 h (d) 160 h

8. The form of energy converted by a generator to electrical energy is
   (a) thermal (b) electrical
   (c) chemical (d) mechanical

9. The form of energy converted by a power supply to electrical energy is
   - (a) thermal
   - (b) electrical
   - (c) chemical
   - (d) mechanical

10. If 200 W are supplied to a power supply and 40 W are given off as heat, the efficiency is
    - (a) 20%
    - (b) 40%
    - (c) 60%
    - (d) 80%

11. A thermistor decreases in resistance as
    - (a) light intensity increases
    - (b) light intensity decreases
    - (c) temperature increases
    - (d) temperature decreases

12. A four-band 360 Ω resistor that has a tolerance of 5% is color-coded
    - (a) red, green, black, gold
    - (b) red, green, brown, gold
    - (c) orange, blue, black, gold
    - (d) orange, blue, brown, gold

13. If you see brown in the fifth position of a five-band resistor, it means
    - (a) the resistor has 1% tolerance
    - (b) the resistor has 2% tolerance
    - (c) the multiplier is 1
    - (d) the multiplier is 10

14. A color that is never in the first position for either a four-band or a five-band resistor is
    - (a) brown
    - (b) gold
    - (c) white
    - (d) gray

15. A meter that requires you to read the position of a needle against a scale is a
    - (a) DMM
    - (b) VOM
    - (c) VMM
    - (d) IBM

Questions

Answers to odd-numbered questions are at the end of the book.

1. Why are solid metals good conductors of electricity?
2. What is the charge carrier in liquids?
3. What are the charge carriers in a mercury vapor lamp?
4. What are the two types of impurities that are used in manufacturing semiconductors?
5. What component allows current in only one direction?
6. What characteristic makes a transistor an active device?
7. What is a plasma?
8. What is the load in a flashlight?
9. How do electrons move in a flashlight?
10. What is meant by a single-pole–single-throw switch?
11. What is the measurement unit for frequency?
12. How is ac compared to dc?
13. What does rms stand for?
14. 100 mA of ac (rms) is equivalent to how much dc?

15. Why is it a misnomer to say "charging a battery"?
16. In an automobile (lead-acid) battery, what are the two ions that carry charge internally in the battery?
17. Approximately how many hours can a 100 Ah battery provide 5 A?
18. What is one application for an ac power supply?
19. Why is it advantageous to have a power supply that is very efficient?
20. What is the measurement unit for resistance?
21. What is a superconductor?
22. Can a potentiometer be connected as a rheostat? If so how?
23. Can a rheostat be connected as a potentiometer? If so how?
24. What is a resistive sensor?
25. What is meant by the term *taper* when applied to a potentiometer?
26. Which band is the multiplier in a five-band resistor?
27. What does red in the fifth band of a five-band resistor mean?
28. What does orange in the third band of a four-band resistor mean?
29. Is silver a color that can appear in the first or second band of a four-band resistor?
30. What does a black band in the third position of a four-band resistor mean?
31. What does AWG stand for?
32. Does wire resistance increase or decrease with higher AWG numbers?
33. What is meant by an autoranging meter?
34. Why is it important to know the frequency response specification for your meter?
35. How is a multimeter connected to measure current in a circuit?
36. What is meant by a 3 1/2 digit display on a DMM?
37. What is different about the resistance scale of most VOMs than the other scales?

Basic Problems

PROBLEMS

Answers to odd-numbered problems are at the end of the book.

1. Determine the direct current (dc) when 20 C passes a point each second.
2. How many coulombs pass a point in 10 s if 15 A of dc is in a wire?
3. How long does it take for 25 mC to pass a point if the current is 12.5 mA?
4. What is the total charge that moves from one terminal of a battery to the other if 50 mA of current are provided to the load for 1 hour?
5. How long can an 80 Ah battery provide 4 A?
6. What is the ampere-hour rating of a battery if it can provide 4 A for 24 hours?
7. If the efficiency of a power supply is 60% when the input power is 25 W, what power is delivered to the load?
8. A power supply has an input power of 200 W and an output power of 165 W.
   (a) What is the efficiency?
   (b) What percentage of the input is lost?
9. What is the power out from a supply that has 500 W of power supplied and is 75% efficient?

10. What is the color code for each of the following four-band resistors?
    (a) 620 kΩ ± 5%   (b) 150 Ω ± 10%
    (c) 2.2 MΩ ± 10%   (d) 91 Ω ± 5%

11. What is the resistance and tolerance of each of the following four-band resistors?
    (a) orange, blue, red, gold   (b) blue, gray, yellow, silver
    (c) red, red, gold, gold   (d) brown, black, blue, silver

12. What is the resistance and tolerance of each of the following five-band resistors?
    (a) yellow, brown, violet, black, brown   (b) green, brown, brown, gold, brown
    (c) brown, blue, black, brown, red   (d) white, gray, gray, silver, brown

13. What is the color code for each of the following five-band resistors?
    (a) 57.6 kΩ ± 1%   (b) 910 Ω ± 2%
    (c) 26.1 Ω ± 1%   (d) 75.0 kΩ ± 2%

14. Determine the resistance of 1 mile (5280 feet) of AWG #26 copper wire.

15. Determine the approximate length of a piece of AWG #14 copper wire if the resistance is 1.2 Ω.

16. Find the approximate length in meters of a piece of #10 aluminum wire that has a resistance of 0.27 Ω.

Basic-Plus Problems

17. If a AWG #12 copper wire replaces the aluminum wire in problem 16, what will be the approximate resistance?

18. The resistance of a 1000 m run of AWG #26 copper wire has 10 Ω of resistance more than is specified for a particular application. Can #24 gauge copper replace the AWG #26 gauge and be within specifications? Justify your answer.

19. Compute the resistivity (in Ω-mm^2/m) of a carbon steel wire if the resistance of a 3.5 m length of AWG #22 wire is 1.94 Ω.

20. Find the resistivity of the carbon steel in Problem 19 in units of Ω-CM/ft.

ANSWERS

Example Questions

3-1: 20 s

3-2: 5 h

3-3: 425 W

3-4: 100 Ω ± 5%

3-5: white, brown, brown, silver

3-6: green, blue, red, brown, brown

3-7: 0.79 Ω

3-8: 7000 Ω

Review Questions

1. Positive ion
2. There is an abundance of electrons that are not bound to a specific atom.

3. Electrolytes

4. Isolation from other conductors and protection against shock hazard

5. A crystalline material that has four electrons in its outer shell and with properties between those of conductors and insulators.

6. A load is a component that uses the power provided by the rest of the circuit.

7. dc and ac

8. ac

9. 120

10. Instantaneous current is any value from the negative peak to the positive peak and changes with time. RMS current is a value that is equivalent to a direct current of the same amount.

11. Oxidation-reduction reactions

12. Lead acid batteries contain acid that can do serious eye damage, and battery gases are explosive.

13. The ampere-hour rating of a battery indicates the length of time that a battery can deliver a certain amount of current to a load at the rated voltage.

14. Mechanical energy

15. 70%

16. A superconductor is a specialized material that at extremely cold temperatures has no resistance.

17. A rheostat is a two-terminal variable resistor for controlling current; a potentiometer is a three-terminal variable resistor for controlling voltage.

18. A linear taper on a potentiometer means that its resistance is proportional to the angular position of the control.

19. The multiplier is 10^{-1}.

20. brown, black, gold, gold

21. dc and ac voltage, resistance, and dc and ac current

22. The meter must be inserted "in line" (series) and the leads must be moved to the current jacks.

23. An incorrect reading will result (or possible damage to the meter).

24. The meter can display the digits from 0000 to 1999; this represents a resolution of 1 part in 2000.

25. The resistance scale on a VOM reads "backward" and is nonlinear. Generally, the zero is on the right side.

Chapter Checkup

1. (a)	2. (c)	3. (b)	4. (d)	5. (c)
6. (a)	7. (d)	8. (d)	9. (b)	10. (d)
11. (c)	12. (d)	13. (a)	14. (b)	15. (b)

CHAPTER 4

OHM'S LAW AND WATT'S LAW

CHAPTER OUTLINE

4–1 Ohm's Law
4–2 Applying Ohm's Law to Basic Circuits
4–3 Electrical Energy and Power
4–4 Watt's Law
4–5 Applying Watt's Law
4–6 Nonlinear Resistance

INTRODUCTION

The basic electrical quantities are voltage, current, resistance, and power. You should be familiar with the definitions for these quantities. In this chapter, you will learn the relationship between these quantities given by two very important laws. The first, Ohm's law, shows the relationship between voltage, current, and resistance. The second, Watt's law, shows the relationship between these same quantities and power. The two laws are closely related and will be applied many times in your work in electronics. This chapter deals strictly with resistive circuits and the application of these two important laws. The chapter ends with a discussion of nonlinear resistance, a concept that is applied in studies of electronic devices.

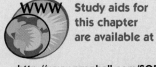

Study aids for this chapter are available at

http://www.prenhall.com/SOE

KEY OBJECTIVES

A section number is given for each objective. After completing this chapter, you should be able to

4–1 State Ohm's law in words and discuss its meaning

4–2 Apply Ohm's law to practical circuits to find current, voltage, or resistance

4–3 Define the kilowatt-hour and apply it to practical situations

4–4 State Watt's law in words and discuss its meaning

4–5 Apply Watt's law to practical circuit problems, including finding the correct wattage required for a resistor

4–6 Explain the difference between dc and ac resistance and calculate the dc and ac resistance for a device with a nonlinear IV curve

COMPUTER SIMULATIONS DIRECTORY

The following figures have Multisim circuit files associated with them. To open a Multisim file, go to the website at http://www.prenhall.com/SOE, click on the cover of this book, choose this chapter, click on "Multisim", and then click on the selected file.

- Figure 4–3 Page 97
- Figure 4–4 Page 98
- Figure 4–8 Page 100

LABORATORY EXPERIMENTS DIRECTORY

The following exercises are for this chapter.

- **Experiment 6** Ohm's Law
- **Experiment 7** Power and Efficiency

KEY TERMS

- Ohm's law
- Conductance
- Kilowatt-hour
- Watt's law
- Horsepower
- Heat sink
- DC resistance
- AC resistance

CHAPTER 4

Sci Hi SCIENCE HIGHLIGHT

An interesting application for lasers is a laser developed at Lawrence Livermore National Laboratory (LLNL) for generating short laser pulses. The world record for the highest pulse power ever achieved is for the Petawatt laser. (The metric prefix peta means quadrillion or 10^{15}.) The Petawatt laser can produce 500 J pulses for 500 femtoseconds (500×10^{-15} s.) What is interesting about these numbers is that they illustrate the difference between energy and power. The energy (500 J) is typical of that used by a television set in two hours. In the case of the Petawatt laser, this amount of energy is delivered in such a short time that the power is extremely high—approximately 1 trillion watts! This is more power (for this extremely short time) than all of the other power generated in the U.S.! For a given amount of energy, a shorter time implies higher power because power is the *rate* that energy is converted and a shorter time corresponds to a higher rate.

Applications for very short pulse lasers like Petawatt include developing a cutting tool that can cut materials with minimal heat damage. It also has applications in energy research and national defense. Even more powerful lasers are planned in the future.

4–1 OHM'S LAW

Georg Simon Ohm formulated the law named in his honor called Ohm's law. Ohm's law expresses the relationship between voltage, current, and resistance.

In this section, you will learn to state Ohm's law in words and discuss its meaning.

Although voltage, current, and resistance were defined in Section 2–3, the definitions are restated here because of their importance.

- Voltage is the work obtained (or energy released) when a unit of charge moves from a more positive point to a more negative point and is measured in volts. Voltage acts like the driving "force" that moves charge through a circuit.
- Current is the rate of charge flow and is measured in amperes.
- Resistance is opposition to current and is measured in ohms.

If the voltage increases across a resistor, current will increase. If voltage decreases, current will decrease. This idea is illustrated in Figure 4–1.

FIGURE 4–1

If the voltage increases, the current increases.

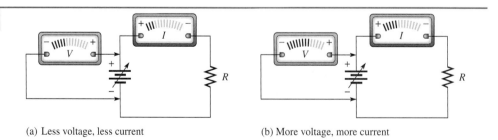

(a) Less voltage, less current (b) More voltage, more current

If the opposition to current (resistance) is increased in a circuit, there will be less current. This idea is illustrated in Figure 4–2.

OHM'S LAW AND WATT'S LAW

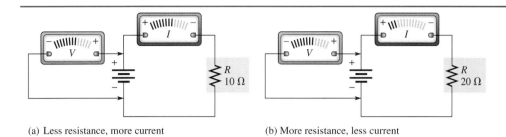

(a) Less resistance, more current (b) More resistance, less current

FIGURE 4–2
If the resistance increases, the current decreases.

HISTORICAL NOTE
The unit of resistance, the ohm, is named in honor of Georg Simon Ohm (1787–1854). Ohm was born in Bavaria and struggled for years to gain recognition for his work in formulating the relationship of current, voltage, and resistance. His work led to the mathematical relationship that is known today as Ohm's law.

Formula for Current

The three quantities of voltage, current, and resistance are related to each other. Ohm recognized the relationship and stated it mathematically as

$$I = \frac{V}{R} \qquad (4\text{–}1)$$

where I is current in amperes (A), V is voltage in volts (V), and R is resistance in ohms (Ω).

Ohm's law shows that current is proportional to voltage and inversely proportional to resistance. This fundamental law is the most important one in electronics. Example 4–1 illustrates how to find the current if you know the voltage and resistance.

EXAMPLE 4–1

Problem
Find the expected reading on the ammeter for the circuit in Figure 4–3.

Solution

$$I = \frac{V}{R} = \frac{15\ \text{V}}{47\ \Omega} = \mathbf{0.319\ A}$$

The calculator sequence for the TI-36X is

The display will show 0.319148936 (if FLO is selected). It is necessary to round the result to the number of digits consistent with your measured precision (typically 3 digits). The TI-36X can be set to keep three digits past the decimal point by pressing the key sequence

This is not quite the same as rounding, as discussed in Section 2–4, but is helpful by eliminating some nonsignificant digits. This setting will be retained for future calculations until changed again. In this text, we will use this mode for all calculator sequences given and round answers to three significant digits.

Question*
What is the current in the resistor if the voltage is raised to 20 V?

FIGURE 4–3

COMPUTER SIMULATION

Open the Multisim file F04-03DC on the website. Using the multimeter, verify the current in the circuit for this example.

---

*Answers are at the end of the chapter.

CHAPTER 4

Although current values of several amperes or more are common in high-power circuits and in appliances, most electronic circuits operate with much smaller current. Example 4–2 illustrates the case where the resistance is much larger, so the current is much smaller. Metric prefixes that are typical in these cases are shown. Notice that the voltage source in this example is drawn with the negative side up. As explained previously, there is no particular significance to the orientation shown, but in actual circuits it is important to observe the polarity so that the meter is connected correctly. The ammeter's negative terminal must be connected to the negative terminal of the voltage source, and it must be in series as shown to avoid damage.

EXAMPLE 4–2

Problem
Find the expected current for the circuit in Figure 4–4.

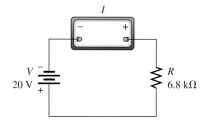

FIGURE 4–4

Solution

$$I = \frac{V}{R} = \frac{20 \text{ V}}{6.8 \text{ k}\Omega} = \mathbf{2.94 \text{ mA}}$$

Since the metric prefix *kilo* represents 10^3, key it into the calculator as [EE] [3]. The complete key sequence is [2] [0] [÷] [6] [.] [8] [EE] [3] [=]. After keying in the sequence, you can select ENG by pressing [3rd] [ENG/+] and limit the display to three digits to the right of the decimal by pressing [2nd] [FIX/CE/C] [3]. The display will show 2.941^{-03}, which is interpreted as 2.94 mA when rounded to three significant digits.

Question
What is the current in the resistor if the voltage is lowered to 10 V?

COMPUTER SIMULATION

Open the Multisim F04-04DC on the website. Using the multimeter, verify the current in the circuit for this example.

Formula for Voltage

If you rearrange Equation 4–1 by multiplying both sides by R and transposing V, you can use Ohm's law to find the voltage when you know the current and resistance.

$$I = \frac{V}{R}$$

$$IR = \frac{V}{R}R$$

OHM'S LAW AND WATT'S LAW

$$V = IR \qquad (4-2)$$

With this formula, you can calculate the voltage in volts if the current is in amps and the resistance is in ohms. In electronic circuits, the current is often stated in milliamps and the resistance is stated in kilohms. In this case, the metric prefixes (*milli* and *kilo*) cancel because $10^{-3} \times 10^{3} = 10^{0} = 1$. Thus, the voltage is in volts (no metric prefix).

EXAMPLE 4–3

Problem
Find the setting of the voltage source that will produce a reading of 5.0 mA on the milliammeter in Figure 4–5. (The arrow through the battery symbol indicates an adjustable voltage source.)

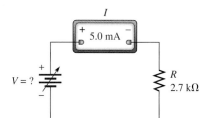

FIGURE 4–5

Solution

$$V = IR = (5.0 \text{ mA})(2.7 \text{ k}\Omega) = \mathbf{13.5 \text{ V}}$$

Because the metric prefixes *milli* and *kilo* cancel, it is not necessary to key in their value in a calculator; however, you may prefer to do so anyway. If you choose to key in the values of the metric prefixes, the key sequence is as follows:

The calculator is still in engineering notation with 3 digits fixed (see Example 4–2), so the display will show 13.500 ^{00}. Because $10^{0} = 1$, this result is read as simply 13.5.

If you mentally cancel the prefix *milli* (m) with the prefix *kilo* (k) in the original problem, the calculator sequence simplifies to

Question
What is the source voltage if the current is 6.0 mA in this circuit?

Formula for Resistance

Sometimes it is necessary to use Ohm's law to find the resistance for a given current and voltage. Ohm's law can be rearranged to find the resistance. Starting with Equation 4–2, divide both sides by I and transpose the terms to obtain the following equivalent form:

$$V = IR$$

$$\frac{V}{I} = \frac{\cancel{I}}{\cancel{I}} R$$

$$R = \frac{V}{I} \qquad (4-3)$$

CHAPTER 4

The resistance is measured in ohms if the voltage is in volts and the current is in amps.

The three equations ($I = V/R$, $V = IR$, and $R = V/I$) are, of course, equivalent. Figure 4–6 may help you remember the relationships of current, voltage, and resistance.

FIGURE 4–6
A memory aid for the three equivalent forms of Ohm's law.

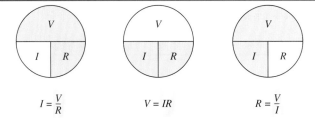

EXAMPLE 4–4

Problem
A small portable hair dryer is plugged into a 12 V circuit and the current is found to be 14 A, as shown in the equivalent circuit in Figure 4–7. Find the resistance for the hair dryer.

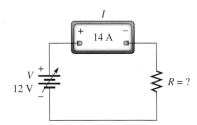

FIGURE 4–7

Solution

$$R = \frac{V}{I} = \frac{12\ \text{V}}{14\ \text{A}} = 0.857\ \Omega$$

The calculator sequence is [1] [2] [÷] [1] [4] [=].

Question
What is the current if the voltage is adjusted to 10 V? Assume the resistance does not change.

The resistance in electrical circuits (such as the hair dryer in Example 4–4) is frequently small. In electronic circuits the resistance is generally much higher. Example 4–5 illustrates a case with very high resistance.

EXAMPLE 4–5

Problem
What is the resistance of the resistor shown in Figure 4–8?

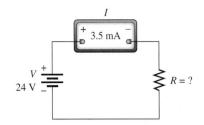

FIGURE 4–8

OHM'S LAW AND WATT'S LAW

Solution

$$R = \frac{V}{I} = \frac{24 \text{ V}}{3.5 \text{ mA}} = 6.86 \text{ k}\Omega$$

The calculator sequence is

[2] [4] [÷] [3] [.] [5] [EE] [+/−] [3] [=]

The display will show $6.857^{\ 03}$ (in engineering notation). This result can be rounded and expressed as either $6.86 \times 10^3\ \Omega$ or $6.86\ \text{k}\Omega$. Although they are equivalent, the metric prefix is generally preferred.

Question
What is the resistance if the voltage is 24 V and the current is 350 μA?

COMPUTER SIMULATION

Open the Multisim file F04-08DC on the website. Using the multimeter, verify the resistance.

A Graph of Current and Voltage

As Ohm's law shows, the voltage and resistance determine the current in a circuit. For many cases, the resistance is a fixed quantity. Let's take a circuit that has a fixed resistance of 330 Ω and vary the voltage from 1 to 10 V in 1 V increments. Starting with 1 V, calculate the current by dividing 1 V by 330 Ω.

$$I = \frac{V}{R} = \frac{1 \text{ V}}{330\ \Omega} = 0.00303 \text{ A} = 3.03 \text{ mA}$$

Doubling the voltage to 2 V will double the current. Thus,

$$I = \frac{V}{R} = \frac{2 \text{ V}}{330\ \Omega} = 6.06 \text{ mA}$$

Each 1 V increase in voltage will increase the current by 3.03 mA. If you do this as an experiment in the lab, the data will be similar to that shown in Table 4–1. You can calculate these points with the TI-36X calculator with the following sequence:

[1] [÷] [3] [3] [0] [=]

The display will show 3.030^{-03}. Then press

in order to add the result to itself and set up the first answer as a constant in the display. For the remaining entries in the table, press the [=] key. This will add the constant to the previous answer.

A plot of the current as a function of voltage is shown in Figure 4–9. Section 2–6 discussed the correct way to plot data. Recall that the dependent variable is plotted on the y-axis; in this case, the dependent variable is the current. The independent variable is plotted on the x-axis; in this case, the independent variable is the voltage. The plot shows a straight line, indicating the linear relationship between current and voltage for a fixed resistor. This is true for all fixed resistors; as long as the resistance is constant, the plot of current as a function of voltage is a straight line.

TABLE 4–1

V (V)	I (mA)
1.0	3.03
2.0	6.06
3.0	9.09
4.0	12.1
5.0	15.2
6.0	18.2
7.0	21.2
8.0	24.2
9.0	27.3
10.0	30.3

CHAPTER 4

FIGURE 4–9

Current versus voltage for a 330 Ω resistor.

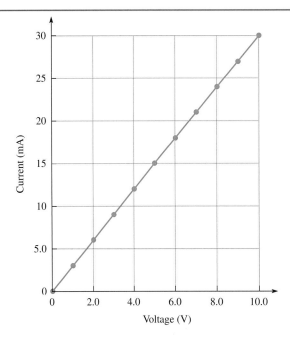

Conductance

Recall that the slope of a line is found by dividing a change (Δ) in the *y* variable by a corresponding change in the *x* variable (Δy/Δx). In the case of a fixed resistor such as the one plotted in Figure 4–9, the slope of the line is constant. Because the slope is constant, it can be calculated by simply dividing the current by the corresponding voltage at any point on the line. The result is called conductance, abbreviated *G*. **Conductance** is the reciprocal of resistance and is a measure of the ability of a substance to pass current. It has units of siemens, abbreviated S. In equation form,

$$G = \frac{1}{R} \tag{4-4}$$

where *G* is the conductance in siemens (S) and *R* is the resistance in ohms (Ω). For example, the 330 Ω resistor plotted in Figure 4–9 has a conductance of 3.0 mS, which is the slope of the line.

Nonlinear Resistance

Fixed resistors are designed to have a specific amount of resistance and have an *IV* characteristic as described in Figure 4–9. As you have seen, the plot is a straight line, meaning that the resistance is constant for all changes in voltage.

Many devices used in electrical and electronic circuits have an *IV* characteristic that is *not* a straight line. The resistance changes when the voltage is changed. Section 4–6 discusses this in more detail and shows how to measure resistance in this case.

Review Questions

Answers are at the end of the chapter.

1. What are the three forms of Ohm's law?
2. If the voltage across a fixed resistor is cut in half, what happens to the current?

OHM'S LAW AND WATT'S LAW

3. If the amount of resistance in a circuit is increased but the voltage remains the same, what happens to the current?
4. Why is current plotted on the *y*-axis in Figure 4–9?
5. What is the relationship between resistance and conductance?

APPLICATIONS OF OHM'S LAW   4–2

Ohm's law is the most important law of electronics. A series of problems and solutions is useful to illustrate applications of Ohm's law. AC and dc sources are shown in examples to emphasize that Ohm's law can be applied to either dc or ac circuits. Try working each problem without looking at the solution and then check how you did.

In this section, you will learn to apply Ohm's law to practical circuits to find current, voltage, or resistance.

Calculating Current, Voltage, or Resistance in a DC Circuit

As you know, you can use Ohm's law in any of three ways to calculate an unknown current, voltage, or resistance when you know the other two quantities. For circuits with only linear elements (such as fixed resistors), you can apply Ohm's law to the entire circuit to calculate the unknown value.

EXAMPLE 4–6

Problem
The current in the circuit in Figure 4–10 is too small for many meters. A typical DMM will show 0.009 DC for the current, which means 0.009 mA. Notice there is only one significant digit. Most analog meters cannot indicate it accurately. Assuming the resistance is the nominal (color-coded) value, what is the calculated current to three significant digits?

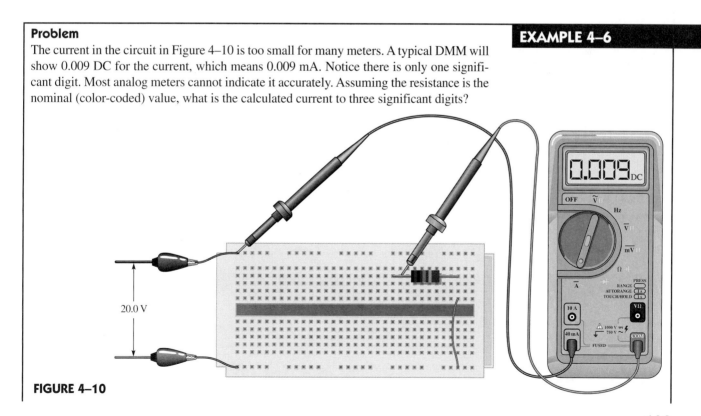

FIGURE 4–10

CHAPTER 4

Solution
First, determine the nominal (color-code) resistance. The resistor color bands are red, red, green, gold. From the color code, the first digit is red = 2, the second digit is red = 2, the multiplier is green = 10^5, and the tolerance is 5%. Therefore, the resistor has a specified value of 2.2 MΩ ± 5%.

$$I = \frac{20.0 \text{ V}}{2.2 \text{ M}\Omega} = \mathbf{9.09 \ \mu A}$$

The calculator sequence is

[2] [0] [÷] [2] [.] [2] [EE] [6] [=]

If you have selected ENG, the display will show 9.091^{-06}. The exponent in the answer is equivalent to the metric prefix *micro* (μ). The answer can be rounded to three significant digits as 9.09 μA. Notice that the prefix *micro* (μ) is the reciprocal of the prefix *mega* (M).

Question
Draw the schematic for the circuit.

EXAMPLE 4–7

Problem
What is the range of expected readings on the ammeter due to the resistor tolerance for the circuit in Figure 4–11? Assume the resistor is within the tolerance specified by the color code.

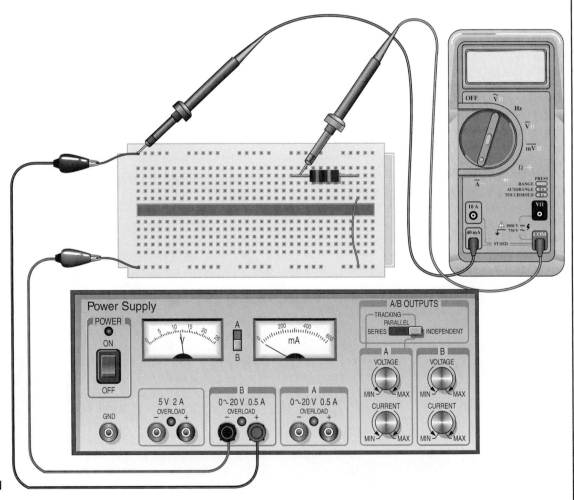

FIGURE 4–11

Solution

First determine the nominal (color-code) resistance. The resistor color bands are red, violet, red, gold. From the color code, the first digit is red = 2, the second digit is violet = 7, the multiplier is red = 10^2, and the tolerance is 5%. Therefore, the resistor has a specified value of 2.7 kΩ ± 5%.

Determine the maximum resistance based on the tolerance.

$$2.7 \text{ k}\Omega + 0.05(2.7 \text{ k}\Omega) = 2.84 \text{ k}\Omega$$

The calculator sequence shows a simplified way for solving this equation by using the percent key.

The power supply is set to +12.0 V, so the minimum current is

$$I = \frac{V}{R} = \frac{12.0 \text{ V}}{2.84 \text{ k}\Omega} = 4.23 \text{ mA}$$

Determine the minimum resistance based on the tolerance.

$$2.7 \text{ k}\Omega - 2.7 \text{ k}\Omega(0.05) = 2.56 \text{ k}\Omega$$

Again, use the percent key on the calculator to simplify the keystrokes.

The maximum current is

$$I = \frac{V}{R} = \frac{12.0 \text{ V}}{2.56 \text{ k}\Omega} = 4.68 \text{ mA}$$

Therefore, the expected range is from **4.23 mA to 4.68 mA.**

Question
What is the range if the tolerance band were silver instead of gold?

EXAMPLE 4–8

Problem
You need to set the power supply to a voltage that will supply 1.0 mA to the resistor in Figure 4–12. What voltage setting is required?

Solution
First determine the resistance. The resistor color bands are brown, green, orange, gold. From the color code, the first digit is brown = 1, the second digit is green = 5, the multiplier is orange = 10^3, and the tolerance is 5%. Therefore, the resistor has a value of 15 kΩ ± 5%. The required current is 1.0 mA, so the required voltage setting is

$$V = IR = (1.0 \text{ mA})(15 \text{ k}\Omega) = \mathbf{15.0 \text{ V}}$$

The calculator sequence is

CHAPTER 4

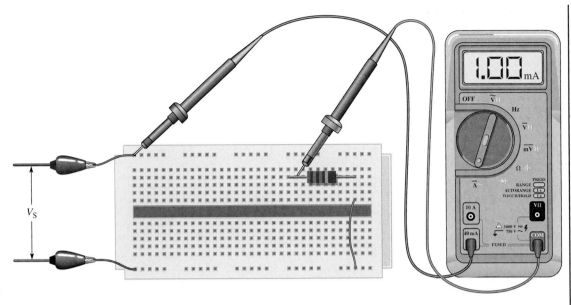

FIGURE 4–12

The result, shown as 15.000 V, is rounded to 15.0 V. Of course, the keystrokes can be simplified by noticing that the product of the prefixes *milli* (m) and *kilo* (k) cancel.

Question
What voltage setting will produce a current of 1.5 mA in the circuit of Figure 4–12?

EXAMPLE 4–9

Problem
You need to select a resistor that will limit the current from a 24 V source to 0.90 mA. What value resistor is required?

Solution
Apply Ohm's law.

$$R = \frac{V}{I} = \frac{24 \text{ V}}{0.90 \text{ mA}} = 26.7 \text{ k}\Omega$$

The calculator sequence is

[2] [4] [÷] [.] [9] [EE] [+/−] [3] [=]

Question
Standard-value resistors are preferred for electronics work. The nearest standard four-band resistor is 27 kΩ. What are the color bands for this resistor? Assume a 5% tolerance.

Calculating Current, Voltage, or Resistance in an AC Circuit

Ohm's law is applied to both dc and ac circuits in exactly the same way as long as the load is resistive. (Later the case of a nonresistive load will be covered.) In ac circuits it is normal that the voltage and current are specified as rms values (Section 3–2). There are other ways to specify an ac waveform; but if not otherwise stated, rms values are assumed in electronics or electrical work. RMS values are used in the examples that follow.

OHM'S LAW AND WATT'S LAW

EXAMPLE 4–10

Problem
The ac ammeter in the circuit for Figure 4–13 has no polarity signs. What is the reading if the voltage source is set to 2.0 Vrms?

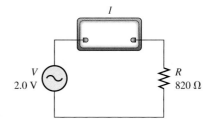

FIGURE 4–13

Solution
Apply Ohm's law.

$$I = \frac{V}{R} = \frac{2.0 \text{ V}}{820 \text{ }\Omega} = \textbf{2.44 mA}$$

Because the voltage was given as an rms value, the current is also an rms value. It is not necessary to state this in the answer because rms is assumed if not otherwise stated.

Question
What is the current if the ac source was replaced with a 2.0 Vdc source (and the ammeter is changed for a dc one)?

EXAMPLE 4–11

Problem
A clamp-on ammeter (described in Section 3–5) is used to measure the current in a circuit connected to 115 Vac, as shown in Figure 4–14. The meter reads 0.87 A. What is the effective resistance of the bulb?

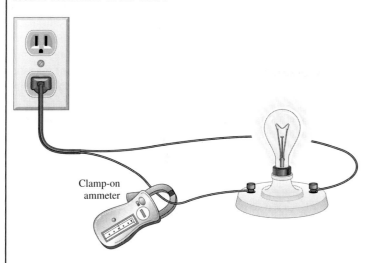

FIGURE 4–14

Solution

$$R = \frac{V}{I} = \frac{115 \text{ V}}{0.87 \text{ A}} = \textbf{132 }\boldsymbol{\Omega}$$

CHAPTER 4

> **Question**
> The most common time for a bulb to burn out is when it is first turned on. From this fact, is the resistance higher or lower when it is cold?

Review Questions

6. What is the current in a 4.7 MΩ resistor that has 1.2 V across it?
7. What voltage is across a 220 Ω resistor if the current through it is 40 mA?
8. What is the value of an unknown resistor with 48 V across it if the current is 7.06 mA?
9. If a 115 Vac source is connected to an electric heater that draws 10 A, what is the resistance of the heater?
10. If a 12 Vdc source is replaced with an equivalent ac source, what voltage is required?

4–3 ELECTRICAL ENERGY AND POWER

Work and energy were introduced in Section 1–3. A joule is a small unit of work or energy. Power is measured in watts (W).

In this section, you will learn to define the kilowatt-hour and apply it to practical situations.

Energy

Energy is defined as the capacity to do work. The unit for energy is the joule, equivalent to the work done when a newton of force is applied over a distance of 1 meter. As discussed in Section 1–3, the three major forms of energy are potential, kinetic, and rest. In turn, each of these major forms is subdivided into basic forms (such as mechanical, chemical, thermal, or light) as previously described. Electrical energy is another basic form of potential energy that can be converted to other energy forms, such as the kinetic energy of a moving motor or the heat from a resistive heater.

Power is defined as the rate energy is transferred or converted to another form such as heat. Although we will say that energy is "used," it is never actually destroyed; generally, this refers to the fact that it is converted to unwanted heat. (See Science Highlight for Chapter 1). Power is typically measured in watts, the standard SI unit. One watt is the power dissipated when 1 J of energy is used in 1 s. Recall that power is defined in equation form as

$$P = \frac{W}{t}$$

where P is power in watts (W), W is energy in joules (J), and t is time in seconds (s). Notice that the unit watts is abbreviated W and the quantity energy is abbreviated W. Rearranging this equation shows that the product of power and time is energy.

$$W = Pt$$

> **SAFETY NOTE**
> Appliances that have three-wire plugs should only be plugged into three-wire receptacles. Extension cords should never be covered when in use and should be regularly inspected for damage. If a cord or plug feels hot, the circuit may be overloaded or the cord may be damaged; corrective action should be taken immediately.

OHM'S LAW AND WATT'S LAW

The Kilowatt-Hour

A much larger unit of energy than the joule is the kilowatt-hour (kWh). In fact, there are 3.6 million joules (J) in one kilowatt-hour! Notice that a kWh is the product of a power unit and a time unit, so the result is energy because $W = Pt$. The **kilowatt-hour** is the amount of energy used if a power level of 1 kilowatt is maintained for one hour. Because it is a much larger unit than the joule, the kilowatt-hour is a useful unit for electric use in your home or other electrical applications. Electrical utility companies charge for kilowatt-hours, not for kilowatts. (An electrical utility company would be more properly called an *energy* company rather than a *power* company.)

EXAMPLE 4–12

Problem
Prove that there are 3.6 million joules in a kilowatt-hour.

Solution

$$1 \text{ kWh} = 1 \text{ kWh} \times \left(\frac{1000 \text{ W}}{\text{kW}}\right)\left(\frac{3600 \text{ s}}{\text{h}}\right) = 3.6 \times 10^6 \text{ Ws} = 3.6 \times 10^6 \text{ J}$$

The quantities in the parentheses are conversion factors. They are set up so that units cancel, leaving Ws or J in place of kWh.

Question
What is the number of kWh in a J?

As mentioned, electrical utility companies charge their customers for energy, not power. Power is the *rate* of energy dissipation. The number of kWh depends on both the rate (kW) and the time (h). Thus, a 1000 W load for 1 hour uses the same energy as a 100 W load for 10 h or a 10 W load for 100 h.

EXAMPLE 4–13

Problem
Find the amount of energy (in kWh) that a 100 W bulb uses in one day (24 h).

Solution

$$W = Pt = 100 \text{ W} \times 24 \text{ h} = 2400 \text{ Wh} = 2.4 \text{ kWh}$$

Question
How many kWh are used by a 200 W bulb burning for 8 h?

Household Appliances

Table 4–2 lists some common household appliances and their typical power ratings in watts, which are based on the rate the appliances can do work. You can determine the maximum kWh used for various appliances by using the power rating in Table 4–2 converted to kilowatts times the number of hours an appliance is used. The cost per kilowatt-hour depends on various factors, including the amount of energy used, type of service (residential, commercial, industrial), and in some cases time of day.

CHAPTER 4

TABLE 4–2
Typical power ratings.

Appliance	Power rating (W)
Clothes dryer	4800
Dishwasher	1200
Fan	75
Lamp	100
Microwave oven	800
Range	12,200
Refrigerator	1800
Television	250
Washing machine	400
Water heater	2500

EXAMPLE 4–14

Problem
You use the following ten appliances for the specified lengths of time during a typical 24-hour period. Determine the energy used and the cost for these appliances in the 24 h period if energy sells for 12 cents/kWh.

Clothes dryer: 1 h
Fan: 24 h
Microwave oven: 15 min
Refrigerator: 12 h
Washing machine: 1 h

Dishwasher: 1 h
Three lamps: 8 h for each
Range: 30 min
Television: 5 h
Water heater: 6 h

Solution
Determine the kWh for each appliance used by converting the watts in Table 4–2 to kilowatts and multiplying by the time in hours.

Clothes dryer:	4.8 kW × 1 h = 4.8 kWh
Dishwasher:	1.2 kW × 1 h = 1.2 kWh
Fan:	0.075 kW × 24 h = 1.8 kWh
Lamps:	3 × 0.100 kWh × 8 h = 2.4 kWh
Microwave:	0.8 kW × 0.25 h = 0.2 kWh
Range:	12.2 kW × 0.5 h = 6.1 kWh
Refrigerator:	1.8 kW × 12 h = 21.6 kWh
Television:	0.25 kW × 5 h = 0.5 kWh
Washing machine:	0.4 kW × 1 h = 0.4 kWh
Water heater:	2.5 kW × 6 h = 15 kWh

Now, add up all the kilowatt-hours to get the total energy for the 24-hour period.

Total energy = (4.8 + 1.2 + 1.8 + 2.4 + 0.2 + 6.1 + 21.6 + 0.5 + 0.4 + 15) kWh
= 54.0 kWh

At 12 cents/kWh, the cost of energy to run the appliances for the 24-hour period is

Energy cost = 54.0 kWh × 0.12 $/kWh = **$6.48**

Question
Suppose you have guests that like hot showers and cause the water heater to run 8 h instead of 6 h. What is your added cost for the 24-hour period for hot water?

Review Questions

11. What are two units for measuring energy?
12. Why is the kWh a more convenient unit than the joule for measuring household energy consumption?
13. What is another name for a joule/second?
14. What is the energy converted to heat (in kWh) if a 5000 W portable heater is operated for 2 hours?
15. If energy costs 15 cents/kWh, what is the cost of operating the heater in Question 14?

WATT'S LAW 4–4

Power is the rate energy is dissipated and is an important parameter for electrical and electronic circuits. The law that relates electrical parameters to power is called Watt's law, named after James Watt.

In this section, you will learn how to state Watt's law in words and discuss its meaning.

Power Dissipation in Resistors

When current is in a resistor, the kinetic energy of electrons colliding with atoms in the resistance is converted to heat. Power is the rate heat is produced and can be expressed as the product of the current and the voltage.

$$P = IV \tag{4-5}$$

where P is the power in watts (W), I is the current in amperes (A), and V is the voltage in volts (V). Notice the units. When current is written as coulombs per second and voltage is written as joules/coulomb, the product is joules per second or watts.

$$P = IV = \left(\frac{\text{coulombs}}{\text{second}}\right)\left(\frac{\text{joules}}{\text{coulombs}}\right) = \frac{\text{joules}}{\text{second}} = \text{watts}$$

By substituting Ohm's law for current ($I = V/R$) in Equation 4–5, you can obtain an equivalent expression.

$$P = IV = \left(\frac{V}{R}\right)V$$
$$P = \frac{V^2}{R} \tag{4-6}$$

You can find another equivalent expression by substituting Ohm's law for voltage ($V = IR$) into Equation 4–5.

$$P = IV = I(IR)$$
$$P = I^2 R \tag{4-7}$$

HISTORICAL NOTE

The unit of power, the watt, is named after James Watt (1736–1819), a Scottish inventor and engineer. Watt was known for his improvements to the steam engine, which made it practical for industrial use. Watt patented several inventions, including the rotary-motion steam engine. This machine enabled steam engines to be applied to many process that required rotary motion.

CHAPTER 4

Collectively, the three equations (4–5, 4–6, and 4–7) are known as **Watt's law**. To calculate the power in a given resistor, you can use any of the three equations, depending on what quantities you know. For example, if you know the current and voltage, use $P = IV$. As in the case of Ohm's law, Watt's law can be applied to both ac and dc for resistive loads. The following examples show how to use the three versions of Watt's law for both ac and dc circuits with resistive loads.

EXAMPLE 4–15

Problem
Calculate the power dissipated in the load resistor, R, for each of the circuits in Figure 4–15.

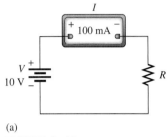
(a)

(b)

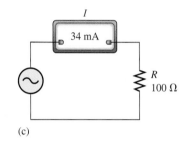
(c)

FIGURE 4–15

Solution
In circuit (a), you know V and I. Determine the power as follows:

$$P = IV = (0.1 \text{ A})(10 \text{ V}) = \mathbf{1 \text{ W}}$$

In circuit (b), you know V and R. Determine the power as follows:

$$P = \frac{V^2}{R} = \frac{20 \text{ V}^2}{270 \text{ }\Omega} = \mathbf{1.48 \text{ W}}$$

The calculator sequence for this calculation is

2 0 x² ÷ 2 7 0 =

In circuit (c), you know I and R. Notice that the source is an ac source. As long as the current is specified as an rms value, the calculation is exactly the same as with a dc source. Determine the power as follows:

$$P = I^2 R = (34 \text{ mA})^2 (100 \text{ }\Omega) = \mathbf{116 \text{ mW}}$$

The calculator sequence is

Question
What is the power in circuit (a) if the voltage is doubled? (*Hint*: Consider the effect on current).

Horsepower

Just as energy has two widely used units (joules and kilowatt-hours), so does power (watts and horsepower). The watt is a relatively small unit. Another commonly used unit for measuring the power of motors, engines, and so forth is the **horsepower (hp)**. One hp is equal to 746 W. The unit has its origins in Scotland. In a test of a very strong horse, the horse was able to deliver 746 W of power in lifting a load. Thus, the horsepower is based on the mechanical work that a horse could do and today is still widely used in mechanics (automobile engines, for example). Although it is not a standard SI unit, you will undoubtedly come across this unit in working with motors. With respect to horsepower ratings, a motor that delivers 746 W is considered to be a 1 hp motor.

Although it delivers 746 W to a load, it actually is consuming about 1000 W from the line (depending on the efficiency of the motor).

Review Questions

16. What are the three forms of Watt's law?
17. What happens to electrical energy when current is in a resistance?
18. What is the current in a 150 W bulb designed for 110 V?
19. How much power is dissipated in a 100 Ω resistor connected to 5.0 V source?
20. What is the power dissipated in a 270 Ω resistor that has a current of 55.6 mA?

APPLICATIONS OF WATT'S LAW 4–5

Watt's law is the second most important law of electronics. A series of practical examples is a useful way to illustrate this law. Some of the examples require you to find a parameter other than the power using Watt's law. Try working each example problem without looking at the solution.

In this section, you will learn to apply Watt's law to practical circuit problems, including finding the correct wattage required for a resistor.

Measuring Power

In many electrical and electronic systems, the measurement of power is extremely important. For example, in communication systems, the power produced at various points of a transmitter is a measurement of its performance. For very high frequencies, a special sensor is used to "pick-off" a small part of the power in the system. The warming of the sensor is a measurement of the power level. High power can be measured using a device called a calorimeter, in which a resistor is submersed in moving fluid. The temperature rise of the fluid between the inlet and outlet is a measure of the power dissipated by the resistor.

In most common applications, a specialized instrument is not required for power measurements. Instead, the power is calculated from application of Watt's law. At least two parameters (voltage, current, or resistance) are needed to apply Watt's law.

CHAPTER 4

EXAMPLE 4–16

Problem
Calculate the power dissipated in the bulb in the ac circuit shown in Figure 4–16. The cord is plugged into an ac outlet that has been measured as 110 Vac. Both the current and the voltage are rms values. Notice that the 10 A scale is selected on the DMM for this measurement.

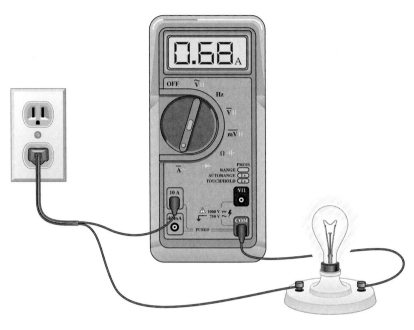

FIGURE 4–16

Solution
The DMM is indicating 0.68 A. Because you know the voltage and current, you can find the power from

$$P = IV = (0.68 \text{ A}) (110 \text{ V}) = \mathbf{74.8 \text{ W}}$$

Question
If the bulb is replaced with a 100 W bulb, what should be the reading on the DMM?

SAFETY NOTE

A consideration for connecting appliances or other electrical loads to a circuit is the size of the wires and circuit breaker for the total load. The National Electrical Code specifies the maximum allowable current in various wire sizes as well as many other conditions for safe operation. The code recognizes two types of circuits for appliances: a dedicated circuit for a single appliance such as an oven and a circuit that allows multiple appliances to be connected to the same circuit.

Power Rating of Appliances

Appliances and tools are given a power rating by their manufacturer that is based on the rate the appliance can do work. A few appliances have heat as the end product (oven or hot water heater), but for many, heat is an unwanted by-product of energy conversion when the electrical energy is converted to a useful form such as motion or stored energy. For example, the electrical energy may be converted to mechanical work (a power saw), or stored energy (air compressor), or may be used to move heat (refrigerator). In each of these cases, some energy is lost to heat.

Power Rating of Resistors

Unlike appliances, the power rating of a resistor is a measure of the amount of unwanted heat that the resistor can get rid of safely. All resistors convert a certain amount of electrical energy to heat when current is present. One of the practical problems that technicians must face is calculating the power that will be dissipated by resistors to assure that the proper replacement is selected if a replacement is necessary. Heat is the great destroyer of electronic components, including resistors. In practice, resistors are extremely reliable components and rarely need to be replaced unless their power rating has been exceeded.

OHM'S LAW AND WATT'S LAW

You should always choose a power rating that is larger than the calculated power to assure a safety factor. Actual circuit conditions may affect the safety factor required, but in general, the next larger standard size is satisfactory. For example, if the power calculated is 0.75 W, you should choose a 1 W resistor.

The amount of power that a resistor can dissipate is related to the physical size of the resistor, not to its ohmic value. A larger surface area can radiate more of the unwanted heat to the surroundings. Metal-film resistors are available in sizes ranging from 1/8 W to 1 W as shown in Figure 4–17. Small 1/8 W and 1/4 W resistors are the most widely used sizes. Other types of resistors are available with much larger power ratings. Figure 4–18 shows some of these resistors.

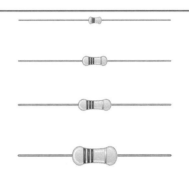

FIGURE 4–17

Relative sizes of metal-film resistors with standard power ratings of 1/8 W, 1/4 W, 1/2 W and 1 W.

Typical resistors with high power ratings. **FIGURE 4–18**

(a) Axial-lead wirewound (b) Adjustable wirewound (c) Radial-lead for PC board insertion

EXAMPLE 4–17

Problem
A 110 V circuit has 24 lamps that can be turned on independently. Each lamp has a 100 W bulb. If the circuit is protected by a 15 A circuit breaker, how many lamps can be turned on before the breaker rating is exceeded?

Solution
Each lamp draws a current of

$$I = \frac{P}{V} = \frac{100 \text{ W}}{110 \text{ V}} = 0.91 \text{ A}$$

Divide the total current of 15 A by the current required for each lamp to determine how many lamps can be on before the breaker limit is exceeded.

$$\frac{15 \text{ A}}{0.91 \text{ A/lamp}} = \mathbf{16 \text{ lamps}}$$

CHAPTER 4

The calculator will show 16.484 for the result. This result is rounded *down* to a whole number; the 17th lamp will exceed the breaker limit.

Question
If all of the lamps had 75 W bulbs instead, how many could be turned on before the current is equal to the breaker capacity?

EXAMPLE 4–18

Problem
A 16 Ω resistor is needed for testing purposes to simulate a speaker in a sound system. Assume the largest power rating available for the test is a resistor rated for 20 W. What is the largest voltage that can be applied to the resistor without exceeding its power rating?

Solution
$$P = \frac{V^2}{R}$$
$$PR = V^2$$

Transpose and take the square root of both sides.

$$V = \sqrt{PR} = \sqrt{(20 \text{ W})(16 \text{ Ω})} = \mathbf{17.9 \text{ V}}$$

The calculator sequence is

Question
If the voltage is set to 15 V, what power is dissipated in the resistor?

EXAMPLE 4–19

Problem
Current-limiting resistors are sometimes used in power supplies. Assume a 0.5 Ω resistor has a power rating of 2 W. What is the maximum current the supply can provide before the power rating is exceeded?

Solution
$$P = I^2 R$$
$$\frac{P}{R} = I^2$$

Transpose and take the square root of both sides.

$$I = \sqrt{\frac{P}{R}} = \sqrt{\frac{2 \text{ W}}{0.5 \text{ Ω}}} = \mathbf{2 \text{ A}}$$

The calculator sequence is

Question
If the current from the supply is 1.5 A, what power is dissipated in the resistor?

OHM'S LAW AND WATT'S LAW

EXAMPLE 4–20

Problem
The maximum power that can be dissipated in a typical small-signal transistor is 0.5 W. If the voltage across the transistor is 15 V, what is the maximum current before the power rating is exceeded?

Solution
$$P = IV$$
$$I = \frac{P}{V} = \frac{0.5 \text{ W}}{15 \text{ V}} = \mathbf{33.3 \text{ mA}}$$

Question
If the current is 25 mA, what power is dissipated in the transistor?

Heat Sinks

Some components such as transistors must dissipate more heat than just the normal packaging can safely dissipate. A common method for dissipating "extra" heat energy is to thermally connect the component to a heat-conductive surface (usually metal) to help conduct heat away and radiate it to the environment. This surface is called a **heat sink**. Many times, heat sinks are equipped with fins to increase the surface area and help radiate heat to the surroundings. Figure 4–19 shows some representative heat sinks.

FIGURE 4–19

Typical small heat sinks.

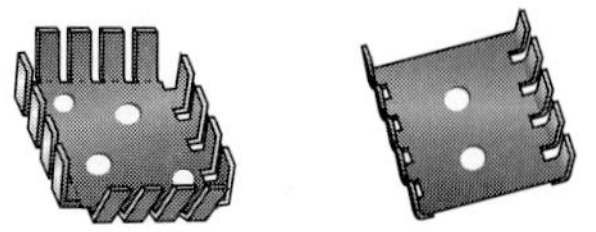

Review Questions

21. What happens to the electrical energy for turning a power tool such as an electric drill?
22. Why does a physically larger resistor have a higher power rating than a smaller one of the same type?
23. What is the range of power ratings for metal-film resistors?
24. What is the ohmic value of a resistor that will dissipate 1 W when the voltage across it is 2 V?
25. What is a heat sink?

NONLINEAR RESISTANCE 4–6

As you have seen, fixed resistors are designed to have a specific amount of resistance that is independent of the voltage across them, provided that their power rating is not exceeded. Many useful devices do not have the linear characteristic of a fixed resistor.

In this section, you will learn the difference between dc and ac resistance and calculate the dc and ac resistance for a device with a nonlinear *IV* curve.

DC Resistance

The common incandescent bulb has much greater resistance as it heats up than when it is cold. The shape of the characteristic *IV* curve is shown in Figure 4–20 for a bulb that has an actual resistance of only 10.5 Ω when it is cold. All tungsten filament bulbs have a similar

CHAPTER 4

FIGURE 4–20

The *IV* characteristic curve for a tungsten light bulb. The dc resistance at each point is found by dividing the voltage by the current.

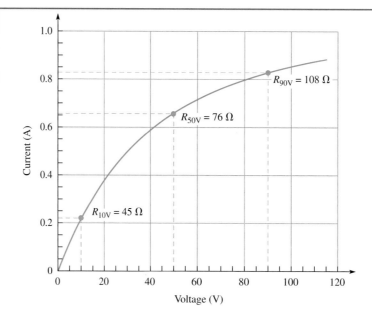

characteristic. As voltage increases, the current causes a heating effect that dramatically reduces the resistance. For the bulb shown, the resistance at 10 V is 45 Ω (10 V/0.22 A). At 90 V, the same bulb has a resistance of 108 Ω (90 V/0.83 A). Although the resistance is not constant, Ohm's law and Watt's can still be applied, provided the voltage is specified at the point where the resistance is determined.

Because the nonlinear resistance is determined by dividing the voltage at some point on the curve by the corresponding current, the result is called the **dc resistance**, shown as *R*. With a fixed resistor, the dc resistance is constant because the *IV* characteristic is a straight line. With a nonlinear resistor, the dc resistance depends on where it is measured because the *IV* characteristic is not a straight line.

EXAMPLE 4–21

Problem
What is the power dissipated by the light bulb whose characteristic *IV* curve is shown in Figure 4–20 for the following applied voltages:

(a) 30 V (b) 60 V (c) 90 V

Solution

(a) At 30 V, the current is 0.50 A (from the graph in Figure 4–20). Applying Watt's law, the power at 30 V (P_{30V}) is

$$P_{30V} = IV = (0.50 \text{ A})(30 \text{ V}) = \mathbf{15.0 \text{ W}}$$

(b) At 60 V, the current is 0.72 A (from the graph). Applying Watt's law, the power at 60 V is

$$P_{60V} = IV = (0.72 \text{ A})(60 \text{ V}) = \mathbf{43.2 \text{ W}}$$

(c) Because you know the resistance from the previous discussion, you can use another of Watt's law formulas.

$$P_{90V} = \frac{V^2}{R} = \frac{90\text{ V}^2}{108\text{ }\Omega} = \mathbf{75\text{ W}}$$

The calculator sequence for answer (c) is

[9][0][x²][÷][1][0][8][=]

Question
What is the wattage rating of the bulb at its rated voltage of 110 V?

AC Resistance

As mentioned previously, the dc resistance is determined by dividing the voltage at a given point on the characteristic curve by the corresponding current at that point. There is another way to specify resistance, which is important when you work with electronic devices, including nonlinear resistors and with ac signals (which will be studied in later courses). The **ac resistance** is defined as the resistance at a point if a small *change* in voltage is divided by a corresponding *change* in current. Figure 4–21 illustrates ac resistance. Notice that the ac resistance is the reciprocal of the slope at the point it is measured and that it is indicated with a lowercase *r*.

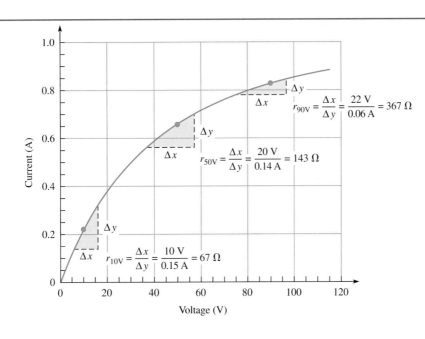

FIGURE 4–21

The ac resistance at each point is found by dividing a small *change* in voltage by the corresponding *change* in current.

Review Questions

26. Why do you think a light bulb is more likely to burn out when it is first turned on?
27. What is dc resistance?
28. What is ac resistance?
29. What does a horizontal line on the *IV* curve represent?
30. What happens to the ac resistance of a tungsten bulb as the voltage across it increases?

CHAPTER REVIEW

Key Terms

AC resistance The resistance at a point on the *IV* curve for a device if a small *change* in voltage is divided by a corresponding *change* in current. AC resistance is designated with the letter *r*.

Conductance The reciprocal of resistance; it is a measure of the ability of a material to conduct current. The unit for conductance is the siemens, S.

DC resistance The resistance of a device found by dividing the voltage at some point by the corresponding current. DC resistance is designated with the letter *R*.

Heat sink A metal surface designed to conduct and radiate heat.

Horsepower (hp) A commonly used unit for measuring the power of motors. One hp is equal to 746 W.

Kilowatt-hour (kWh) An amount of energy equivalent to a power level of one kilowatt maintained for one hour. It is equal to 3.6 million joules (J).

Ohm's law A fundamental law of electricity that states that the current in a circuit is proportional to the applied voltage and inversely proportional to the resistance.

Watt's law A law that states the relationships of power to current, voltage, and resistance.

Important Facts

- Ohm's law gives the relationship between voltage, current, and resistance.
- Conductance is the reciprocal of resistance and is measured in siemens.
- Fixed resistors have a linear *IV* characteristic.
- Electrical energy is a form of potential energy that can be converted to other energy forms.
- Watt's law describes the relationship between power and voltage, current, and resistance.
- The power rating of a resistor is a measure of the amount of unwanted heat that the resistor can get rid of safely.
- The amount of power that a resistor can dissipate is related to the size of the resistor, not to its ohmic value.
- A heat sink is a metal surface designed to conduct and radiate heat.
- DC resistance is determined by dividing the voltage at some point on the *IV* curve by the corresponding current.
- AC resistance is defined as the resistance at a point if a small *change* in voltage is divided by a corresponding *change* in current.

Formulas

Ohm's law:

$$I = \frac{V}{R} \tag{4-1}$$

$$V = IR \qquad (4\text{-}2)$$

$$R = \frac{V}{I} \qquad (4\text{-}3)$$

Conductance (definition):

$$G = \frac{1}{R} \qquad (4\text{-}4)$$

Watt's law:

$$P = IV \qquad (4\text{-}5)$$

$$P = \frac{V^2}{R} \qquad (4\text{-}6)$$

$$P = I^2R \qquad (4\text{-}7)$$

Chapter Checkup

Answers are at the end of the chapter.

1. Ohm's law for current is
   (a) voltage times resistance
   (b) voltage divided by resistance
   (c) resistance divided by voltage
   (d) voltage divided by power

2. Ohm's law for resistance is
   (a) voltage times current
   (b) current times power
   (c) current divided by voltage
   (d) voltage divided by current

3. Ohm's law for voltage is
   (a) current times resistance
   (b) current squared times resistance
   (c) current divided by resistance
   (d) resistance divided by current

4. If an electric heater with a resistance of 15 Ω is connected to a 110 V circuit, the current will be
   (a) 133 mA   (b) 1.33 A
   (c) 7.33 A   (d) 15 A

5. If the voltage across a resistor is doubled, the current will be
   (a) halved   (b) unchanged
   (c) doubled   (d) quadrupled

6. If the resistance in a circuit is doubled, the current will be
   (a) halved   (b) unchanged
   (c) doubled   (d) quadrupled

7. Conductance is the reciprocal of
   (a) power
   (b) current
   (c) voltage
   (d) resistance
8. Watt's law can be written:
   (a) $P = I^2V$
   (b) $P = IR$
   (c) $P = \dfrac{V^2}{R}$
   (d) all of these answers
9. The current in a 50 W light bulb rated for a 110 V circuit is approximately
   (a) 0.45 A
   (b) 2.2 A
   (c) 4.5 A
   (d) 22 A
10. If the voltage across a resistor is doubled, the power dissipated in the resistor will be
    (a) halved
    (b) unchanged
    (c) doubled
    (d) quadrupled
11. Electric utility companies charge customers for
    (a) power
    (b) energy
    (c) kilowatts
    (d) electrons
12. A unit for power is
    (a) horsepower
    (b) joule
    (c) kilowatt-hour
    (d) all of these answers
13. The *IV* characteristic curve for a fixed resistor is
    (a) a straight line
    (b) a line that has an increasing slope
    (c) a line that has a decreasing slope
14. The resistance of a tungsten light bulb
    (a) is constant for any voltage
    (b) decreases with voltage
    (c) increases with voltage
15. AC resistance is the same as dc resistance for
    (a) all resistive devices
    (b) light bulbs
    (c) fixed resistors

Questions

Answers to odd-numbered questions are at the end of the book.

1. What happens to the current in a circuit if the voltage is increased?
2. How much current is in a circuit with a 3 V source and a total resistance of 25 Ω?
3. What is the voltage across a 330 Ω resistor if the current is 15 mA?
4. How much resistance is needed to limit the current from a 12 V battery to 1 mA?
5. If a 3.6 kΩ resistor is placed across a 5.0 V source, what current is in the resistor?
6. What current is in a 2.2 MΩ resistor that has 10 V across it?
7. What is the voltage across a 470 kΩ resistor if the current is 35 μA?

8. What is the resistance of a bulb if it has 24 V across it with a current of 1.5 A?
9. How much resistance is needed to limit the current from a 110 V source to 10 μA?
10. What is the current in a 68 kΩ resistor that has 25 V across it?
11. What is the conductance of a 10 Ω resistor?
12. What is the conductance of a 270 kΩ resistor?
13. If a resistor has a conductance of 21.3 mS, what is the value of the resistance?
14. If a resistor has a conductance of 3.03 μS, what is the value of the resistance?
15. What is a kilowatt-hour?
16. How many joules are in 1 kWh?
17. What is horsepower?
18. If electricity costs 12 cents per kWh, what is the cost of the electricity for operating a 7 watt night light for 24 hours?
19. If electricity costs 12 cents per kWh, what is the cost of the electricity for operating a 250 watt TV set for 6 hours?
20. What is the power rating of a saw that uses 4.0 A when connected to a 110 V line?
21. If a clothes dryer uses 4.8 kW when it is connected to a 220 V line, what is the current?
22. What is the smallest resistance that can be used in a 5 V circuit if the resistors available are rated for 1/4 W?
23. Assume a 1/2 Ω resistor is rated for 5 W. What is the maximum current that can be in the resistor?
24. How many watts are dissipated by a 20 Ω resistor connected across a 5 V source?
25. What happens to the resistance of a tungsten filament light bulb as it heats up?
26. If the *IV* curve of a resistor is a straight line, what can you say about its dc and ac resistance?

Basic Problems

PROBLEMS

Answers to odd-numbered problems are at the end of the book.

1. Find the current for each of the following cases:
    (a) $V = 10$ V, $R = 270$ Ω
    (b) $V = 110$ V, $R = 50$ Ω
    (c) $V = 20$ V, $R = 82$ kΩ
    (d) $V = 6.1$ V, $R = 4.7$ MΩ
2. Find the current for each of the following cases:
    (a) $V = 15$ V, $R = 10$ kΩ
    (b) $V = 30$ V, $R = 2.2$ MΩ
    (c) $V = 110$ V, $R = 2.0$ kΩ
    (d) $V = 5.0$ V, $R = 560$ Ω
3. Find the voltage for each of the following cases:
    (a) $I = 4.0$ mA, $R = 1.5$ kΩ
    (b) $I = 2.94$ μA, $R = 5.1$ MΩ
    (c) $I = 0.2$ A, $R = 550$ Ω
    (d) $I = 121$ mA, $R = 82$ Ω

4. Find the voltage for each of the following cases:
   (a) $I = 60\ \mu A$, $R = 220\ k\Omega$
   (b) $I = 12\ \mu A$, $R = 1.5\ M\Omega$
   (c) $I = 4.2\ A$, $R = 52.4\ \Omega$
   (d) $I = 1.11\ mA$, $R = 27\ k\Omega$

5. Find the resistance for each of the following cases:
   (a) $I = 90\ nA$, $V = 90\ mV$
   (b) $I = 66.7\ \mu A$, $V = 18\ V$
   (c) $I = 22\ A$, $V = 220\ V$
   (d) $I = 2.56\ mA$, $V = 10\ V$

6. Find the resistance for each of the following cases:
   (a) $I = 8.2\ mA$, $V = 22\ V$
   (b) $I = 550\ \mu A$, $V = 36\ V$
   (c) $I = 40\ mA$, $V = 5.0\ V$
   (d) $I = 55\ mA$, $V = 66\ V$

7. What is the conductance of each resistor?
   (a) $10\ \Omega$
   (b) $2.7\ k\Omega$
   (c) $10\ M\Omega$

8. What is the conductance of each resistor?
   (a) $5.1\ \Omega$
   (b) $200\ k\Omega$
   (c) $1.0\ M\Omega$

9. What happens to the current in a circuit containing a fixed resistor for each of the following conditions? (Assume each condition is independent of the others.)
   (a) The voltage is reduced to 1/3 of its original value.
   (b) The voltage is tripled.
   (c) The fixed resistor is replaced with one twice as large.
   (d) The fixed resistor is replaced with one half as large.

10. For each circuit in Figure 4–22, determine the expected reading on the ammeter.

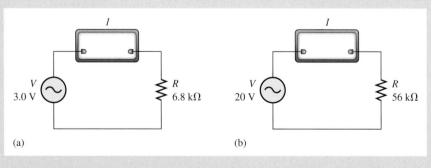

FIGURE 4–22

11. For each circuit in Figure 4–23, determine the voltage.

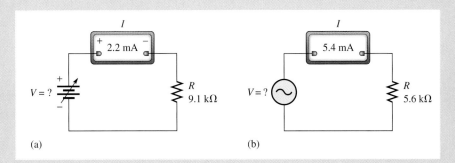

FIGURE 4–23

12. For each circuit in Figure 4–24, determine the resistance.

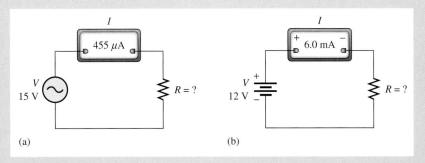

FIGURE 4–24

13. For the circuit in Figure 4–25, determine the maximum and minimum current, assuming the variation is due to the resistor's tolerance. The power supply is set for +7.0 V.

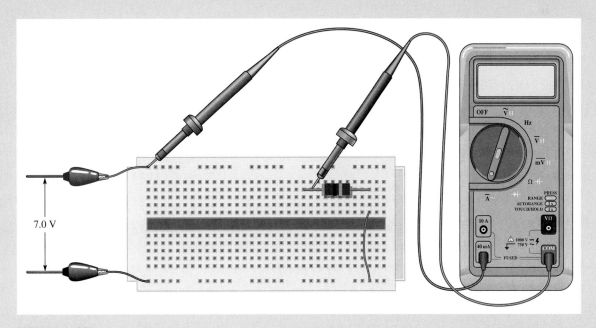

FIGURE 4–25

14. Refer to Figure 4–25. What is the minimum wattage rating the resistor should have?
15. The taillight for a trailer is rated for 12 V at 10 W. What current do you expect in the bulb?
16. A 220 V electric clothes dryer is rated for 4.0 kW. What is the current when the dryer is on?
17. Find the power dissipated in a resistor for each of the following cases:
    (a) $I = 10$ A, $V = 110$ V (b) $I = 18$ mA, $R = 330$ Ω
    (c) $V = 10$ V, $R = 27$ Ω (d) $I = 30$ mA, $R = 560$ Ω
18. Find the power dissipated in a resistor for each of the following cases:
    (a) $I = 1.5$ A, $V = 12$ V (b) $I = 120$ mA, $R = 15$ Ω
    (c) $V = 5.0$ V, $R = 33$ Ω (d) $V = 16$ V, $R = 270$ Ω
19. What is the current in a 60 W bulb designed for 110 V?
20. What is the current in a 250 W bulb designed for 110 V?
21. What is the energy in kWh used when a 200 W bulb is on for 10 hours?
22. The maximum power that can be dissipated in a small LED is about 40 mW (depending on the LED). If the voltage across the LED is 2.0 V, what is the maximum current before the power rating is exceeded?
23. Calculate the total cost of operating an electric water heater rated at 2500 W for 12 h if electricity is 12 cents per kWh.
24. Calculate the total cost of operating an electric range rated at 12,200 W for 2 h if electricity is 12 cents per kWh.
25. Assume that you use the following eight appliances for the specified lengths of time during a typical 24-hour period. Determine the energy used and the cost for these appliances in the 24 h period if energy sells for 12 cents/kWh. Refer to Table 4–2 for power ratings of the appliances.

    Clothes dryer: 45 minutes   Dishwasher: 2 hour
    Microwave oven: 30 minutes   Range: 1 hour
    Refrigerator: 12 hours   Television: 2 hours
    Washing machine: 30 minutes   Water heater: 8 hours

26. A motor is rated at 1/2 hp. What is this rating in watts?
27. Draw the *IV* curve for a 10 kΩ resistor.

Basic-Plus Problems

28. What is the maximum current that a 100 Ω, 2 W resistor can have without exceeding the rated power?
29. What is the maximum voltage that can be across a 330 Ω, 1/2 W resistor?
30. What is the cost of operating a 1 hp motor that has an efficiency of 75% for 2 hours if the cost of electricity is 10 cents per kWh?
31. A field-effect transistor (FET) is a solid-state device that can be used as an "electronic variable resistor" for certain applications. The resistance is controlled by a voltage called V_{GS}. As V_{GS} is made more negative, the resistance increases. Figure 4–26 shows a family of four *IV* curves for an actual FET with very low voltage across the FET. From each curve, calculate the dc resistance. (*Hint*: Recall than the slope on an *IV* curve represents the conductance).

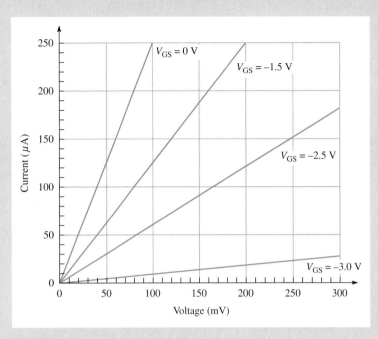

FIGURE 4–26

Example Questions

4–1: 0.425 A

4–2: 1.47 mA

4–3: 16.2 V

4–4: 11.7 A

4–5: 68.6 kΩ

4–6: See Figure 4–27.

4–7: 4.04 mA to 4.94 mA

4–8: 22.5 V

4–9: red, violet, red, gold

4–10: 2.44 mA

4–11: 16.4 V to 20.0 V

4–12: It is the reciprocal of 3.6×10^6 J/kWh (or 278×10^{-9} kWh/J)

4–13: 1.6 kWh

4–14: $0.60

4–15: 4.0 W

4–16: 0.909 A

4–17: 22 bulbs

4–18: 14 W

4–19: 1.12 W

4–20: 375 mW

4–21: 100 W

ANSWERS

FIGURE 4–27

Review Questions

1. $I = \dfrac{V}{R}, V = IR, R = \dfrac{V}{I}$
2. Current is one-half also.
3. Decreases
4. Current is the dependent variable.
5. Conductance is the reciprocal of resistance.
6. 255 nA
7. 8.8 V
8. 6.8 kΩ
9. 11.5 Ω
10. 12 Vrms
11. Joules and kilowatt-hours
12. The kWh is a much larger unit than the joule; typical household electrical energy usage would involve many millions of joules.
13. Watt
14. 10 kWh
15. $1.50
16. $P = IV, P = V^2/R, P = I^2R$
17. It is converted to heat.
18. 1.36 A
19. 0.25 W
20. 834 mW
21. It is converted to kinetic energy of the drill and heat.
22. It has more surface area to radiate heat to the atmosphere.
23. 1/8 W to 1 W
24. 4 Ω
25. A surface (usually metal) designed to conduct heat away from an electronic device and radiate it to the surroundings.
26. Because the resistance is low when it is cold and the current is high
27. Voltage divided by current
28. Change in voltage divided by change in current
29. Infinite resistance
30. It increases.

Chapter Checkup

1. (b) 2. (d) 3. (a) 4. (c) 5. (c)
6. (a) 7. (d) 8. (c) 9. (a) 10. (d)
11. (b) 12. (a) 13. (a) 14. (c) 15. (c)

CHAPTER 5

SERIES AND PARALLEL CIRCUITS

CHAPTER OUTLINE

- 5–1 Resistors in Series
- 5–2 Applying Ohm's Law in Series Circuits
- 5–3 Kirchhoff's Voltage Law
- 5–4 Voltage Dividers
- 5–5 Resistors in Parallel
- 5–6 Applying Ohm's Law in Parallel Circuits
- 5–7 Kirchhoff's Current Law

Study aids for this chapter are available at

http://www.prenhall.com/SOE

INTRODUCTION

When a circuit has a single path for current, it is called a series circuit. When a circuit has more than one path connected to a common voltage source, it is called a parallel circuit. In a series circuit, the same current is in all parts of the circuit. In a parallel circuit, the same voltage is across each component. In this chapter, you will learn to deal with both series and parallel resistive circuits. Resistive circuits are those that contain one or more voltage sources and resistors or components that behave as resistors. In addition, important circuit laws are introduced.

KEY OBJECTIVES

A section number is given for each objective. After completing this chapter, you should be able to

5–1 Identify series circuits and calculate the total resistance of resistors in series

5–2 Use Ohm's law to calculate an unknown current, voltage, or resistance in a series circuit

5–3 Apply Kirchhoff's voltage law

5–4 Apply the voltage-divider formula to find voltages in a series circuit

5–5 Recognize parallel circuits and calculate the total resistance of parallel resistors

5–6 Use Ohm's law to calculate an unknown current, voltage, or resistance in a parallel circuit

5–7 Apply Kirchhoff's current law

COMPUTER SIMULATIONS DIRECTORY

The following figures have Multisim circuit files associated with them.

◆ Figure 5–6 Page 136
◆ Figure 5–13 Page 140
◆ Figure 5–18 Page 144
◆ Figure 5–31 Page 154

LABORATORY EXPERIMENTS DIRECTORY

The following exercises are for this chapter.

◆ **Experiment 8** Voltage Dividers
◆ **Experiment 9** Parallel Circuits

KEY TERMS

- Series circuit
- Kirchhoff's voltage law
- Voltage divider
- Parallel circuit
- Branch
- Node
- Kirchhoff's current law
- Current divider

CHAPTER 5

The science of surveying involves measuring a portion of the earth's surface. One important measurement in surveying is called differential leveling. Differential leveling is used to find the elevation of an unknown point. The surveyor starts at a known reference point and makes a series of elevation measurements along a path. After finding the elevation of the unknown point, the surveyor "closes the loop" by making additional measurements back to the starting point. As an accuracy check, the total rise in elevation must equal the total fall. In ordinary surveying work, the accuracy is typically better than 0.5 inch in a 1 mile path.

Another common surveying problem is to traverse a parcel of land for the purpose of describing the property. For example, if the parcel is a rectangle, each of the four corners is located by measuring the distance and the direction of each side. When analyzing the data, the surveyor must again "close the loop" by making sure that the last side's end point is at the first side's starting point. This is equivalent to saying that the starting point and the ending point must be the same.

An analogy to these surveying problems occurs in circuit analysis. The voltage rises and drops must have the same magnitude if you make a loop and must return to the starting point. The path you take doesn't matter—only that you return to the starting point over a single path. Although the concept is basic, it is an important idea in circuit analysis.

5–1 RESISTORS IN SERIES

A **series circuit** is one in which there is a single complete path (forming a string) from the voltage source and through the load (or loads) and back. When two or more resistors are connected together in series, the total resistance is the sum of the individual resistors.

In this section, you will learn to identify series circuits and calculate the total resistance of resistors in series.

A series circuit can always be identified by the fact that it has only one current path. Figure 5–1 shows three resistors (R_1, R_2, and R_3) in series. All three circuits are electrically the same but have been drawn differently. Of course, there are other ways of drawing these three resistors in series. The important point to remember with a series connection is that there is only *one* current path from the source through the resistors and back to the source.

In a series circuit, the current is the same in all components.

This idea is extremely useful for solving problems with series circuits, and you will apply it many times in your work in electronics. Figure 5–2 shows four ammeters placed in a series circuit. All ammeters have identical readings, which indicates the current is the same throughout the circuit.

FIGURE 5–1

Three resistors in series. Although drawn differently, each resistive circuit is electrically the same.

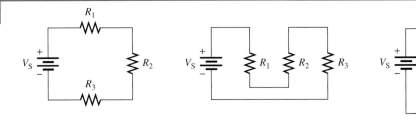

132

SERIES AND PARALLEL CIRCUITS

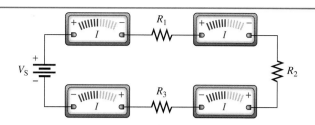

FIGURE 5–2

Ammeters in a series circuit have identical readings. Notice the polarity of the ammeters.

The voltage source maintains a certain current in any resistive circuit. The magnitude of this current depends on the source voltage and the total resistance of the circuit. In the case of a series circuit, the total resistance is the sum of the individual resistors, no matter how many are in series. As an equation this is expressed as

$$R_T = R_1 + R_2 + R_3 + \ldots + R_n \qquad (5\text{–}1)$$

where R_T represents the total resistance of the series resistors and R_n is the resistance of the last resistor in the string. The subscript n is commonly applied to electronics equations to indicate a count. It can stand for any positive integer. In Equation 5–1, n is equal to the number of resistors in series. For example, if there are five resistors in series, $n = 5$ and Equation 5–1 could be written as

$$R_T = R_1 + R_2 + R_3 + R_4 + R_5$$

In this case, the total resistance is the sum of the five resistors.

EXAMPLE 5–1

Problem
Find the total resistance of a series circuit with the following resistors: 1.0 kΩ, 2.7 kΩ, 3.3 kΩ, and 10 kΩ.

Solution
The total resistance is the sum of the four resistors.

$$R_T = R_1 + R_2 + R_3 + R_4 = 1.0 \text{ k}\Omega + 2.7 \text{ k}\Omega + 3.3 \text{ k}\Omega + 10 \text{ k}\Omega = \mathbf{17 \text{ k}\Omega}$$

The calculator sequence for the TI-36X is

[1][.][0][EE][3][+][2][.][7][EE][3][+][3][.][3][EE][3][+][1][0][EE][3][=]

Because all of the resistor values are in kΩ units, you may prefer to key in the numbers without the exponents. The answer will have the same kΩ units that each resistor value has.

Question*
What is the total resistance if the 10 kΩ resistor is replaced with a 5.6 kΩ resistor?

Measuring the Resistance of Series Resistors

Any time you measure resistance, you must disconnect all voltage sources to assure they do not affect the reading. To measure the total resistance of series resistors, place one lead from a multimeter on each end of the string of resistors. Example 5–2 illustrates this method.

Answers are at the end of the chapter.

CHAPTER 5

EXAMPLE 5–2

Problem
Show how to connect the resistors in Figure 5–3 in series on a protoboard and how to measure their total resistance with a DMM. (Protoboard wiring is discussed in the lab manual.)

FIGURE 5–3

Solution
The resistors can be connected in series in many ways, but all require that a single path exists. One way to connect them is shown in Figure 5–4. The DMM is connected to the ends of the string of resistors to read the total resistance. It is important to disconnect voltage sources when you make the resistance measurement.

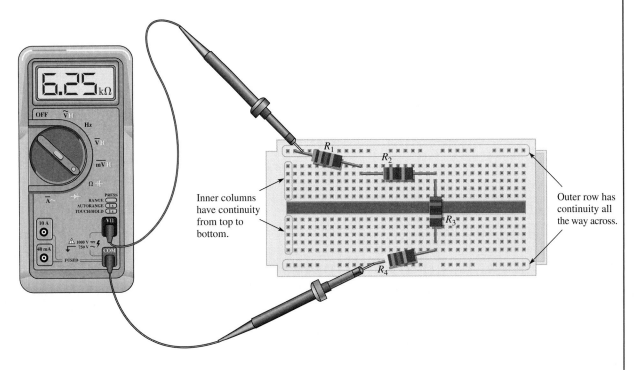

FIGURE 5–4

Question
Is the reading on the DMM the expected value?

Equal-Value Resistors in Series

Occasionally, you will find a circuit with two or more resistors of the same value connected in series. To find the total resistance of equal-value resistors, simply multiply the number of resistors by the resistance of one of the resistors. Of course, this is equivalent to summing the values. In equation form,

$$R_T = nR \tag{5–2}$$

where n is the number of equal-value resistors and R is the resistance of one of the resistors.

EXAMPLE 5–3

Problem
What is the total resistance of five 330 Ω resistors in series?

Solution
Substitute the given values into Equation 5–2. The total resistance is

$$R_T = nR = 5(330 \, \Omega) = \mathbf{1.65 \, k\Omega}$$

Question
What is the resistance of one resistor if three equal-value resistors have a total resistance of 14.1 kΩ?

Review Questions

Answers are at the end of the chapter.

1. What is true about the current in a series circuit?
2. The following resistors are in series: 470 Ω, 560 Ω, 1.2 kΩ and 1.5 kΩ. What is the total resistance?
3. Suppose you have three resistors that are 5.6 kΩ each. If you wanted to get a total series resistance of 25 kΩ with one more resistor, what resistance must it have?
4. What is the total resistance of six 220 Ω resistors connected in series?
5. What is the resistance of a 1.5 MΩ resistor that is in series with a 560 kΩ resistor?

APPLYING OHM'S LAW IN SERIES CIRCUITS 5–2

Ohm's law can be applied to series circuits for the total circuit or for individual components in the circuit. Try working the examples given before you look at the solutions.

In this section, you will learn to use Ohm's law to calculate an unknown current, voltage, or resistance in a series circuit.

Subscript Notation

A series circuit consists of a voltage source and two or more components connected to form a single current path. For example, the circuit may consist of a voltage source, an ammeter, and two resistors as shown in Figure 5–5.

The ammeter indicates the current in the circuit. It could be identified as the total current (I_T) or as the current in either resistor (I_1 or I_2). The subscripts refer to the specific component; in this case, R_1 has current I_1 and R_2 has current I_2. As you know, the total current is identical to the current in either resistor.

$$I_T = I_1 = I_2 \qquad (5\text{–}3)$$

When a voltage difference appears across a component, it is referred to as a *voltage drop*. The voltage drop across a component is also indicated with a subscript. For example, the voltage across R_1 is indicated as V_1 and the voltage across R_2 is indicated as V_2. The total voltage can be shown as V_T, but it is more commonly shown on schematics as V_S (for source voltage).

FIGURE 5–5

A series circuit consisting of a voltage source, an ammeter, and two resistors.

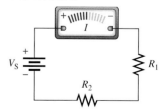

CHAPTER 5

When you apply Ohm's law to a series circuit, you can use the total current, total voltage, or total resistance. If you are interested in finding the total current, you can write Ohm's law as

$$I_T = \frac{V_T}{R_T}$$

Notice that the same subscripts are used in writing Ohm's law. The source voltage is shown here as V_T to emphasize the common subscript for all variables. This is exactly the same as writing Ohm's law as

$$I_T = \frac{V_S}{R_T}$$

Because V_S is more widely used than V_T, we will use V_S in the examples that follow. Keep in mind that V_S is the total applied voltage.

The idea of writing a common subscript on each variable can be applied to a single component such as R_1. Thus, the current in R_1 is given as

$$I_1 = \frac{V_1}{R_1}$$

Of course, I_1 is the same as I_T because it is a series circuit. In fact, the current in any component is the same as the total current. In the case of the series circuit in Figure 5–5, you can express this current as

$$I_T = \frac{V_S}{R_T} = \frac{V_1}{R_1} = \frac{V_2}{R_2}$$

You can apply Ohm's law in a series circuit to the circuit as a whole or to an individual component. Also, you can find the total resistance of series resistors by adding the individual resistors. Armed with these two ideas, you can easily find other unknown quantities in the circuit. The following examples illustrate the process. As in many problems in electronics, there is more than one starting point for solving these examples. Keep in mind that the solutions shown are not the only way to approach the problem.

EXAMPLE 5–4

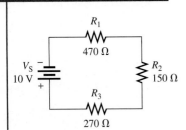

FIGURE 5–6

Problem
For the circuit of Figure 5–6, what is the total current and the voltage across R_1?

Solution
Step 1: Calculate the total resistance.

$$R_T = R_1 + R_2 + R_3 = 470 \ \Omega + 150 \ \Omega + 270 \ \Omega = 890 \ \Omega$$

Step 2: Apply Ohm's law to find the total current.

$$I_T = \frac{V_S}{R_T} = \frac{10 \text{ V}}{890 \ \Omega} = 11.2 \text{ mA}$$

Because it is a series circuit, this is the current in each of the resistors.

$$I_T = I_1 = I_2 = I_3 = 11.2 \text{ mA}$$

Step 3: To find the voltage across R_1, rearrange Ohm's law to solve for V_1.

$$V_1 = I_1 R_1 = (11.2 \text{ mA})(470 \text{ }\Omega) = \mathbf{5.28 \text{ V}}$$

Question
What are V_2 and V_3?

COMPUTER SIMULATION

Open the Multisim file F05-06DC on the website. Using the multimeter, verify the current.

EXAMPLE 5-5

Problem
For the circuit of Figure 5–7, the given values are listed in Table 5–1. Calculate the unknown values indicated in the table.

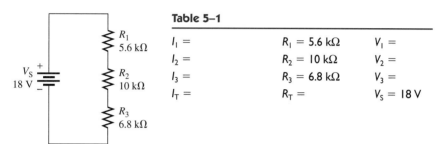

Table 5–1

$I_1 =$	$R_1 = 5.6 \text{ k}\Omega$	$V_1 =$
$I_2 =$	$R_2 = 10 \text{ k}\Omega$	$V_2 =$
$I_3 =$	$R_3 = 6.8 \text{ k}\Omega$	$V_3 =$
$I_T =$	$R_T =$	$V_S = 18 \text{ V}$

FIGURE 5–7

Solution
Step 1: Calculate the total resistance by summing the resistance of the three resistors.

$$R_T = R_1 + R_2 + R_3 = 5.6 \text{ k}\Omega + 10 \text{ k}\Omega + 6.8 \text{ k}\Omega = 22.4 \text{ k}\Omega$$

Step 2: Use Ohm's law with the total voltage and total resistance to calculate the total current.

$$I_T = \frac{V_S}{R_T} = \frac{18 \text{ V}}{22.4 \text{ k}\Omega} = 0.804 \text{ mA}$$

Because this is a series circuit, the current is identical in all components. Enter this value for I_1, I_2, and I_3.

Step 3: Use Ohm's law to calculate V_1, V_2, and V_3.

$$V_1 = I_1 R_1 = (0.804 \text{ mA})(5.6 \text{ k}\Omega) = 4.50 \text{ V}$$
$$V_2 = I_2 R_2 = (0.804 \text{ mA})(10 \text{ k}\Omega) = 8.04 \text{ V}$$
$$V_3 = I_3 R_3 = (0.804 \text{ mA})(6.8 \text{ k}\Omega) = 5.46 \text{ V}$$

The completed table is shown as Table 5–2.

Table 5–2

$I_1 = \mathbf{0.804 \text{ mA}}$	$R_1 = 5.6 \text{ k}\Omega$	$V_1 = \mathbf{4.50 \text{ V}}$
$I_2 = \mathbf{0.804 \text{ mA}}$	$R_2 = 10 \text{ k}\Omega$	$V_2 = \mathbf{8.04 \text{ V}}$
$I_3 = \mathbf{0.804 \text{ mA}}$	$R_3 = 6.8 \text{ k}\Omega$	$V_3 = \mathbf{5.46 \text{ V}}$
$I_T = \mathbf{0.804 \text{ mA}}$	$R_T = \mathbf{22.4 \text{ k}\Omega}$	$V_S = 18 \text{ V}$

Question
If a 1.0 kΩ resistor were accidentally used for R_2 instead of a 10 kΩ resistor, what will happen to the voltage across each of the resistors?

CHAPTER 5

EXAMPLE 5–6

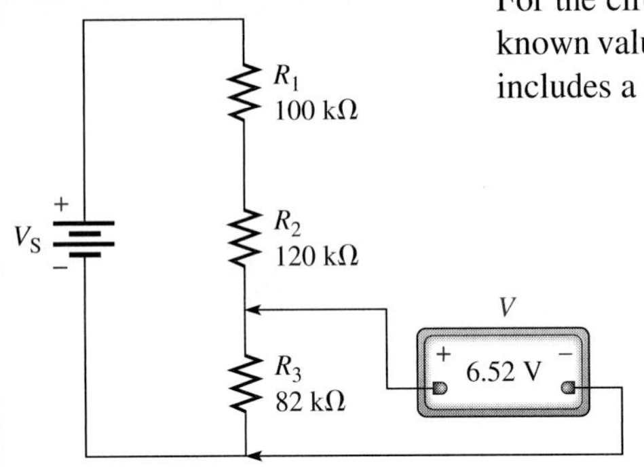

Problem
For the circuit of Figure 5–8, the given values are listed in Table 5–3. Calculate the unknown values in the table. Notice that the source voltage is not given. In addition, the table includes a column for the power dissipated in each resistor and the total power.

FIGURE 5–8

Table 5–3

$I_1 =$	$R_1 = 100\ k\Omega$	$V_1 =$	$P_1 =$
$I_2 =$	$R_2 = 120\ k\Omega$	$V_2 =$	$P_2 =$
$I_3 =$	$R_3 = 82\ k\Omega$	$V_3 = 6.52\ V$	$P_3 =$
$I_T =$	$R_T =$	$V_S =$	$P_T =$

Solution

Step 1: Calculate the total resistance by summing the resistance of the three resistors. (*Note:* Alternatively, you could calculate the current first.)

$$R_T = R_1 + R_2 + R_3 = 100\ k\Omega + 120\ k\Omega + 82\ k\Omega = 302\ k\Omega$$

Step 2: Use Ohm's law with the known V_3 and known resistance of R_3 to calculate the current.

$$I_3 = \frac{V_3}{R_3} = \frac{6.52\ V}{82\ k\Omega} = 79.5\ \mu A$$

Because this is a series circuit, the same current is in all components.

$$I_3 = I_T = I_1 = I_2 = 79.5\ \mu A$$

Step 3: Use Ohm's law to calculate V_1 and V_2.

$$V_1 = I_1 R_1 = (79.5\ \mu A)(100\ k\Omega) = 7.95\ V$$
$$V_2 = I_2 R_2 = (79.5\ \mu A)(120\ k\Omega) = 9.54\ V$$

V_S is left as a question exercise.

Step 4: Use Watt's law to calculate P_1, P_2, and P_3.

$$P_1 = I_1 V_1 = (79.5\ \mu A)(7.95\ V) = 632\ \mu W$$
$$P_2 = I_2 V_2 = (79.5\ \mu A)(9.54\ V) = 758\ \mu W$$
$$P_3 = I_3 V_3 = (79.5\ \mu A)(6.52\ V) = 518\ \mu W$$

P_T is left as a question exercise.

The completed table, except for V_S and P_T, is shown as Table 5–4.

Table 5–4

$I_1 = 79.5\ \mu A$	$R_1 = 100\ k\Omega$	$V_1 = 7.95\ V$	$P_1 = 632\ \mu W$
$I_2 = 79.5\ \mu A$	$R_2 = 120\ k\Omega$	$V_2 = 9.54\ V$	$P_2 = 758\ \mu W$
$I_3 = 79.5\ \mu A$	$R_3 = 82\ k\Omega$	$V_3 = 6.52\ V$	$P_3 = 518\ \mu W$
$I_T = 79.5\ \mu A$	$R_T = 302\ k\Omega$	$V_S =$	$P_T =$

Question
What are the values of V_S and P_T?

SERIES AND PARALLEL CIRCUITS

Review Questions

6. A 12 V battery is connected to three 1.0 kΩ resistors in series. What is the current in each resistor?
7. What is the total current in the circuit shown in Figure 5–9?

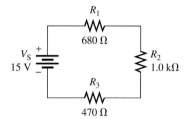

FIGURE 5–9

8. What is V_1 for the circuit shown in Figure 5–9?
9. Four equal-value resistors are connected in series to a 20 V source. The current is 50 mA. What is the value of each resistor?
10. What is the value of R_2 in the circuit of Figure 5–10 that will limit current to 100 mA?

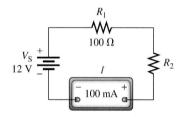

FIGURE 5–10

KIRCHHOFF'S VOLTAGE LAW 5–3

Kirchhoff was a physicist who extended the work of Ohm in circuit theory. He developed two fundamental laws for circuits and made important contributions to other sciences including astronomy.

In this section, you will learn to apply Kirchhoff's voltage law.

In a simple resistive circuit, the polarity of the voltage drop across a resistor is always opposite to the polarity of the source voltage, as illustrated in Figure 5–11(a). With several resistors, the polarity across each resistor is still opposite to the source, as shown in Figure 5–11(b).

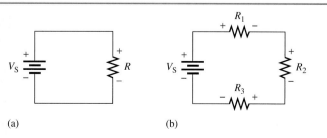

FIGURE 5–11

The polarity of the source voltage and the voltage drop across a load resistor are always opposite.

139

CHAPTER 5

One way to view the polarity of the voltages across the components of Figure 5–11 is in terms of "rises" and "drops." If we define the source voltage as a "rise," then the voltages across the resistors represent "drops." Kirchhoff observed that the sum of the drops was equal to the source voltage in a closed path. From this observation, he developed a law that is widely applied to solving circuit problems. **Kirchhoff's voltage law** is generally stated as

The sum of all the voltage drops around a single closed path in a circuit is equal to the total source voltage in that closed path.

Expressed in equation form, Kirchhoff's voltage law (KVL) is written

$$V_S = V_1 + V_2 + V_3 + \ldots + V_n \quad (5\text{–}4)$$

where n represents the number of voltage drops.

Both circuits in Figure 5–12 illustrate Equation 5–4. The polarity of the source does not matter. KVL is shown here for a three-resistor series circuit, but it applies to any circuit as long as you follow a single path when you write the equation.

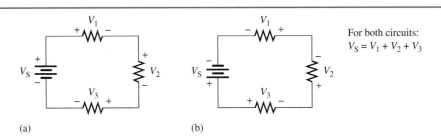

FIGURE 5–12

The sum of n voltage drops is equal to the total source voltage in a closed path. In this circuit, $n = 3$.

Another way to consider Kirchhoff's voltage law is in terms of an algebraic sum. An algebraic sum simply means to add numbers while keeping track of the sign. A voltage rise is assigned one sign (generally positive), and a voltage drop is assigned the opposite sign. Voltages are either positive or negative and are always measured *across* a component. In this form, Kirchhoff's voltage law is stated as

The algebraic sum of the voltages around a single closed path in a circuit equals zero.

In this form of KVL, the algebraic sign of the source is opposite to the sign of the drops across the resistors. If there are more than one source in the closed path, then the total source voltage depends on whether they are aiding or opposing each other. Multiple sources will be discussed in Section 6–5.

EXAMPLE 5–7

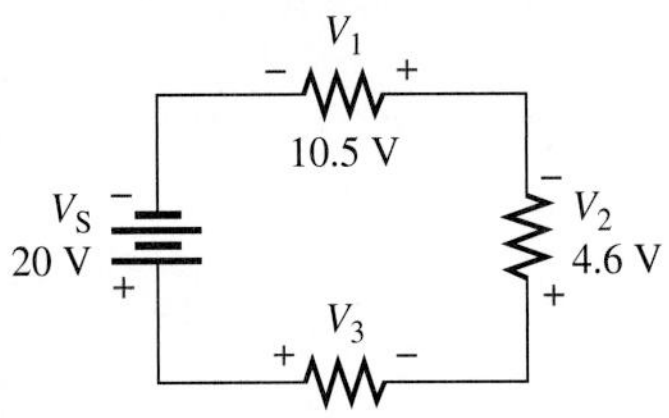

Problem
For the circuit of Figure 5–13, calculate V_3.

Solution
One of Kirchhoff's voltage law statements says that the sum of the voltage drops around a single closed path is equal to the source voltage in that closed path.

FIGURE 5–13

$$V_S = V_1 + V_2 + V_3$$
$$V_3 = V_S - (V_1 + V_2) = 20\text{ V} - (10.5\text{ V} + 4.6\text{ V}) = \mathbf{4.9\text{ V}}$$

The equivalent voltage law statement says that the algebraic sum of the voltages around a single closed path equals zero.

$$V_S - V_1 - V_2 - V_3 = 20\text{ V} - 10.5\text{ V} - 4.6\text{ V} - V_3 = 0$$
$$V_3 = \mathbf{4.9\text{ V}}$$

Question
If the source voltage is doubled, what are V_1, V_2 and V_3?

COMPUTER SIMULATION

Open the Multisim file F05-13DC on the website. Using the multimeter, measure the voltage drops and confirm that KVL is true.

EXAMPLE 5–8

Problem
You can use both Ohm's law and Kirchhoff's voltage law to find the solution to a problem. Table 5–5 lists known values for the circuit of Figure 5–14. Complete the table by finding the unknown quantities.

Table 5–5

$I_1 =$	$R_1 = 330\ \Omega$	$V_1 =$
$I_2 =$	$R_2 =$	$V_2 =$
$I_3 =$	$R_3 = 270\ \Omega$	$V_3 = 1.35\text{ V}$
$I_T =$	$R_T =$	$V_S = 8.0\text{ V}$

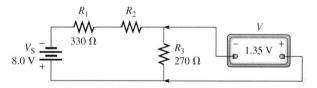

FIGURE 5–14

Solution
Step 1: Use Ohm's law to calculate the current in R_3.

$$I_3 = \frac{V_3}{R_3} = \frac{1.35\text{ V}}{270\ \Omega} = 5.00\text{ mA}$$

Because this is a series circuit, the same current is in all components.

$$I_3 = I_1 = I_2 = I_T = 5.00\text{ mA}$$

Step 2: Calculate the voltage across R_1.

$$V_1 = I_1 R_1 = (5.00\text{ mA})(330\ \Omega) = 1.65\text{ V}$$

Step 3: Apply Kirchhoff's voltage law to calculate the voltage across R_2.

$$V_S = V_1 + V_2 + V_3$$
$$V_2 = V_S - (V_1 + V_3) = 8.0\text{ V} - (1.65\text{ V} + 1.35\text{ V}) = 5.0\text{ V}$$

Step 4: Use Ohm's law to calculate the value of R_2.

$$R_2 = \frac{V_2}{I_2} = \frac{5.0 \text{ V}}{5.0 \text{ mA}} = 1.0 \text{ k}\Omega$$

Step 5: Calculate the total resistance by summing the resistors.

$$R_T = R_1 + R_2 + R_3 = 330 \, \Omega + 1.0 \text{ k}\Omega + 270 \, \Omega = 1.60 \text{ k}\Omega$$

The results are shown in Table 5–6.

Table 5–6

I_1 = 5.0 mA	R_1 = 330 Ω	V_1 = **1.65 V**
I_2 = 5.0 mA	R_2 = 1.0 kΩ	V_2 = **5.00 V**
I_3 = 5.0 mA	R_3 = 270 Ω	V_3 = **1.35 V**
I_T = 5.0 mA	R_T = **1.60 kΩ**	V_S = 8.0 V

Question
What mathematical check can you make to verify that the solution is correct?

Interestingly, Kirchhoff's voltage law works in an open circuit. An open circuit means the path is incomplete and there is no current. If there is an open path because of a failed component (such as a blown fuse), the source voltage will appear across the open.

Review Questions

11. What is Kirchhoff's voltage law?
12. Assume three equal-value resistors are connected in series with a 12 V source. What is the voltage across each resistor?
13. Assume voltages are measured across the resistors of Figure 5–15 as shown. What is V_S?
14. In a series circuit with three resistors, the source voltage is 20 V. If two of the voltage drops are 8.0 V and 6.0 V, what is the third voltage drop?
15. If a series circuit with a 100 V source is open, will there be a voltage across the open? If so what is the voltage?

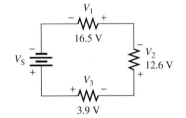

FIGURE 5–15

5–4 VOLTAGE DIVIDERS

For circuit analysis purposes, the resistors in a resistive series circuit can be thought of as "dividing" the source voltage between them. The term *voltage divider* comes from this idea. By thinking of series resistors as voltage dividers, you can compute the voltage drop across a resistor.

In this section, you will learn how to apply the voltage-divider formula to find voltages in a series circuit.

If two unequal resistors are in a series circuit, the larger resistor will have a proportionately larger voltage drop. For example, if one resistor is twice the size of a second resistor, the larger resistor will have twice the voltage of the smaller one. The reason is clear when you examine Ohm's law as applied to a series circuit. Because both resistors have the same current, the voltage drop is proportional to the resistance. This can be shown mathematically.

The current in R_1 is I_1 and the current in R_2 is I_2.

$$I_1 = I_2$$

Substituting Ohm's law for the current,

$$\frac{V_1}{R_1} = \frac{V_2}{R_2}$$

Rearranging terms by multiplying both sides by R_1 and dividing both sides by V_2 gives

$$\frac{V_1}{V_2} = \frac{R_1}{R_2}$$

This last equation shows that in a series circuit the ratio of the voltage drops (V_1/V_2) is equal to the ratio of the resistances (R_1/R_2). An equivalent statement is to say that the voltage drop is proportional to the resistance. The circuit in Figure 5–16 illustrates this idea. The resistance of R_1 is twice that of R_2. The total resistance is 12 kΩ, so the current in each resistor is

$$I_T = \frac{V_S}{R_T} = \frac{12\ \text{V}}{12\ \text{k}\Omega} = 1.0\ \text{mA}$$

By Ohm's law, the voltage across R_1 is

$$V_1 = I_1 R_1 = (1.0\ \text{mA})(8.0\ \text{k}\Omega) = 8.0\ \text{V}$$

and the voltage across R_2 is

$$V_2 = I_2 R_2 = (1.0\ \text{mA})(4.0\ \text{k}\Omega) = 4.0\ \text{V}$$

The voltage drop on the larger resistor is twice that of the smaller resistor (2:1), in proportion to the resistance ratio (2:1).

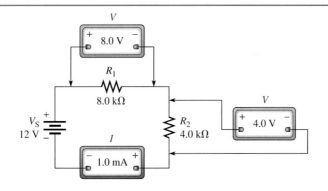

FIGURE 5–16

A series circuit with two resistors. R_1 is twice the value of R_2 and will have twice the voltage across it. (Nonstandard resistor values are used for convenience).

Figure 5–16 illustrates another important point. The total resistance is 12 kΩ. Resistor R_1 represents 2/3 of the total resistance, and resistor R_2 represents 1/3 of the total resistance. The voltage across R_1 is 2/3 of the source voltage; the voltage across R_2 is 1/3 of the source voltage. This observation is a direct result of Ohm's law again.

CHAPTER 5

FIGURE 5-17

A series circuit with two resistors.

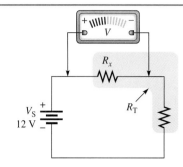

Figure 5-17 shows another series circuit with two resistors. R_x represents the resistance of one of the two resistors; R_T is the total resistance (sum of the resistors). The current in R_x is the same as the total current in the circuit. Therefore, from Ohm's law,

$$I_x = I_T$$

Substituting Ohm's law for the current,

$$\frac{V_x}{R_x} = \frac{V_S}{R_T}$$

Multiplying both sides by R_x gives

$$V_x = \left(\frac{V_S}{R_T}\right)R_x$$

Rearranging terms,

$$V_x = \left(\frac{R_x}{R_T}\right)V_S \qquad (5\text{-}5)$$

Equation 5-5 is the general voltage-divider formula. The portion in the parentheses represents the fraction that R_x is of the total resistance. A **voltage divider** consists of two or more series resistors across which one or more voltages are taken. Although the voltage-divider formula was developed here for two resistors, it can be applied to any number of resistors in series. The formula can be stated as

The voltage drop across any given resistor in a series circuit is equal to the ratio of that resistor to the total resistance, multiplied by the source voltage.

EXAMPLE 5-9

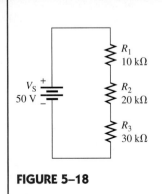

FIGURE 5-18

Problem

For each resistor in Figure 5-18,

(a) Determine the fraction of the total resistance represented by that resistor.

(b) Determine the voltage drop across each resistor using the voltage-divider formula.

Solution

(a) The total resistance is

$$R_T = R_1 + R_2 + R_3 = 10 \text{ k}\Omega + 20 \text{ k}\Omega + 30 \text{ k}\Omega = 60 \text{ k}\Omega$$

Determine the fraction of the total resistance represented by each resistor.

For R_1,
$$\left(\frac{R_1}{R_T}\right) = \left(\frac{10 \text{ k}\Omega}{60 \text{ k}\Omega}\right) = \frac{1}{6}$$

For R_2,
$$\left(\frac{R_2}{R_T}\right) = \left(\frac{20 \text{ k}\Omega}{60 \text{ k}\Omega}\right) = \frac{2}{6} = \frac{1}{3}$$

For R_3,
$$\left(\frac{R_3}{R_T}\right) = \left(\frac{30 \text{ k}\Omega}{60 \text{ k}\Omega}\right) = \frac{3}{6} = \frac{1}{2}$$

Thus, R_1 represents 1/6 of the total resistance, R_2 represents 1/3 of the total resistance and R_3 represents 1/2 of the total resistance.

(b) Find the voltage drops across each resistor by substituting into the voltage-divider equation. In each case R_x is replaced with the appropriate resistor.
The voltage across R_1 is
$$V_1 = \left(\frac{R_1}{R_T}\right)V_S = \left(\frac{10 \text{ k}\Omega}{60 \text{ k}\Omega}\right)50 \text{ V} = \mathbf{8.33 \text{ V}}$$

The voltage across R_2 is
$$V_2 = \left(\frac{R_2}{R_T}\right)V_S = \left(\frac{20 \text{ k}\Omega}{60 \text{ k}\Omega}\right)50 \text{ V} = \mathbf{16.7 \text{ V}}$$

The voltage across R_3 is
$$V_3 = \left(\frac{R_3}{R_T}\right)V_S = \left(\frac{30 \text{ k}\Omega}{60 \text{ k}\Omega}\right)50 \text{ V} = \mathbf{25.0 \text{ V}}$$

The calculator sequence for the voltage across R_1 is

The parentheses do not need to be keyed into the calculator in this case but are shown to clarify the resistance ratio. In addition, you can simplify the calculator entry if you cancel the prefixes *kilo* (k) before entering the numbers. With these simplifications, the calculator sequence is

1 0 ÷ 6 0 × 5 0 =

Question
If four equal-value resistors are in series, what fraction of the source voltage is across each resistor?

COMPUTER SIMULATION

Open the Multisim file F05-18DC on the website. Using the multimeter, verify the voltage drops.

CHAPTER 5

EXAMPLE 5–10

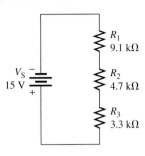

FIGURE 5–19

Problem
Use the voltage-divider formula to calculate the voltage across R_1 and R_2 in Figure 5–19.

Solution
The total resistance is

$$R_T = R_1 + R_2 + R_3 = 9.1\ \text{k}\Omega + 4.7\ \text{k}\Omega + 3.3\ \text{k}\Omega = 17.1\ \text{k}\Omega$$

The voltage across R_1 is

$$V_1 = \left(\frac{R_1}{R_T}\right)V_S = \left(\frac{9.1\ \text{k}\Omega}{17.1\ \text{k}\Omega}\right)15\ \text{V} = \mathbf{7.98\ V}$$

The voltage across R_2 is

$$V_2 = \left(\frac{R_2}{R_T}\right)V_S = \left(\frac{4.7\ \text{k}\Omega}{17.1\ \text{k}\Omega}\right)15\ \text{V} = \mathbf{4.12\ V}$$

Question
What is the voltage across R_3?

The Potentiometer as a Variable Voltage Divider

Recall from Section 3–4 that a potentiometer is a three-terminal variable resistor. A potentiometer connected to a dc voltage source is shown in Figure 5–20(a). Notice that the two end-terminals are labeled 1 and 2. The adjustable terminal or wiper is labeled 3. The potentiometer acts as two resistors in series, which can be illustrated by separating the total resistance into two parts as shown in Figure 5–20(b). The resistance between terminal 1 and terminal 3 (R_{13}) is one part, and the resistance between terminal 3 and terminal 2 (R_{32}) is the other part. The addition of R_{13} and R_{32} always produces the total resistance of the potentiometer. Therefore, the potentiometer is actually an adjustable two-resistor voltage divider.

FIGURE 5–20

The potentiometer as a voltage divider.

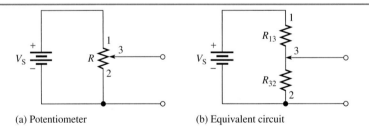

(a) Potentiometer   (b) Equivalent circuit

Figure 5–21 shows what happens when you move the wiper contact (3) of a potentiometer. A 10 kΩ potentiometer is used as an example with a 20 V source connected across terminals 1 and 2. The output voltage measurement is taken across terminals 2 and 3. In part (a), the wiper is exactly centered, making R_{13} and R_{32} equal to 5.0 kΩ each. If you measure the voltage across the output as indicated, you have one-half of the total source voltage or 10 V. When you move the wiper up, as in part (b), the resistance between terminals 2 and 3 increases, and the resistance between terminals 1 and 3 decreases. In this case, the 10 kΩ resistor is divided between 1.0 kΩ and 9.0 kΩ. The output voltage is now 90% of the input voltage or 18 V. When you move the wiper down, as in part (c), the resistance between terminals 3 and 2 decreases, and the voltage decreases proportionally. Now the resistances shown are 9.0 kΩ and 1.0 kΩ, and the output voltage is 2.0 V.

FIGURE 5–21

Adjusting the voltage divider. The position of the center wiper determines how the total resistance is divided.

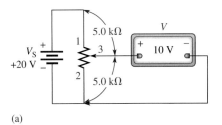
(a)

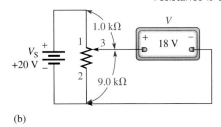
(b)

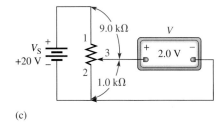
(c)

A potentiometer used as a voltage-divider has a number of applications in electronics. One common application is as a volume control for a sound system. The input signal is an audio (ac) signal that is applied to the outer terminals of the potentiometer. The output is taken from the variable terminal and one side of the potentiometer, as shown in Figure 5–22. As the potentiometer is adjusted, the output amplitude varies in proportion to the setting.

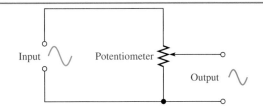

FIGURE 5–22

A potentiometer used as a volume control. The input is an audio signal. The output is the same signal but may be reduced in amplitude by the adjustment of the potentiometer.

Review Questions

16. What is the voltage-divider formula?
17. If two unequal resistors are in a series circuit, which one will have the larger voltage drop?
18. If a 1.0 kΩ and 9.0 kΩ resistor are connected in series across a 100 V source, what is the voltage across each one?
19. How is a potentiometer connected as a voltage divider?
20. Assume a 10 kΩ potentiometer is in series with a fixed 10 kΩ resistor and a 10 V source, as shown in Figure 5–23. What are the smallest and the largest output voltages that can be obtained?

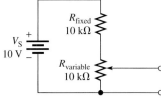

FIGURE 5–23

RESISTORS IN PARALLEL 5–5

A **parallel circuit** is one in which two or more resistors (or other devices) are connected across a common voltage source. Each resistor provides a separate path for current. When two or more resistors are connected in parallel, the total resistance is less than the value of the smallest-value resistor.

In this section, you will learn to recognize parallel circuits and calculate the total resistance of parallel resistors.

CHAPTER 5

With a parallel circuit, you can always trace a complete path from the voltage source to any resistor (or load) and back to the voltage source without going through another resistor. Figure 5–24 shows three examples of parallel circuits. Figure 5–24(a) shows two resistors connected across a dc source. Figure 5–24(b) shows two resistors across an ac source. Figure 5–24(c) shows four resistors. Ideally, there is no limit to the number of resistors that can be in a parallel circuit, but in practice, there is a limit to how much current a source can provide. Notice that in each circuit, the voltage source is connected directly across two or more components.

FIGURE 5–24

Examples of parallel circuits.

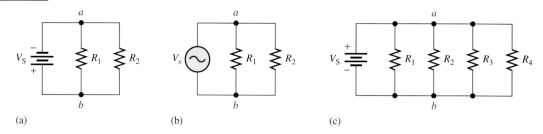

The circuits shown in Figure 5–24 are drawn in a traditional manner. Each parallel current path is called a **branch**. The first two circuits have two branches; the third has four branches. Sometimes, in a parallel circuit, the components will not appear to be in parallel at first. Figure 5–25 shows some more examples, but they are drawn "nontraditionally." Study each and convince yourself that each is a parallel circuit. Of course, there are many other variations that could be shown.

FIGURE 5–25

Additional examples of parallel circuits.

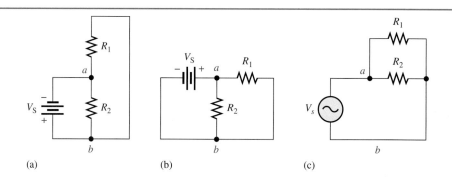

Nodes

A connection point in a circuit that has a common path for two or more circuit elements (e.g., resistors, voltage sources) is called a **node**. A node is sometimes thought of as a *junction* where two or more conductors come together, but in electrical work it has a broader meaning. Any points connected together with a common conductive path form a single node. Notice that each circuit in Figures 5–24 and 5–25 has exactly two nodes, labeled a and b. In fact, all parallel circuits have exactly two nodes. We'll look more at nodes in Section 5–7.

When you say two or more resistors are in parallel, then you are implying that they are connected between the same two nodes. Example 5–11 shows how you could connect resistors in parallel in the lab. Notice that all of the resistors are connected between the same two nodes.

SERIES AND PARALLEL CIRCUITS

EXAMPLE 5–11

Problem
Show how to connect the resistors shown in Figure 5–26 in parallel on a protoboard and how to measure their total resistance with a DMM. (Protoboard connections are discussed in the lab book.)

FIGURE 5–26

Solution
Although there are many ways to connect the resistors in parallel, all require that the resistors be connected across two nodes. One way to connect them is shown in Figure 5–27. Notice that the tops of the resistors are connected together through the protoboard. The bottoms of the resistors are connected together with wire jumpers. The DMM is shown connected across R_1 to read the total resistance.

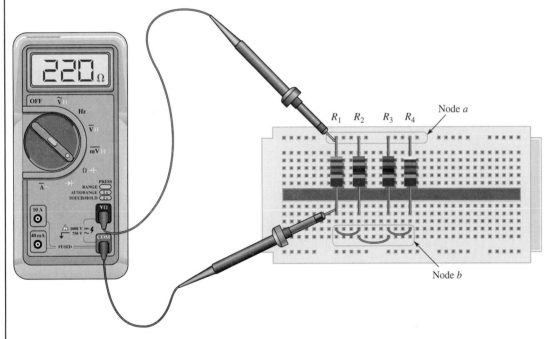

FIGURE 5–27

Question
Can the DMM be connected across a different resistor to read the total resistance?

Total Resistance

In the case of a parallel circuit, the total resistance is *less* than the smallest-value individual resistor in the circuit. To understand why this is true, consider a circuit with one resistor connected across a voltage source as shown in Figure 5–28(a). When a second resistor is connected to the source in parallel as in Figure 5–28(b), the source current must increase because there is an additional path for current. Ohm's law shows that this increase in total

149

CHAPTER 5

FIGURE 5-28

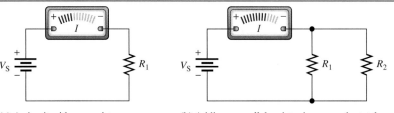

(a) A circuit with one resistor (b) Adding a parallel resistor increases the total current.

current implies that the total resistance has decreased (for constant voltage). That is, with constant voltage,

$$R\downarrow = \frac{V}{I\uparrow}$$

An increase in total I implies that the total R must have decreased.

It is useful to develop the idea of parallel resistance from the viewpoint of conductance. Recall that conductance is the reciprocal of resistance given by the equation

$$G = \frac{1}{R}$$

Resistors in parallel can be thought of in terms of conductive paths. When resistors are connected in parallel, the total conductance is the sum of the individual conductances given by the equation

$$G_T = G_1 + G_2 + G_3 + \ldots + G_n \tag{5-6}$$

where G_T represents the total conductance of the parallel resistors and G_n is the conductance of the last (nth) resistor. Recall that the subscript n is equal to the number of resistors.

Notice the similarity between Equation 5–6 for parallel resistors and Equation 5–1 for series resistors. In a series circuit, you find the total resistance by adding the individual resistors; in a parallel circuit, you find the total conductance by adding the individual conductances.

To calculate resistance in a parallel circuit, begin by substituting $1/R$ for the conductances. Thus, Equation 5–6 becomes

$$\frac{1}{R_T} = \frac{1}{R_1} + \frac{1}{R_2} + \frac{1}{R_3} + \ldots + \frac{1}{R_n} \tag{5-7}$$

Many people prefer to solve for total resistance of parallel resistors by solving the right side of Equation 5–7, then taking the reciprocal of the answer. This is illustrated for a calculator with the following general procedure:

Step 1: Enter the value of R_1, and then press the [1/x] key (this is a second function key on some calculators). The notation x^{-1} also means 1/x.

Step 2: Press the [+] key; then enter the value of R_2 and press the [1/x] key.

Step 3: Repeat step 2 until all of the resistor values have been entered and the reciprocal of each has been added.

Step 4: Press the [=] key to complete the right side operation. The display shows the total conductance.

Step 5: To compute the total resistance, take the reciprocal of the answer by pressing the [1/x] key again. The display will now show the total resistance.

Equation 5–7 can be solved for the total resistance by simply taking the reciprocal of both sides of the equation. This procedure gives the general equation for parallel resistors.

$$R_T = \frac{1}{\dfrac{1}{R_1} + \dfrac{1}{R_2} + \dfrac{1}{R_3} + \ldots + \dfrac{1}{R_n}} \qquad (5\text{--}8)$$

Equation 5–8 can be used with a calculator to compute the total resistance of a group of resistors but requires extra keystrokes over the method shown for Equation 5–7. Example 5–12 illustrates both methods for solving total resistance.

EXAMPLE 5–12

Problem
Find the total resistance of the resistors in Figure 5–29.

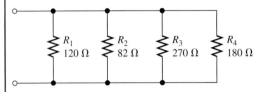

FIGURE 5–29

Solution
Substitute the resistor values into Equation 5–8 (for four resistors).

$$R_T = \frac{1}{\dfrac{1}{R_1} + \dfrac{1}{R_2} + \dfrac{1}{R_3} + \dfrac{1}{R_4}} = \frac{1}{\dfrac{1}{120\,\Omega} + \dfrac{1}{82\,\Omega} + \dfrac{1}{270\,\Omega} + \dfrac{1}{180\,\Omega}} = 33.6\,\Omega$$

The total resistance is smaller than the smallest resistor value.
    The calculator sequence is

[1] [÷] [(] [1] [2] [0] [1/x] [+] [8] [2] [1/x] [+] [2] [7] [0] [1/x] [+] [1] [8] [0] [1/x] [)] [=]

Parentheses must enclose the denominator to force the correct sequence of operation.
    The same calculation can be done with fewer keystrokes by applying Equation 5–7, then taking the reciprocal of the answer as described previously. The keystrokes for this method are

[1] [2] [0] [1/x] [+] [8] [2] [1/x] [+] [2] [7] [0] [1/x] [+] [1] [8] [0] [1/x] [=] [1/x]

Question
If a 100 Ω resistor is connected in parallel with the other resistors, what is the new total resistance?

Special Cases of Parallel Resistors

There are two special cases of parallel resistors that we will discuss. The first is the case of *exactly* two resistors in parallel. Equation 5–8 can be rearranged for this special case with the formula

$$R_T = \frac{R_1 R_2}{R_1 + R_2} \qquad (5\text{--}9)$$

This equation is known as the product-over-sum rule for two resistors. It does not apply to cases where there are more than two resistors in parallel. For example, if a 4.7 kΩ resistor is in parallel with a 10 kΩ resistor, the total resistance is

$$R_T = \frac{R_1 R_2}{R_1 + R_2} = \frac{(4.7 \text{ k}\Omega)(10 \text{ k}\Omega)}{4.7 \text{ k}\Omega + 10 \text{ k}\Omega} = 3.20 \text{ k}\Omega$$

A less-frequent special case arises when the parallel resistors have the same resistance. In this case, Equation 5–8 can be simplified to

$$R_T = \frac{R}{n}$$

EXAMPLE 5–13

Problem
A car's rear window defroster uses 15 strips of resistive wire in a parallel arrangement. If the total resistance is 1.4 Ω, what is the resistance of one wire?

Solution

$$R_T = \frac{R}{n}$$
$$R = nR_T = (15)(1.4 \text{ }\Omega) = \textbf{21 }\boldsymbol{\Omega}$$

Question
What is the total power dissipated in the defroster if 12 V is applied to it?

Notation for Parallel Resistors

A shorthand way of showing parallel resistors is to place two parallel vertical lines between the labels (or values). Thus, to show that R_1, R_2, and R_3 are all in parallel, you can write $R_1 \| R_2 \| R_3$. You can also show the vertical lines between values. For example, if the values of R_1, R_2, and R_2 are 1.0 kΩ, 2.0 kΩ and 3.3 kΩ, you could show that they are in parallel by writing 1.0 kΩ ∥ 2.0 kΩ ∥ 3.3 kΩ.

Review Questions

21. How many nodes are present in a parallel circuit?
22. What happens to the total conductance as additional resistors are connected in parallel?
23. What happens to the total resistance as additional resistors are connected in parallel?
24. What is the total resistance if a 1.5 MΩ resistor is in parallel with a 1.0 MΩ resistor?
25. What is the resistance of three 150 Ω resistors in parallel?

5–6 APPLYING OHM'S LAW IN PARALLEL CIRCUITS

Ohm's law can be applied to parallel circuits for the total circuit or for individual components. Try working each of the examples given before you look at the solutions.

In this section, you will learn to use Ohm's law to calculate an unknown current, voltage, or resistance in a parallel circuit.

SERIES AND PARALLEL CIRCUITS

A parallel circuit consists of a voltage source and two or more components connected directly across it. As a result, each component has exactly the same voltage across it as the source voltage. In the case of a parallel circuit, this is the key idea for applying Ohm's law. You can express this idea as

$$V_S = V_1 = V_2 = V_n$$

where n is the last component in the circuit.

Because the voltage is the same across each branch in a parallel circuit, the current in a given branch is independent of current in other branches. In parallel circuits, you can find the current in any branch by dividing the source voltage by the resistance in that branch. The following examples illustrate solving for an unknown voltage, current, or resistance in a parallel circuit.

EXAMPLE 5–14

Problem
Calculate the voltage and current in each resistor for the circuit shown in Figure 5–30. Complete Table 5–7.

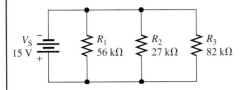

FIGURE 5–30

Table 5–7

$I_1 =$	$R_1 = 56\ k\Omega$	$V_1 =$
$I_2 =$	$R_2 = 27\ k\Omega$	$V_2 =$
$I_3 =$	$R_3 = 82\ k\Omega$	$V_3 =$
$I_T =$	$R_T =$	$V_S = 15\ V$

Solution

Step 1: Calculate the total resistance by substituting into Equation 5–8, the parallel resistance formula. (*Note:* Alternatively, you could determine the voltage values first).

$$R_T = \frac{1}{\dfrac{1}{R_1} + \dfrac{1}{R_2} + \dfrac{1}{R_3}} = \frac{1}{\dfrac{1}{56\ k\Omega} + \dfrac{1}{27\ k\Omega} + \dfrac{1}{82\ k\Omega}} = 14.9\ k\Omega$$

Step 2: Because this is a parallel circuit, the voltage across all components is the same as the source voltage.

$$V_1 = V_2 = V_3 = V_S = 15\ V$$

Step 3: To find each current, apply Ohm's law to the individual components and to the total.

$$I_1 = \frac{V_1}{R_1} = \frac{15\ V}{56\ k\Omega} = 0.268\ mA$$

$$I_2 = \frac{V_2}{R_2} = \frac{15\ V}{27\ k\Omega} = 0.556\ mA$$

$$I_3 = \frac{V_3}{R_3} = \frac{15\ V}{82\ k\Omega} = 0.183\ mA$$

$$I_T = \frac{V_S}{R_T} = \frac{15\ V}{14.9\ k\Omega} = 1.007\ mA$$

CHAPTER 5

The completed table is shown as Table 5–8.

Table 5–8

$I_1 = 0.268$ mA	$R_1 = 56$ kΩ	$V_1 = 15$ V
$I_2 = 0.556$ mA	$R_2 = 27$ kΩ	$V_2 = 15$ V
$I_3 = 0.183$ mA	$R_3 = 82$ kΩ	$V_3 = 15$ V
$I_T = 1.007$ mA	$R_T = 14.9$ kΩ	$V_S = 15$ V

Question
Of the four currents listed in Table 5–8, which one will change if another resistor is connected in parallel? Explain your answer.

EXAMPLE 5–15

Problem
Calculate the unknown quantities in the parallel circuit shown in Figure 5–31. Complete Table 5–9.

Table 5–9

$I_1 = 15.2$ mA	$R_1 = 330$ Ω	$V_1 =$
$I_2 =$	$R_2 = 470$ Ω	$V_2 =$
$I_3 =$	$R_3 = 270$ Ω	$V_3 =$
$I_T =$	$R_T =$	$V_S =$

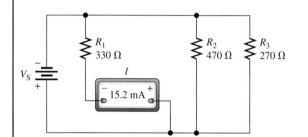

FIGURE 5–31

Solution

Step 1: Calculate the total resistance by substituting into Equation 5–8. (*Note:* You could begin by calculating *either* R_T or V_1.)

$$R_T = \frac{1}{\frac{1}{R_1} + \frac{1}{R_2} + \frac{1}{R_3}} = \frac{1}{\frac{1}{330\ \Omega} + \frac{1}{470\ \Omega} + \frac{1}{270\ \Omega}} = 113\ \Omega$$

Step 2: Calculate V_1 from Ohm's law.

$$V_1 = I_1 R_1 = (15.2\ \text{mA})(330\ \Omega) = 5.02\ \text{V}$$

Step 3: Because this is a parallel circuit, the voltage across all components, including the source itself, is the same as V_1.

$$V_S = V_2 = V_3 = V_1 = 5.02\ \text{V}$$

Step 4: Use Ohm's law to calculate the remaining currents.

$$I_2 = \frac{V_2}{R_2} = \frac{5.02\ \text{V}}{470\ \Omega} = 10.7\ \text{mA}$$

$$I_3 = \frac{V_3}{R_3} = \frac{5.02\ \text{V}}{270\ \Omega} = 18.6\ \text{mA}$$

$$I_T = \frac{V_S}{R_T} = \frac{5.02\ \text{V}}{113\ \Omega} = 44.4\ \text{mA}$$

SERIES AND PARALLEL CIRCUITS

The completed table is shown as Table 5–10.

Table 5–10

$I_1 = 15.2$ mA	$R_1 = 330\ \Omega$	$V_1 = 5.02$ V
$I_2 = 10.7$ mA	$R_2 = 470\ \Omega$	$V_2 = 5.02$ V
$I_3 = 18.6$ mA	$R_3 = 270\ \Omega$	$V_3 = 5.02$ V
$I_T = 44.4$ mA	$R_T = 113\ \Omega$	$V_S = 5.02$ V

Question
What is the total current if the 470 Ω resistor is open?

COMPUTER SIMULATION

Open the Multisim file F05-31DC on the website. Using the multimeter, verify the current in each leg of the parallel circuit.

EXAMPLE 5–16

Problem
Calculate the unknown quantities in the parallel circuit shown in Figure 5–32. For this problem, the total resistance is 3.19 kΩ but you do not know the value of R_3. Complete Table 5–11.

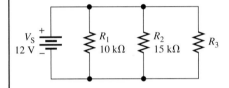

Table 5–11

$I_1 =$	$R_1 = 10$ kΩ	$V_1 =$
$I_2 =$	$R_2 = 15$ kΩ	$V_2 =$
$I_3 =$	$R_3 =$	$V_3 =$
$I_T =$	$R_T = 3.19$ kΩ	$V_S = 12$ V

FIGURE 5–32

Solution

Step 1: Recall the equation for the reciprocal of the total resistance of parallel resistors (Equation 5–7). For three resistors, it is written as

$$\frac{1}{R_T} = \frac{1}{R_1} + \frac{1}{R_2} + \frac{1}{R_3}$$

Rearranging the terms,

$$\frac{1}{R_3} = \frac{1}{R_T} - \left(\frac{1}{R_1} + \frac{1}{R_2}\right)$$

and taking the reciprocal of both sides,

$$R_3 = \frac{1}{\frac{1}{R_T} - \left(\frac{1}{R_1} + \frac{1}{R_2}\right)} = \frac{1}{\frac{1}{3.19\ \text{k}\Omega} - \left(\frac{1}{10\ \text{k}\Omega} + \frac{1}{15\ \text{k}\Omega}\right)} = 6.81\ \text{k}\Omega$$

The calculator sequence for this calculation is

`1 ÷ ( 1 ÷ 3 . 1 9 EE 3 - ( 1 ÷ 1 0 EE 3 + 1 ÷ 1 5 EE 3 ) ) =`

Step 2: Because this is a parallel circuit, the voltage across all components is the same as V_S.

$$V_1 = V_2 = V_3 = V_S = 12\ \text{V}$$

155

Step 3: Apply Ohm's law to find the current in each resistor.

$$I_1 = \frac{V_1}{R_1} = \frac{12\text{ V}}{10\text{ k}\Omega} = 1.20\text{ mA}$$

$$I_2 = \frac{V_2}{R_2} = \frac{12\text{ V}}{15\text{ k}\Omega} = 0.80\text{ mA}$$

$$I_3 = \frac{V_3}{R_3} = \frac{12\text{ V}}{6.81\text{ k}\Omega} = 1.76\text{ mA}$$

The completed table, except for the total current, is shown as Table 5–12.

Table 5–12

I_1 = 1.20 mA	R_1 = 10 kΩ	V_1 = 12 V
I_2 = 0.80 mA	R_2 = 15 kΩ	V_2 = 12 V
I_3 = 1.76 mA	R_3 = 6.81 kΩ	V_3 = 12 V
I_T =	R_T = 3.19 kΩ	V_S = 12 V

Question
What is the total current?

Review Questions

26. For the circuit in Figure 5–33, what is the source voltage?

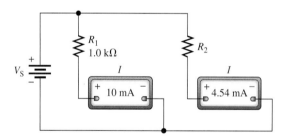

FIGURE 5–33

27. For the circuit in Figure 5–33, what is the resistance of R_2?
28. For the circuit in Figure 5–33, what is the total resistance of the circuit?
29. The total resistance of two parallel resistors is 9.65 kΩ. One of the resistors is 15 kΩ. What is the resistance of the other resistor?
30. Why are parallel circuits used in home wiring?

5–7 KIRCHHOFF'S CURRENT LAW

In addition to his voltage law, Kirchhoff developed a second fundamental law for current known as Kirchhoff's current law.

In this section, you will learn to apply Kirchhoff's current law.

A node was defined earlier as a connection point in a circuit that is common to two or more components. Figure 5–34(a) shows an example of nodes in a series circuit, and Figure 5–34(b) shows an example of the nodes in a parallel circuit. Notice that in the series circuit,

SERIES AND PARALLEL CIRCUITS

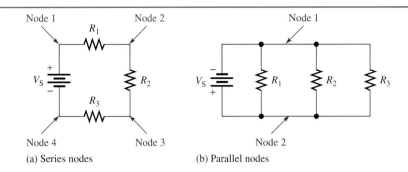

FIGURE 5–34

Examples of nodes in series and parallel circuits.

any number of nodes may be present but only two components can join at a node. In a parallel circuit, there can only be two nodes in the entire circuit and all components are connected between these two nodes.

Kirchhoff's current law (KCL) is stated as follows:

> **The sum of the currents entering a node is equal to the sum of the currents leaving the node.**

In a series circuit, a node cannot have more than two components connected to it. For this reason, the current from one component must be the same as the current to the next component. In effect, KCL shows that the current must be the same anywhere in a series circuit.

In a parallel circuit, any number of components can be connected to the two nodes. Current enters one node from the source and leaves through the various branches. Current enters the other node from the branches and leaves that node by returning to the source. Figure 5–35 illustrates the currents in a parallel circuit with three branches. (The ammeters themselves do not affect the nodes). Notice that the total source current (to and from the source) is equal to the sum of the branch currents.

Kirchhoff's current law can aid in solving problems in electronics and is an excellent check when you use other methods to arrive at the answer to a problem.

SAFETY NOTE

Liquids and electricity don't mix! You should never place a drink or other liquid where it can be accidentally spilled onto an electrical ciruit.,

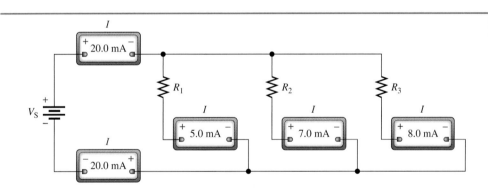

FIGURE 5–35

Current to and from the source is equal to the currents in the parallel branches.

EXAMPLE 5–17

Problem

For the parallel circuit shown in Figure 5–36, find I_3.

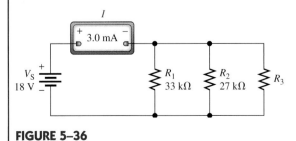

FIGURE 5–36

157

CHAPTER 5

Solution

Step 1: $V_1 = V_2 = V_3 = V_S = 18$ V because it is a parallel circuit.

Step 2: Use Ohm's law to calculate the current in R_1 and R_2.

$$I_1 = \frac{V_1}{R_1} = \frac{18 \text{ V}}{33 \text{ k}\Omega} = 0.545 \text{ mA}$$

$$I_2 = \frac{V_2}{R_2} = \frac{18 \text{ V}}{27 \text{ k}\Omega} = 0.667 \text{ mA}$$

Step 3: Apply KCL to find I_3. The current into one of the nodes is the source current. The currents from this node are the branch currents.

$$I_S = I_1 + I_2 + I_3$$
$$I_3 = I_S - (I_1 + I_2) = 3.0 \text{ mA} - (0.545 \text{ mA} + 0.667 \text{ mA}) = \mathbf{1.79 \text{ mA}}$$

Question
What is the value of R_3?

EXAMPLE 5–18

Problem
For the parallel circuit shown in Figure 5–37, find I_3 by KCL and the other values indicated in Table 5–13.

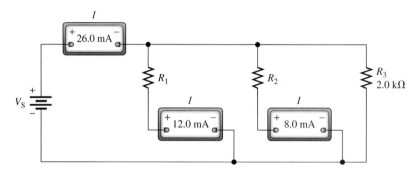

FIGURE 5–37

Table 5–13

$I_1 = 12.0$ mA	$R_1 =$	$V_1 =$
$I_2 = 8.0$ mA	$R_2 =$	$V_2 =$
$I_3 =$	$R_3 = 2.0$ kΩ	$V_3 =$
$I_T = 26.0$ mA	$R_T =$	$V_S =$

Solution

Step 1: Calculate the current in R_3 by KCL.

$$I_S = I_1 + I_2 + I_3$$
$$I_3 = I_S - (I_1 + I_2) = 26.0 \text{ mA} - (12.0 \text{ mA} + 8.0 \text{ mA}) = 6.0 \text{ mA}$$

Step 2: Calculate V_3 from Ohm's law.

$$V_3 = I_3 R_3 = (6.0 \text{ mA})(2.0 \text{ k}\Omega) = 12 \text{ V}$$

Step 3: Observe that because it is a parallel circuit,

$$V_S = V_1 = V_2 = V_3 = 12 \text{ V}$$

Step 4: Calculate the resistances by Ohm's law.

$$R_1 = \frac{V_1}{I_1} = \frac{12 \text{ V}}{12 \text{ mA}} = 1.0 \text{ k}\Omega$$

$$R_2 = \frac{V_2}{I_2} = \frac{12 \text{ V}}{8.0 \text{ mA}} = 1.5 \text{ k}\Omega$$

$$R_T = \frac{V_S}{I_T} = \frac{12 \text{ V}}{26 \text{ mA}} = 0.462 \text{ k}\Omega$$

These results are shown in Table 5–14.

Table 5–14

I_1 = **12.0 mA**	R_1 = **1.0 kΩ**	V_1 = **12 V**
I_2 = **8.0 mA**	R_2 = **1.5 kΩ**	V_2 = **12 V**
I_3 = **6.0 mA**	R_3 = **2.0 kΩ**	V_3 = **12 V**
I_T = **26.0 mA**	R_T = **0.462 kΩ**	V_S = **12 V**

Question
Prove that the total resistance is correct by the parallel resistance formula.

Current Dividers

Recall that in a resistive series circuit the current is the same in all resistors and the voltage is divided between them. In a parallel resistive circuit, the voltage is the same across all resistors and the current divides between them. All parallel circuits divide the current entering a node into the individual branch currents. A **current divider** consists of two or more parallel resistors that provide a smaller current in each branch than the input current. Because the voltage is the same across parallel resistors, the current is proportionately larger in a smaller-value resistor. The current-divider rule can be expressed as

> **The total current in a parallel circuit divides into branch currents that are inversely proportional to the branch resistance.**

As you know, the total resistance of a parallel circuit is always smaller than the smallest resistance. If a parallel circuit has a total resistance of R_T, then the fraction of the total current that will be in any one resistor is simply R_T/R_x, where R_x is the resistance of the resistor in question. The current in R_x can be given in terms of the total current as

$$I_x = \left(\frac{R_T}{R_x}\right) I_T \qquad (5\text{–}10)$$

Notice the similarity of Equation 5–10 to the voltage-divider equation (Equation 5–5). Keep in mind that the quantity in parentheses is always a fraction less than 1.

EXAMPLE 5–19

Problem
Assume a source current of 10 mA enters a node. Connected to the node are three parallel resistors with values of R_1 = 1.5 kΩ, R_2 = 2.7 kΩ, and R_3 = 3.3 kΩ. What is the current in R_1 and R_2?

Solution

Find the total resistance.

$$R_T = \frac{1}{\frac{1}{R_1} + \frac{1}{R_2} + \frac{1}{R_3}} = \frac{1}{\frac{1}{1.5\text{ k}\Omega} + \frac{1}{2.7\text{ k}\Omega} + \frac{1}{3.3\text{ k}\Omega}} = 0.746\text{ k}\Omega$$

Apply Equation 5–10.

$$I_1 = \left(\frac{R_T}{R_1}\right)I_T = \left(\frac{0.746\text{ k}\Omega}{1.5\text{ k}\Omega}\right)10\text{ mA} = \mathbf{4.97\text{ mA}}$$

$$I_2 = \left(\frac{R_T}{R_2}\right)I_T = \left(\frac{0.746\text{ k}\Omega}{2.7\text{ k}\Omega}\right)10\text{ mA} = \mathbf{2.76\text{ mA}}$$

Question

What is the current in R_3?

Review Questions

31. What is a node?
32. State Kirchhoff's current law.
33. For what type of circuits does Kirchhoff's current law apply?
34. If a two-branch parallel circuit has 120 mA in one branch and 48 mA in the other, what is the source current?
35. What is the current-divider rule?

CHAPTER REVIEW

Key Terms

Branch One current path in a parallel circuit.

Current divider A parallel circuit in which the current entering a node is divided into different branches.

Kirchhoff's current law A circuit law stating that the total current into a node is equal to the total current out of the node.

Kirchhoff's voltage law A circuit law stating that the sum of all the voltage drops around a single closed path in a circuit is equal to the total source voltage in that closed path. An equivalent statement for Kirchhoff's voltage law is *The algebraic sum of the voltages around a single closed path in a circuit equals zero.*

Node A connection point in a circuit that has a common path for two or more components.

Parallel circuit A circuit that has two or more current paths connected to a common voltage source.

Series circuit A circuit with only one current path.

Voltage divider A circuit consisting of series resistors across which one or more output voltages are taken.

Important Facts

- A series circuit always can be identified by the fact that it has only one current path.
- In a series circuit, the current is the same in all components.
- In a series circuit, the total resistance is the sum of the individual resistors.
- A voltage divider consists of two or more series resistors across which one or more voltages are taken.
- The voltage-divider rule states that the voltage drop across any given resistor in a series circuit is equal to the ratio of that resistor to the total resistance, multiplied by the source voltage.
- A potentiometer can be used as a variable voltage divider.
- A parallel circuit is one in which the same voltage is applied to two or more components.
- In a parallel circuit, each parallel current path is called a branch.
- A node is a junction in a circuit where two or more components are joined.
- Parallel circuits have exactly two nodes.
- The voltage across all components in a parallel circuit is the same.
- The total conductance of parallel resistors is the sum of the individual conductances.
- The total resistance of parallel resistors is smaller than the smallest individual resistor value.
- For exactly two resistors in parallel, the total resistance can be found by the product-over-sum rule.
- A shorthand way of showing parallel resistors is to place two parallel vertical lines between the labels (or values).
- The current-divider rule states that the total current in a parallel circuit divides into branch currents that are inversely proportional to the branch resistance.

Formulas

Total resistance of series resistors:

$$R_T = R_1 + R_2 + R_3 + \ldots + R_n \tag{5-1}$$

Total resistance of equal-value series resistors:

$$R_T = nR \tag{5-2}$$

Current in series components:

$$I_T = I_1 = I_2 \tag{5-3}$$

Kirchhoff's voltage law:

$$V_S = V_1 + V_2 + V_3 + \ldots + V_n \tag{5-4}$$

Voltage-divider rule:

$$V_x = \left(\frac{R_x}{R_T}\right) V_S \tag{5-5}$$

Total conductance of parallel resistors:

$$G_T = G_1 + G_2 + G_3 + \ldots + G_n \tag{5-6}$$

Parallel resistance rule:

$$\frac{1}{R_T} = \frac{1}{R_1} + \frac{1}{R_2} + \frac{1}{R_3} + \ldots + \frac{1}{R_n} \tag{5-7}$$

Total resistance of parallel resistors:

$$R_T = \frac{1}{\dfrac{1}{R_1} + \dfrac{1}{R_2} + \dfrac{1}{R_3} + \ldots + \dfrac{1}{R_n}} \tag{5-8}$$

Total resistance of exactly two parallel resistors:

$$R_T = \frac{R_1 R_2}{R_1 + R_2} \tag{5-9}$$

Current-divider rule:

$$I_x = \left(\frac{R_T}{R_x}\right) I_T \tag{5-10}$$

Chapter Checkup

Answers are at the end of the chapter.

1. A series circuit always has
   (a) more than one path
   (b) the same voltage across all components
   (c) exactly two nodes
   (d) the same current through all components

2. Three equal-value resistors are connected in series to a voltage source. The voltage across each resistor is
   (a) the same as the source voltage
   (b) three times the source voltage
   (c) one-half of the source voltage
   (d) one-third of the source voltage

3. A 2.0 kΩ resistor and a 1.0 kΩ resistor are connected in series with a voltage source. The current in the 2.0 kΩ resistor is
   (a) one-half that of the 1.0 kΩ resistor
   (b) the same as the 1.0 kΩ resistor
   (c) twice that of the 1.0 kΩ resistor
   (d) none of the above

4. If a number of equal-value resistors are connected in series, the total resistance is
   (a) the product of the number of resistors times the resistance of one resistor
   (b) the resistance of one resistor divided by the number of resistors
   (c) the product of all of the resistances divided by the number of resistors
   (d) the sum of all of the resistances divided by the number of resistors

5. If one of four series resistors is removed from a series circuit and the circuit is reconnected, the current from the source will be
   (a) smaller than before
   (b) the same as before
   (c) larger than before

6. Four resistors are connected in series to a voltage source of 15 V. If each of the first three resistors drops 3 V, then the fourth resistor will drop

   (a) 3 V   (b) 6 V
   (c) 9 V   (d) not enough information to tell

7. If you measure all of the voltage drops and the source voltage in a series circuit and add them together, taking into account their polarities, you will get a total

   (a) of zero   (b) equal to the source voltage
   (c) equal to the smallest drop (d) equal to the largest drop

8. A parallel circuit always has

   (a) more than one path

   (b) the same voltage across all components

   (c) exactly two nodes

   (d) all of the above

9. Three equal-value resistors are connected in parallel to a voltage source. The voltage across each resistor will be

   (a) the same as the source voltage

   (b) three times the source voltage

   (c) one-half of the source voltage

   (d) one-third of the source voltage

10. A 2.0 kΩ resistor and a 1.0 kΩ resistor are connected in parallel with a voltage source. The current in the 2.0 kΩ resistor is

    (a) one-half that of the 1.0 kΩ resistor

    (b) the same as the 1.0 kΩ resistor

    (c) twice that of the 1.0 kΩ resistor

    (d) none of the above

11. If a number of equal-value resistors are connected in parallel, the total resistance is

    (a) the product of the number of resistors times the resistance of one resistor

    (b) the resistance of one resistor divided by the number of resistors

    (c) the product of all of the resistances divided by the number of resistors

    (d) the sum of all of the resistances divided by the number of resistors

12. If one of four series resistors is removed from a parallel circuit, the current from the source will be

    (a) smaller than before   (b) the same as before

    (c) larger than before

13. Four equal-value resistors are connected in parallel with a voltage source of 12 V. The voltage across the fourth resistor will be

    (a) 3 V   (b) 6 V
    (c) 12 V   (d) 48 V

14. If you measure all of the current into a junction and all of the current from the same junction, the current into the junction will be

    (a) smaller than the current out

    (b) equal to the current out

    (c) larger than the current out

15. If a 33 kΩ resistor, a 27 kΩ resistor, and a 1.5 kΩ resistor are connected in parallel, the total resistance is

    (a) smaller than 1.5 kΩ

    (b) larger than 1.5 kΩ but smaller than 33 kΩ

    (c) larger than 33 kΩ

Questions

Answers to odd-numbered questions are at the end of the book.

1. What is the total resistance when a 1.2 MΩ resistor is connected in series with a 470 kΩ resistor?
2. What is the total resistance when a 330 Ω resistor is connected in series with a 1.0 kΩ resistor?
3. What is the total resistance of a series circuit with the following resistors: 270 kΩ, 560 kΩ, 1.0 MΩ?
4. What is the total resistance of a series circuit with the following resistors: 910 Ω, 1.0 kΩ, 1.5 kΩ?
5. Six 4.7 kΩ resistors are in series. What is the total resistance?
6. Three 330 Ω resistors are in series with five 470 Ω resistors. What is the total resistance?
7. Suppose you have four resistors that are 1.0 kΩ each. If you wanted to get a total series resistance of 4.27 kΩ with one more resistor, what resistance must it have?
8. Why is it necessary to disconnect a voltage source when you measure resistance?
9. If a source voltage of 15 V is applied to two 10 kΩ resistors in series, what current is in each resistor?
10. What is Kirchhoff's voltage law?
11. Two resistors are in series with a 5.0 V source. The first resistor drops 3.0 V. What is the voltage drop across the second resistor?
12. In a series resistive circuit, if the source voltage drops to one-half of its original value, what happens to the current in the circuit?
13. Three equal-value resistors are in series with a voltage source. If one of the resistors has a voltage across it of 6 V, what is the source voltage?
14. A 10 kΩ resistor is in series with a 27 kΩ resistor. If 2.0 V is across the 10 kΩ resistor, what voltage will be across the 27 kΩ resistor?
15. How is the voltage-divider formula stated in words?
16. If a certain resistance represents 1/3 of the total resistance of a series circuit, what voltage will be across it if the source voltage is 12 V?
17. How is a potentiometer connected to form a variable voltage divider?
18. With reference to circuits, what is a branch?
19. What is the total conductance of a 270 Ω resistor in parallel with a 470 Ω resistor?
20. What is the total resistance of a 270 Ω resistor in parallel with a 470 Ω resistor?
21. What is the total resistance of a 680 kΩ resistor in parallel with a 1.0 MΩ resistor?
22. What is a shorthand way of showing that two resistors are in parallel?
23. What is Kirchhoff's current law?

24. If the current leaving a source is 300 mA and 200 mA goes through one branch of a two-resistor parallel circuit, what current is in the other branch?

25. For question 24, what current returns to the source?

26. When can the product-over-sum rule for parallel resistors be applied?

Basic Problems

PROBLEMS

Answers to odd-numbered problems are at the end of the book.

1. Find the total resistance of the resistors in Figure 5–38.

2. What is the total resistance in Figure 5–38 if R_2 is replaced with an orange, white, orange, gold resistor?

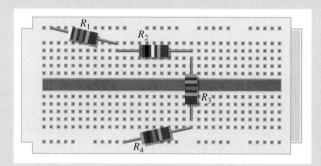

FIGURE 5–38

3. The total resistance in Figure 5–39 is 88 kΩ. What is the resistance of R_3?

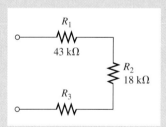

FIGURE 5–39

4. Resistor R_3 in Figure 5–39 is color-coded brown, red, orange, gold. What is R_T?

5. Calculate R_T for the resistors in Figure 5–40.

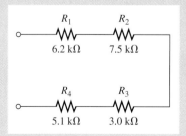

FIGURE 5–40

165

6. Calculate R_T for the resistors in Figure 5–41.

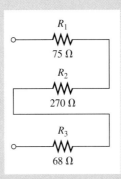

FIGURE 5–41

7. A 10 V source is connected between the terminals of Figure 5–40. What is the current in the circuit?

8. A 15 V source is connected between the terminals of Figure 5–41. What is the current in the circuit?

9. Three equal-value resistors are connected in series with a 10 V source and a source current of 1.0 mA is measured. What is the value of each resistor?

10. Table 5–15 lists the known values for the circuit in Figure 5–42. Calculate the remaining unknown values listed in the table.

Table 5–15

$I_1 =$	$R_1 = 2.7$ kΩ	$V_1 =$
$I_2 =$	$R_2 = 6.2$ kΩ	$V_2 = 6.64$ V
$I_3 =$	$R_3 = 5.1$ kΩ	$V_3 =$
$I_T =$	$R_T =$	$V_S =$

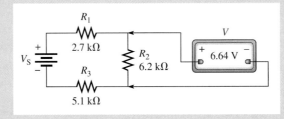

FIGURE 5–42

11. For the circuit of Figure 5–43, calculate V_3.

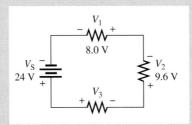

FIGURE 5–43

12. The total resistance of a series circuit is 1000 Ω. What fraction of the source voltage will be dropped across a 330 Ω resistor?

13. Use the voltage-divider formula to calculate the output voltage in Figure 5–44.

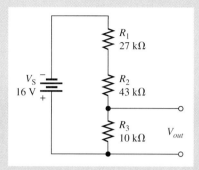

FIGURE 5–44

14. Calculate the minimum and maximum output voltage for the voltage divider in Figure 5–45.

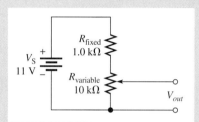

FIGURE 5–45

15. Find the total resistance between points A and B of the resistors in Figure 5–46.

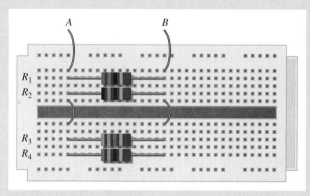

FIGURE 5–46

16. What is the total resistance read between points A and B in Figure 5–46 if R_4 is replaced with a violet, green, brown, gold resistor?

17. What is R_T for the circuit in Figure 5–47?

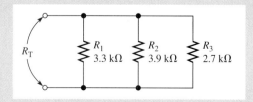

FIGURE 5–47

18. The total resistance in Figure 5–48 is 64 Ω. What is the resistance of R_3?

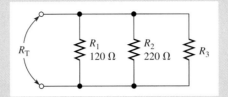

FIGURE 5–48

19. Assume R_3 in Figure 5–48 is open (infinite resistance). What will an ohmmeter read for R_T?

20. A heating unit uses twelve 300 Ω resistive wires in parallel. What is the total resistance of the heating unit?

21. What is the total source current for the circuit in Figure 5–49?

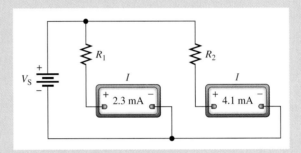

FIGURE 5–49

22. Table 5–16 lists the known values for the circuit in Figure 5–50. Calculate the remaining unknown values listed in the table.

Table 5–16

$I_1 =$	$R_1 = 82\ \Omega$	$V_1 =$	$P_1 =$
$I_2 =$	$R_2 = 150\ \Omega$	$V_2 =$	$P_2 =$
$I_3 =$	$R_3 = 100\ \Omega$	$V_3 =$	$P_3 =$
$I_T =$	$R_T =$	$V_S = 3.0\ V$	$P_T =$

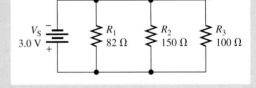

FIGURE 5–50

23. For the parallel circuit shown in Figure 5–51, find I_1 and the value of R_1.

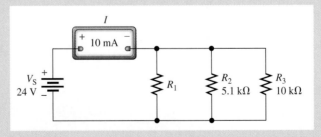

FIGURE 5–51

24. Table 5–17 lists the known values for the circuit in Figure 5–52. Calculate the remaining unknown values listed in the table.

Table 5–17

$I_1 =$	$R_1 = 6.2\ k\Omega$	$V_1 =$
$I_2 =$	$R_2 = 15\ k\Omega$	$V_2 =$
$I_3 =$	$R_3 = 10\ k\Omega$	$V_3 =$
$I_T = 3.94\ mA$	$R_T =$	$V_S =$

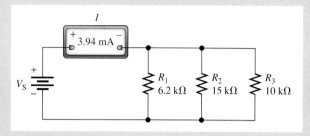

FIGURE 5–52

Basic-Plus Problems

25. Four equal-value resistors are connected in series to a 12 V source. If the current from the source is 100 mA, what is the value of one of the resistors?

26. Draw a two-resistor voltage divider that has a 6.0 V output from a 24 V source. The larger-value resistor is required to be 10 kΩ.

27. Draw a variable voltage divider using a 10 kΩ potentiometer that divides a 12 V source between 4.0 V minimum and 8.0 V maximum.

28. Assume two resistors are in parallel. The value of the second resistor is twice the value of the first. The total resistance is 100 Ω. What is the value of the first resistor?

29. Calculate I_3 in Figure 5–53.

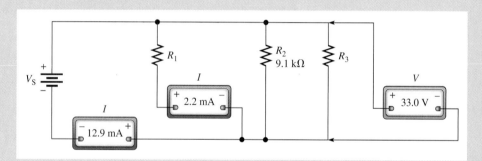

FIGURE 5–53

30. Calculate the value of R_3 in Figure 5–53.

31. The total resistance of a parallel circuit is 200 Ω. What is the current through a 680 Ω resistor that makes up part of the parallel circuit if the total current is 40 mA?

32. The circuit in Figure 5–54 is fused for 0.1 A. What value of R_2 will just cause excessive current?

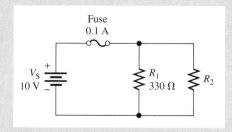

FIGURE 5–54

33. The wiring for household lights and plugs is always done in parallel. Assume a 115 V source voltage is applied to three 100 W lamps. (a) What is the current in each lamp? (b) What is the total current?
34. Two 1000 W heaters are to be connected in parallel to a 115 V source. Will a 20 A circuit breaker be able to carry the total current without tripping? Justify your answer.

ANSWERS

Example Questions

5–1: 12.6 kΩ

5–2: yes

5–3: 4.7 kΩ

5–4: $V_2 = 1.69$ V and $V_3 = 3.03$ V

5–5: The current in the circuit will increase to 1.34 mA. Then $V_1 = 7.52$ V, $V_2 = 1.34$ V and $V_3 = 9.13$. Notice the sum is still 18 V.

5–6: 24.0 V, 1.91 mW

5–7: $V_1 = 21$ V, $V_2 = 9.2$ V, $V_3 = 9.8$ V

5–8: One way is to check KVL that the computed voltage drops add up to the source voltage.

5–9: 1/4

5–10: 2.89 V

5–11: Yes; the DMM will read the same across any of the resistors.

5–12: 25.1 Ω

5–13: 6.86 W

5–14: I_T is the only one affected because branch currents are independent of each other.

5–15: The new total current is the sum of I_1 and $I_3 = 33.8$ mA.

5–16: 3.76 mA

5–17: 10 kΩ

5–18: The proof is

$$R_T = \frac{1}{\frac{1}{R_1} + \frac{1}{R_2} + \frac{1}{R_3}} = \frac{1}{\frac{1}{1.0 \text{ k}\Omega} + \frac{1}{1.5 \text{ k}\Omega} + \frac{1}{2.0 \text{ k}\Omega}} = 462 \text{ }\Omega$$

5–19: 2.26 mA

Review Questions

1. Current is the same throughout the circuit.
2. 3.73 kΩ
3. 8.2 kΩ
4. 1.32 kΩ
5. 2.06 MΩ
6. 4.0 mA
7. 6.98 mA
8. 4.74 V
9. 100 Ω

10. 20 Ω

11. There are two equivalent ways of stating it: (1) The sum of all the voltage drops around a single closed path in a circuit is equal to the total source voltage in that closed path; or (2) The algebraic sum of the voltages around a single closed path in a circuit equals zero.

12. 4.0 V

13. 33.0 V

14. 6.0 V

15. Yes, the voltage will be equal in magnitude to the source voltage (100 V). (Note that this can be shown with KVL.)

16. $V_x = \left(\dfrac{R_x}{R_T}\right) V_S$

17. The larger resistor

18. 10 V across the 1.0 kΩ resistor; 90 V across the 9.0 kΩ resistor

19. The input voltage is connected across the fixed terminals; the output voltage is taken across the variable terminal and one of the fixed terminals.

20. The smallest voltage is 0 V; the largest voltage is 5.0 V.

21. Exactly two.

22. It increases.

23. It decreases.

24. 0.60 MΩ (or 600 kΩ)

25. 50 Ω

26. 10 V

27. 2.2 kΩ

28. 688 Ω

29. 27 kΩ

30. With a parallel circuit, the current in each load is independent of other loads.

31. A junction in a circuit

32. The sum of the currents entering a node is equal to the sum of the currents leaving the node.

33. All circuits

34. 168 mA

35. The total current in a parallel circuit divides into branch currents that are inversely proportional to the branch resistance.

Chapter Checkup

1. (d)	2. (d)	3. (b)	4. (a)	5. (c)
6. (b)	7. (a)	8. (d)	9. (a)	10. (a)
11. (b)	12. (a)	13. (c)	14. (b)	15. (a)

CHAPTER 6

COMBINATIONAL SERIES/PARALLEL CIRCUITS

INTRODUCTION

Most circuits in electronics are neither series circuits nor parallel circuits. Instead, they are combinations of series and parallel elements called combinational circuits. In most cases, combinational circuits can be analyzed by various simplifying methods. In this chapter, methods to solve various types of combinational circuits will be covered, and you will learn to apply them to specific circuits.

CHAPTER OUTLINE

- 6–1 Equivalent Circuits
- 6–2 Analysis of Combinational Circuits
- 6–3 Thevenin's Theorem
- 6–4 Loading Effects
- 6–5 Multiple Sources
- 6–6 The Wheatstone Bridge
- 6–7 Troubleshooting Combinational Circuits

Study aids for this chapter are available at

http://www.prenhall.com/SOE

KEY OBJECTIVES

A section number is given with each objective. After completing this chapter, you should be able to

6–1 Explain how to break down a basic combinational circuit into an equivalent series or parallel form and solve for current, voltage, and resistance

6–2 Solve for current and voltage in complex circuits that contain resistors in various arrangements

6–3 Use Thevenin's theorem to replace a linear, combinational circuit with an equivalent circuit from the standpoint of the load

6–4 Calculate the loading effect of various resistors on a voltage divider and discuss the resistive loading effect of electronic instruments on circuits

6–5 Apply the superposition method to circuits to find the currents and voltages in a circuit with multiple sources

6–6 Describe the Wheatstone bridge and calculate the load current in an unbalanced Wheatstone bridge using Thevenin's theorem

6–7 In a general way, discuss the analysis, planning, and measurement process for troubleshooting a faulty circuit

COMPUTER SIMULATIONS DIRECTORY

The following figures have Multisim circuit files associated with them.

- Figure 6–14 page 183
- Figure 6–21 page 186
- Figure 6–42 page 202

LABORATORY EXPERIMENTS DIRECTORY

The following exercises are for this chapter.

- **Experiment 10** Series-Parallel Combination Circuits
- **Experiment 11** Thevenin's Theorem
- **Experiment 12** The Wheatstone Bridge

KEY TERMS

- Equivalent circuit
- Thevenin's theorem
- Terminal equivalency
- Superposition method
- Wheatstone bridge
- Transducer

ON THE JOB . . .

Technical skills include competency in the area of electronics in which you are working. You should learn as much as possible about any product on which you work. You must be able to apply your knowledge to solve problems on the job and be able to acquire more advanced knowledge. Also, you must be proficient in the use of tools and instruments that are used to perform the job duties. Look for ways to improve your technical skills by independent study and taking training offered by your employer and area schools.

173

CHAPTER 6

Sci Hi SCIENCE HIGHLIGHT

All scientific measurements involve a "loading effect." It is not possible to make a measurement that does not affect the quantity being measured in some way. Obviously, for a measurement to be accurate, the effect of loading should be extremely small. At first glance, you may think of a measurement that does not seem to be affected at all by the process of making a measurement. For example, how could measuring the temperature of a room be affected by the presence of the thermometer?

A way to consider this problem is to realize that a thermometer actually takes its own temperature and must exchange energy with its surroundings in order to come into equilibrium with the surroundings. The total amount of energy exchanged with a thermometer is very tiny compared to the total energy in the room, and the loading effect is of no consequence. However, if we take the problem to the extreme condition of using a thermometer to measure the temperature of a nearly perfect vacuum, we see that the molecules in the vacuum exchanging energy with the thermometer are clearly affected by the presence of the thermometer.

A more bizarre example is the effect of a radar gun on the speed of a car. Radar possesses energy, so an imperceptibly small amount of the radar's energy will go toward speeding up (or slowing down) the car (depending if it is approaching or going away from the source). This defense has never been known to get a motorist out of a speeding ticket, however.

In many cases of making measurements, loading has utterly no significance, as in the case of the radar gun; nevertheless, loading is still present. Electrical and electronic measurements are no exception. Most electrical instruments must extract a certain amount of energy from the circuit in order to make a measurement. Occasionally, this loading effect can have a serious effect on the circuit or the measurement. The key to good electrical measurements is to recognize when loading is a factor and be able to choose an appropriate instrument for the job.

6–1 EQUIVALENT CIRCUITS

Many electronic circuits involve series or parallel combinations of circuit elements. Breaking them down to an equivalent form simplifies finding the current, voltage, or resistance of these circuits.

In this section, you will learn how to break down a basic combinational circuit into an equivalent series or parallel form and solve for current, voltage, and resistance.

Equivalent Series Circuits

Many complicated circuits can be broken down into simpler circuits. If a series or parallel combination of resistors is part of a more complicated circuit, you can replace this combination with one equivalent resistor and proceed with an analysis. Sometimes the process of simplifying a circuit requires several replacements. An **equivalent circuit** is one that has characteristics that are electrically the same as another circuit and is generally simpler than the original circuit. Equivalent circuits frequently are used in electronics for analysis purposes.

To begin an analysis, identify any series or parallel combination of resistors within the circuit. Then simplify the combination by finding a single equivalent replacement resistor. The resistors in Figure 6–1(a) are not in a basic series or basic parallel arrangement but rather form a series-parallel combination. Resistors R_2 and R_3 are in parallel with each other

COMBINATIONAL SERIES/PARALLEL CIRCUITS

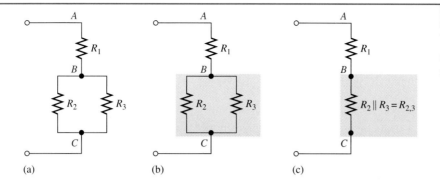

FIGURE 6–1

The blue box in part (b) contains an equivalent resistance as shown in part (c).

because they are connected between the same two nodes (labeled B and C). The blue area of Figure 6–1(b) indicates the parallel combination of these resistors. In Figure 6–1(c), the parallel combination of R_2 and R_3 has been replaced with a single equivalent resistor labeled $R_{2,3}$. This equivalent resistor is in series with R_1.

When a voltage source is connected to the resistors as in Figure 6–2(a), begin the analysis procedure by replacing R_2 and R_3 with the equivalent resistance, as shown in the yellow shaded box. The resulting equivalent circuit is shown in Figure 6–2(b). This series circuit can be analyzed by the methods given in the last chapter. All results outside of the yellow box are identical for both circuits. Values found in the equivalent series circuit can then be applied to the original circuit to complete the analysis.

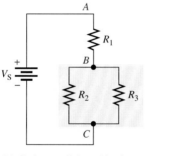
(a) Series-parallel combination

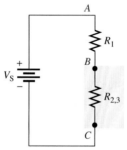
(b) Equivalent series circuit

FIGURE 6–2

The resistance for each yellow box is the same.

EXAMPLE 6–1

Problem
Figure 6–3 shows a series-parallel combinational circuit, and the given values are listed in Table 6–1. Complete the table by finding the unknown values.

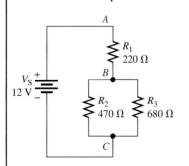

Table 6–1

$I_1 =$	$R_1 = 220\ \Omega$	$V_1 =$
$I_2 =$	$R_2 = 470\ \Omega$	$V_2 =$
$I_3 =$	$R_3 = 680\ \Omega$	$V_3 =$
$I_T =$	$R_T =$	$V_S = 12.0\ V$

FIGURE 6–3

CHAPTER 6

Solution
The following steps illustrate one way to solve the problem. Other methods are often good to use as checks on the solution. Notice that subscripts from the original resistors are combined for the equivalent resistor. This idea for subscripts should help you apply circuit laws in a consistent manner and reduce errors.

Step 1: Calculate the resistance of the parallel combination of $R_2 \parallel R_3$, as shown in Figure 6–4(a). The equivalent resistance, called $R_{2,3}$ and shown in the green box, is the resistance between nodes B and C of the circuit.

$$R_{2,3} = \frac{1}{\frac{1}{R_2} + \frac{1}{R_3}} = \frac{1}{\frac{1}{470\ \Omega} + \frac{1}{680\ \Omega}} = 278\ \Omega$$

This replacement resistance value is shown in Figure 6–4(b).

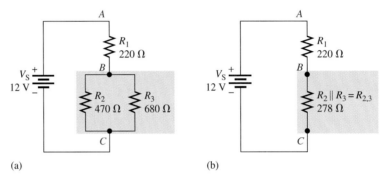

(a)　　　　　　　　　　(b)

FIGURE 6–4

Step 2: Calculate the total resistance of the circuit between nodes A and C. Because the circuit in Figure 6–4(b) is a series circuit, the total is simply the sum of R_1 and $R_{2,3}$.

$$R_T = R_1 + R_{2,3} = 220\ \Omega + 278\ \Omega = 498\ \Omega$$

Step 3: Apply Ohm's law to find the total current from the source.

$$I_T = \frac{V_S}{R_T} = \frac{12\ \text{V}}{498\ \Omega} = 24.1\ \text{mA}$$

Step 4: Calculate current in the equivalent circuit. Apply Ohm's law to each resistor in the equivalent series circuit of Figure 6–4(b). Because the equivalent circuit is a series circuit, the same current is in R_1 and $R_{2,3}$. (Of course, this is true *only* for the series equivalent circuit, not the original circuit.) As you know, the same current is in series components.

$$I_1 = I_{2,3} = I_T = 24.1\ \text{mA}$$

Step 5: Calculate voltage drops in the equivalent circuit. Substitute the current value into Ohm's law for V_1 and $V_{2,3}$.

$$V_1 = I_1 R_1 = (24.1\ \text{mA})(220\ \Omega) = 5.30\ \text{V}$$

and

$$V_{2,3} = I_{2,3}R_{2,3} = (24.1 \text{ mA})(278 \text{ }\Omega) = 6.70 \text{ V}$$

A good check at this point is to verify that KVL is valid for the equivalent circuit. Notice that $V_1 + V_{2,3} = V_S$. ✓

Step 6: You now know the current, voltage, and resistance in the equivalent circuit. Return to the original circuit in Figure 6–3. The voltage between nodes B and C is the same as the voltage across R_2 and R_3 because they are in parallel. Therefore,

$$V_2 = V_3 = V_{2,3} = 6.70 \text{ V}$$

Apply Ohm's law to R_2 and R_3 to find the current in each.

$$I_2 = \frac{V_2}{R_2} = \frac{6.70 \text{ V}}{470 \text{ }\Omega} = 14.3 \text{ mA}$$

and

$$I_3 = \frac{V_3}{R_3} = \frac{6.70 \text{ V}}{680 \text{ }\Omega} = 9.9 \text{ mA}$$

A good check for this step is to apply KCL to node B. (You will note a slight round-off error if you do this.)

The overall results are shown in Table 6–2.

Table 6–2

I_1 = 24.1 mA	R_1 = 220 Ω	V_1 = 5.30 V
I_2 = 14.3 mA	R_2 = 470 Ω	V_2 = 6.70 V
I_3 = 9.9 mA	R_3 = 680 Ω	V_3 = 6.70 V
I_T = 24.1 mA	R_T = 498 Ω	V_S = 12.0 V

Question*
What is the current entering and leaving node B?

Equivalent Parallel Circuits

Another example of a series-parallel combination of resistors is shown in Figure 6–5(a). This time, the simplification will lead to an equivalent parallel arrangement of the resistors. In this case, R_1 and R_2 are in series, as shown by the blue box in Figure 6–5(b) and can be replaced by an equivalent resistor, $R_{1,2}$, equal to their sum. The equivalent resistor $R_{1,2}$ is in parallel with R_3, as shown in Figure 6–5(c).

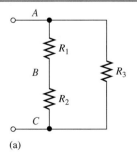
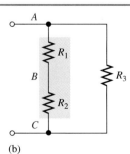
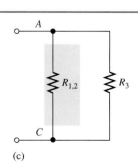

FIGURE 6–5

The blue box in (b) contains an equivalent resistor in part (c).

Answers are at the end of the chapter.

CHAPTER 6

When a voltage source is connected to the three resistors as in Figure 6–6(a), begin the analysis procedure by simplifying the circuit with the equivalent resistance. In Figure 6–6(b), note that the resistors in the yellow box are in series and can be combined (added) to form one resistor, as shown in the yellow box in Figure 6–6(c). This equivalent circuit is a pure parallel circuit; notice that there are only two nodes in the equivalent circuit (A and C) as required for all parallel circuits. As before, results from outside of the yellow box are identical for both circuits. Values found in the equivalent parallel circuit can then be applied back to the original circuit to complete the analysis.

FIGURE 6–6

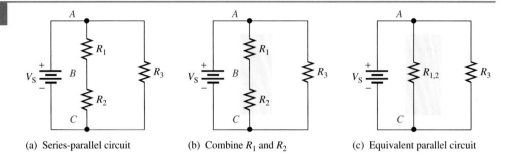

(a) Series-parallel circuit (b) Combine R_1 and R_2 (c) Equivalent parallel circuit

EXAMPLE 6–2

Problem

Figure 6–7 shows a series-parallel combinational circuit driven with an ac source. The given values for this circuit are listed in Table 6–3. Find the current in each resistor and the voltage across each resistor.

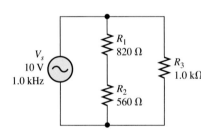

Table 6–3

$I_1 =$	$R_1 = 820\ \Omega$	$V_1 =$
$I_2 =$	$R_2 = 560\ \Omega$	$V_2 =$
$I_3 =$	$R_3 = 1.0\ k\Omega$	$V_3 =$
$I_{tot} =$	$R_{tot} =$	$V_s = 10.0\ V$

FIGURE 6–7

Solution

The fact that the source is an ac source does not affect how you approach the solution in a resistive circuit. All currents and voltages will be rms values. As mentioned before, there is more than one way to solve a combinational circuit problem. The solution given here is one way.

Step 1: Calculate the resistance of the series combination in Figure 6–8(a) by adding R_1 and R_2 (shown in the purple box).

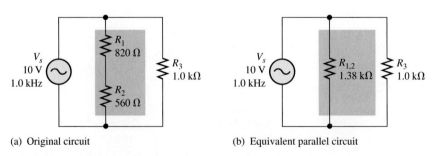

(a) Original circuit   (b) Equivalent parallel circuit

FIGURE 6–8

178

$$R_{1,2} = R_1 + R_2 = 820 \, \Omega + 560 \, \Omega = 1.38 \, k\Omega$$

The value is shown in the equivalent parallel circuit in Figure 6–8(b).

Step 2: Determine the voltage in the equivalent parallel circuit. The equivalent circuit can be analyzed like any parallel circuit. Note that the source voltage is the same across all components because it is a parallel circuit.

$$V_{1,2} = V_3 = V_s = 10.0 \, V$$

Step 3: Calculate the currents in the equivalent circuit.

$$I_{1,2} = \frac{V_{1,2}}{R_{1,2}} = \frac{10.0 \, V}{1.38 \, k\Omega} = 7.25 \, mA$$

and

$$I_3 = \frac{V_3}{R_3} = \frac{10.0 \, V}{1.0 \, k\Omega} = 10.0 \, mA$$

Step 4: Apply the known parameters from the equivalent parallel circuit to the original circuit. The equivalent resistor $R_{1,2}$ is composed of two series resistors; therefore, the current in each of the series resistors is the same as that found for the equivalent resistor.

$$I_1 = I_2 = I_{1,2} = 7.25 \, mA$$

Step 5: Apply Ohm's law to each of the series resistors in the original circuit.

$$V_1 = I_1 R_1 = (7.25 \, mA)(820 \, \Omega) = 5.94 \, V$$

and

$$V_2 = I_2 R_2 = (7.25 \, mA)(560 \, \Omega) = 4.06 \, V$$

The current and voltage for each resistor are shown in Table 6–4.

Table 6–4

$I_1 = 7.25 \, mA$	$R_1 = 820 \, \Omega$	$V_1 = 5.94 \, V$
$I_2 = 7.25 \, mA$	$R_2 = 560 \, \Omega$	$V_2 = 4.06 \, V$
$I_3 = 10 \, mA$	$R_3 = 1.0 \, k\Omega$	$V_3 = 10 \, V$
$I_{tot} =$	$R_{tot} =$	$V_s = 10.0 \, V$

Question
What are the total current and the total resistance in the circuit in Figure 6–7?

The two examples in this section have illustrated how to reduce a basic series-parallel combinational circuit into either an equivalent series or parallel circuit. Practical series-parallel circuits usually involve more components, but the concept is the same. In some cases, more than one equivalent circuit is necessary to solve a problem.

Review Questions

Answers are at the end of the chapter.

1. What is an equivalent circuit?
2. For the circuit in Figure 6–9, what is the total resistance seen by the source?
3. For the circuit in Figure 6–9, what is the total current from the source?

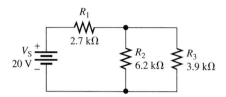

FIGURE 6–9

4. For the circuit in Figure 6–10, what is the total resistance seen by the source?
5. For the circuit in Figure 6–10, what is the total current from the source?

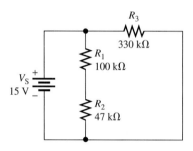

FIGURE 6–10

6–2 ANALYSIS OF COMBINATIONAL CIRCUITS

The methods of simplifying a complex circuit introduced in the last section can be applied to complicated circuits by forming more than one equivalent circuit. Sometimes applying the voltage-divider rule or using one of Kirchhoff's laws is the most efficient way to arrive at a solution for more complicated circuits.

In this section, you will learn to solve for current and voltage in complex circuits that contain resistors in various arrangements.

When you analyze a complex circuit by using equivalent circuits, certain results from the equivalent circuit are applied back to the original circuit to complete the analysis. With resistive circuits, you will generally work with known resistors, so it is useful to start by finding the total resistance. When you know the total resistance, you can calculate the total current from Ohm's law and apply the result back to the original circuit. Many times, you will see that an equivalent circuit has series resistors, so you can apply the voltage-divider rule to the series portion. This is particularly useful if you are not concerned with finding current.

Examples 6–3 and 6–4 show how to find the total resistance and voltages in series-parallel combinational circuits with five or six resistors. The method illustrated in these two

COMBINATIONAL SERIES/PARALLEL CIRCUITS

examples involves simplification by forming more than one equivalent circuit and applying the voltage-divider rule, as well as applying other basic rules for series and parallel circuits.

EXAMPLE 6–3

Problem
Determine the total resistance and the voltage across each resistor in the circuit shown in Figure 6–11. Complete Table 6–5.

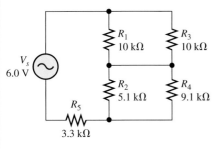

Table 6–5

$R_1 = 10\ k\Omega$	$V_1 =$
$R_2 = 5.1\ k\Omega$	$V_2 =$
$R_3 = 10\ k\Omega$	$V_3 =$
$R_4 = 9.1\ k\Omega$	$V_4 =$
$R_5 = 3.3\ k\Omega$	$V_5 =$
$R_{tot} =$	$V_s = 6.0\ V$

FIGURE 6–11

Solution

Step 1: Find the total resistance of the circuit. Sometimes it helps to visualize the series and parallel combinations by redrawing the circuit. Figure 6–12 is the same circuit redrawn to emphasize the parallel connections. There are four nodes in the circuit, labeled A, B, C, and D. Resistors R_1 and R_3 are connected between the same two nodes (A and B) and are therefore in parallel. Likewise, resistors R_2 and R_4 are connected between the same two nodes (B and C) and are also in parallel. The parallel combinations are shown in the two colored boxes in Figure 6–13(a). Figure 6–13(b) shows the circuit redrawn with series-equivalent resistors. Calculate the values of $R_{1,3}$ and $R_{2,4}$. Note that because R_1 and R_3 are equal, you can find the total resistance with the simpler equation for equal-value resistors.

$$R_{1,3} = \frac{R}{n} = \frac{10\ k\Omega}{2} = 5.0\ k\Omega$$

and

$$R_{2,4} = \frac{1}{\dfrac{1}{R_2} + \dfrac{1}{R_4}} = \frac{1}{\dfrac{1}{5.1\ k\Omega} + \dfrac{1}{9.1\ k\Omega}} = 3.27\ k\Omega$$

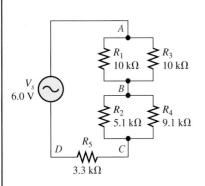

FIGURE 6–12

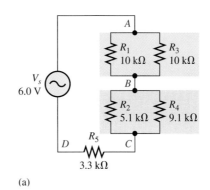 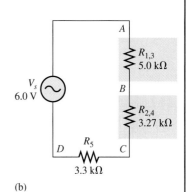

(a) (b)

FIGURE 6–13

181

The three resistors in the equivalent circuit are all in series. Notice that the four nodes from the original circuit are also shown.

Find the total resistance by adding the values of the resistors in the equivalent series circuit.

$$R_{tot} = R_{1,3} + R_{2,4} + R_5 = 5.0 \text{ k}\Omega + 3.27 \text{ k}\Omega + 3.3 \text{ k}\Omega = 11.6 \text{ k}\Omega$$

Starting from the original circuit, you can find the total resistance with the following calculator sequence (for the TI-36X). Intermediate values are stored in two memory locations for this calculation.

Step 2: Calculate the voltage across each equivalent resistor. The equivalent circuit is a series circuit, which means you can apply the voltage-divider rule. Equation 5–5 is the voltage-divider rule.

$$V_x = \left(\frac{R_x}{R_{tot}}\right) V_s$$

Substituting,

$$V_{1,3} = \left(\frac{5.0 \text{ k}\Omega}{11.6 \text{ k}\Omega}\right) 6.0 \text{ V} = 2.59 \text{ V}$$

$$V_{2,4} = \left(\frac{3.27 \text{ k}\Omega}{11.6 \text{ k}\Omega}\right) 6.0 \text{ V} = 1.69 \text{ V}$$

$$V_5 = \left(\frac{3.3 \text{ k}\Omega}{11.6 \text{ k}\Omega}\right) 6.0 \text{ V} = 1.71 \text{ V}$$

Step 3: You now know the voltages across the original resistors. Because $R_{1,3}$ is the parallel combination of R_1 and R_3, the voltage is the same for the individual resistors as you found for the equivalent resistor.

$$V_1 = V_3 = V_{1,3}$$
$$V_1 = 2.59 \text{ V}$$
$$V_3 = 2.59 \text{ V}$$

The same idea applies to the voltage across R_2 and R_4.

$$V_2 = V_4 = V_{2,4}$$
$$V_2 = 1.69 \text{ V}$$
$$V_4 = 1.69 \text{ V}$$

The results are summarized in Table 6–6.

Table 6–6

$R_1 = 10 \text{ k}\Omega$	$V_1 = $ **2.59 V**
$R_2 = 5.1 \text{ k}\Omega$	$V_2 = $ **1.69 V**
$R_3 = 10 \text{ k}\Omega$	$V_3 = $ **2.59 V**
$R_4 = 9.1 \text{ k}\Omega$	$V_4 = $ **1.69 V**
$R_5 = 3.3 \text{ k}\Omega$	$V_5 = $ **1.71 V**
$R_{tot} = 11.6 \text{ k}\Omega$	$V_s = 6.0 \text{ V}$

Question
What is the current in each resistor?

COMBINATIONAL SERIES/PARALLEL CIRCUITS

EXAMPLE 6-4

Problem

Determine the total resistance and the voltage across each resistor for the circuit shown in Figure 6–14. Complete Table 6–7.

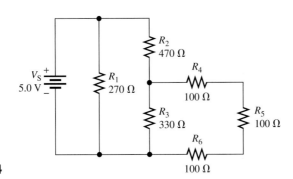

FIGURE 6-14

Table 6-7

$R_1 = 270\ \Omega$	$V_1 =$
$R_2 = 470\ \Omega$	$V_2 =$
$R_3 = 330\ \Omega$	$V_3 =$
$R_4 = 100\ \Omega$	$V_4 =$
$R_5 = 100\ \Omega$	$V_5 =$
$R_6 = 100\ \Omega$	$V_6 =$
$R_T =$	$V_S = 5.0\ V$

Solution

Step 1: Determine the equivalent resistance for the series combination of R_4, R_5, and R_6, in Figure 6–14. There is only one place to start the simplification process.

$$R_{4,5,6} = nR = (3)(100\ \Omega) = 300\ \Omega$$

The circuit with this equivalent resistor is shown in Figure 6–15.

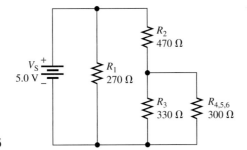

FIGURE 6-15

Step 2: Determine the resistance of R_3 in parallel with the equivalent resistor $R_{4,5,6}$. Notice that they are both connected between the same two nodes. This combination gives a resistance of

$$R_{3,4,5,6} = R_3 \parallel R_{4,5,6} = \frac{1}{\dfrac{1}{R_3} + \dfrac{1}{R_{4,5,6}}} = \frac{1}{\dfrac{1}{330\ \Omega} + \dfrac{1}{300\ \Omega}} = 157\ \Omega$$

This resistance has the subscripts of all four resistors that were used to calculate it. It is redrawn in Figure 6–16 in the yellow box.

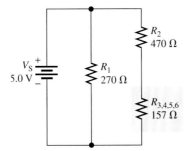

FIGURE 6-16

183

Step 3: The equivalent resistance $R_{3,4,5,6}$ is in series with R_2, so add $R_{3,4,5,6}$ and R_2 to find the resistance of the right-hand branch. This forms a new equivalent resistance $R_{2,3,4,5,6}$ as shown in Figure 6–17.

$$R_{2,3,4,5,6} = R_2 + R_{3,4,5,6} = 470\ \Omega + 157\ \Omega = 627\ \Omega$$

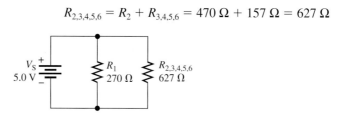

FIGURE 6–17

The original circuit has been reduced to a basic parallel circuit like those you studied in Chapter 5. The total resistance of the parallel circuit is

$$R_T = \cfrac{1}{\cfrac{1}{R_{2,3,4,5,6}} + \cfrac{1}{R_1}} = \cfrac{1}{\cfrac{1}{627\ \Omega} + \cfrac{1}{270\ \Omega}} = 189\ \Omega$$

Step 4: In Example 6–3, you used the voltage-divider rule to calculate all of the voltages. A similar approach is shown in this example. The key to applying the rule correctly is to apply it only to resistors in the equivalent circuit that are in series and the voltage across them is known. Although it is not part of a voltage-divider, R_1 is connected directly across the source. Therefore, the voltage across R_1 is

$$V_1 = V_S = 5.0\ \text{V}$$

Refer to Figure 6–16. Determine V_2 and $V_{3,4,5,6}$ by applying the voltage-divider rule to this branch in the equivalent circuit. Although the circuit itself is not a series circuit, you can still apply the voltage-divider rule to the two series resistors.

$$V_2 = \left(\frac{R_2}{R_2 + R_{3,4,5,6}}\right)V_S = \left(\frac{470\ \Omega}{470\ \Omega + 157\ \Omega}\right)5.0\ \text{V} = 3.75\ \text{V}$$

$$V_{3,4,5,6} = \left(\frac{R_{3,4,5,6}}{R_2 + R_{3,4,5,6}}\right)V_S = \left(\frac{157\ \Omega}{470\ \Omega + 157\ \Omega}\right)5.0\ \text{V} = 1.25\ \text{V}$$

The resistance in the denominators of both fractions is the total resistance in the series path, *not* the total resistance in the circuit.

Step 5: At this point, you know the voltage across $R_{3,4,5,6}$. Figure 6–18 shows the portion of the original circuit left to be solved. From step 4, you know that 1.25 V is directly across R_3.

$$V_3 = V_{3,4,5,6} = 1.25\ \text{V}$$

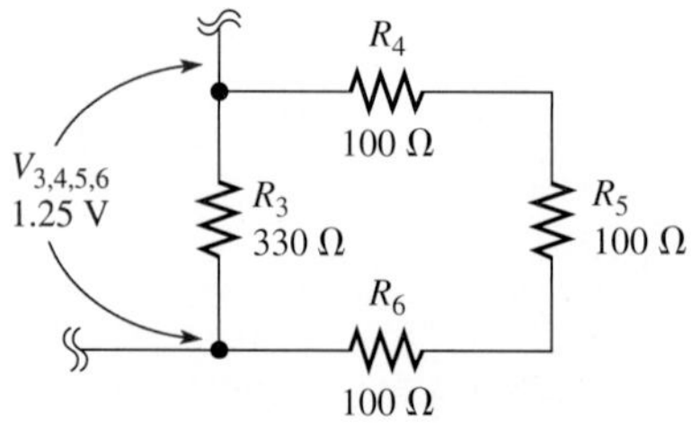

FIGURE 6–18

A voltage divider can be applied to the series branch composed of R_4, R_5, and R_6 because there is a known voltage (1.25 V) across it. The total resistance of this branch is 300 Ω. Applying the voltage-divider rule,

$$V_4 = \left(\frac{R_4}{R_4 + R_5 + R_6}\right) V_{3,4,5,6} = \left(\frac{100\ \Omega}{100\ \Omega + 100\ \Omega + 100\ \Omega}\right) 1.25\ \text{V} = 0.417\ \text{V}$$

Because R_4, R_5, and R_6 are equal values, the voltage across each of these resistors is the same.

$$V_5 = 0.417\ \text{V}$$
$$V_6 = 0.417\ \text{V}$$

Table 6–8 summarizes the results.

Table 6–8

$R_1 = 270\ \Omega$	$V_1 = 5.0\ \text{V}$
$R_2 = 470\ \Omega$	$V_2 = 3.75\ \text{V}$
$R_3 = 330\ \Omega$	$V_3 = 1.25\ \text{V}$
$R_4 = 100\ \Omega$	$V_4 = 0.417\ \text{V}$
$R_5 = 100\ \Omega$	$V_5 = 0.417\ \text{V}$
$R_6 = 100\ \Omega$	$V_6 = 0.417\ \text{V}$
$R_T = 189\ \Omega$	$V_S = 5.0\ \text{V}$

Question
The voltage-divider rule was developed for series circuits. Under what circumstances can it be applied to a series-parallel circuit?

COMPUTER SIMULATION

Open the Multisim file F06-14DC on the website. Using the multimeter, verify the voltages across all of the resistors.

Applying Kirchhoff's Laws to Series-Parallel Combinational Circuits

Kirchhoff's voltage and current laws (KVL and KCL) apply to all circuits and the combinational circuits given previously are no exception. For example, in Figure 6–19, the circuit from Example 6–4 is shown with "footprints" representing an arbitrary closed path around the outside of the circuit. The direction you traverse the circuit and the starting point don't matter. The only requirement is that the path is a single path and that the finish point is the same as the starting point. As you pass the voltage source, you would observe a 5.0 V change—in this case a rise in voltage because you are walking from the negative terminal toward the positive terminal. As you pass each resistor, you observe a drop in voltage as you move closer to the negative end of the source and further from the positive end. As indicated in Table 6–8, the drops would be V_2 (3.75 V), V_4 (0.417 V), V_5 (0.417 V), and V_6 (0.417 V). The rise and drops are equal as expected. You might try another closed path to verify that KVL works in any closed path that you choose.

CHAPTER 6

FIGURE 6-19

Any path around the circuit will have a sum of zero volts. The path shown and direction traversed are arbitrary.

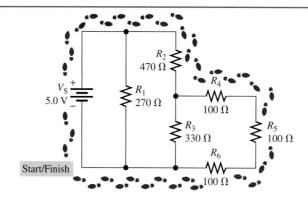

KCL can also be applied to any junction in this circuit. KCL basically states that the currents entering and leaving any node are equal. For example, if ammeters are inserted as shown in Figure 6-20, the readings show that current into and out of node A are the same.

FIGURE 6-20

KCL can be applied to any circuit junction.

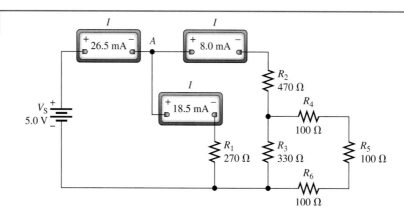

Depending on what is known, Kirchhoff's voltage and current laws may simplify finding the solution for a particular circuit. The next example illustrates the importance of laws like KVL and KCL.

EXAMPLE 6-5

Problem
Determine the current in R_3 in Figure 6-21.

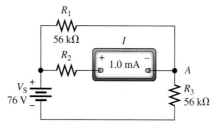

FIGURE 6-21

Solution
R_1 and R_2 are in parallel and R_3 is in series with this combination. There is no immediate place to calculate a voltage drop by Ohm's law because none of the resistors has *both* a known current and resistance. Neither can the total resistance be obtained because R_2 is unknown. Find the current in R_3 by applying KCL to node A.

The total current must go through R_3 because it is in series with the source. The total current is divided between R_1 and R_2.

$$I_1 + I_2 = I_T$$

The current in R_2 is 1.0 mA. Therefore,

$$I_1 = I_T - I_2 = I_T - 1.0 \text{ mA}$$

Now write KVL for the outside loop.

$$V_1 + V_3 = V_S$$

Apply Ohm's law and substitute the known quantities.

$$I_1 R_1 + I_3 R_3 = V_S$$
$$(I_T - 1.0 \text{ mA})56 \text{ k}\Omega + I_T 56 \text{ k}\Omega = 76 \text{ V}$$
$$I_T 56 \text{ k}\Omega - 56 \text{ V} + I_T 56 \text{ k}\Omega = 76 \text{ V}$$

Solving this equation for I_T,

$$I_T 56 \text{ k}\Omega + I_T 56 \text{ k}\Omega = 76 \text{ V} + 56 \text{ V} = 132 \text{ V}$$
$$I_T = \frac{132 \text{ V}}{112 \text{ k}\Omega} = 1.179 \text{ mA}$$

Because $I_3 = I_T$, the current through R_3 is **1.179 mA**.

Question
What is the current in R_1?

COMPUTER SIMULATION

Open the multisim file F06-21DC on the website. Using the multimeter, verify the voltages across all of the resistors.

Review Questions

6. Which resistors are in series and which are in parallel for the circuit in Figure 6–22?

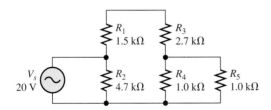

FIGURE 6–22

7. What is the total resistance as seen by the source for the circuit in Figure 6–22?
8. What is the total current from the source in Figure 6–22?
9. What is the voltage across R_1, R_3, and R_5 in Figure 6–22?
10. Is Kirchhoff's voltage law satisfied for the outside loop in Figure 6–22? Prove your answer.

CHAPTER 6

6-3 THEVENIN'S THEOREM

Thevenin's theorem is a powerful method for simplifying a circuit looking from the output terminals back toward the source. In some cases, circuits that cannot be solved using equivalent series or parallel circuits can be solved by applying Thevenin's theorem.

In this section, you will learn to use Thevenin's theorem to replace a linear, combinational circuit with an equivalent circuit from the standpoint of the load.

Thevenin Equivalent Circuit

Thevenin's theorem states that any two-terminal, resistive circuit can be replaced with a simple equivalent circuit when viewed from two output terminals. The Thevenin equivalent circuit is a simple circuit, indeed. It is a voltage source (V_{TH}) and a resistance (R_{TH}) in series, as shown in Figure 6–23. The values of the Thevenin equivalent voltage and Thevenin equivalent resistance depend on the values in the original circuit. This is a useful idea for determining the effect of different loads on a given circuit.

FIGURE 6–23

The general form of a Thevenin equivalent circuit. It can replace any resistive circuit from the perspective of the output terminals.

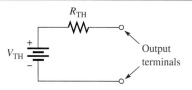

A Thevenin equivalent circuit is much easier to use for calculating the load voltage or current than the original circuit. In some cases, Thevenin's theorem makes it possible to solve circuits that could not be solved using equivalent series or parallel circuits.

The equivalent voltage, V_{TH}, is one part of the complete Thevenin equivalent circuit. The other part is R_{TH}. V_{TH} is defined as the no-load voltage between the two output terminals of a circuit. *No load* means the load is removed and the circuit is open at the output.

$$V_{TH} = V_{NL} \qquad (6-1)$$

The most common method to calculate R_{TH} is to look back into the original circuit from the output terminals and calculate the resistance from that point of view. R_{TH} is the total resistance appearing between the two output terminals when all sources have been replaced by their internal resistances. The internal resistance of a voltage source is usually taken as zero ohms. Example 6–6 illustrates how to find a Thevenin equivalent circuit.

EXAMPLE 6–6

Problem
Determine the Thevenin equivalent circuit for the circuit in Figure 6–24.

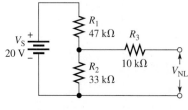

FIGURE 6–24

188

COMBINATIONAL SERIES/PARALLEL CIRCUITS

Solution
The voltage across the output terminals with no load is the Thevenin voltage. There can be no current in R_3 if there is no load because there is no path for current. Therefore, the no-load output voltage, V_{NL}, is the same as the voltage across R_2 (because there is no voltage drop across R_3). Ignoring R_3, the voltage-divider theorem can be applied to R_1 and R_2.

$$V_{NL} = V_{TH} = V_S\left(\frac{R_2}{R_1 + R_2}\right) = 20\text{ V}\left(\frac{33\text{ k}\Omega}{47\text{ k}\Omega + 33\text{ k}\Omega}\right) = \mathbf{8.25\text{ V}}$$

Determine the Thevenin resistance by replacing the source with its internal resistance and calculating the resistance between the output terminals. As mentioned earlier, the internal resistance of a voltage source will be assumed to be zero for now. Figure 6–25(a) shows the situation looking from the output into the circuit with the voltage source replaced with a short. Figure 6–25(b) is the same circuit redrawn to emphasize the series and parallel arrangements.

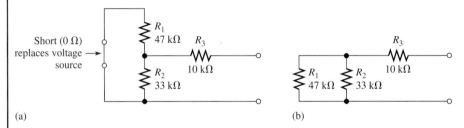

FIGURE 6–25

From examining Figure 6–25, you should see that the resistance seen from the output terminals is

$$R_{TH} = R_3 + R_1 \parallel R_2 = 10\text{ k}\Omega + \frac{1}{\dfrac{1}{47\text{ k}\Omega} + \dfrac{1}{33\text{ k}\Omega}} = \mathbf{29.4\text{ k}\Omega}$$

The values for the Thevenin equivalent circuit are shown in Figure 6–26.

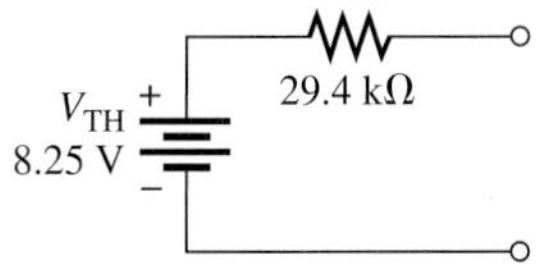

FIGURE 6–26

Recall that the notation ∥ is used between resistors to indicate they are in parallel and can be solved with the reciprocal of the sum of the reciprocals (see Section 5–5). The equation for the Thevenin resistance can be solved with the following calculator sequence:

| 1 | 0 | EE | 3 | + | 1 | ÷ | (| 4 | 7 | EE | 3 | 1/x | + | 3 | 3 | EE | 3 | 1/x |) | = |

Question
If R_3 is 15 kΩ instead of 10 kΩ, what effect will this have on the Thevenin voltage? On the Thevenin resistance?

CHAPTER 6

Another method for calculating the Thevenin resistance, R_{TH}, is to visualize a short on the output terminals and calculate the current in the short. Then apply Ohm's law.

$$R_{TH} = \frac{V_{NL}}{I_{SHORT}} \qquad (6\text{--}2)$$

EXAMPLE 6–7

Problem
Determine the Thevenin resistance for the circuit in Figure 6–24 in Example 6–6 by applying the equation

$$R_{TH} = \frac{V_{NL}}{I_{SHORT}}$$

Solution
V_{NL} is V_{TH}; therefore, $V_{NL} = 8.25$ V.

To calculate the current in a shorted load, begin by finding the resistance seen by the source with the short in place. The circuit is drawn in Figure 6–27; the ammeter indicates the current in the shorted load (I_{SL}), which is identical to the current in R_3.

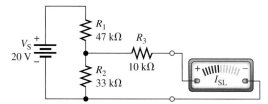

FIGURE 6–27

The total resistance seen by the source is

$$R_T = R_1 + R_2 \parallel R_3 = 47 \text{ k}\Omega + 33 \text{ k}\Omega \parallel 10 \text{ k}\Omega = 54.7 \text{ k}\Omega$$

The total current from the source is

$$I_T = \frac{V_S}{R_T} = \frac{20 \text{ V}}{54.7 \text{ k}\Omega} = 0.366 \text{ mA}$$

The voltage drop across R_3 is

$$V_3 = V_S - I_T R_1 = 20 \text{ V} - (0.366 \text{ mA})(47 \text{ k}\Omega) = 2.81 \text{ V}$$

The current in the shorted load is

$$I_{SL} = \frac{V_3}{R_3} = \frac{2.81 \text{ V}}{10 \text{ k}\Omega} = 0.281 \text{ mA}$$

The Thevenin resistance is

$$R_{TH} = \frac{V_{NL}}{I_{SHORT}} = \frac{V_{TH}}{I_{SL}} = \frac{8.25 \text{ V}}{0.281 \text{ mA}} = \mathbf{29.4 \text{ k}\Omega}$$

This is the same result obtained in Example 6–6.

Question
A load resistor of 29.4 kΩ is connected to the circuit in Figure 6–24. What is the current in the load?

Equivalency in Thevenin's Theorem

Although a Thevenin equivalent circuit is not the same as its original circuit, it acts the same in terms of the output voltage and current. For example, place a resistive circuit of any complexity in a box with only the output terminals exposed. Then place the Thevenin equivalent in a different box with, again, only the output terminals exposed. Connect identical load resistors across the output terminals of each box. Next connect a voltmeter and an ammeter to each output to measure the voltage and current for each load, as shown in Figure 6–28. The measured values will be identical and you will not be able to determine which box contains the original circuit and which contains the Thevenin equivalent from any electrical measurement made at the output terminals. That is, in terms of your observations, both circuits are the same. This condition is sometimes known as **terminal equivalency** because both circuits respond identically from the viewpoint of the two output terminals.

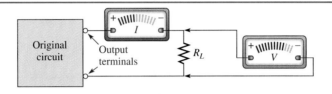

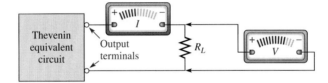

FIGURE 6–28
Which box contains the original circuit and which contains the Thevenin equivalent circuit? You cannot tell by observing the meters.

Measuring the Thevenin Resistance Indirectly

Sometimes it is not possible to observe the internal circuitry to calculate the Thevenin resistance. One simple way to measure the Thevenin resistance of a linear circuit is to use a variable resistor for a load. Adjust the variable resistor until one half of the Thevenin (no-load) voltage is dropped across the output. Then measure the variable resistor. It will have the same resistance as the Thevenin resistance! The Thevenin voltage has been divided equally between the Thevenin resistance and the load resistor. From the voltage-divider rule, if the voltage drops across two series resistors are identical, the resistors are identical.

Review Questions

11. What is Thevenin's theorem?
12. In general, how would you measure the Thevenin voltage of a circuit?
13. In general, how would you measure the Thevenin resistance of a circuit indirectly?
14. What is meant by terminal equivalency?
15. If $R_L = R_{TH}$ in a given circuit, what can you say about V_L?

HISTORICAL NOTE

Leon Charles Thevenin 1857-1926 was born in Paris, France. He graduated from the Ecole Polytechnique in 1876 and, in 1878, joined the Corps of Telegraph Engineers where he initially worked on the development of long-distance underground telegraph lines. During his career, Thevenin became increasingly interested in the problems of measurements in electrical circuits and later developed his now famous theorem which made calculations involving complex circuits possible.

LOADING EFFECTS 6–4

When a resistive load is connected to the output of a circuit, it tends to lower the output voltage of that circuit. This change in output voltage is referred to as a loading effect. The amount of loading effect depends on the load itself and the circuit to which it is connected.

In this section, you will learn to calculate the loading effect of various resistors on a voltage divider and discuss the resistive loading effect of electronic instruments on circuits.

CHAPTER 6

Voltage-Divider Loading

As you learned in Section 5–4, a voltage divider is a circuit that may have two or more resistors in series with a source. When another resistor called a load resistor is connected to the output of the voltage divider, the new circuit is now a combinational circuit. The output voltage of the original circuit decreases as a result of the load resistor. Although we will look specifically at the effect of loading a voltage divider with a resistor, the effect of loading is important in many aspects of electronics and includes components other than resistors. In cases where the effect is small, it usually can be ignored. In some cases, the effect is large and may lead to adverse effects on a circuit or erroneous measurements.

To understand the effect of electrical loading and when it is important, it is helpful to simplify a circuit to its Thevenin equivalent. Then the loading effect can immediately be determined by applying the voltage-divider rule to the equivalent circuit.

EXAMPLE 6–8

Problem
Determine the output voltage for the voltage divider in Figure 6–29 with no load. Then calculate the output voltage if a 1.0 kΩ and a 10 kΩ load are connected to the output terminals. Only one load is connected at a time.

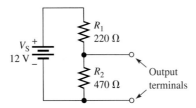

FIGURE 6–29

Solution
The easiest way to evaluate the effect of different loads is to Thevenize the circuit. Recall from Section 6–3 that the Thevenin equivalent voltage appears across the output terminals with no load. For this circuit, apply the voltage-divider theorem to determine the Thevenin voltage.

$$V_{TH} = V_S\left(\frac{R_2}{R_1 + R_2}\right) = +12\text{ V}\left(\frac{470\text{ }\Omega}{220\text{ }\Omega + 470\text{ }\Omega}\right) = 8.17\text{ V}$$

The no-load output voltage is the Thevenin voltage.

$$V_{NL} = V_{TH} = \mathbf{8.17\text{ V}}$$

Determine the Thevenin resistance by replacing the voltage source with its internal resistance and calculating the resistance seen at the output terminals. The internal resistance of a voltage source is assumed to be 0 Ω (a short). Looking back from the output terminals, there are two paths—one through R_2 and the other through R_1 and the short. This means that R_1 is in parallel with R_2 from the perspective of the output. Therefore,

$$R_{TH} = R_1 \parallel R_2 = 220\text{ }\Omega \parallel 470\text{ }\Omega = 150\text{ }\Omega$$

Figure 6–30 shows the Thevenin equivalent circuit.
You can readily determine the effect of various loads using the voltage-divider rule. For a 1.0 kΩ load, the output voltage is

$$V_{L(1.0\,k\Omega)} = V_{TH}\left(\frac{R_L}{R_{TH} + R_L}\right) = 8.17\,\text{V}\left(\frac{1.0\,\text{k}\Omega}{150\,\Omega + 1.0\,\text{k}\Omega}\right) = \mathbf{7.11\,V}$$

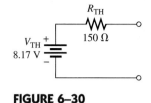

FIGURE 6–30

The output has dropped by almost 13%.
For a 10 kΩ load, the output voltage is

$$V_{L(10\,k\Omega)} = V_{TH}\left(\frac{R_L}{R_{TH} + R_L}\right) = 8.17\,\text{V}\left(\frac{10\,\text{k}\Omega}{150\,\Omega + 10\,\text{k}\Omega}\right) = \mathbf{8.05\,V}$$

With the larger load resistor, the output has dropped by only 1.5%. The amount of current the load draws determines the loading effect. Therefore, a smaller resistance has a larger loading effect on the circuit because it draws more current.

Question
What is the output voltage if a 100 kΩ load is connected to the output?

Stiff Voltage Dividers

Voltage dividers are widely used in electronics; most voltage dividers should have an output voltage that drops no more than 10% when the load is connected. A divider that meets this criterion is said to be "stiff." As Example 6–8 has shown, the resistive loading effects of a voltage divider are determined by two main parameters: the Thevenin resistance of the divider itself and the load resistor. To keep the loaded output voltage within 10% of the unloaded voltage, the load resistance should be at least ten times larger than the Thevenin resistance of the divider.

Calculating the Thevenin equivalent circuit for a voltage divider is not the only way to determine loading effects. When you are troubleshooting, you may estimate the output of a loaded voltage divider by comparing the divider resistor's values with the load resistance. Keep in mind that the output voltage will be nearly the same as the unloaded voltage as long as the current to the load is small compared to the current in the divider.

EXAMPLE 6–9

Problem
Compare the output voltage for each divider in Figure 6–31 for no load and for a 10 kΩ load resistor. Is either or both of the dividers considered to be stiff with this load?

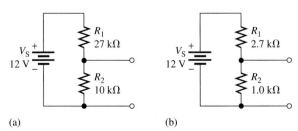

FIGURE 6–31   (a)   (b)

Solution
The output voltage with no load is the same for each voltage divider because the resistance ratio is the same. Applying the voltage-divider rule for the first circuit, you get the output voltage with no load.

$$V_{NL} = V_2 = V_S\left(\frac{R_2}{R_1 + R_2}\right) = 12\text{ V}\left(\frac{10\text{ k}\Omega}{27\text{ k}\Omega + 10\text{ k}\Omega}\right) = \mathbf{3.24\text{ V}}$$

As you saw in Example 6–8, you can apply Thevenin's theorem to find the output voltage. However, let's use an alternative method here. Consider each loaded voltage divider as a series-parallel combinational circuit by combining the load resistor, R_L, with R_2 as a parallel combination that is in series with R_1. Figure 6–32(a) shows the parallel combination in the yellow box, and Figure 6–32(b) shows the equivalent series circuit. Then apply the voltage-divider rule to this equivalent circuit.

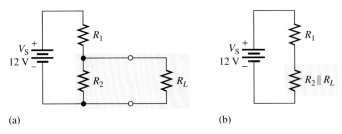

(a)   (b)

FIGURE 6–32

For the circuit in Figure 6–32(a), the parallel combination of R_L and R_2 results in a 5.0 kΩ equivalent resistor. Apply the voltage-divider rule to this equivalent circuit; the voltage across the parallel combination of R_L and R_2 is

$$V_{2,L} = V_S\left(\frac{R_2 \| R_L}{R_1 + R_2 \| R_L}\right) = 12\text{ V}\left(\frac{5.0\text{ k}\Omega}{27\text{ k}\Omega + 5.0\text{ k}\Omega}\right) = \mathbf{1.88\text{ V}}$$

Because the equivalent resistor is a parallel combination, the load voltage, V_L, in the original circuit is the same. This divider is clearly not stiff because the output dropped by more than 40% as a result of connecting the 10 kΩ load.

For the circuit in Figure 6–31(b), the parallel combination of a R_L and R_2 results in a 909 Ω equivalent resistor. Apply the voltage-divider theorem to this equivalent circuit; the voltage across the parallel combination of R_L and R_2 is

$$V_{2,L} = V_S\left(\frac{R_2 \| R_L}{R_1 + R_2 \| R_L}\right) = 12\text{ V}\left(\frac{909\text{ }\Omega}{2.7\text{ k}\Omega + 909\text{ }\Omega}\right) = \mathbf{3.02\text{ V}}$$

Again, this voltage is the same as V_L in the original circuit. The divider in Figure 6–31(b) is considered to be stiff because the output voltage dropped by only 7% as a result of connecting the 10 kΩ load. Notice that the stiff divider has smaller resistors in the divider string.

Question
Is the first voltage divider in Figure 6–31(a) stiff if an 82 kΩ resistor is the load?

Resistive Loading Effect of Instruments

Whenever an electronic instrument such as an oscilloscope (a device for measuring voltage as a function of time) or a voltmeter is connected to a circuit, the instrument represents a "load" on the circuit. The circuit must supply current to the instrument, so the process of connecting the instrument changes the circuit in some way. If the instrument's resistance is much larger than the circuit resistance, then the resistive loading effect is small and can be ignored. For example, a DMM used as a voltmeter may have several megohms of internal resistance. The loading effect of this meter on most circuits is negligible except in cases where you are measuring in a very high resistance circuit.

In circuits with a very high resistance, instrument loading will change the circuit voltages and, in severe cases, may cause the circuit to operate incorrectly or not at all. One example occurs when measuring a voltage in certain transistor amplifiers called field-effect transistors. The transistor's input resistance can be as high as 10 MΩ. Most instruments are unable to measure a voltage across this size resistance accurately; some instruments will change the reading enough to make it completely useless. An instrument such as a VOM has much lower internal resistance than a DMM. A VOM should never be used in very high resistance circuits because it will seriously change the voltage to be measured.

Oscilloscopes also have a loading effect on circuits. The signal input to a typical oscilloscope will have a 1.0 MΩ resistance to ground. At first glance, this would appear to be the resistive load that would be seen when the oscilloscope is connected to a circuit; however, manufacturers nearly always package a probe with the scope that attenuates (reduces) the input signal. When the probe is used for a measurement, the resistance to ground is increased by some fixed amount (typically a factor of 10). Manufacturers accomplish this by adding a series 9.0 MΩ resistor internally in the probe, thus increasing the apparent input resistance of the probe to 10 MΩ. This is one reason (of several) that an attenuating probe should be used for almost all oscilloscope work. An exception is when you need to measure a very small signal in a low-resistance circuit; then a $\times 1$ probe is satisfactory.

EXAMPLE 6–10

Problem

A student exclaims, "I have a circuit for which Kirchhoff's voltage law doesn't work. I made three measurements and the voltage drops across the resistors don't add up to the source voltage." The circuit is shown in Figure 6–33 with three measurements made by the student. Explain what is happening. Try to figure this one out before you look at the solution!

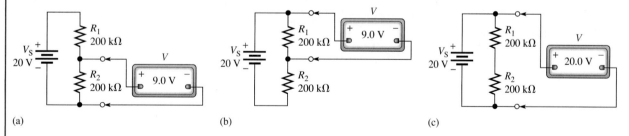

FIGURE 6–33

Solution

The internal meter resistance is loading the circuit in (a) and (b) but not in (c). The reason can be seen if each circuit is Thevenized. The Thevenin circuits for (a) and (b) are identical as shown in Figure 6–34(a). The meter reads 90% of the source voltage so must have 90% of the circuit resistance. An internal meter resistance of 900 kΩ will give the reading shown. The Thevenin circuit in Figure 6–34(b) for the circuit in part (c) of Figure 6–33 is different because the output terminals have changed; the meter is placed directly across

CHAPTER 6

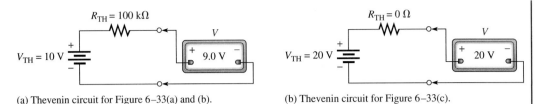

(a) Thevenin circuit for Figure 6–33(a) and (b).   (b) Thevenin circuit for Figure 6–33(c).

FIGURE 6–34

the source. The internal resistance of the source is assumed to be zero, so the Thevenin resistance is now zero, and the meter reads the full source voltage.

Question
Assume the resistors in Figure 6–33 were replaced with 20 kΩ resistors. What voltage reading would you expect in each case if the same meter were used to make the measurements?

Review Questions

16. What is a general rule to determine of a voltage divider is "stiff"?
17. In general, when should you be concerned that an instrument will cause resistive loading of a circuit?
18. Why is a VOM a poor choice for measuring a voltage in a high resistance circuit?
19. How does an oscilloscope probe reduce the loading effect of the scope on a circuit?
20. What load resistor in Figure 6–35 will cause the loaded output voltage to be 9.0 V?

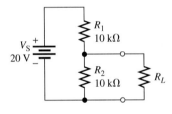

FIGURE 6–35

6–5 MULTIPLE SOURCES

Some circuits require more than one dc voltage source. For example, certain types of amplifiers require both a positive and a negative voltage source for proper operation. These circuits are solved with a method called the superposition method.

In this section, you will learn to apply the superposition method to find the currents and voltages in a circuit with multiple sources.

The Superposition Method

The **superposition method** is a way to determine currents and voltages in a linear circuit that has multiple sources by taking one source at a time and algebraically summing the results. The other sources are replaced by their internal resistances. The ideal voltage source has a zero internal resistance, so it is replaced with a short. An ideal current source has infinite resistance (an open). In this section, only voltage sources will be treated and they are all considered ideal in order to simplify the coverage.

The steps in applying the superposition method are as follows:

Step 1: Take one voltage source at a time and replace each of the other voltage sources with its internal resistance (ideally zero).

Step 2: Determine the current or voltage that you need just as if there were only that one source in the circuit.

Step 3: Repeat Steps 1 and 2 for each source in turn.

Step 4: To find an actual current or voltage, algebraically sum the currents or voltages due to each individual source. The term *algebraically sum* means you need to add positive and negative numbers according to the rules for signed numbers.

The superposition method is demonstrated in Example 6–11. It is important to keep a bookkeeping method for recording signs of current properly. Notice in the example how the ammeters keep track of algebraic signs "automatically" as they are left in place as the problem is solved.

EXAMPLE 6–11

Problem
Calculate the current that should be displayed on each ammeter in Figure 6–36. Assume the ammeters can read either a positive or a negative current but are placed in the circuit with the polarity shown on the schematic.

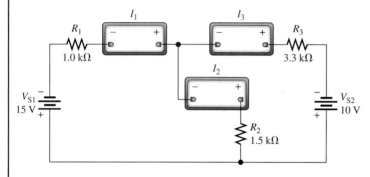

FIGURE 6–36

Because it is necessary to keep track of several numbers, tabulate the current as you calculate it. Use Table 6–9.

Table 6–9

Source	Reading of I_1	Reading of I_2	Reading of I_3
V_{S1} only:			
V_{S2} only:			
V_{S1} and V_{S2}:			

Solution
Step 1: Replace V_{S2} with a short (the internal resistance of a power supply) and calculate the current in each ammeter as if V_{S1} acted alone. The new circuit is shown in Figure 6–37. Resistor R_1 is in series with the parallel combination of R_2 and R_3. The total resistance seen by V_{S1} is

$$R_{T(S1)} = R_1 + R_2 \| R_3 = 1.0 \text{ k}\Omega + 1.5 \text{ k}\Omega \| 3.3 \text{ k}\Omega = 2.03 \text{ k}\Omega$$

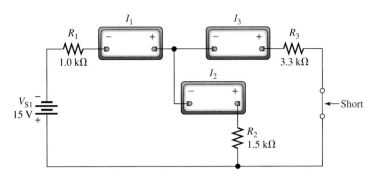

FIGURE 6–37

The total current is

$$I_{T(S1)} = \frac{V_{S1}}{R_{T(S1)}} = \frac{15 \text{ V}}{2.03 \text{ k}\Omega} = 7.39 \text{ mA}$$

This current is the same as I_1. Calculate the voltage across R_2 and R_3 by applying KVL to the circuit.

$$V_{2,3} = V_{S1} - I_1 R_1 = 15 \text{ V} - (7.39 \text{ mA})(1.0 \text{ k}\Omega) = 7.61 \text{ V}$$

Use Ohm's law to calculate the current in R_2 and R_3.

$$I_2 = \frac{V_2}{R_2} = \frac{7.61 \text{ V}}{1.5 \text{ k}\Omega} = 5.08 \text{ mA}$$

$$I_3 = \frac{V_3}{R_3} = \frac{7.61 \text{ V}}{3.3 \text{ k}\Omega} = 2.31 \text{ mA}$$

Enter these three currents into the first row of Table 6–10. The ammeters all have the correct polarity; so indicate all values as positive.

Step 2: Replace V_{S1} with a short and activate the second source. This is illustrated in Figure 6–38. Notice that the ammeter polarities are not changed. This time R_3 is in series with the source and R_1 is in parallel with R_2. The total resistance seen by V_{S2} is

$$R_{T(S2)} = R_3 + R_1 \| R_2 = 3.3 \text{ k}\Omega + 1.5 \text{ k}\Omega \| 1.0 \text{ k}\Omega = 3.9 \text{ k}\Omega$$

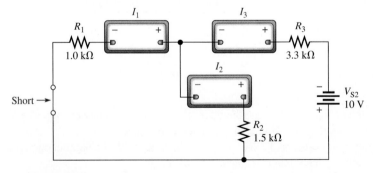

FIGURE 6–38

The total current from V_{S2} is

$$I_{T(S2)} = \frac{V_{S2}}{R_{T(S2)}} = \frac{10 \text{ V}}{3.9 \text{ k}\Omega} = 2.56 \text{ mA}$$

Notice that this current has the same magnitude as I_3 and will register on the I_3 meter as a negative current. For this reason, record it as a negative current.

Apply KVL to obtain the voltage across R_1 and R_2.

$$V_{1,2} = V_{S2} - I_3 R_3 = 10 \text{ V} - (2.56 \text{ mA})(3.3 \text{ k}\Omega) = 1.54 \text{ V}$$

Use Ohm's law to find I_1 and I_2.

$$I_1 = \frac{V_1}{R_1} = \frac{1.54 \text{ V}}{1.0 \text{ k}\Omega} = 1.54 \text{ mA}$$

$$I_2 = \frac{V_2}{R_2} = \frac{1.54 \text{ V}}{1.5 \text{ k}\Omega} = 1.03 \text{ mA}$$

The I_1 ammeter will show a negative value because of the meter polarity. Therefore, record this current as a negative current also. The I_2 ammeter has the correct polarity; therefore record I_2 as a positive value. Add the values from V_{S2} to Table 6–10.

Step 3: Algebraically add the columns to find the current with both sources acting together. The results are shown in Table 6–10. The final result on I_3, indicating a negative value, shows that the initial assumed polarity of the ammeter is reversed.

Table 6–10

Source	Reading of I_1	Reading of I_2	Reading of I_3
V_{S1} only:	+7.39 mA	+5.08 mA	+2.31 mA
V_{S2} only:	−1.54 mA	+1.03 mA	−2.56 mA
V_{S1} and V_{S2}:	+5.85 mA	+6.11 mA	−0.25 mA

Question
Show that KCL is valid for the circuit with both sources on.

While there are other methods for dealing with multiple sources, the superposition theorem is one of the easiest to apply. The previous example showed how to calculate current in a multiple-source problem using the superposition theorem. It is just as easy to find voltages in a problem by the superposition theorem.

EXAMPLE 6–12

Problem
Calculate the voltage across each resistor in the previous problem with both sources active.

Solution
Use the data from Table 6–10 to calculate voltages with Ohm's law.

$$V_1 = I_1 R_1 = (5.85 \text{ mA})(1.0 \text{ k}\Omega) = \mathbf{5.85 \text{ V}}$$
$$V_2 = I_2 R_2 = (6.11 \text{ mA})(1.5 \text{ k}\Omega) = \mathbf{9.17 \text{ V}}$$

Notice that I_3 is negative. This means that it is opposite to the direction that the ammeter shows in the circuit.

$$V_3 = I_3 R_3 = (-0.25 \text{ mA})(3.3 \text{ k}\Omega) = \mathbf{-4.95 \text{ V}}$$

CHAPTER 6

> **Question**
> What does the negative result for V_3 indicate?

Review Questions

21. When the superposition method is applied to a circuit, how do you "deactivate" an ideal voltage source?
22. When the superposition method is applied to a circuit, how do you "deactivate" an ideal current source?
23. When the superposition method is applied to a circuit, what is the meaning of a negative current in the answer?
24. If a current is calculated to be in one direction from a source and the opposite direction from a different source, what determines the direction of the resulting current?
25. What is the current in R_2 for the circuit in Figure 6–39?

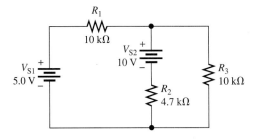

FIGURE 6–39

6–6 THE WHEATSTONE BRIDGE

The Wheatstone bridge is an important combinational circuit that has application in electronic measuring circuits. It is widely used in instruments that need to measure a very small change in resistance accurately.

In this section, you will learn to calculate the load current in an unbalanced Wheatstone bridge using Thevenin's theorem.

The Balanced Wheatstone Bridge

A **Wheatstone bridge** circuit is shown in Figure 6–40(a). This traditional Wheatstone bridge circuit consists of four resistive arms, a dc voltage source, and a special current detector called a galvanometer. The galvanometer (indicated with the circled G) is a sensitive analog ammeter with a needle in the center that can deflect either left or right depending on the current direction. The Wheatstone bridge is usually drawn with the diamond shape to identify it, but it is equivalent to two voltage dividers with an output taken between the dividers. At one time this type of bridge was widely used as a laboratory instrument for precise resistance measurements, but today it is more frequently used in automated measuring instruments, such as certain temperature- and force-measuring instruments. Most electronic scales use an automated Wheatstone bridge.

COMBINATIONAL SERIES/PARALLEL CIRCUITS

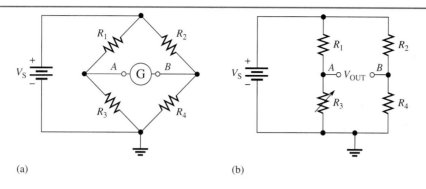

FIGURE 6–40

The Wheatstone bridge.

In many applications the point of interest is the output voltage. The output voltage is measured between the voltage dividers, as indicated in Figure 6–40(b). One of the bridge resistors may be a variable resistor as indicated by the arrow through R_3. If it is adjusted until the voltage across the output terminals is zero, the bridge is said to be **balanced**. The variable resistor may be composed of a series of precision resistors that are inserted into the circuit. Modern instruments that use the Wheatstone bridge can be electronically balanced or calibrated to read the small output voltage between A and B. They will often have additional circuitry for converting the output to a usable number.

A balanced Wheatstone bridge is easy to analyze. Because the output voltage is zero, the voltage ratios on each side of the bridge must be equal.

$$\frac{V_1}{V_3} = \frac{V_2}{V_4}$$

Substituting Ohm's law yields

$$\frac{I_1 R_1}{I_3 R_3} = \frac{I_2 R_2}{I_4 R_4}$$

At balance, there is no current in the detector, so $I_1 = I_3$ and $I_2 = I_4$. The currents in the previous equation cancel, giving

$$\frac{R_1}{R_3} = \frac{R_2}{R_4}$$

Rearranging terms,

$$R_1 R_4 = R_2 R_3 \qquad (6\text{–}3)$$

Equation 6–3 holds the key to the balanced Wheatstone bridge and can be summarized as follows:

In a balanced Wheatstone bridge, the products of the opposite diagonal resistances are equal.

In a typical application to determine an unknown resistor, one of the resistors is unknown and the others are known. One of the bridge arm resistors is a precision adjustable resistor and can be varied until the output is zero volts. Assuming R_1 is the unknown,

$$R_1 = R_3 \left(\frac{R_2}{R_4} \right) \qquad (6\text{–}4)$$

Keep in mind that this equation is *only* valid if the bridge is balanced.

EXAMPLE 6–13

Problem
What is the value of R_1 in the Wheatstone bridge shown in Figure 6–41?

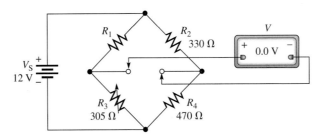

FIGURE 6–41

Solution
The bridge is balanced because the output voltage is zero. Therefore, apply Equation 6–4 to find the unknown value of R_1.

$$R_1 = R_3\left(\frac{R_2}{R_4}\right) = 305\,\Omega\left(\frac{330\,\Omega}{470\,\Omega}\right) = \mathbf{214\,\Omega}$$

Question
For the circuit in Figure 6–41, what is the voltage drop across R_1?

The Unbalanced Wheatstone Bridge

In many applications the basic bridge is modified so that each of the arms of the bridge contains a resistive transducer that converts a physical quantity to a certain amount of resistance. A **transducer** is a device that converts energy from one form to another. The energy can be in the form of a force moving a resistance by a small amount, thus converting mechanical work to an electrical quantity. In the case of a resistive transducer, the transducers are placed in the bridge arms. This method is widely used in many measurements including most electronic scales. Usually, the bridge is unbalanced, but it still is very useful for making precise measurements.

An example of an unbalanced Wheatstone bridge is shown in Figure 6–42. To simplify the discussion, a ground is shown at the bottom of the bridge and lettered points (A and B) are shown at the output. A load resistance (R_L) is placed across the output terminals. When a load resistor is added to the Wheatstone bridge, the resulting circuit does not contain any pair of resistors that are in parallel with each other or any resistors that are in series with each other. Consequently, you cannot simplify the circuit with equivalent parallel or series combinations. Applying Thevenin's theorem is one method for solving this type of circuit, as Example 6–14 illustrates.

FIGURE 6–42

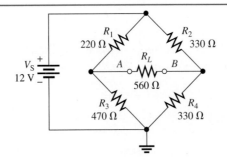

COMBINATIONAL SERIES/PARALLEL CIRCUITS

EXAMPLE 6–14

Problem
What is the current in R_L for the Wheatstone bridge shown in Figure 6–42?

Solution
You are interested in the load current, so Thevenin's theorem will allow you to replace the rest of the circuit with an equivalent circuit. The simplest way to find an equivalent Thevenin circuit is to break the problem into two parts. Thevenize between point A and ground and then between point B and ground. Then combine the two Thevenin circuits. Recall that the Thevenin voltage and resistance are determined with the load resistor removed. Start by visually removing the load resistor, as shown in Figure 6–43.

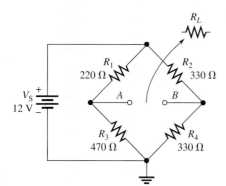

FIGURE 6–43

Part 1: As you know, the Thevenin equivalent circuit consists of a voltage source driving a series resistance. The Thevenin voltage is the output voltage between point A and ground with no load and can be found immediately with the voltage-divider rule. The left side of the bridge is the same basic circuit (except for resistor numbers) given earlier in Example 6–8. In this case, the Thevenin voltage is the same as the voltage across R_3.

$$V_{TH} = V_3 = V_S\left(\frac{R_3}{R_1 + R_3}\right) = +12\text{ V}\left(\frac{470\text{ }\Omega}{220\text{ }\Omega + 470\text{ }\Omega}\right) = 8.17\text{ V}$$

Determine the Thevenin resistance in the same way as you did in Example 6–8. The voltage source is replaced with a short that represents its internal resistance (0 Ω), as shown in Figure 6–44. This connects a direct path from the top of R_2 to the bottom of R_4, so they are not part of the resistance between point A and ground. When you look back into the driving circuit between A and ground, as shown by the "eye" in the figure, there are two separate paths—one through R_3 and the other through R_1 and the short. This means that the Thevenin resistance between A and ground is simply the parallel combination of R_1 and R_3.

$$R_{TH} = R_1 \parallel R_3 = 220\text{ }\Omega \parallel 470\text{ }\Omega = 150\text{ }\Omega$$

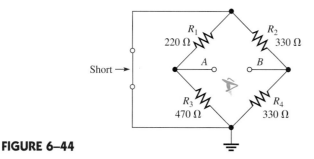

FIGURE 6–44

203

Part 2: The Thevenin equivalent circuit between point B and ground is analogous to the Thevenin circuit between point A and ground. Determine the Thevenin voltage by applying the voltage-divider rule to R_4. Use a prime with the variable V_{TH} to differentiate the voltage from the Thevenin circuit found in part 1.

$$V'_{TH} = V_4 = V_S \left(\frac{R_4}{R_2 + R_4} \right) = +12 \text{ V} \left(\frac{330 \text{ }\Omega}{330 \text{ }\Omega + 330 \text{ }\Omega} \right) = 6.0 \text{ V}$$

The resistance seen between point B and ground is the parallel combination of R_2 and R_4. Resistors R_1 and R_3 are not part of the Thevenin resistance on the right side of the bridge because the short bypasses them. The Thevenin resistance on the right side is

$$R'_{TH} = R_2 \parallel R_4 = 330 \text{ }\Omega \parallel 330 \text{ }\Omega = 165 \text{ }\Omega$$

All that remains is to put the two Thevenin equivalent circuits in place of the bridge and place the load between points A and B. The results of this are shown in Figure 6–45.

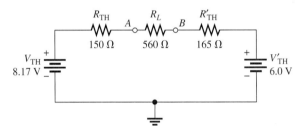

FIGURE 6–45

You can now find the current in R_L from the equivalent circuit. Notice that the two Thevenin sources are in series but opposing each other. The net driving voltage is the difference in the two sources, or 2.17 V. This difference is across three resistors with a total resistance of

$$R_T = R_{TH} + R_L + R'_{TH} = 150 \text{ }\Omega + 560 \text{ }\Omega + 165 \text{ }\Omega = 875 \text{ }\Omega$$

The current in the equivalent circuit is the same in all three resistors and is equal to the load current. Therefore, by Ohm's law,

$$I_L = I_T = \frac{2.17 \text{ V}}{875 \text{ }\Omega} = \textbf{2.48 mA}$$

Question
What is the voltage across the load resistor?

COMPUTER SIMULATION

Open the Multisim file F06-42DC on the website. Using the multimeter, verify the voltage across the load resistor.

Review Questions

26. What are the conditions for balance in a Wheatstone bridge?
27. What happens to the output voltage of a balanced Wheatstone bridge if the source voltage is halved?
28. What is a transducer?
29. Why can't an unbalanced Wheatstone bridge with a load resistor be solved by using combinational circuit analysis?
30. How does Thevenin's theorem simplify solving an unbalanced Wheatstone bridge?

TROUBLESHOOTING COMBINATIONAL CIRCUITS 6–7

Troubleshooting is the application of logical thinking to correcting a malfunctioning circuit or system.

In this section, you will learn the analysis, planning, and measurement process for troubleshooting a faulty circuit.

Analysis, Planning, and Measuring

When you troubleshoot any circuit, the first step is to analyze the clues (symptoms) of a failure. Begin the analysis by determining the answer to several questions. Examples of questions you might ask are: Has the circuit ever worked? If so, under what conditions did it fail? If not, have component values been checked? What are the symptoms of a failure? What are possible causes of this failure? Questions of this type can lead you toward an efficient plan for finding the problem.

The suggested questions are part of the analysis process. In every failure are clues to the failure; the process of troubleshooting begins with analyzing these clues, then forming a logical plan for troubleshooting. Sometimes a clue is visual (a burned component for example); more often it is not. You can save a lot of time by planning the steps you will take. As part of this plan, it is important to have a working understanding of the circuit you are troubleshooting.

Logical thinking is one of the most important tools of good troubleshooting but rarely can solve the problem by itself. The next step is to narrow the possible failures by making carefully thought out measurements. These measurements usually either confirm the direction you are taking in solving the problem or point to a new direction.

The thinking process that is part of analysis, planning, and measuring is best illustrated with an example. Suppose you have a string of 16 decorative lamps connected in series to a 120 V source, as shown in Figure 6–46. Let's further assume that this circuit worked at one time and stopped after moving it to a new location. When plugged in, the lamps fail to turn on. How would you go about finding the trouble?

SAFETY NOTE

When working in the electronics lab, be aware of the location of the emergency cut off switch. You should never work alone. If you suspect an electrical problem with a piece of equipment, remove it from service and tag it with a label identifying the suspected problem. Write "Do Not Use" on the label.

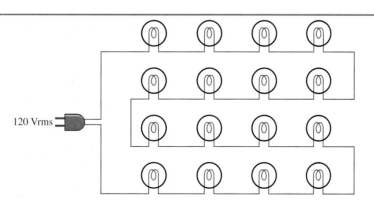

FIGURE 6–46

A series of lights. Is one of them open?

You might think like this: Since the circuit worked before it was moved, the problem could be that there is no voltage at this location. Perhaps the wiring was loose and pulled apart when moved. It's possible a bulb burned out or became loose.

Your reasoning should consider possible causes and failures that could have occurred. The fact that the circuit was once working eliminates the possibility that the original circuit may have been incorrectly wired. In a series circuit, the possibility of two open paths occurring together is unlikely. You have analyzed the problem and now you are ready to plan the troubleshooting approach.

The first part of your plan is to measure (or test) for voltage at the new location. If voltage is present, then the problem is in the light string. If voltage is not present, check the circuit breakers at the input panel to the house. Before resetting breakers, you should think about why a breaker may be tripped.

The second part of your plan assumes voltage is present and the string is bad. You can disconnect power from the string and make resistance checks to begin isolating the problem. Alternatively, you could apply power to the string and measure voltage at various points. The decision whether to measure resistance or voltage is a toss-up and can be made based on the ease of making the test. Seldom is a troubleshooting plan developed so completely that all possible contingencies are included. The troubleshooter will frequently need to modify the plan as tests are made. You are ready to make measurements and work your way to the fault.

Suppose you have a digital multimeter (DMM) handy. You check the voltage at the source and find 120 V present. You know the problem is in the string, so you proceed. You might think: "Since I have voltage across the entire string, and apparently no current in the circuit (since no bulb is on), there is almost certainly an open in the path—either a bulb or a connection." To eliminate testing each bulb, you decide to break the circuit in the middle, and check the resistance of each half of the circuit.

The technique you are using is a common troubleshooting procedure called *half-splitting*. By measuring the resistance of half the bulbs at once, you can reduce the effort required to find the open. Continuing along these lines, by half-splitting again, will lead to the problem in a few tests. Some circuits don't lend themselves to the technique but for those that do, it simplifies the effort to find a problem.

Troubleshooting a combinational circuit is generally more difficult than this example given with the series light string. Still, if a circuit is not working correctly, the troubleshooter needs to have a plan for testing based on the analysis step. Example 6–15 illustrates the process with a combinational circuit.

EXAMPLE 6–15

Problem
Does the reading on the voltmeter indicate a problem with the circuit in Figure 6–47? If so, what steps would you take to isolate the problem?

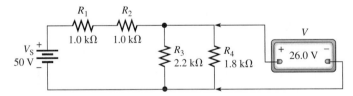

FIGURE 6–47

Solution
Analysis of the circuit is the first step. R_3 and R_4 are in parallel, and a quick calculation of their combined resistance gives 990 Ω. You can estimate that the voltmeter should read about 1/3 of the source voltage or about 16.7 V. Because the voltmeter is reading 26 V, a problem is indicated. Keep in mind that the reading may vary from ideal due to component variations or meter limitations.

The reading itself does not point to the specific problem; you need to develop a troubleshooting *plan* for finding the specific problem. A troubleshooting plan doesn't need to be written down (and seldom is), but it should have logical steps. You might think, "The fact that the voltage is present indicates a path for current exists. Either R_1 or R_2 could be shorted or R_3 or R_4 could be open." You might also consider that the source voltage could be set too high. These are not the only possible problems, but others (such as a wrong-value resistor) are less likely, particularly for a circuit that had been working correctly.

The only way to isolate the problem is to take a few *measurements*. In this case, voltage measurements can lead to the problem directly. Check the source voltage, then the voltage drops across R_1 and R_2. If the series path through R_1 and R_2 is okay, you would expect to see about 13 V across each. If measurements indicate this voltage for each, then remove power and check the resistance of R_3 and R_4 to isolate the problem.

Question
If R_4 is open, what is the ideal reading you would expect on the voltmeter in Figure 6–47?

Review Questions

31. What are three important questions a troubleshooter might ask in the analysis phase of a faulty circuit?

32. What kind of problem might be present in a circuit that is new and has never worked before that would not be likely in a previously working circuit?

33. What is *half-splitting*?

34. What reading would you expect on the voltmeter in Figure 6–48 if the circuit is working correctly?

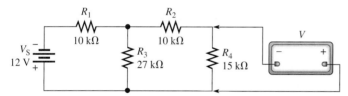

FIGURE 6–48

35. What reading would you expect on the voltmeter in Figure 6–48 if R_4 is open? Explain your answer.

CHAPTER REVIEW

Key Terms

Equivalent circuit A circuit that has characteristics that are electrically the same as another circuit and is generally simpler than the original circuit.

Superposition method A way to determine currents and voltages in a linear circuit that has multiple sources by taking one source at a time and algebraically summing the results.

Terminal equivalency Two different circuits that respond identically from the viewpoint of the two output terminals.

Thevenin's theorem A theorem that states that any two-terminal, resistive circuit can be replaced with a voltage source (V_{TH}) and a series resistance (R_{TH}) when viewed from two output terminals.

Transducer A device that converts energy from one form to another.

Wheatstone bridge A circuit that is used for precise resistance measurements. It consists of four resistive arms, a dc voltage source, and a detector.

Important Facts

❑ Many combinational circuits can be solved by finding an equivalent circuit that is electrically the same but simpler than the original circuit.

❑ Kirchhoff's voltage and current laws are frequently useful for solving combinational circuits.

❑ The Thevenin voltage is the open circuit voltage between the two output terminals of a circuit.

❑ The Thevenin resistance is the total resistance appearing between the two output terminals when all sources have been replaced by their internal resistances.

❑ Loading effects occur when an instrument or a load resistor is connected to a circuit.

❑ All measurements involve a loading effect. Loading effects may be insignificant or may seriously affect the measured quantity, depending on the circumstances.

❑ The Wheatstone bridge is a circuit that is used for precise resistance measurements. It consists of four resistive arms, a dc voltage source, and a detector.

❑ In a balanced Wheatstone bridge, the products of the diagonal resistances are equal.

❑ Wheatstone bridges are widely used with resistive transducers in the bridge arms; an example is most electronic scales.

❑ Thevenin's theorem can be applied to calculate the current in the load of an unbalanced Wheatstone bridge.

❑ Troubleshooting requires logical thinking. It generally requires analysis, planning, and measurements to find a fault. Measured results may affect the troubleshooting plan.

Formulas

Thevenin voltage:

$$V_{TH} = V_{NL} \quad (6\text{--}1)$$

Thevenin resistance:

$$R_{TH} = \frac{V_{NL}}{I_{SHORT}} \quad (6\text{--}2)$$

Balanced Wheatstone bridge:

$$R_1 R_4 = R_2 R_3 \quad (6\text{--}3)$$

$$R_1 = R_3 \left(\frac{R_2}{R_4} \right) \quad (6\text{--}4)$$

Chapter Checkup

Answers are at the end of the chapter.

1. For the circuit in Figure 6–49, the resistor with the greatest voltage across it is
   (a) R_1 (b) R_2
   (c) R_3 (d) both R_2 and R_3

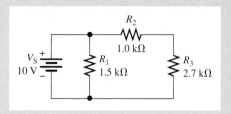

FIGURE 6–49

2. For the circuit in Figure 6–49, the resistor with the smallest current is
   (a) R_1 (b) R_2
   (c) R_3 (d) both R_2 and R_3

3. For the circuit in Figure 6–49, which statement is true?
   (a) The sum of the voltages across all of the resistors is equal to the source voltage.
   (b) The current in each resistor is the same as the source current.
   (c) The current in R_1 and R_2 is the same.
   (d) None of the above is true.

4. For the circuit in Figure 6–49, assume R_1 fails open. Which of the following will happen as a result?
   (a) source voltage decreases
   (b) total current decreases
   (c) voltage across R_2 decreases
   (d) voltage across R_1 decreases

5. For the circuit in Figure 6–50, the resistor with the greatest current is
   (a) R_1 (b) R_2
   (c) R_3 (d) not enough information to tell

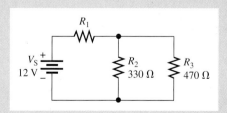

FIGURE 6–50

6. For the circuit in Figure 6–50, the resistor with the greatest voltage across it is
   (a) R_1 (b) R_2
   (c) R_3 (d) not enough information to tell

7. For the circuit in Figure 6–50, which one of the following statements is true?
   (a) R_1 and R_2 are in series.
   (b) R_1 and R_3 are in series.
   (c) R_2 and R_3 are in parallel.
   (d) None of the above is true.

8. An ideal voltage source has an internal resistance of
   (a) 0 Ω (b) 50 Ω
   (c) 600 Ω (d) infinity

9. A voltage divider consisting of two 10 kΩ resistors has a load resistor connected across one of the resistors. Of the following load resistors, which will have the greatest effect on the output voltage?
   (a) 100 kΩ (b) 10 kΩ
   (c) 1.0 kΩ (d) All have the same effect.

10. When a load resistor is connected to the output of a voltage divider, the output voltage will
    (a) not change (b) always decrease
    (c) always increase (d) increase or decrease depending on R_L

11. A Thevenin equivalent circuit consists of a
    (a) voltage source and a parallel resistor
    (b) voltage source and a series resistor
    (c) voltage source, parallel resistor, and a load resistor
    (d) voltage source, series resistor, and a load resistor

12. If the output voltage source is +5.0 V with no load, the Thevenin voltage is
    (a) +2.5 V (b) +5.0 V
    (c) +10 V (d) not enough information to tell

13. If you apply the superposition method to a circuit with two voltage sources, you begin by replacing one of the voltage sources with
    (a) an open circuit (b) a Thevenin circuit
    (c) the load resistor (d) its internal resistance

14. Using the superposition method, the total current in R_1 of Figure 6–51 is
    (a) 0 mA (b) 1 mA
    (c) 3 mA (d) none of these answers

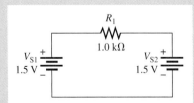

FIGURE 6–51

15. Suppose the polarity of V_{S2} is reversed in Figure 6–51. The current in R_1 will be
    (a) 0 mA (b) 1 mA
    (c) 3 mA (d) none of these answers

16. In a two-source circuit, one source acting alone produces 3.0 mA in a certain branch. The second source acting alone produces −2 mA in the same branch. The current due to both sources is

    (a) 1.0 mA (b) 2.0 mA
    (c) 3.0 mA (d) 5.0 mA

17. The output voltage of a balanced Wheatstone bridge is

    (a) dependent on the size of the resistors in the arms of the bridge
    (b) dependent on the size of the load resistor
    (c) equal to the source voltage
    (d) equal to zero

18. A troubleshooter should start troubleshooting a faulty circuit by

    (a) applying power to the faulty circuit
    (b) analyzing the symptoms
    (c) making as many measurements as possible
    (d) calling for help

Questions

Answers to odd-numbered questions are at the end of the book.

1. What sequence of steps would you take to calculate the total resistance for the circuit shown in Figure 6–52?

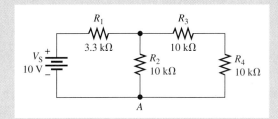

FIGURE 6–52

2. The node marked A in Figure 6–52 has current entering and leaving. According to Kirchhoff's current law, what is true about these currents?

3. If a path is traced around the outside of the circuit in Figure 6–52, what is true about the voltage rises and drops?

4. If a path is traced around resistors R_2, R_3, and R_4 of the circuit in Figure 6–52, what is true about the voltage rises and drops?

5. If a path is traced around V_S, R_1, and R_2 of the circuit in Figure 6–52, what is true about the voltage rises and drops?

6. If an equivalent resistor is in parallel with another resistance, what statement can be made about the voltage across the combination?

7. If an equivalent resistor is in series with another resistance, what statement can be made about the current in the combination?

8. For the circuit in Figure 6–53, are any two resistors in parallel with each other? If so, which ones?

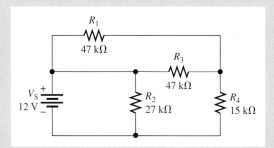

FIGURE 6–53

9. For the circuit in Figure 6–53, are any two resistors in series with each other? If so, which ones?

10. For the circuit in Figure 6–53, does the total source current go through any one resistor? If so, which one?

11. For the circuit in Figure 6–53, can any one current be determined without knowing the total resistance? If so, which one?

12. Assume the source in Figure 6–53 is changed for an ac source set for 12 Vrms. How does this change affect the currents and voltages in the circuit?

13. What are the two components in a Thevenin circuit?

14. If a Thevenin equivalent circuit is constructed for a resistive circuit and the Thevenin circuit is compared to the original circuit, what electrical difference can you measure at the *external* output terminals?

15. If a Thevenin equivalent circuit is constructed for a resistive circuit and the Thevenin circuit is compared to the original circuit, can there be a difference *internally?*

16. When a load resistor is connected to a resistive circuit, the output voltage will drop by one-half if the load resistor is equal to the Thevenin resistance. Explain why.

17. For the combinational circuit in Figure 6–54, R_1 has no effect on the Thevenin voltage and no effect on the Thevenin resistance. Explain why not.

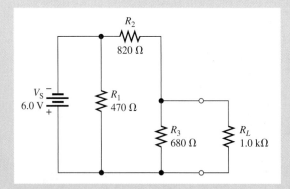

FIGURE 6–54

18. When a load is connected across one of the resistors in a voltage divider, what happens to the voltage across that resistor?

19. Three voltage dividers are shown in Figure 6–55 that divide V_S by two. Are any of the dividers "stiff" if the load resistor is a 10 kΩ resistor? If so, which?

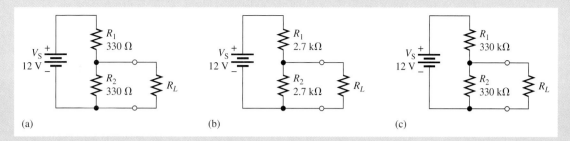

FIGURE 6–55

20. For the three voltage dividers in Figure 6–55, are any of the dividers "stiff" if the load resistor is a 1.0 MΩ resistor? If so, which?

21. If the load resistor in Figure 6–55 represents the internal meter resistance of a voltmeter, which of the three circuits will produce the most accurate voltage reading of the unloaded voltage divider? Explain your answer.

22. For what type of circuits can the superposition method be applied?

23. Why is it important to keep track of the signs of voltages and currents when applying the superposition method?

24. What is true about the diagonal resistors in a balanced Wheatstone bridge?

25. What are the steps for Thevenizing a Wheatstone bridge?

26. When troubleshooting, why should you analyze the symptoms of a nonworking circuit first?

27. A series light string contains 12 bulbs and one of the bulbs is burned out. When the string is plugged into 110 Vac, what voltage do you expect to read across the burned out bulb? Why?

Basic Problems

PROBLEMS

Answers to odd-numbered problems are at the end of the book.

1. Calculate the unknown values listed in Table 6–11 for the circuit in Figure 6–52.

Table 6–11

$I_1 =$	$R_1 = 3.3\ \text{k}\Omega$	$V_1 =$
$I_2 =$	$R_2 = 10\ \text{k}\Omega$	$V_2 =$
$I_3 =$	$R_3 = 10\ \text{k}\Omega$	$V_3 =$
$I_4 =$	$R_4 = 10\ \text{k}\Omega$	$V_4 =$
$I_T =$	$R_T =$	$V_S = 10.0\ \text{V}$

2. Calculate the unknown values listed in Table 6–12 for the circuit in Figure 6–53.

Table 6–12

$I_1 =$	$R_1 = 47\ \text{k}\Omega$	$V_1 =$
$I_2 =$	$R_2 = 27\ \text{k}\Omega$	$V_2 =$
$I_3 =$	$R_3 = 47\ \text{k}\Omega$	$V_3 =$
$I_4 =$	$R_4 = 15\ \text{k}\Omega$	$V_4 =$
$I_T =$	$R_T =$	$V_S = 12.0\ \text{V}$

3. For the circuit shown in Figure 6–55(a), calculate the value of R_L that will cause the output voltage to be exactly 5.0 V.

4. For the circuit shown in Figure 6–55(b), calculate the value of R_L that will cause the output voltage to be exactly 5.0 V.

5. For the circuit in Figure 6–56, calculate the unknown values listed in Table 6–13.

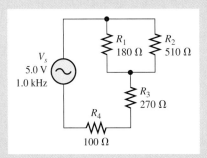

Table 6–13

$I_1 =$	$R_1 = 180\ \Omega$	$V_1 =$
$I_2 =$	$R_2 = 510\ \Omega$	$V_2 =$
$I_3 =$	$R_3 = 270\ \Omega$	$V_3 =$
$I_4 =$	$R_4 = 100\ \Omega$	$V_4 =$
$I_{tot} =$	$R_{tot} =$	$V_s = 5.0\ V$

FIGURE 6–56

6. For the circuit in Figure 6–57, calculate the unknown values listed in Table 6–14.

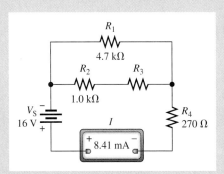

Table 6–14

$I_1 =$	$R_1 = 4.7\ k\Omega$	$V_1 =$
$I_2 =$	$R_2 = 1.0\ k\Omega$	$V_2 =$
$I_3 =$	$R_3 =$	$V_3 =$
$I_4 =$	$R_4 = 270\ \Omega$	$V_4 =$
$I_T = 8.41\ mA$	$R_T =$	$V_S = 16\ V$

FIGURE 6–57

7. A voltage divider consists of two 10 kΩ series resistors and a 24 V source. Calculate the output voltage if a 270 kΩ load resistor is connected across one of the resistors.

8. Repeat Problem 7 for a 56 kΩ load.

9. A meter with an internal resistance, R_M, of 200 kΩ is connected across the output voltage of the voltage divider in Figure 6–58. What is the output voltage as a result of connecting the meter to the circuit?

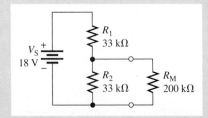

FIGURE 6–58

10. Draw the Thevenin equivalent circuit for the voltage divider in Figure 6–58. The output terminals are the open circles. The meter resistance, R_M, represents the load.

11. For the circuit in Figure 6–59, the output terminals are the open circles.
    (a) Calculate the Thevenin voltage.
    (b) Calculate the Thevenin resistance.
    (c) Determine the value of load resistor that will drop one-half the Thevenin voltage across it.
    (d) Determine the power that will be dissipated in the load resistor found in part (c).

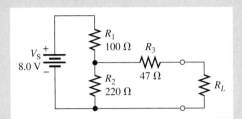

FIGURE 6–59

12. Assume the load resistor in Figure 6–59 is 220 Ω. What is the voltage across the load?
13. Assume the load resistor in Figure 6–59 is replaced with a short. What is the current in the short?
14. Use the superposition method to calculate the currents that should be shown on the ammeters in Figure 6–60 with each source acting alone and with the sources acting together. Set up a data table like Table 6–15 for your results. Notice the polarity of the meters and the sources; assume the meters can read a positive or negative current.

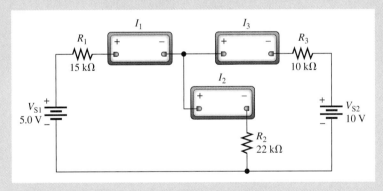

FIGURE 6–60

Table 6–15

Source	Reading of I_1	Reading of I_2	Reading of I_3
V_{S1} only:			
V_{S2} only:			
V_{S1} and V_{S2}:			

15. Calculate the voltage between the A and B terminals for the Wheatstone bridge shown in Figure 6–61.

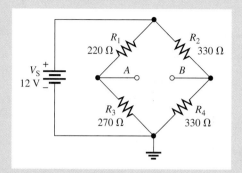

FIGURE 6–61

16. A 100 Ω load resistor is connected between the A and B terminals of the Wheatstone bridge shown in Figure 6–61. Calculate the current in the load using Thevenin's theorem.

17. To what value must R_3 be set in order to balance the Wheatstone bridge in Figure 6–62?

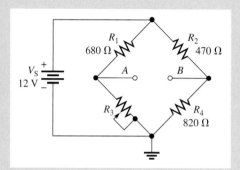

FIGURE 6–62

18. Use Thevenin's theorem to calculate the current in the load of the Wheatstone bridge shown in Figure 6–63.

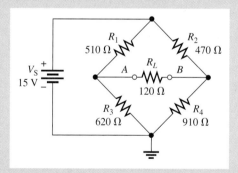

FIGURE 6–63

19. The voltmeter reading in Figure 6–64 is 2.97 V. What is the most likely problem with the circuit?

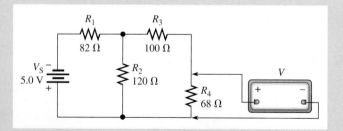

FIGURE 6–64

20. The voltmeter reading in Figure 6–64 is 0 V. What are at least three possible problems with the circuit?

Basic-Plus Problems

21. Determine the value of R_2 in Figure 6–65. *Hint:* Start by writing KVL around the outside path.

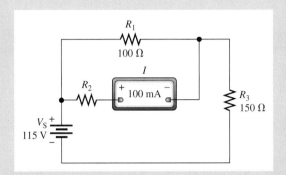

FIGURE 6–65

22. The voltage divider in Figure 6–66 has a no-load voltage of 5.9 V. When a load resistor is connected across the output terminals, the voltage drops to 5.1 V. Find the value of R_2 and R_L.

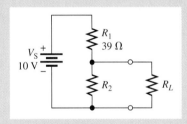

FIGURE 6–66

23. Resistor R_2 in Figure 6–66 has a value of 39 Ω. What is the smallest load resistor that can be connected to the output with a minimum voltage of 4.5 V?

24. What is the minimum power rating for the divider resistors in Problem 23?

25. Use the superposition method to find the current in R_2 of Figure 6–67.

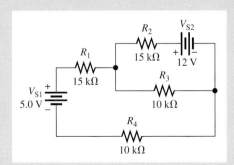

FIGURE 6–67

26. For the Wheatstone bridge in Figure 6–68, what value of R_L will limit the load current to 1.0 mA?

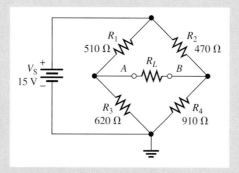

FIGURE 6–68

27. For the circuit in Figure 6–68, what is the current between A and B if R_L is replaced with a short?

Example Questions

6–1: 24.1 mA

6–2: I_{tot} = 17.25 mA; R_{tot} = 580 Ω

6–3: $I_1 = I_3$ = 0.259 mA, I_2 = 0.331 mA, I_4 = 0.186 mA, $I_5 = I_{tot}$ = 0.517 mA

6–4: It can be applied to series resistors within a combinational circuit or to equivalent resistors that are in series.

6–5: I_1 = 0.179 mA

6–6: R_3 has no effect on the Thevenin voltage. If it is 15 kΩ instead of 10 kΩ, the Thevenin resistance would be 34.4 kΩ.

6–7: I_L = 0.140 mA

6–8: $V_{L(100 kΩ)}$ = 8.16 V

6–9: Yes, because the load voltage drops by less than 10%.

6–10: 9.89 V

6–11: $I_1 = I_2 + I_3$. 5.85 mA = 6.11 mA + (−0.25 mA)

6–12: The assumed polarity is opposite the actual polarity.

6–13: 4.95 V

6–14: 1.39 V

6–15: 26.2 V

Review Questions

1. A circuit that is electrically the same as another circuit and is generally simpler than the original circuit.
2. 5.09 kΩ
3. 3.93 mA
4. 102 kΩ
5. 197 μA
6. R_1 and R_3 are in series; R_4 and R_5 are in parallel.
7. 2.35 kΩ
8. 8.51 mA
9. V_1 = 6.38 V, V_3 = 11.49 V, V_5 = 2.13 V

10. $-20\text{ V} + 6.38\text{ V} + 11.49\text{ V} + 2.13\text{ V} = 0\text{ V}$
11. Any two-terminal, resistive circuit can be replaced with a voltage source (V_{TH}) in series with a resistance (R_{TH}).
12. Measure the voltage on the output terminals with no load.
13. Measure indirectly by placing a variable resistor across the output and adjusting it until one-half the Thevenin (no load) voltage is across the variable resistor. Then measure the variable resistor.
14. Terminal equivalency occurs when two different circuits respond identically from the viewpoint of the two output terminals.
15. V_L is one-half of the V_{TH}.
16. If the load voltage changes by less than 10% when a resistor is connected to the output, the voltage divider is stiff.
17. The instrument's internal resistance should be very much larger than circuit resistances to be measured. If not, consider loading effects.
18. The internal resistance of a VOM is generally too low to measure voltage accurately in a high resistance circuit.
19. An oscilloscope probe increases the resistance of the scope's input circuitry by the attenuation factor of the probe.
20. $R_L = 45\text{ k}\Omega$
21. Replace the voltage source with a short (its internal resistance).
22. Replace the current source with an open (its internal resistance).
23. The actual current is in the opposite direction of the assumed current.
24. The net current is in the direction of the larger of the two.
25. 0.773 mA
26. The products of the opposite diagonal resistances are equal.
27. Nothing, it remains at zero.
28. A device that converts a quantity from one form to another.
29. In an unbalanced Wheatstone bridge with a load resistor there are no series or parallel combinations.
30. Thevenin's theorem allows the circuit to be reduced to a simple series form.
31. Has the circuit ever worked? Under what conditions did it fail? What were the symptoms of the failure?
32. Component values may be incorrect, a wrong part may be substituted, wiring may be open or a component may not be inserted in the board correctly (bent pin, etc.).
33. Half-splitting is the process of dividing a troubleshooting problem in half and measuring at that point to try to isolate the problem.
34. 4.07 V
35. 8.76 V. Because there is no path through R_2 and R_4, the calculated voltage is the same as the voltage across R_3 and can be found by applying the voltage-divider rule.

Chapter Checkup

1. (a)   2. (d)   3. (d)   4. (b)   5. (a)
6. (d)   7. (c)   8. (c)   9. (c)   10. (b)
11. (b)   12. (b)   13. (d)   14. (a)   15. (c)
16. (a)   17. (d)   18. (b)

CHAPTER 7

MAGNETISM AND MAGNETIC CIRCUITS

CHAPTER OUTLINE

7–1 Magnetic Quantities
7–2 Magnetic Materials
7–3 Magnetic Circuits
7–4 Transformers
7–5 Solenoids and Relays
7–6 Programmable Logic Controllers

Study aids for this chapter are available at

http://www.prenhall.com/SOE

INTRODUCTION

Although advances in electronics are well documented, the importance of magnetism is often overlooked. Energy stored in magnetic fields is converted to other forms in motors, generators, and transformers. Magnetism goes hand in hand with electricity; in fact, magnetism is really an aspect of electricity rather than something separate. One of the fascinating things about permanent magnets is that when you break one in two, each piece is also a magnet. If you could continue bisecting all the way to the atomic level, you would find that the electrons and protons in the atom itself act as tiny magnets. As you know, current is the movement of charge; this movement produces a magnetic field. In this chapter, you will learn about magnetism and its relationship to electricity. The chapter introduces solenoids and relays. It concludes with a discussion of the widely used ladder diagrams as a means of showing how relay logic diagrams are used in control applications with specialized computers called programmable logic controllers.

KEY OBJECTIVES

A section number is given for each objective. After completing this chapter, you should be able to

7–1 Define key terms used for magnetic quantities and state the SI unit for measuring each one

7–2 Describe how magnetic materials become magnetized and describe properties of magnetic materials

7–3 Contrast basic laws for electrical circuits with those of magnetic circuits and apply magnetic laws to problem solving

7–4 Describe how a transformer is constructed and how it works

7–5 Describe the principles of operation of solenoids, solenoid valves, switches, and relays

7–6 Read and explain a basic ladder diagram

LABORATORY EXPERIMENTS DIRECTORY

The following exercises are for this chapter.

◆ **Experiment 13**
Magnetic Fields

◆ **Experiment 14**
Magnetic Devices

KEY TERMS

- Flux
- Flux density
- Magnetomotive force
- Magnetic field intensity
- Magnetic domains
- Permeability
- Relative permeability
- Reluctance
- Hysteresis
- Retentivity
- Magnetization curve
- Ladder diagram

CHAPTER 7

It has been known for centuries that the earth has a magnetic field. The Chinese are credited with inventing the first compass over 2000 years ago, but there is no evidence that they understood that the early compass was responding to the earth's magnetic field. The Chinese compass consisted of a small piece of magnetic rock, called a lodestone, on a piece of wood. When the wood and lodestone were allowed to float in water, it would align itself in a north-south direction. Later, European explorers used a compass to point approximately to the north despite the fact that the earth's magnetic pole is a thousand miles from its axis of rotation (the true pole). Scientists at the time of Columbus speculated on how the earth might have acquired its magnetic field. Some scientists thought that a permanent mass of iron within the earth's interior formed the magnetic field—like a giant bar magnet.

Primarily from the study of earthquakes, geologists today believe that the earth's interior contains an iron-rich molten core that surrounds a solid inner core. The solid inner core is very hot—about the temperature of the surface of the sun—and heats the molten material surrounding it. This outer core is in motion due to thermal convection within the earth's interior and the rotation of the earth. As this fluid moves, current is generated because the conducting material passes weak magnetic fields generated by the sun or sources within the earth. The current produces the earth's magnetic field.

Recently, scientists Gary Glatzmaier, U.C.-Santa Cruz, and Paul Roberts, UCLA, have produced complicated mathematical equations that help explain the earth's magnetic field. Their models even show how the earth's magnetic field has reversed a number of times in the distant past. Evidence of these magnetic reversals is found in ancient rocks that retain residual magnetism.

7–1 MAGNETIC QUANTITIES

Whenever an electrical current is present, a magnetic field exists. Today, this fact is well known; but early in the 19th century, it was not. To describe magnetic fields, a whole new set of terms and units had to be introduced. Unfortunately, there are many magnetic units; the SI units are the most commonly used set.

In this section, you will learn key terms for magnetic quantities and learn the SI unit for measuring each one.

A compass needle is a small magnet that aligns itself with the earth's own magnetic field. When another magnet is brought near a compass, the compass needle will deflect due to the stronger magnet. A compass needle will also deflect when a sufficiently strong electric current is present in a nearby wire as shown in Figure 7–1. Danish professor Hans Christian Oersted first noticed this effect when he was lecturing on electricity. Oersted's discovery showed that electric current produces a magnetic field. This important discovery showed that magnetic fields and electricity were related.

It is logical to ask if an electric current can produce a magnetic field, can a magnetic field produce a current. A basic experiment to test the idea is illustrated in Figure 7–2. If a galvanometer (a sensitive bidirectional ammeter) is connected to a coil as shown and a magnet is moved in the coil, a voltage appears across the coil. This induces a current that can be detected by the galvanometer. The relationship between the magnetic field and electric current is thus shown to be complementary.

Both Oersted's experiment and the moving magnet involve *change*. In Oersted's experiment with the compass, current is basically moving charge, producing the magnetic field. In the experiment with the magnet in the coil, no current is generated unless the magnet is

FIGURE 7–1

The effect on a compass needle of a current in a wire. The compass needle follows the magnetic field lines, which surround the current-carrying conductor in concentric circles.

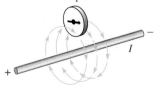

222

MAGNETISM AND MAGNETIC CIRCUITS

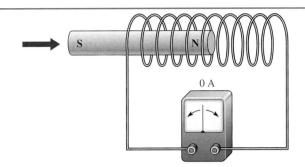

FIGURE 7-2

A moving magnet induces a current in a coil. The direction of the induced current depends on whether the magnet is moving in or out. When the magnet is not moving, the current is zero.

moved, creating a changing magnetic field. A magnetic field surrounds a stationary magnet because of the motion of electrons in the magnetic material. Moving electrons constitute electric current, creating the magnetic field.

Motors and generators exploit this link between electricity and magnetism. Motors convert electrical current into rotational motion by moving a magnetic field. Conversely, a rotating magnetic field can generate current, as is the case in alternators and generators. These devices are but two that fall into the category of electromagnetic components. An introduction to magnetic circuits, which are similar in many respects to electric circuits, will help you understand the operation of electromagnetic devices.

Flux and Flux Density

Magnetic fields are described by drawing lines that represent locations where a magnetic force is present. To visualize these lines, you can sprinkle iron filings in the region of a magnetic field. If the magnetic field is strong enough, the iron filings align themselves with the field. Although they do not show the direction of the field, they will line up in a pattern that reveals the basic shape of the magnetic field. The direction of a magnetic field is the direction a compass needle will point. Figure 7–3 shows photographs of the patterns formed from iron filings sprinkled in two different magnetic fields.

HISTORICAL NOTE

The unit of magnetic flux, the weber, is named in honor of a German physicist, Wilhelm Eduard Weber (1804–1891). Weber established a system of absolute electrical units and also performed work that was crucial to the later development of the electromagnetic theory of light.

Patterns of magnetic field lines are shown by iron filings that align with the magnetic field.

FIGURE 7-3

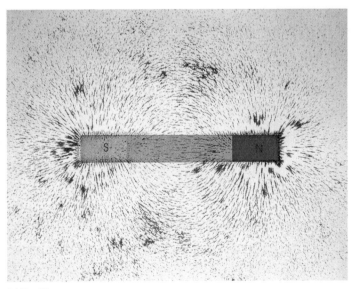

(a) Bar Magnet

(b) Current-carrying wire. Notice the magnetic field is stronger closer to the wire.

223

CHAPTER 7

HISTORICAL NOTE

The unit of magnetic flux density, the tesla, is named in honor of Nikola Tesla (1856–1943). Tesla was born in Croatia (then Austria-Hungary). He was an electrical engineer who invented the ac induction motor, polyphase ac systems, the Tesla coil transformer, wireless communications, and fluorescent lights. He worked for Edison when he first came to the U.S. in 1884 and later for Westinghouse.

Imaginary field lines called **flux** are used to describe both the magnitude and the shape of a magnetic field. In SI units, the unit of flux is the weber (Wb). One weber represents 10^8 lines of flux, a very large unit. Thus, the milliweber and the microweber are commonly used units. Flux is represented with the Greek letter ϕ (lowercase phi—pronounced "fee").

When flux lines are concentrated, the flux density is said to be high. **Flux density** is a value that is proportional to the number of lines of flux that are perpendicular to a given area, as illustrated in Figure 7–4. Flux density is represented by the letter B. As a defining equation, flux density is

$$B = \frac{\phi}{A} \qquad (7\text{--}1)$$

where B is flux density in Wb/m² or tesla, T, ϕ is flux in Wb, and A is area in m². The SI unit for flux density is the tesla, T, after Nikola Tesla, which is the same as the unit Wb/m². Both units are in common use.

FIGURE 7–4

The flux density is determined by the number of lines perpendicular to an area.

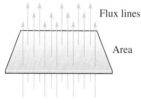
(a) Higher flux density

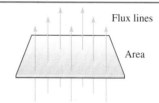
(b) Lower flux density

EXAMPLE 7–1

Problem
What is the flux density in a core that has a rectangular cross section that is 1.00 cm long and 0.50 cm wide and has a flux of 2.0 mWb?

Solution
Find the cross-sectional area by converting the lengths of the sides from centimeters to meters and calculating the product. 1.00 cm = 0.0100 m and 0.50 cm = 0.0050 m.

$$A = lw = (0.0100 \text{ m})(0.0050 \text{ m}) = 5.0 \times 10^{-5} \text{ m}^2$$

Substitute into Equation 7–1 and convert mWb to Wb.

$$B = \frac{\phi}{A} = \frac{0.002 \text{ Wb}}{5.0 \times 10^{-5} \text{ m}^2} = 40 \text{ Wb/m}^2 = \mathbf{40 \text{ T}}$$

Question*
What happens to the flux density if the same number of lines is in an area that is twice as large?

Answers are at the end of the chapter.

MAGNETISM AND MAGNETIC CIRCUITS

EXAMPLE 7–2

Problem
Compare the flux and the flux density in the two magnetic cores shown in Figure 7–5 by completing Table 7–1. The diagram represents the cross section of a magnetized material. Assume that each dot represents 100 field lines or 1 μWb.

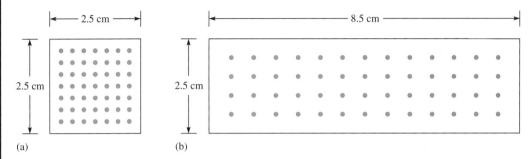

FIGURE 7–5

Table 7–1

	Flux (μWb)	Area (m²)	Flux density (T)
Figure 7–5(a)			
Figure 7–5(b)			

Solution
The flux is simply the number of field lines. In Figure 7–5(a) there are 49 dots. Each represents 1 μWb, so the flux is 49 μWb. In Figure 7–5(b) there are 52 dots, so the flux is 52 μWb.

To calculate the flux density, first calculate the area in m². For Figure 7–5(a), the area is

$$A = lw = (0.025 \text{ m})(0.025 \text{ m}) = 6.25 \times 10^{-4} \text{ m}^2$$

For Figure 7–5(b), the area is

$$A = lw = (0.025 \text{ m})(0.085 \text{ m}) = 2.12 \times 10^{-3} \text{ m}^2$$

Use Equation 7–1 to calculate the flux density. For Figure 7–5(a), the flux density is

$$B = \frac{\phi}{A} = \frac{49 \ \mu\text{Wb}}{6.25 \times 10^{-4} \text{ m}^2} = 78.4 \times 10^{-3} \text{ Wb/m}^2 = 78.4 \times 10^{-3} \text{ T}$$

For Figure 7–5(b), the flux density is

$$B = \frac{\phi}{A} = \frac{52 \ \mu\text{Wb}}{2.12 \times 10^{-3} \text{ m}^2} = 24.5 \times 10^{-3} \text{ Wb/m}^2 = 24.5 \times 10^{-3} \text{ T}$$

These results are entered in Table 7–2. This illustrates that the core with the larger flux does not necessarily have the higher flux density.

Table 7–2

	Flux (μWb)	Area (m²)	Flux density (T)
Figure 7–5(a)	49	6.25×10^{-4}	78.4×10^{-3}
Figure 7–5(b)	52	2.12×10^{-3}	24.5×10^{-3}

Question
What happens to the flux density if the same flux shown in Figure 7–5(a) is in a core that is 5.0 cm × 5.0 cm?

CHAPTER 7

Magnetomotive Force

All magnetic fields originate from moving electric charge. At the atomic level, the motion of an electron in orbit about the nucleus gives rise to a magnetic field. This magnetic field is an intrinsic property of the electron. For most materials, the magnetic field that arises from one electron's motion is often canceled out by another electron moving in the opposite direction on the same atom. Permanent magnets have a magnetic field due to the addition of these magnetic fields that are not cancelled.

In the case of a current-carrying wire, magnetic field lines are generated by a huge number of electrons in motion together; the field lines surround the wire due to their collective motion as previously shown. When the wire is formed into a coil, the flux increases even more, due to a multiplying effect of the turns in the coil. Current in an electric circuit and flux in a magnetic circuit are both *effects*. Just as voltage causes current in an electrical circuit, a quantity called **magnetomotive force** (mmf) causes flux in a magnetic circuit. *Magnetomotive force* is something of a misnomer because in the physics sense, it is not really a force; rather it is a direct result of the movement of charge (current). Magnetomotive force is given by the equation

$$F_m = NI \qquad (7\text{--}2)$$

where F_m is magnetomotive force (At), N is number of turns of wire in a coil, and I is current (A). Although the number of turns of wire is a dimensionless number, it is common to show the ampere-turn as a unit for magnetomotive force; however, the ampere alone is also commonly used.

The simplest form of a magnetic circuit is a coil of wire wrapped around a ferromagnetic material such as an iron core, as illustrated in Figure 7–6. As mentioned, only the current in the wire and the number of turns determine the magnetomotive force. It is analogous to the voltage in an electric circuit.

FIGURE 7–6

A simple magnetic circuit. Current in the coil causes flux in the iron core.

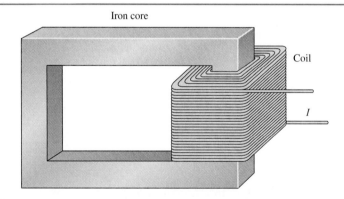

EXAMPLE 7–3

Problem
For the coil shown in Figure 7–6, calculate F_m if the number of turns is 100 and a current of 2 A is in the coil.

Solution
Apply Equation 7–2.

$$F_m = NI = (100 \text{ t})(2 \text{ A}) = \mathbf{200 \text{ At}}$$

Question
What happens to the magnetomotive force if the current is doubled and the number of turns is halved?

MAGNETISM AND MAGNETIC CIRCUITS

Magnetic Field Intensity

Another important quantity is the **magnetic field intensity** (sometimes called the magnetic force) that is defined as the magnetomotive force per unit length. The magnetic field intensity is shown with the letter H. You can think of magnetic field intensity as the independent variable that represents the effort that a given current must put into establishing a certain flux density in a material. The defining equation for magnetic field intensity is

$$H = \frac{F_m}{l} \tag{7-3}$$

where F_m is magnetomotive force (At) and l is average length of the path (m).

By substituting the definition for F_m into Equation 7–3, you obtain an alternate formula for H.

$$H = \frac{NI}{l} \tag{7-4}$$

EXAMPLE 7–4

Problem
For the rectangular-shaped core shown in Figure 7–7, calculate F_m and H. The number of turns is 250, and a current of 1.5 A is in the coil. The average length and width of the core are 8.0 cm by 5.0 cm as shown.

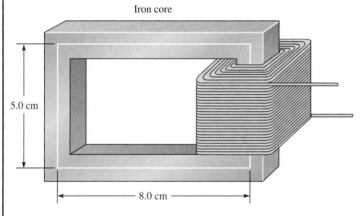

FIGURE 7–7

Solution
Apply Equation 7–2 to calculate F_m.

$$F_m = NI = (250 \text{ t})(1.5 \text{ A}) = \textbf{375 At}$$

Add the lengths around the perimeter to find the average length of the core.

$$l = 2 \times 8.0 \text{ cm} + 2 \times 5.0 \text{ cm} = 26 \text{ cm} = 0.26 \text{ m}$$

Apply Equation 7–4.

$$H = \frac{F_m}{l} = \frac{375 \text{ At}}{0.26 \text{ m}} = \textbf{1.44} \times \textbf{10}^3 \textbf{ At/m}$$

Question
What is the magnetic field intensity if the number of windings is increased to 400?

227

CHAPTER 7

Review Questions

Answers are at the end of the chapter.

1. What is the difference between flux and flux density?
2. What is the measurement unit for flux?
3. What is the measurement unit for flux density?
4. What is magnetomotive force?
5. What is the measurement unit for magnetomotive force?

7–2 MAGNETIC MATERIALS

Many materials, including iron, nickel, and cobalt as well as some ceramics, are magnetic materials. Magnetic materials are used in a number of electrical and electronic products including motors and generators, computer disks, video or audio tapes, and computer monitors.

In this section, you will learn how magnetic materials become magnetized and the properties of magnetic materials.

All materials can be classified according to their magnetic properties into one of four groups:

- Ferromagnetic
- Ferrites
- Paramagnetic
- Diamagnetic

Both ferromagnetic materials and ferrites are attracted to magnets and can be made into magnets themselves. Ferromagnetic materials include iron, cobalt, and certain alloys such as steel. Ferrites are compounds that are made from a ferromagnetic material and a ceramic or plastic material (an example is the common refrigerator magnet). Both ferromagnetic materials and ferrites are good conductors of magnetic flux; however, ferromagnetic materials will also conduct electricity, whereas ferrites are poor conductors of electricity.

Paramagnetic and diamagnetic materials are both considered nonmagnetic materials. The primary difference is that paramagnetic materials have a very weak attraction to magnets whereas diamagnetic materials are slightly repelled by magnets. Diamagnetic materials include carbon graphite, bismuth, water, and gold. Because of their slight repulsion to magnetic fields, diamagnetic materials are interesting materials for levitation experiments and in applications such as sensors.

Magnetic Domains

Ferromagnetic materials have minute magnetic domains created within their atomic structure by the orbital motion and spin of electrons. **Magnetic domains** consist of millions of atoms that have naturally aligned their magnetic fields. The domains can be viewed as very small bar magnets with north and south poles. When a ferromagnetic material is not exposed to an external magnetic field, its magnetic domains are randomly oriented, as shown in Figure 7–8(a). When the material is placed in a magnetic field, the domains align themselves, as shown in part (b). Thus, the object itself effectively becomes a magnet. This is how materials like a string of paper clips are able to hang together from a single magnet.

One of the most interesting properties of certain iron alloys is the ability of the alloy to retain the direction of the aligned magnetic domains. When the domains remain aligned af-

MAGNETISM AND MAGNETIC CIRCUITS

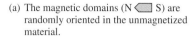

(a) The magnetic domains (N ⬅ S) are randomly oriented in the unmagnetized material.

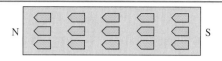

(b) The magnetic domains become aligned when the material is magnetized.

FIGURE 7–8

Ferromagnetic domains in (a) an unmagnetized and (b) a magnetized material.

ter removing the material from a magnetic field, a permanent magnet is formed. Nonmagnetic materials do not have domains, so they cannot form magnets.

Permeability

An important parameter for a material that defines the ease with which a magnetic field can be established in a given material is called **permeability**. The permeability (μ) of a material indicates the amount of flux density (B) that will occur for a given magnetic field intensity (H). Think of permeability as a number assigned to a material that relates the cause (H) to the effect (B). Flux density is related to the permeability and the magnetic field intensity by

$$B = \mu H \quad (7\text{--}5)$$

Notice that Equation 7–5 is in the form of a straight line, where H is the independent variable, B is the dependent variable, and μ is the slope of the line. The most common unit for permeability is the Wb/At-m (weber/ampere-turn-meter).

Permeability for nonmagnetic materials such as air, wood, or water is a constant, so a plot of the flux density, B, that occurs due to a given magnetomotive force for these materials is a straight line, as the red line illustrates in Figure 7–9. The flux density for a magnetic material is proportional to the magnetic field intensity for small values of flux density but falls off for increasing magnetomotive force as the blue line shows in Figure 7–9. For a given magnetic field intensity (H), the flux density (B) in a magnetic material is significantly larger than in a nonmagnetic material.

SAFETY NOTE

Strong magnets, such as neodymium magnets, can slam together rapidly and pinch fingers (like pliers!), Neodymium and NIB (Neodymium-Iron-Boron) magnets are brittle and prone to shatter when this happens. When working with strong magnets, you should wear safety glasses because magnetic chips can be propelled at high velocities if they accidentally snap together! Be careful of surrounding materials (magnetic disks, watches, hearing aids) when handling strong magnets. Strong magnets should never be given to children for play.

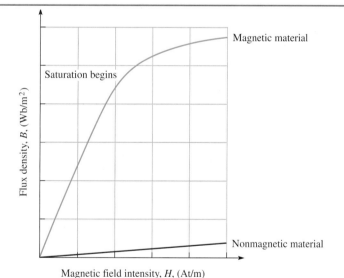

FIGURE 7–9

Comparison of the *B-H* curve for a nonmagnetic and a magnetic material.

Figure 7–9 illustrates another important difference between magnetic materials and nonmagnetic materials. For nonmagnetic materials the permeability is small but constant as shown by the straight-line relationship between H and B. By contrast, the relationship between H and B for magnetic materials is not constant but depends to some extent on the flux density. At the point where nonlinearity begins, the material is beginning to become **saturated.** This means that the flux density tends to remain constant even as the magnetomotive force increases. Equation 7–5 is valid only in the first (linear) portion of the curve.

CHAPTER 7

Because permeability of magnetic materials is much larger than nonmagnetic materials, the B-H curve is much steeper for these materials than for nonmagnetic materials. The curves shown in Figure 7–9 are only representative of the two types of materials; there are considerable differences between various materials.

Relative Permeability

As you know, permeability depends on the type of material. The permeability of a vacuum (μ_0) is $4\pi \times 10^{-7}$ Wb/At-m and can be used as a reference for other materials. Ferromagnetic materials typically have permeabilities hundreds and even thousands of times larger than that of a vacuum, which indicates that a magnetic field can be set up with relative ease in these materials. For a vacuum, Equation 7–5 is modified to read

$$B = \mu_0 H = (4\pi \times 10^{-7} \text{ Wb/At-m})H$$

The **relative permeability** (μ_r) of a material is the ratio of its absolute permeability (μ) to the permeability of a vacuum (μ_0). Since μ_r is a ratio, it has no units.

$$\mu_r = \frac{\mu}{\mu_0}$$

Reluctance

Recall that in electrical circuits, resistance is the opposition to current. A similar unit for magnetic circuits is called reluctance. **Reluctance** is the opposition to the establishment of a magnetic field in a material. The SI unit of reluctance is the ampere-turn/weber (At/Wb). The value of reluctance ($\mathcal{R}$) is directly proportional to the length (l) of the magnetic path and inversely proportional to the permeability (μ) and to the cross-sectional area (A) of the material. The defining equation for reluctance is

$$\mathcal{R} = \frac{l}{\mu A} \qquad (7\text{–}6)$$

where $\mathcal{R}$ is reluctance (At/Wb), μ is permeability (Wb/At-m), and A is area (m²). Equation 7–6 for reluctance is similar to Equation 3–5 (used for determining wire resistance). Recall Equation 3–5 is

$$R = \frac{\rho l}{A}$$

The reciprocal of resistivity (ρ, Greek rho) is conductivity (σ, Greek sigma). By substituting $1/\sigma$ for ρ, the equation for wire resistance can be written as

$$R = \frac{l}{\sigma A}$$

Compare this last equation for wire resistance with Equation 7–6. The length (l) and the area (A) have the same meaning in both equations. The conductivity (σ) in electrical circuits is analogous to permeability (μ) in magnetic circuits. Also, resistance (R) in electric circuits is analogous to reluctance ($\mathcal{R}$) in magnetic circuits; both are oppositions. Typically, the reluctance of a magnetic circuit is 50,000 At/Wb or more, depending on the size and type of material.

EXAMPLE 7–5

Problem
Calculate the reluctance of a torus (a doughnut-shaped core) made of iron. The inner radius of the torus is 1.75 cm, and the outer radius of the torus is 2.25 cm. The permeability of iron is 2×10^{-4} Wb/At-m.

Solution
You must convert cm to m before calculating the area and length. From the dimensions given, the thickness (diameter) is 0.5 cm = 0.005 m. Thus, the cross-sectional area is

$$A = \pi r^2 = \pi(0.0025)^2 = 1.96 \times 10^{-5} \text{ m}^2$$

The length is equal to the circumference of the torus measured at the average radius of 2.0 cm or 0.020 m.

$$l = C = 2\pi r = 2\pi(0.020 \text{ m}) = 0.125 \text{ m}$$

Substitute values into Equation 7–6. The reluctance is

$$\mathcal{R} = \frac{l}{\mu A} = \frac{0.125 \text{ m}}{(2 \times 10^{-4} \text{ Wb/At-m})(1.96 \times 10^{-5} \text{ m}^2)} = \mathbf{31.9 \times 10^6 \text{ At/Wb}}$$

Question
What happens to the reluctance if cast steel with a permeability of 5×10^{-4} Wb/At-m is substituted for the iron core?

Hysteresis

Another property of magnetic materials, called **hysteresis**, is the effect in a magnetic material that occurs when the flux density lags behind the application of the field intensity, (H). Hysteresis comes from the root word *history*. A material with hysteresis behaves as if it had a memory. Imagine a core material wrapped with a coil such as the one shown in Figure 7–6. The magnetomotive force (mmf) can be increased or decreased by varying the current in a coil. Reversing the direction of current in the coil will reverse the direction of the mmf.

Figure 7–10 extends the idea introduced in Figure 7–9 by including the effect of hysteresis in a previously unmagnetized core. As the magnetic field intensity (H) is increased

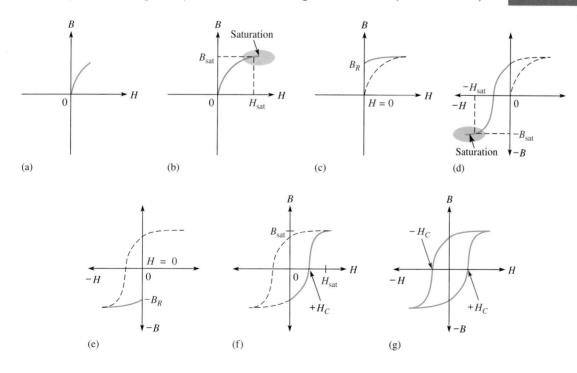

FIGURE 7–10 Development of a magnetic hysteresis curve. H is magnetic field intensity; B is flux density.

from zero, the flux density (B) increases proportionally, as indicated by the curve in Figure 7–10(a). When H reaches a certain value, the value of B begins to level off. As H continues to increase, B reaches a saturation value (B_{sat}) when H reaches a value (H_{sat}), as illustrated in Figure 7–10(b). Once saturation is reached, a further increase in H will not increase B (as you saw in Figure 7–9 for the magnetic material).

Now, if H is decreased to zero, B will fall back along a different path to a residual value (B_R), as shown in Figure 7–10(c). This indicates that the material continues to be magnetized (referred to as *residual magnetism*) even with the magnetic field intensity removed (H = 0). In effect, the material is now a "permanent" magnet. **Retentivity** is the property of a material to retain some amount of residual magnetism without the presence of a magnetizing force. The retentivity of a material is indicated by the ratio of B_R to B_{sat}.

Reversal of the magnetic field intensity is represented by negative values of H on the curve and is achieved by reversing the current in the coil of wire. An increase in H in the negative direction causes saturation to occur at a value ($-H_{sat}$) where the flux density is at its maximum negative value, as indicated in Figure 7–10(d).

When the magnetic field intensity is removed (H = 0), the flux density goes to its negative residual value ($-B_R$), as shown in Figure 7–10(e). From the $-B_R$ value, the flux density follows the curve indicated in part (f) back to its maximum positive value when the magnetic field intensity equals H_{sat} in the positive direction.

The complete B-H curve is shown in Figure 7–10(g) and is called the hysteresis curve. The magnetic field intensity required to make the flux density zero is called the coercive force, H_C.

Materials with a low retentivity do not retain a magnetic field very well while those with high retentivities exhibit values of B_R very close to the saturation value of B. Depending on the application, retentivity in a magnetic material can be an advantage or a disadvantage. In permanent magnets and in magnetic tape, for example, high retentivity is required to maintain a magnetic field in the absence of a magnetizing force. In ac motors, retentivity is undesirable because the residual magnetic field must be overcome each time the current reverses, thus wasting energy. A narrower hysteresis curve implies less energy loss, which is desirable in motors.

Summary of Magnetic Terms

Table 7–3 lists several terms and units that have been introduced in our discussion of magnetism so far.

TABLE 7–3

Magnetic terms, definitions, and units.

Term	Definition	Symbol	SI unit
Flux	A quantity related to the number of lines in a magnetic field. One weber represents 10^8 lines.	ϕ	Weber (Wb)
Flux density	Flux per area for which the flux lines are perpendicular to the area	B	Tesla (T)
Magnetomotive force	The cause of magnetic flux	F_m	Ampere-turn (At)
Magnetic field intensity	Magnetomotive force per unit length	H	Ampere-turn/meter (At/m)
Permeability	Ability of a given material to establish a magnetic field	μ	Weber/ampere-turn-meter (Wb/At-m)
Relative permeability	Ability of a given material to establish a magnetic field compared to a vacuum	μ_r	Dimensionless
Reluctance	Property of a material to oppose the establishment of flux	$\mathcal{R}$	Ampere-turn/weber (At/Wb)

Review Questions

6. How do magnetic and nonmagnetic materials differ?
7. What does it mean if a material has high permeability?
8. What electrical unit is similar to reluctance?
9. What is retentivity?
10. What does hysteresis refer to with a magnetic material?

MAGNETIC CIRCUITS 7–3

Magnetic circuits have important similarities to electrical circuits but also have some important differences. Comparing the similarities and the differences between electrical and magnetic circuits leads to a better understanding of both types.

In this section, you will learn to compare electrical and magnetic circuits and apply basic laws for magnetic circuits to problem solving.

Ohm's Law for Magnetic Circuits

As you know from Ohm's law for an electrical circuit, you can express current as the ratio of voltage to resistance.

$$I = \frac{V}{R}$$

An analogous law for magnetic circuits is

$$\phi = \frac{F_m}{\mathcal{R}} \qquad (7\text{--}7)$$

Equation 7–7 is called Ohm's law for magnetic circuits because the flux (ϕ) is analogous to current, the magnetomotive force (F_m) is analogous to voltage, and the reluctance ($\mathcal{R}$) is analogous to resistance. Just as voltage in an electric circuit is the cause for current, magnetomotive force is the cause for magnetic flux. Similarly, the opposition to current is resistance, and the opposition to flux is reluctance.

Although these are important similarities between electrical and magnetic circuits, there are also several important differences. One difference is that in an electrical circuit current cannot be present if there is no voltage. In a magnetic circuit, flux can be present (as in a permanent magnet) without a magnetomotive force. Another difference is that in an electrical circuit, there are excellent insulators that block current. There are no true insulators for magnetic flux—flux can be established in air for example. Another difference is that electrical resistance is essentially independent of the current, but in a magnetic circuit, reluctance tends to increase with higher flux due to saturation. Equation 7–7 does not include this effect so the equation is valid only up to the point where the reluctance begins to increase (and saturation begins).

EXAMPLE 7–6

Problem
What flux is established in the iron core of Example 7–5 if coil of 150 turns is wrapped on the core and a current of 2 A is in the coil?

Solution
The reluctance of the core is 31.9×10^6 At/Wb (see Example 7–5). The mmf is

$$F_m = NI = (150 \text{ turns})(2 \text{ A}) = 300 \text{ At}$$

CHAPTER 7

Substitute values into Equation 7–7. The flux is

$$\phi = \frac{F_m}{\mathcal{R}} = \frac{300 \text{ At}}{31.9 \times 10^6 \text{ At/Wb}} = \mathbf{9.4\ \mu Wb}$$

Question
What happens to the flux if the coil is wound on an identically-sized core but with higher permeability?

Finding Magnetic Flux from a Graph

The relationship between magnetic field intensity (H) and flux density (B) was introduced in the last section. As the magnetic intensity is increased, the core eventually reaches a point where an additional increase in the magnetic intensity produced only a small increase in flux density. Figure 7–11 shows an example of a B-H curve for a common ferromagnetic material used in transformers, annealed iron. A B-H curve like this is generally referred to as a **magnetization curve** for a particular material that is initially unmagnetized. A magnetization curve like this is useful to establish the flux density in a specific core; the flux density value can be read directly from the graph for a given magnetic field intensity.

FIGURE 7–11

Magnetization curve for annealed iron.

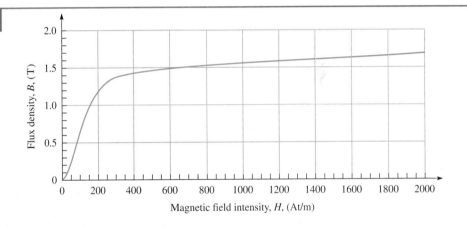

EXAMPLE 7–7

Problem
What is the flux density in the annealed iron core shown in Figure 7–12? The cross-sectional area is uniform and dimensions shown are to the center of the core.

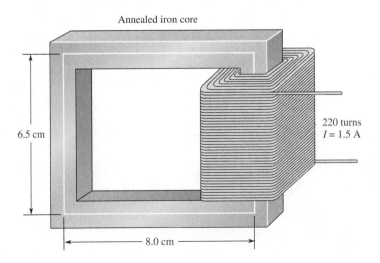

FIGURE 7–12

MAGNETISM AND MAGNETIC CIRCUITS

Solution
The average length of the core in meters is

$$l = 2(0.080 \text{ m} + 0.065 \text{ m}) = 0.29 \text{ m}$$

From Equation 7–4, the field intensity is

$$H = \frac{NI}{l} = \frac{(220 \text{ t})(1.5 \text{ A})}{0.29 \text{ m}} = 1.14 \times 10^3 \text{ At/m}$$

The flux density, B, is read from the graph in Figure 7–11 as **1.6 T**. Notice that because this is in the saturation region, Equation 7–7 is *not* valid for this case.

Question
If the same coil is wired on a 20% larger core, and the current is the same, what will happen to the flux density?

The graphical method in Example 7–7 can be used to solve for the required current to produce a certain flux density. In addition, the method can be extended to include more than one material. Two or more materials have a total reluctance that is additive, just as resistances in a series electrical circuit. In addition, the magnetomotive forces in a series magnetic circuit add like voltages around a closed path of an electrical circuit. For example, if there are two different materials in the magnetic path (such as iron and air), each material contributes to the total mmf. Air gaps are common in the magnetic paths of motors and generators because the magnetic field must cross between a fixed part and a rotating part. The following example illustrates how having two materials in the magnetic path affects the total magnetic circuit.

EXAMPLE 7–8

Problem
What is the required current to produce a flux density of 1.0 T for the core in Figure 7–13? Notice that the core has a 1.0 mm air gap at one point.

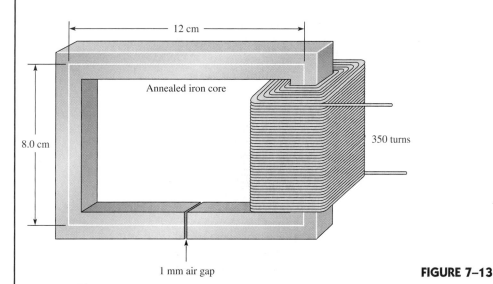

FIGURE 7–13

CHAPTER 7

Solution

An approach to the solution is to list the required variables in a table. Table 7–4 is set up with the required variables in the order they are found. The second column contains the required flux density of 1.0 T that is given in the problem. Like current in a circuit, the flux density is the same throughout (except for minor fringing effects at the gap), so enter this value into the table for each material. It is also the flux density for the last row, reserved for the total.

Table 7–4

Material	Flux density, (B)	Magnetic field intensity, (H)	Path length, (l)	Magnetomotive force, (F_m)
Annealed iron	1.0 T			
Air	1.0 T			
Total	1.0 T	—	—	

You can now find the field intensity for each part. Determine the magnetic field intensity for the annealed iron by simply reading the graph in Figure 7–11. For 1.0 T, the magnetic field intensity is 150 At/m. Enter this value into Table 7–4.

Use Equation 7–5 to calculate the magnetic field intensity for the air gap. A vacuum and air have nearly identical values of permeability.

$$B = \mu_0 H$$

$$H = \frac{B}{\mu_0} = \frac{1.0}{4\pi \times 10^{-7} \text{ Wb/At-m}} = 7.96 \times 10^5 \text{ At/m}$$

Next, enter the path lengths into the table by converting the given units to meters. For the annealed iron, ignoring the very small air gap, the total length is

$$l = 2(0.12 \text{ m} + 0.08 \text{ m}) = 0.40 \text{ m}$$

The length of the air gap, in meters, is 0.001 m, as given in the problem.

Determine the magnetomotive force, F_m, by rearranging the terms in Equation 7–3

$$H = \frac{F_m}{l}$$

$$F_m = Hl$$

The magnetomotive force for the annealed iron is

$$F_m = Hl = (150 \text{ At/m})(0.40 \text{ m}) = 60 \text{ At}$$

The magnetomotive force for the air gap is

$$F_m = Hl = (7.96 \times 10^5 \text{ At/m})(0.001 \text{ m}) = 796 \text{ At}$$

As mentioned, the magnetomotive forces in a series magnetic circuit add, just as voltages around a closed path of an electric circuit. The total magnetomotive force required is

$$F_m = F_{m(\text{iron})} + F_{m(\text{air})} = 60 \text{ At} + 976 \text{ At} = 856 \text{ At}$$

Interestingly, the annealed iron has approximately 400 times longer path length, yet the magnetomotive force is much higher for the air gap than for the iron. This illustrates why it is important to keep air gaps as small as possible in magnetic circuits.

The total number of turns given in Figure 7–13 is 350. Apply Equation 7–2 to find the current necessary to establish the required flux density.

$$F_m = NI$$
$$I = \frac{F_m}{N} = \frac{856}{350} = \mathbf{2.45\ A}$$

The completed table is shown as Table 7–5.

Table 7–5

Material	Flux density, (B)	Magnetic field intensity, (H)	Path length, (l)	Magnetomotive force, (F_m)
Annealed iron	1.0 T	150 At/m	0.40 m	60 At
Air	1.0 T	7.96 × 10⁵ At/m	0.001 m	796 At
Total	1.0 T	—	—	856 At

Question
How much current is required for the same flux if the air gap is widened to 2 mm?

Review Questions

11. What is Ohm's law for magnetic circuits?
12. What is a magnetization curve?
13. How are resistance and reluctance similar?
14. How are resistance and reluctance different?
15. Why is it important to have only a small air gap in a motor or generator?

TRANSFORMERS 7–4

One of the most interesting and important applications of magnetic cores is found in transformers. Transformers can change an ac voltage from one value to another.

In this section, you will learn how transformers are constructed and how they work.

A **transformer** is a device formed by two or more windings that are magnetically coupled to each other and provide for transfer of power electromagnetically from one winding to the other. A basic transformer is shown for a small power transformer in Figure 7–14. A schematic for an iron core transformer is shown in Figure 7–15. One coil is called the **primary winding**, and the other coil is called the **secondary winding**. An ac source voltage is applied to the primary winding, and a load is connected to the secondary winding, as shown. The primary winding establishes the flux in the core.

Transformer Action

Recall that a changing magnetic field can produce an electric current. This is the basic idea behind transformer action. A changing current on the primary side induces a flux in the core, which follows the changing current. This changing magnetic flux induces a voltage across the secondary coil. Keep in mind that a transformer can *only* work with an ac input

FIGURE 7–14

A small power transformer has two coils wound on a common core.

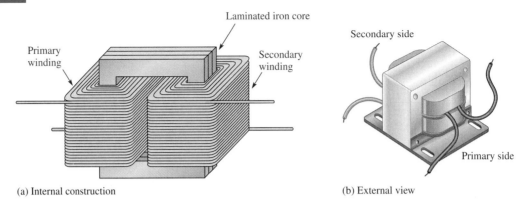

(a) Internal construction (b) External view

FIGURE 7–15

Schematic of a transformer. The horizontal lines between the coils indicate an iron core.

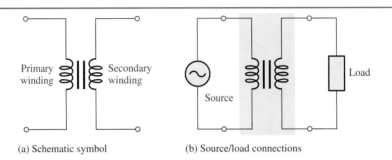

(a) Schematic symbol (b) Source/load connections

because it is necessary for the flux in the core to change in order to induce the voltage in the secondary winding.

The secondary voltage may be larger, smaller, or equal to the primary voltage; it depends on the ratio of the number of turns in the secondary winding to the number of turns in the primary winding. In some cases, there may be more than one secondary. The number of turns in a given secondary divided by the number of turns in the primary is called the **turns ratio**[*], given by the equation

$$n = \frac{N_{sec}}{N_{pri}}$$

where n is turns ratio, N_{sec} is number of turns in the secondary winding, and N_{pri} is number of turns in the primary winding.

When the transformer has a larger number of turns in the secondary winding than in the primary winding, the output voltage is greater than the input voltage; this transformer is a **step-up transformer**. The turns ratio will be greater than 1. When the transformer has a smaller number of turns in the secondary winding than in the primary winding, the output voltage is smaller than the input voltage; this transformer is a **step-down transformer**. The turns ratio will be less than 1. Although the idea of the turns ratio is useful to understanding how a transformer operates, most transformers are marked by the required input voltage and the corresponding output voltage rather than the turns ratio.

---

[*] This definition is from the IEEE Standard Dictionary of Electrical and Electronic Terms for electronic power transformers. Power and distribution transformers are defined differently.

MAGNETISM AND MAGNETIC CIRCUITS

The relationship between the turns ratio and the voltage ratio is a simple one given by the equation

$$n = \frac{N_{sec}}{N_{pri}} = \frac{V_{sec}}{V_{pri}} \qquad (7\text{–}8)$$

where V_{sec} is voltage across secondary winding and V_{pri} is voltage across primary winding. Equation 7–8 shows an important relationship for transformers and is simply stated as the voltage ratio is equal to the turns ratio.

A transformer is a passive device, so the power taken from the secondary cannot exceed the power applied to the primary. For many basic calculations of transformers, the ideal transformer (100% efficient) is assumed (no power loss). This ideal transformer is one in which load power is equal to power delivered to the primary, which is expressed as

$$P_L = P_{pri}$$

where P_L is power delivered to the load and P_{pri} is power delivered to the primary. The load power is the product of the secondary voltage and secondary current; the power delivered to the primary is the product of the primary voltage and primary current. By substitution,

$$I_{pri} V_{pri} = I_{sec} V_{sec}$$

where I_{sec} is current from the secondary winding and I_{pri} is current to the primary winding. This equation implies if the voltage is stepped up, the current must be stepped down and vice-versa.

SAFETY NOTE

Transformers should be tested and inspected before using them. There should be complete isolation between the primary and secondary windings and between any winding and the case or ground. A cracked power cord should be replaced immediately. When operating, a transformer should not get hot. The output voltage should be near its rated value when operating under load.

EXAMPLE 7–9

Problem
What is the turns ratio of a transformer that has a primary voltage of 120 V and a secondary voltage of 24 V?

Solution

$$n = \frac{V_{sec}}{V_{pri}} = \frac{24 \text{ V}}{120 \text{ V}} = \mathbf{0.2}$$

The turns ratio is less than 1, which is true when the output voltage is less than the input voltage for electronic power transformers.

Question
Is the current higher in the primary or the secondary? Justify your answer.

Applications

Transformers are primarily used to convert an ac voltage from one value to another. Because ac could be transformed, power companies standardized on ac rather than dc early in the development of power grids. Power transformers range in size from small wall units, common with electronic appliances, to huge ones the size of a car used by power companies.

Another important application for transformers is in high-frequency circuits. Air-core and ferrite-core transformers are used in coupling high-frequency signals from one amplifier stage to another and are usually much smaller than power transformers. The windings

CHAPTER 7

are on an insulating shell that is hollow in some cases or constructed of ferrite, as depicted in Figure 7–16. A varnish-type coating covers the wire in the transformer to prevent the windings from shorting together. The entire assembly may be housed in a small can, and many can be "tuned" to a specific high frequency.

FIGURE 7–16

High-frequency transformers.

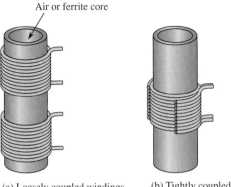

(a) Loosely coupled windings

(b) Tightly coupled windings. Cutaway view shows both windings.

Another type of transformer, called an isolation transformer, has a turns ratio of 1. In this case, the primary and secondary voltages are the same. This type of transformer isolates two circuits from each other so that they do not have a common ground. This is useful in certain measurement applications.

Review Questions

16. How is an iron core indicated on the schematic of a transformer?
17. Why aren't transformers used for changing dc voltages?
18. What is turns ratio?
19. What type of transformer has a turns ratio that is less than 1?
20. What are air-core transformers used for?

7–5 SOLENOIDS AND RELAYS

Solenoids and relays use a coil to convert an electrical current into a magnetic force that moves a ferromagnetic material. Solenoids move the material to perform an action, such as opening a valve. Relays are electrically controlled switches.

In this section, you will learn more about the operation of solenoids, solenoid valves, switches, and relays.

Solenoids

A **solenoid** is a device designed to convert an electrical signal into a mechanical motion. It is basically an electromagnet with a coil wrapped about a magnetic core with a movable portion called a plunger. The plunger is attached to something that needs to be moved. Examples of devices that use solenoids include the gear on a starter motor in a car, change machines, electronic locks, speakers, and electronic valves (called solenoid valves). The basic structure of a solenoid and its schematic symbol are shown in Figure 7–17.

MAGNETISM AND MAGNETIC CIRCUITS

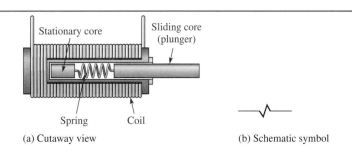

FIGURE 7–17

A basic solenoid.

In the at-rest (or unenergized) state, the plunger is pushed out by the spring. Current through the coil energizes the solenoid. The current sets up an electromagnetic field that magnetizes both the stationary core and the sliding core. This causes the magnetic cores to attract each other, thus retracting the plunger and compressing the spring. As long as there is current in the coil, the plunger remains retracted by the attractive force of the magnetic fields. When the current is cut off, the magnetic field collapses and the force of the compressed spring pushes the plunger back out.

In speakers, the solenoid is a coil that has a current proportional to the signal (music or voice) from a power amplifier. The coil produces a magnetic field that interacts with a permanent magnet and moves back and forth with the signal. This vibrates a paper cone assembly to turn the electrical signal into sound.

Solenoids are rated for the coil voltage, dc resistance, pull-in strength, and wattage rating of the coil. The coil voltage can be either ac or dc. The current depends on the coil resistance but typically is higher when first energized due to inductive effects (described in Section 11–4).

Solenoid Valves

In industrial controls, **solenoid valves** are widely used to control the flow of air, water, steam, oils, refrigerants, and other fluids. Solenoid valves are used in both pneumatic (air) and hydraulic (oil) systems, common in machine controls. Solenoid valves are also common in the aerospace and medical fields. Solenoid valves can either move a plunger to open or close a port or can rotate a blocking flap a fixed amount.

A solenoid valve consists of two functional units: a solenoid coil that provides the magnetic field to provide the required movement to open or close the valve and a valve body, which is isolated from the coil assembly via a leakproof seal and includes a pipe and butterfly valve. Figure 7–18 shows a cutaway of one type of solenoid valve. When the solenoid is energized, the butterfly valve is turned to open a normally closed (NC) valve or to close a normally open (NO) valve.

Solenoid valves are available with a wide variety of configurations including normally open or normally closed valves. They are rated for different types of fluids (for example, gas or water), pressures, number of pathways, sizes, and more. The same valve may control more than one line and may have more than one solenoid to move.

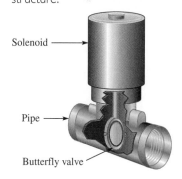

FIGURE 7–18

A basic solenoid valve structure.

Switches

A **switch** is a control element that can open or close one or more lines of a circuit. Each line that is controlled represents a **pole** of the switch; thus, a single-pole switch can open or close one line and a double-pole switch can open or close two lines. The term *pole* represents the movable arm of the switch. The contact points of a switch represent **throws**. The number of throws (contacts) are specified for each pole of the switch. Figure 7–19 shows some basic switches and illustrates these definitions.

241

FIGURE 7–19

Switch definitions.

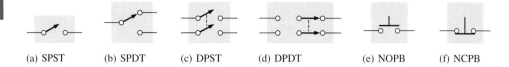

(a) SPST (b) SPDT (c) DPST (d) DPDT (e) NOPB (f) NCPB

The Relay

A **relay** is an electrically controlled switch. Relays differ from solenoids in that the coil is used to open or close electrical contacts rather than provide mechanical movement. All relays have at least one movable contact and one or more fixed contacts. The contacts generally carry a high current to a load. A control voltage energizes the coil; this voltage is generally lower than the voltage that is switched so that electronic switching is simplified.

Figure 7–20 shows the operation of a basic relay with one normally open (NO) fixed contact, one normally closed (NC) fixed contact, and a movable contact, which is usually referred to as the pole (single pole–double throw). When there is no coil current, the armature is held against the upper contact by the spring, thus providing continuity from terminal 1 to terminal 2, as shown in Figure 7–20(a). When the control voltage is applied, the relay is energized and the attractive force of the electromagnetic field pulls down the armature. It makes connection with the lower contact to provide continuity from terminal 1 to terminal 3, as shown in Figure 7–20(b).

FIGURE 7–20

A basic relay structure.

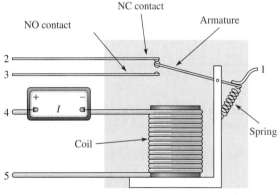
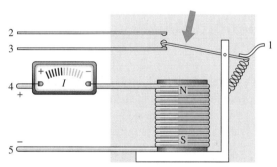

(a) Unenergized: continuity from terminal 1 to terminal 2

(b) Energized: continuity from terminal 1 to terminal 3

A typical small relay is shown in Figure 7–21(a). Parts (b) and (c) of the figure show two ways of drawing the same relay. The symbol shown in Figure 7–21(c), used in many industrial schematics, is called a ladder diagram (covered in the next section). The coil is drawn as a circle. The letters *CR* stand for Control Relay and a number designator is generally shown. The coil is physically separated from the contacts, so a common method of designating a particular set of contacts is to use a two number system. The first number indicates the relay coil associated with the contacts, and the second number designates the contact set. One of the contacts in a set will always be movable. Thus, the designation CR2-1 refers to the first set of contacts on relay number 2.

The voltage and current rating of both the coil and the relay contacts are important specifications for relays. Small relays, such as the one shown in Figure 7–21, are rated for about 5–10 A of contact current. Larger relays, called contactors, are designed to switch high

MAGNETISM AND MAGNETIC CIRCUITS

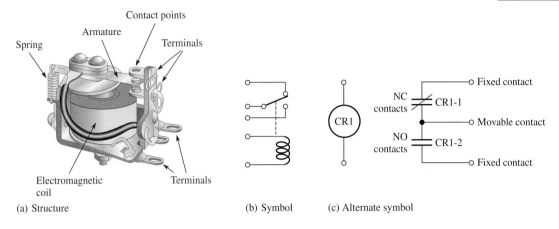

FIGURE 7–21 A typical relay.

(a) Structure   (b) Symbol   (c) Alternate symbol

voltages and high currents. Sometimes a control relay is used to close a contactor, which may be handling a very high current or voltage. The number of movable contacts and number of fixed contacts must be specified for any type of relay. The number of movable contacts is given as the number of poles; the number of fixed contacts on each switch is called the throws. Thus, the relay diagrammed in Figure 7–21(b) and (c) is a single-pole, double-throw (SPDT) relay. If a second movable contact with two more fixed contacts were added to this relay, it would be a double-pole, double-throw (DPDT) relay.

Solid-State Relays

For many applications, solid-state relays perform the same function as electromechanical relays but without the coil and movable contact. A solid-state relay is a transistorized relay with no moving parts. Solid-state relays have certain advantages such as speed, resistance to vibration, and life expectancy. However, electromechanical relays have the advantage of better isolation, ability to withstand high temperatures and surge currents, and capability of switching both ac and dc with the same relay.

Review Questions

21. What is the purpose of a solenoid?
22. What are the two parts to a solenoid valve?
23. If a NO solenoid valve is energized, is it open or closed?
24. What are the two parts of a relay?
25. How many fixed contacts are on each switch of a DPDT relay?

PROGRAMMABLE LOGIC CONTROLLERS 7–6

For many years, industrial control was done by electromechanical relays working in logic circuits. The relay circuits were drawn with a ladder diagram, so named because of its resemblance to a ladder, with sidebars and rungs. Ladder drawings are still used extensively but are associated with modern computerized controllers called programmable logic controllers.

In this section, you will learn how to read and explain a basic ladder diagram.

CHAPTER 7

A **programmable logic controller** or PLC is a specialized computer that is designed for industrial control applications. When PLCs were introduced in the late 1960s, they were designed to replace large blocks of relay control circuits, which were widely used in industry at the time. Although a PLC does not have any relays inside it, the transition from relay logic to electronic control was simplified by simulating these relay circuits in software.

The programs that run PLCs are designed to look exactly like the relay control logic they replaced. Today, PLCs are still programmed by simulating relay logic because it is simple and readily understood by the factory technicians that use it. Different manufacturers of PLCs have various ways of showing logic diagrams; the diagrams shown in this section are generic and do not represent any specific manufacturer. Whole books are devoted to this topic, so the drawings in this section are intended only to introduce the topic. Keep in mind that references to relays are for simulated relays when a PLC is used, but the logic will work with mechanical relays as well.

Ladder Diagrams

A **ladder diagram**, sometimes called a line diagram, is the most commonly used method for showing the logic associated with relay control circuits. Figure 7–22 shows a basic ladder diagram. The two vertical legs of the ladder, labeled L1 and L2 for line inputs, are connected to the supply voltage that operates the relay coils or any loads that are connected across the legs of the ladder. Each "rung" on the ladder is across this control voltage and must have a relay coil (labeled CR) or a load that operates on this supply voltage. Each coil and relay are designed for this supply voltage, so coils and loads cannot be in series. (However, switches and contacts can be in either series or parallel with other switches but never across a load or relay coil.) Notice that each rung of the ladder is numbered. In this case, rung 2 has a light (labeled L1) and rung 3 has a motor (labeled M1). In keeping with standard practice for ladder drawings, switches and other control elements are always drawn on the left and coils or loads are always shown on the right side of a rung. To read a ladder diagram, start at the top rung and work your way down.

FIGURE 7–22

A basic ladder drawing of a relay circuit.

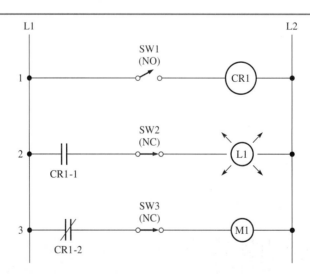

The contacts for a given relay are numbered with two numbers as described in the last section—one for the relay and the second for the contacts. Notice that the line through the contacts indicates they are normally closed (in the unenergized position) whereas the absence of a line indicates that they are normally open.

For the diagram shown in Figure 7–22, when SW1 is closed, the coil for relay CR1 is energized. This action causes normally open (NO) contacts CR1-1 to close, completing a path through SW2 to turn on the light, L1. At the same time, normally closed (NC) contacts CR1-2 is open, which breaks the path through SW3 and turns off the motor, M1. SW2 and SW3 are manually operated switches that are in series with the relay contacts in this example.

EXAMPLE 7–10

Problem
Figure 7–23 shows a partial ladder diagram for a door control. Explain what happens when the OPEN pushbutton switch is momentarily pressed. Assume that contacts CR2-1 remain closed. The coil for relay 2 is not shown.

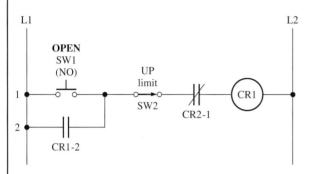

FIGURE 7–23

Solution
The pushbutton switch completes a path through the UP limit switch and closed contacts CR2-1 to the coil of CR1 and it is energized. This causes NO contacts CR1-2 to close. This is called *latching action* because when the pushbutton is released, a path still exists to the relay coil through contacts CR1-1.

Question
What are the two ways that CR1 is no longer energized after pushbutton SW1 is pressed?

In Example 7–10, switches and contacts are shown in series and in parallel. If two switches are in series, both must be closed to provide a path to the relay coil or the load. If switches and contacts are in parallel, either can be closed to provide a path to the load. Combinations of series and parallel switches and relay contacts allow various logical possibilities and are frequently part of the logic required for controlling complicated operations.

Multiple loads (lights, motors) can be shown on a single rung of a ladder diagram, but the loads must be connected in parallel, as illustrated in Figure 7–24. In the case shown,

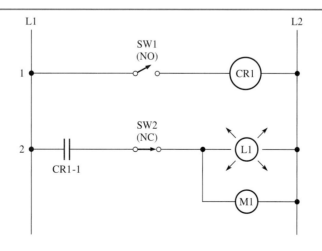

FIGURE 7–24

Two loads on the same rung of a ladder diagram.

both loads will be active together or both will be off. The light and the motor will be active together when SW1 is closed. Of course, the current rating of the contacts must be sufficient to handle both loads.

EXAMPLE 7–11

Problem
Explain how to apply power to the light, L1, in the partial ladder diagram in Figure 7–25. For simplicity, the three coils are not shown.

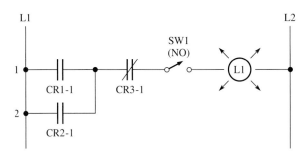

FIGURE 7–25

Solution
Normally open contacts CR1-1 and CR2-1 are in parallel. Therefore, either relay 1 or relay 2 can provide a path to normally closed contacts CR3-1. Relay 3 must *not* be energized or the path will be broken. Normally open SW1 must be closed.

Question
If the circuit is changed so that contacts CR2-1 are normally closed, how is the light turned on?

Review Questions

26. Why is it necessary to have a load or relay coil on every rung of a ladder diagram?
27. How are normally closed contacts indicated?
28. If two switches are shown in parallel, what does this indicate?
29. If two switches are shown in series, what does this indicate?
30. How should you show two loads connected to one rung of a ladder diagram?

CHAPTER REVIEW

Key Terms

Flux Imaginary lines used to describe the magnitude and shape of a magnetic field.

Flux density A value that is proportional to the number of lines of flux that are perpendicular to a given area.

Hysteresis The effect in a magnetic material that occurs when the flux density lags behind the application of the magnetic field intensity.

Ladder diagram A drawing that shows the logic associated with relay control circuits. Two vertical legs are connected to a voltage source with relay coils or other loads connected to the "rungs" of the ladder.

Magnetic domain A group of millions of atoms that have naturally aligned their magnetic fields within certain types of magnetic materials.

Magnetic field intensity The magnetomotive force per unit length.

Magnetization curve A B-H curve for a particular material showing the flux density as a function of the magnetic field intensity.

Magnetomotive force The cause of magnetic flux. It is the product of the number of turns of a conductor and the current in that conductor.

Permeability A parameter that defines the ease with which a magnetic field can be established in a given material.

Relative permeability For a given material, the ratio of absolute permeability (μ) to the permeability of a vacuum (μ_o).

Reluctance The opposition to the establishment of a magnetic field in a material.

Retentivity The ability of a material, once magnetized, to maintain a magnetized state without the presence of a magnetizing force.

Important Facts

❏ Magnetic field lines are used to describe the magnitude and shape of the magnetic field. The imaginary lines are called flux.

❏ Magnetic flux is caused by magnetomotive force, which is proportional to the number of turns of a conductor and the current in the conductor.

❏ Ferromagnetic materials have minute magnetic domains created within their atomic structure by the orbital motion and spin of electrons.

❏ Permeability is a measure of how easily a magnetic field can be established in a material for a given magnetomotive force.

❏ A B-H curve describes the flux that is set up in a material as a function of the magnetic field intensity.

❏ Reluctance is the opposition to the establishment of a magnetic field in a material.

❏ The ability of a magnetized material to maintain a magnetized state without the presence of a magnetizing force is called retentivity.

❏ An equivalent law to Ohm's law for electric circuits operates with magnetic circuits. It states that the flux is directly proportional to the magnetomotive force and inversely proportional to the reluctance.

❏ A transformer is used with ac circuits to change a voltage from one value to another.

❏ Two important devices that use magnetic fields to operate are the solenoid and the relay.

Formulas

Flux density:

$$B = \frac{\phi}{A} \qquad (7-1)$$

Magnetomotive force:

$$F_m = NI \qquad (7-2)$$

Magnetic field intensity:

$$H = \frac{F_m}{l} \tag{7-3}$$

or

$$H = \frac{NI}{l} \tag{7-4}$$

Flux density for a given magnetic field intensity:

$$B = \mu H \tag{7-5}$$

Reluctance:

$$\mathcal{R} = \frac{l}{\mu A} \tag{7-6}$$

Ohm's law for magnetic circuits:

$$\phi = \frac{F_m}{\mathcal{R}} \tag{7-7}$$

Turns ratio and voltage ratio for a transformer:

$$n = \frac{N_{sec}}{N_{pri}} = \frac{V_{sec}}{V_{pri}} \tag{7-8}$$

Chapter Checkup

Answers are at the end of the chapter.

1. Flux density depends on the
    (a) area of a material   (b) number of lines of force
    (c) both (a) and (b)   (d) neither (a) or (b)

2. If the current in a coil is doubled, the magnetomotive force will be
    (a) halved   (b) unchanged
    (c) doubled   (d) quadrupled

3. The quantity that is abbreviated with the letter H is
    (a) flux   (b) flux density
    (c) magnetomotive force   (d) magnetic field intensity

4. The unit for relative permeability is
    (a) weber   (b) weber/At-m
    (c) At   (d) dimensionless

5. A property of a material that opposes a magnetic field is
    (a) reluctance   (b) resistance
    (c) hysteresis   (d) retentivity

6. A magnetization curve is a plot of flux density as a function of
    (a) magnetomotive force   (b) magnetic field intensity
    (c) permeability   (d) path length

7. The ampere-turn is the unit for
   (a) magnetomotive force (b) magnetic field intensity
   (c) flux (d) flux density

8. The unit for flux density is the
   (a) weber (b) weber/meter^2
   (c) tesla (d) Both (b) and (c) are correct.

9. An electronic power transformer that has a turns ratio of 5 is a
   (a) step-up (b) step-down
   (c) not enough information to tell

10. If an air gap is cut into an iron core that has a flux established in it, the flux will
    (a) increase (b) not change
    (c) decrease

11. A device that is a magnetically controlled switch is a
    (a) relay (b) solenoid
    (c) solenoid valve (d) electric door lock

12. On a ladder diagram, two relay coils in series
    (a) may be operated independently
    (b) must have different current
    (c) are numbered the same
    (d) are not allowed

13. On a ladder diagram, two switches in parallel
    (a) may be operated independently
    (b) are controlled by the same action
    (c) are numbered the same
    (d) are not allowed

Questions

Answers to odd-numbered questions are at the end of the book.

1. What is the basic cause of magnetic fields?
2. What magnetic quantity is analogous to the voltage in an electrical circuit?
3. What is the unit for magnetic field intensity?
4. What is a magnetic domain?
5. What is the difference between permeability and relative permeability?
6. What does it mean when a core is said to be saturated?
7. How is reluctance similar to resistance?
8. What is the unit for reluctance?
9. How does residual magnetism occur?
10. When is a high retentivity desirable?
11. What is flux analogous to in an electrical circuit?
12. What is a magnetization curve?
13. What advantage does a smaller air gap have in a motor or generator?
14. What is the schematic symbol for a solenoid?

15. What is a solenoid valve?
16. What is the movable arm of a switch called?
17. What is a contactor?
18. What are advantages of a solid-state relay over an electromechanical relay?
19. Where are switches shown on a ladder diagram?
20. What is meant by the number 3-2 associated with a set of relay contacts on a ladder diagram?
21. What is another name for a ladder diagram?
22. What do the letters CR stand for on a ladder drawing?

PROBLEMS

Basic Problems

Answers to odd-numbered problems are at the end of the book.

1. Calculate the flux in a rectangular-shaped core if the flux density is 75 T and the core is 1.0 cm by 1.0 cm.

2. Figure 7–26 shows the cross section of a magnetic core. Assume each dot represents 1 μWb. Calculate the flux and the flux density for the core.

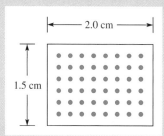

FIGURE 7–26

3. Assume a 600 turn coil is wrapped around a rectangular core. The coil has a current of 650 mA of current. What is the magnetomotive force?

4. If the core in Problem 3 is 5.0 cm by 3.0 cm (average). What is the magnetic field intensity?

5. A rectangular annealed iron core has a 1.0 cm^2 cross-sectional area, a length of 6.0 cm, and a height of 4.0 cm as shown in Figure 7–27. The permeability of annealed iron is 2×10^{-4} Wb/At-m. What is the reluctance of the core?

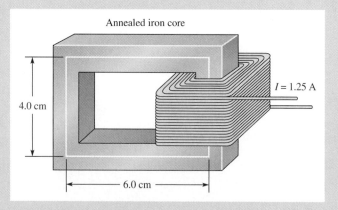

FIGURE 7–27

6. What is the relative permeability of a ferromagnetic core whose absolute permeability is 600×10^{-6} Wb/At-m?

7. For the uniform rectangular annealed iron core shown in Figure 7–27, assume a 300 turn coil is wrapped around it with 1.25 A of current in the coil. The average length and width of the core is 4.0 cm by 6.0 cm. Calculate F_m and H.

8. For the core in Figure 7–27, what flux density is established? The magnetization curve for annealed iron was given in Figure 7–11.

9. Assume you want to establish a flux of 140 μWb in the annealed iron core shown in Figure 7–27. The core has a uniform cross-sectional area of 1 cm^2.

    (a) Determine the mmf required.

    (b) If the current is 1.25 A, how many turns are required?

10. A coil with 500 turns is wound on a closed core that is 0.12 m long. If the current in the coil is 0.8 A, what are the mmf and the magnetic field intensity?

11. A torus made from annealed iron has an inner radius of 1.6 cm and an outer radius of 2.0 cm. Assume the permeability of annealed iron is 2×10^{-4} Wb/At-m. Calculate the reluctance of the torus.

12. A 200 turn coil is wrapped on the torus described in Problem 11 and a current of 1.5 A is in the coil. Find the flux density in the torus.

13. What is the permeability of a material that has a flux density of 1.4 T when the magnetic field intensity is 2000 At/m?

14. What is the relative permeability of the material in Problem 13?

15. What is the flux in a core that has 300 turns with 1.5 A of current if reluctance of the core is 22.5×10^6 At/Wb?

16. What flux is established in a core that has 400 turns with 2 A if the reluctance is 12×10^6 At/Wb?

17. A core consists of a cast steel section and an air gap. To establish a certain flux in the cast steel section requires an mmf of 220 At and an additional mmf of 560 At to establish the flux across the air gap. If the core is wrapped with 500 turns of wire, what current is required?

18. For the ladder circuit in Figure 7–28, list all of the conditions necessary to cause the light (on line 4) to turn on.

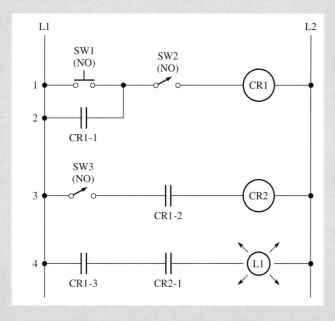

FIGURE 7–28

19. Draw a ladder diagram that uses a switch labeled SW1 to energize CR1. CR1 is to turn off a light on a second line and turn on a light on line 3. The light on line 3 can also be with a manual switch labeled SW2.

Basic-Plus Problems

20. What is the required current to produce a flux density of 1.5 T for the annealed-iron core in Figure 7–29? Notice that the core has a 2.5 mm air gap at one point.

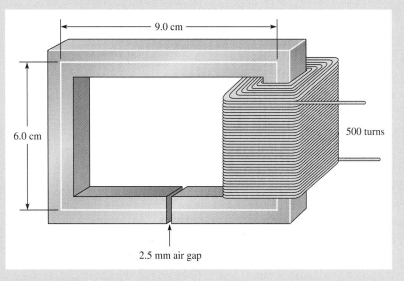

FIGURE 7–29

21. Explain how the CLOSE pushbutton works on the garage door ladder diagram in Figure 7–30. Motor contacts are not shown. When CR2 is active, a motor turns to close the door.

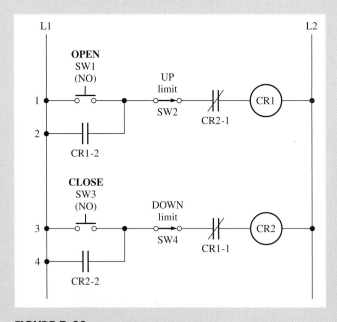

FIGURE 7–30
Garage door control.

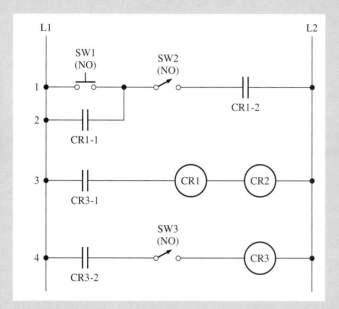

FIGURE 7-31

22. Name three errors in the ladder diagram of Figure 7–31.

23. Draw a ladder diagram that has a pressure switch and temperature switch connected in a way that both switches must be closed to energize a motor starter. Include a manual override switch for the pressure switch but not the temperature switch. Also include a normally closed overload switch that will remove power from the motor if it opens.

24. Draw a ladder diagram that will turn on a lamp if any of three NO pushbuttons are pressed momentarily. The light is to be latched on after any one of the pushbuttons is pressed. The light is turned off with a fourth NC pushbutton.

25. Isolation transformers were briefly mentioned in Section 7–4. Use the Internet or other resources to find more information about isolation transformers, particularly with respect to medical-grade isolation transformers. Prepare a short report on your research.

Example Questions

ANSWERS

7–1: Flux density is halved.

7–2: Flux density is one-fourth.

7–3: No change in the mmf.

7–4: The magnetic field intensity is increased to 2.31×10^3 At/m (assuming the current is maintained at 1.5 A).

7–5: The reluctance is reduced to 12.8×10^6 At/Wb.

7–6: The flux will increase for the same magnetomotive force.

7–7: The flux density will be only slightly lower because the core is saturated.

7–8: The required mmf is 1652 At; the new current is 4.72 A.

7–9: The current is higher in the secondary because the voltage is lower.

7–10: The upper limit switch can be open or relay 2 (not shown) is energized.

7–11: Either relay 1 or relay 2 should be energized. In addition, relay 3 should be unenergized and SW1 closed.

Review Questions

1. Flux is the number of lines; flux density is the number of lines perpendicular to a given area.
2. Weber
3. Weber/meter^2 or tesla
4. Magnetomotive force is the cause of flux in a magnetic circuit; it is the product of number of turns of a conductor and current in those turns.
5. Ampere-turn
6. Magnetic materials have domains, regions where their atoms have aligned magnetic fields; nonmagnetic materials do not.
7. A material with high permeability is relatively easy to set up a magnetic field.
8. Resistance
9. The ability of a material, once magnetized, to maintain a magnetized state.
10. The property of a material to keep a magnetic field after the initiating magnetic force has been removed.
11. Flux is equal to magnetomotive force divided by reluctance.
12. A plot of flux density, B, as a function of magnetic field intensity, H, for a given material.
13. Both quantities are oppositions. Resistance opposes current; reluctance opposes flux. Also, in a series electrical or a series magnetic circuit, oppositions add.
14. Resistance is generally independent of current; reluctance tends to increase with higher flux. Also, there are no true insulators for flux although there are for resistance.
15. The air gap in a magnetic circuit adds a significant amount of reluctance to the path.
16. Two vertical lines between the coils
17. A voltage is induced across the secondary only by a changing flux.
18. The number of turns of the secondary winding divided by the number of turns of the primary winding
19. Step-down
20. High-frequency applications
21. A solenoid converts an electrical signal into mechanical motion.
22. A coil and a valve body
23. Closed
24. The coil and a set of switch contacts
25. Two
26. If there is no load or relay coil, a direct short would exist across the voltage source.
27. A slash is drawn across the contacts.
28. Either switch can be closed to complete the path.
29. Both switches must be closed to complete the path.
30. In parallel

Chapter Checkup

1. (c) 2. (c) 3. (d) 4. (d) 5. (a)
6. (b) 7. (a) 8. (d) 9. (a) 10. (c)
11. (a) 12. (d) 13. (a)

CHAPTER 8

MOTORS AND GENERATORS

CHAPTER OUTLINE

- 8–1 Voltage Induced Across a Moving Conductor
- 8–2 Force on a Current-Carrying Conductor
- 8–3 Generators
- 8–4 DC Motors
- 8–5 AC Motors

INTRODUCTION

In the last chapter, you learned how a magnetic field is produced and were introduced to the basic magnetic quantities of flux, flux density, magnetomotive force, and reluctance. The connection between electricity and magnetism was established early in the nineteenth century with the experiments of Oersted, Faraday, and others. These experiments are basic to understanding the fundamental principles of generators and motors. A generator is a device that converts mechanical energy into electrical energy, and a motor is a device that converts electrical energy into mechanical energy.

Two related principles underlie the operation of all generators and motors. The first principle is that of electromagnetic induction. If there is relative motion between a conductor and a magnetic field such that magnetic field lines are crossed by the conductor, a voltage is induced across the conductor, a process known as *generator action*. The second principle, which is the converse of the first principle, is that if a current is passed through a conductor that is in a magnetic field, a force is induced across the conductor, a process known as *motor action*. These two key ideas are inseparable—all generators and motors are simultaneously generating voltages and experiencing motor forces.

Nearly all of the electrical energy in the United States comes directly from generators. In response, about two-thirds of this electrical energy goes to turning motors. For example, in households, the motors in air conditioners, heat pumps, refrigerators, washing machines, hair dryers, dishwashers, and many other appliances are common applications. In this chapter, you will learn about the principles that underlie generators and motors.

Study aids for this chapter are available at

http://www.prenhall.com/SOE

KEY OBJECTIVES

A section number is given for each objective. After completing this chapter, you should be able to

8–1 Apply Faraday's law to calculate the induced voltage in a moving conductor

8–2 Calculate the force on a conductor moving in a magnetic field

8–3 Explain the operating principles for dc generators and ac generators (alternators)

8–4 Explain how a dc motor converts electrical energy to mechanical motion

8–5 Compare synchronous motors with induction motors and explain how each converts electrical energy to mechanical motion

LABORATORY EXPERIMENTS DIRECTORY

The following exercise is for this chapter.

◆ **Experiment 15**
Constructing a Reed Switch Motor

KEY TERMS

- Faraday's law
- Generator
- Alternator
- Stator
- Rotor
- Armature
- Slip ring
- Brushes
- Armature reaction
- Commutator
- Synchronous motor
- Induction motor
- Squirrel cage
- Slip

CHAPTER 8

Sci Hi SCIENCE HIGHLIGHT

In 1959, physicist Richard Feynman gave an after-dinner talk in which he discussed miniaturization. He related that students at Los Angeles High School had sent a pin to students at Venice High School on which the words "How's this?" were written. V.H.S. students returned the pin; in the dot of the *i* it said "Not so hot."

Today, the science of miniaturization is working with much smaller processes than required for writing on the head of a pin that Feynman described. Nanotechnology is the science of miniaturization on a molecular scale with parts constructed from individual molecules. One interesting area of research involves living cells, called ATPase, that transport energy across cell membranes. The ATPase is a molecular "motor" made from proteins and found in mitochondria, microscopic bodies in the cells of nearly all living organisms. A central protein is equivalent to an electric motor's rotor (the moving part), and three protein channels are equivalent to the stator coils (fixed part) of an electric motor. The central shaft rotates in one direction for converting ATP (adenosine triphosphate) fuel into ADP (adenosine diphosphate) for the cell. It rotates in the opposite direction when it synthesizes ATP from ADP.

Many exciting prospects for nanotechnology have been envisioned. In the future, it may be possible for molecular motors to manipulate tiny amounts of fluids for testing specific compositions of cells or for working with new methods for releasing medications within the body. Some scientists envision nanoelectronic devices to replace silicon electronics, in the relentless pursuit of even smaller and faster electronic devices.

8-1 VOLTAGE INDUCED ACROSS A MOVING CONDUCTOR

Prior to 1831, the only known method for producing a continuous electric current was with batteries. In 1831, Michael Faraday discovered that a voltage is induced across a conductor that moves through a magnetic field. This profound discovery led to the laws for electromagnetic induction.

In this section, you will learn Faraday's law and apply it to calculate the induced voltage in a moving conductor.

Relative Motion

When a conductor such as a wire is moved across a magnetic field, there is a relative motion between the conductor and the magnetic field. Likewise, when a magnetic field is moved past a stationary conductor, there is also relative motion. In either case, this relative motion results in an induced voltage (v_{ind}) in the conductor, as Figure 8–1 indicates. The lowercase *v* stands for instantaneous voltage. The amount of the induced voltage depends on the rate at which the conductor and the magnetic field move with respect to each other: The faster the relative motion, the greater the induced voltage.

Recall from Chapter 7 that a magnetic field can be described in terms of lines of flux. When a conductor is moved in a magnetic field, a voltage is induced *only* if the conductor moves in such a way as to cross magnetic field lines. No voltage is induced when the conductor moves parallel to the lines. Figure 8–2 illustrates the relative motion necessary to induce a voltage in the conductor. As the conductor is moved up or down, it "cuts" magnetic field lines and a voltage is induced across it, as illustrated in Figure 8–2(a). If the conductor moves side to side, parallel to the magnetic field, no voltage is induced, as illustrated in Figure 8–2(b). Of course, if the wire is moved in and out, no lines are cut and no voltage is induced in it in this case.

MOTORS AND GENERATORS

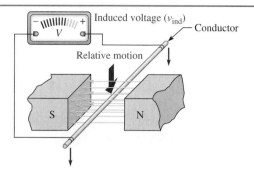

FIGURE 8–1

Relative motion between a conductor and a magnetic field.

(a) Conductor moving downward; magnetic field stationary

(b) Magnetic field moving upward; conductor stationary

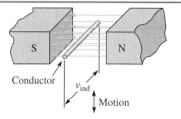 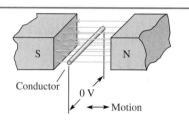

FIGURE 8–2

The relative motion of the conductor must be perpendicular to the magnetic field to induce a voltage.

(a) A voltage is induced across the conductor.

(b) No voltage is induced across the conductor.

Polarity of the Induced Voltage

If a conductor is moved downward, a voltage is induced with the polarity indicated in Figure 8–3(a). As the conductor is moved upward, the polarity is as indicated in part (b) of the figure. If a conductor is first moved down and then up in the magnetic field, a reversal of the polarity of the induced voltage will be observed.

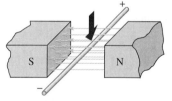 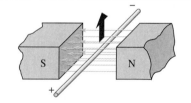

FIGURE 8–3

Polarity of induced voltage depends on direction of motion.

(a) Downward relative motion

(b) Upward relative motion

Induced Current

If a closed path and a load are provided for the moving conductor, as in Figure 8–4(a), then current will be in the load. The polarity of the induced voltage across the load determines the polarity of the current.

Sometimes it is useful to determine the direction of current for a moving conductor. Fleming formulated his right-hand rule when current was assumed to be from positive to negative before the discovery of the electron. Keep in mind that it involves *motion*. Figure 8–4(b) illustrates the right-hand rule. The thumb is in the direction of the motion of the conductor and the index finger points in the direction of the magnetic field (north pole to south pole). The middle finger will point in the direction of conventional current as indicated. If you prefer to think in terms of electron flow, you can use your left hand; your middle finger will represent current electron flow instead of conventional current.

CHAPTER 8

FIGURE 8–4
Relationship between a moving conductor and induced current. The conductor acts like a voltage source because of the induced current.

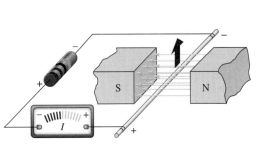
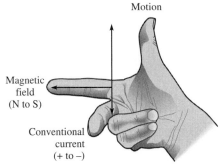

(a) Induced current (i_{ind}) in a load as the conductor moves through the magnetic field.

(b) The right-hand generator rule. The right hand can be used to show the right-angle relationship between the motion of a conductor, the magnetic field, and the induced conventional current (plus to minus).

Faraday's Law for a Straight Conductor

HISTORICAL NOTE

The unit of capacitance, the farad, was named for English physicist and chemist Michael Faraday (1791–1867). He is best remembered for his contribution to the understanding of electromagnetism. He discovered that electricity could be produced by moving a magnet inside a coil of wire, and he was able to build the first electric motor. He later built the first electromagnetic generator and transformer. The statement of the principles of electromagnetic induction is known today as Faraday's law.

Faraday's study of magnetic fields led to a law that bears his name. **Faraday's law** allows you to calculate the voltage that is induced across a wire that is moving relative to a magnetic field. Four factors determine the voltage induced in the conductor:

1. The strength of the magnetic field
2. The length of the conductor exposed to the magnetic field
3. The speed that the conductor is moving relative to the magnetic field
4. The angle the conductor is moving relative to the magnetic field

The mathematical law expressed by Equation 8–1 was developed by Neuman in response to Faraday's observations. The factors listed above can be summarized in equation form as

$$v_{ind} = Blv \sin \theta \qquad (8\text{--}1)$$

where v_{ind} is instantaneous voltage induced across the conductor in V, B is flux density in T, l is length of the conductor that is subjected to the magnetic field in m, v is relative velocity between the conductor and the magnetic field in m/s, and θ is angle between the motion of the conductor and the magnetic field.

The sine function (abbreviated sin) shown in Equation 8–1 is a trigonometric function that is developed from a right triangle. You can find the value of the sine for any angle directly on your calculator (as shown in Example 8–2). This term will be zero if the conductor is moving parallel to the magnetic field lines ($\sin 0° = 0$) and no voltage is induced in the conductor. When the conductor is moving at an angle of 90° to the magnetic field, the induced voltage is at its maximum value ($\sin 90° = 1$). In the special case where the motion is perpendicular ($\perp$) to the direction of the field, Equation 8–1 reduces to

$$v_{ind} = B_\perp lv \qquad (8\text{--}2)$$

where $B_\perp$ indicates that the magnetic field is perpendicular to the direction of motion.

EXAMPLE 8–1

Problem
Assume the conductor in Figure 8–2 is 15 cm long, and the pole face of the magnet is 10 cm long. The flux density is 0.75 T, and the conductor is moved upward at 90° with respect to the magnetic field at a velocity of 1.0 m/s. What voltage is induced across the conductor?

Solution
Although the conductor is 15 cm long, only 10 cm (0.1 m) is in the magnetic field because of the size of the pole faces. Substitute into Equation 8–2.

$$v_{ind} = B_\perp lv = (0.75 \text{ T})(0.1 \text{ m})(1.0 \text{ m/s}) = 0.075 \text{ V} = \mathbf{75 \text{ mV}}$$

Question*
What voltage is induced in the wire if the velocity is 2.0 m/s?

EXAMPLE 8–2

Problem
A wire is moved with a constant 0.8 m/s at an angle of 30° to a magnetic field. The poles of the magnet are 5 cm across. If a voltage of 10 mV is induced across the conductor, what is the flux density of the magnetic field?

Solution
Rearrange Equation 8–1.

$$v_{ind} = Blv \sin \theta$$

$$B = \frac{v_{ind}}{lv \sin \theta} = \frac{10 \text{ mV}}{(0.05 \text{ m})(0.8 \text{ m/s})(\sin 30)} = \mathbf{0.5 \text{ Wb/m}^2 = 0.5 \text{ T}}$$

The calculator keystrokes for the TI-36X are

[1] [0] [EE] [+/–] [3] [÷] [(] [.] [0] [5] [×] [.] [8] [×] [3] [0] [SIN] [)] [=]

Like many calculators, the TI-36X performs the sin function on the displayed value. That is why you press the angle (30°) before the [SIN] key. Some calculators use a different order or set of keystrokes for solving this problem; check your manual if you are not sure.

Question
Assume the wire in this example is moved perpendicular to the magnetic field. What voltage will be induced across the wire?

When a resistive load is connected across a conductor that is cutting magnetic field lines, the induced voltage causes a current in the load in accordance with Ohm's law. The current that results from this action is fundamental to electrical generators as will be seen in Section 8–3. The current at any instant in time is shown as a lowercase italic *i*. This current is a "snapshot" of the current at an instant in time; it depends on the instantaneous voltage that causes it at that instant. If the conductor is moved up and down, the polarity of the induced voltage and current reverses.

---

*Answers are at the end of the chapter.

CHAPTER 8

Faraday's Law for Coils

Faraday's law for a single conductor shows that the induced voltage is related to the length of the conductor exposed to the magnetic field. A simple way to increase the length of the conductor exposed to the field is to form a coil. Rotate the coil in the magnetic field or move the magnetic field in the coil and a voltage is induced across the coil. The idea of a rotating coil in a magnetic field is fundamental to electrical generators.

In 1831, Michael Faraday experimented with generating current by moving a magnet inside a coil of wire and with moving a coil in a fixed magnetic field. In both cases, magnetic field lines are cut by a conductor because of the relative motion, producing a voltage between the leads of the coil. The amount of voltage induced across a coil is determined by two factors:

1. The rate of change of the magnetic flux with respect to the coil
2. The number of turns of wire in the coil

Although these factors are specifically for a coil, they are related to Faraday's law for a straight conductor, given as Equation 8–1. Recall that the induced voltage is determined by the strength of the field (B), the length of the conductor (l), the velocity of the conductor (v), and the angle it is moved (θ). Except for the length of the conductor, these factors are all related to the rate of change of the magnetic field for coils.

The effect of a changing magnetic field can be demonstrated easily in the lab. In Figure 8–5, a bar magnet moves through a coil, thus creating a changing magnetic flux with respect to the coil. In part (a), the magnet is moved at a certain rate, and a certain induced voltage is produced as indicated. In part (b), the magnet is moved at a faster rate through the coil, creating a greater induced voltage.

FIGURE 8–5

Demonstration that the rate magnetic field lines are cut is proportional to the induced voltage.

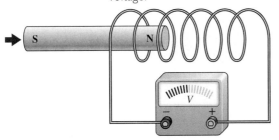
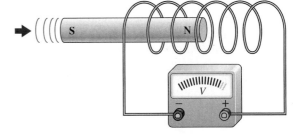

(a) As the magnet moves slowly to the right, its magnetic field is changing with respect to coil, and a voltage is induced.

(b) As the magnet moves more rapidly to the right, its magnetic field is changing more rapidly with respect to coil, and a greater voltage is induced.

The number of turns in a coil is related to the length of the conductor exposed to the magnetic field because the longer the conductor, the more turns you can have. As Equation 8–1 showed, the induced voltage is proportional to the length, l, of the conductor. It follows that the induced voltage for a coil will be related directly to the number of turns in the coil as stated in the second factor for coils. The simple experiment shown in Figure 8–6 demonstrates this idea. In part (a), the magnet moves through the coil and a voltage is induced. In part (b), the magnet moves at the same speed through a coil that has a greater number of turns. The greater number of turns creates a greater induced voltage. Thus for a coil, Faraday's law can be stated as follows:

The voltage induced across a coil of wire that is moving in a magnetic field is proportional to the number of turns in the coil times the rate of change of the magnetic flux.

The number of turns in a coil is related to the induced voltage.

FIGURE 8–6

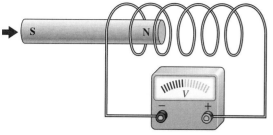
(a) Magnet moves through a coil and induces a voltage.

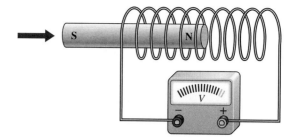
(b) Magnet moves at same rate through a coil with more turns (loops) and induces a greater voltage.

At the time of Faraday, there was no immediate practical application of the law that bears his name. For many years, Faraday's discovery was no more than a scientific curiosity. However, the fact that a voltage can be induced in a conductor moving in a magnetic field had profound implications later for the development of generators and motors. Essentially, all generators convert mechanical motion into electric current by the relative motion between a conductor and a magnetic field. Faraday's law is one of the fundamental laws of all rotating electrical machines.

Review Questions

Answers are at the end of the chapter.

1. What are two requirements for inducing a voltage in a conductor?
2. What are the four factors that determine the amount of voltage induced in a moving conductor?
3. How can a voltage be induced in a stationary conductor?
4. What happens to the polarity of an induced voltage or current if the motion of a conductor in a magnetic field is reversed?
5. What voltage is induced in a conductor moving parallel to magnetic field lines?

FORCE ON A CURRENT-CARRYING CONDUCTOR 8–2

The motion of a conductor cutting lines in a magnetic field causes a voltage to be induced across the conductor. When a complete path is provided, a current is induced in the moving conductor and the conductor experiences a force, which is fundamental to motors.

In this section, you will learn how to calculate the force on a conductor moving in a magnetic field.

Magnetic Field Around a Conductor

When a conductor carries a current, a weak magnetic field is induced by the moving charge, which is perpendicular (at a right angle) to the field. Figure 8–7 shows a cross-sectional view of a conductor. (The complete path is not shown.) The field lines are concentric circles surrounding the wire; there are no north or south poles. Electrons moving in the conductor create the magnetic field shown. The magnetic field is stronger near the conductor, as indicated by the closer spacing of the lines. The arrowheads shown on the magnetic field lines point in the direction that a compass will point if placed near the conductor.

CHAPTER 8

FIGURE 8–7

Magnetic field lines surrounding a current-carrying conductor. The conductor is drawn as a cross-sectional view.

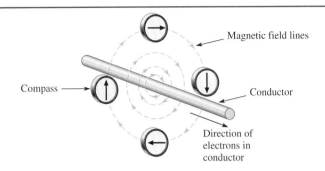

If the conductor is placed in a stationary magnetic field, the stationary magnetic field will interact with the magnetic field produced by the current. This is illustrated in Figure 8–8, which shows a cross-sectional view of the conductor. The result of this interaction causes the field lines to concentrate on one side of the conductor because they reinforce, producing a stronger magnetic field. On the other side of the conductor, the field lines tend to cancel each other, producing a weaker field. Because of this interaction, a force is produced on the conductor itself. The force is downward because the stronger field is above the conductor. This case shows current as the cause and force as the effect.

FIGURE 8–8

The magnetic field lines surrounding a current-carrying conductor interact with the magnetic field from a stationary magnet and produce a force on the conductor. Conventional current is perpendicular to the magnetic field and is into the page.

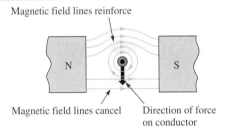

Essentially, all motors convert electrical current into a force to enable work to be done. This is possible because of the force produced along a current-carrying conductor in a magnetic field. The amount of force is determined by the flux density (B), the current in the conductor (I), the length of the conductor exposed to the magnetic field (l), and the angle at which the conductor cuts the magnetic field (θ). The relationship of these factors is known as Biot-Savart's law. Expressed as an equation,

$$F = BIl \sin \theta \tag{8-3}$$

where F is force in newtons (N), B is flux density in teslas (T), I is current in amperes (A), l is length of the conductor in the magnetic field (m), and θ is angle between the conductor and the magnetic field lines.

In the special case where the conductor is perpendicular to the magnetic field lines, Equation 8–3 reduces to

$$F = B_\perp Il \tag{8-4}$$

where $B_\perp$ indicates that the magnetic field is perpendicular to the conductor.

The relationship between the force, the magnetic field, and the current in Equation 8–4 can be shown with the left-hand rule for motors. Figure 8–9 shows the relationship using the left hand. The three variables are always at right angles to each other as indicated. Notice that two of the variables (magnetic field and current) are assigned to the same fingers as the generator law shown in Figure 8–4(b). For the motor law, the thumb represents force, whereas with generators it represents motion. The magnetic field is again from north pole to south pole. Also, current is shown as conventional current, plus to minus. If you prefer to show electron flow, use the right hand.

MOTORS AND GENERATORS

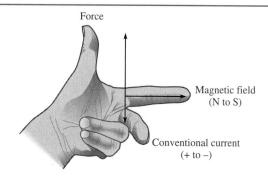

FIGURE 8–9
The left-hand motor rule. The left hand can be used to show the right-angle relationship between the force on a conductor, the magnetic field, and the induced conventional current flow (plus to minus).

EXAMPLE 8–3

Problem
A 10 cm section of wire is exposed to a magnetic field from a strong permanent magnet of 0.5 T at a right angle to the wire. If the wire carries a current of 25 A, what is the force due to the magnetic field?

Solution
You can use Equation 8–4 because the field is perpendicular to the wire.

$$F = B_\perp I l = (0.5 \text{ T})(25 \text{ A})(0.1 \text{ m}) = \mathbf{1.25 \text{ N}}$$

Notice that the force (1.25 N) on a single conductor is small, despite the fact that the magnetic field is fairly strong (0.5 T) and the current is high (25 A).

Question
Assume the wire is moved such that it is at an angle of 60° to the magnetic field lines. What is the force on the wire?

One of the applications of Equation 8–4 is to use it to determine the flux density, B, of a magnet. The force, current, and length of a conductor can be measured directly. The equation is rearranged to solve for B in the next example.

EXAMPLE 8–4

Problem
A 0.20 m wire is placed at a right angle to a uniform magnetic field. The wire carries a current of 40 A, and a force of 15 N is measured. What is the flux density?

Solution
Use Equation 8–4 because the field is perpendicular to the wire. Rearranging the equation,

$$B_\perp = \frac{F}{Il} = \frac{15 \text{ N}}{(40 \text{ A})(0.20 \text{ m})} = \mathbf{1.88 \text{ T}}$$

This is a relatively high flux density that forms a strong electromagnet such as found in a small generator.

Question
Assume the current is reduced to 30 A. What is the force on the wire?

Magnetic Field Around Parallel Conductors

For most practical electrical circuits, the magnetic field surrounding a wire is too small to be noticed. Although the magnetic field is small near a single conductor, it can be made stronger by increasing the current in the conductor. Another way to increase the magnetic field near a wire is to place it parallel to another current-carrying conductor. This causes the magnetic field from one conductor to interact with the magnetic field from the other. If the currents in the two conductors are in the same direction, the field surrounding the conductors is stronger because of the addition of the fields from each conductor; however, between the conductors, the fields tend to cancel. This is illustrated in Figure 8–10(a). If the current is in opposite directions in the two conductors, the field between the conductors tends to intensify. The intensified field produces a magnetic force that pushes the conductors apart, as illustrated in Figure 8–10(b).

FIGURE 8–10

The magnetic fields from parallel conductors interact with each other. The dot represents the point of an arrow coming out of the paper; the x represents the feathers of an arrow going into the paper.

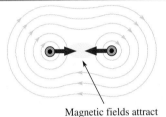
Magnetic fields attract
(a) When current is the same direction, the magnetic field between the conductors tends to cancel.

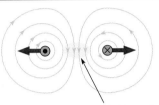
Magnetic fields repel
(b) When current is the opposite direction, the magnetic field between the conductors is reinforced, causing the conductors to be pushed apart.

As in the case of the stationary field, the force is caused by the interaction of the magnetic fields. You may have heard the noise from high-tension wires vibrating. The changing force is due to alternating current; this creates the vibrations and noise you hear.

Magnetic Field Around a Coil

As you know, a current in a conductor has a magnetic field associated with it that is proportional to the current. For most practical applications, the magnetic field around a single wire conductor that carries a current is generally not strong enough. However, by wrapping the wire into coil, the coil acts as a series of parallel conductors with the current in the same direction. As a result, the magnetic field surrounding the coil intensifies, depending on the number of turns. Each turn adds to the total magnetic field of the coil, so with more turns the field gets stronger. The coil will have a north pole and a south pole, similar to a bar magnet, as illustrated in Figure 8–11.

FIGURE 8–11

The magnetic field around a coil.

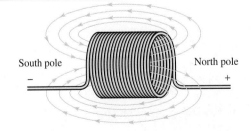

If a piece of soft iron is placed inside the coil, then the magnetic domains in the iron align with the field lines of the coil, creating an even stronger magnet for as long as the current is switched on. This is how an **electromagnet** is formed and is the reason solenoids and other

MOTORS AND GENERATORS

magnetic devices almost always have a magnetic core material. Small electromagnets are used in manufacturing, especially with robotic controls because they can hold or release ferromagnetic materials easily. Figure 8–12 shows small electromagnets useful in material handling.

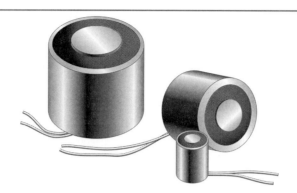

FIGURE 8–12
Round electromagnets for material handling.

Review Questions

6. What is the shape of the magnetic field lines around a single conductor?
7. What determines the strength of a magnetic field around a single conductor?
8. What is the direction of the forces that act on two parallel conductors that have current in the same direction?
9. If the current in a coil reverses direction, what happens to the poles of the electromagnet?
10. What are three factors that control the strength of an electromagnet?

GENERATORS 8–3

A voltage is induced across a conductor when there is relative motion between the conductor and a magnetic field. Electrical generators are practical applications of this principle.

In this section, you will learn about the operating principles for dc generators and ac generators (alternators).

SAFETY NOTE
The National Electrical Code (NEC) includes many standards for generators and motors. For example, because motors draw much more current at startup than when running, motor circuits are often designed to allow a temporary overload at startup. Overload protection devices are but one of many specifications for motors and generators written into the NEC.

An electrical **generator** is a device that converts mechanical work to electricity. Recall that a voltage is induced in a conductor when there is relative motion between the conductor and a magnetic field. This relative motion is the operating principle behind all electrical generators; for almost all generators the motion is rotational. The output can be either dc or ac, but ac generators are usually referred to as alternators. The word *alternator* is a contraction of the words *altern*ating current gener*ator*.

In practical generators, either the magnetic field is rotated or the conductor is rotated. In many smaller generators, the magnetic field is stationary and the conductor rotates. This type is called a *stationary field generator* or *stationary field alternator*. A second type, called a *rotating field generator* or *rotating field alternator*, is more widely used for higher power applications and in cases where the application requires dc (as in cars). In this type, the magnetic field rotates and the conductor is stationary. From a physics standpoint, it does not matter if the magnetic field or the conductor moves; either way a voltage is induced across the conductor. From a practical construction point of view, there are certain advantages to one type or the other, as you will see.

Circular Motion of a Conductor in a Stationary Field

When a conductor rotates at a constant speed within a magnetic field, a sinusoidal voltage is generated, as illustrated in Figure 8–13(a). Circular motion is found in virtually all electrical generators. Starting at the bottom (position *A*), the conductor is moving parallel to the magnetic field lines, so no voltage is induced across it. As it moves toward position *B*, it cuts field lines at a faster rate until it is moving perpendicular to the field lines. At this point, the maximum voltage is induced across it. As it rotates toward position *C*, lines are cut at a slower rate until it is moving parallel to the lines and again induces 0 V. As the conductor moves to position *D*, it is again cutting lines at a faster rate; but the motion is opposite to the direction between positions *A* and *C*, so the polarity is reversed. When it reaches position *D*, it is again traveling perpendicular to the field lines, so the induced voltage is a maximum with the opposite polarity. Then it returns to the starting point, and the process repeats. A graph of the voltage traces out a voltage waveform that has the same shape as the sine wave. This is illustrated in Figure 8–13(b).

FIGURE 8–13

A conductor moving in a circular path in a magnetic field has a sinusoidal voltage induced across it.

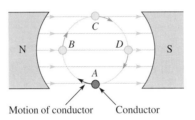
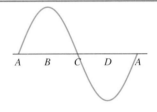

(a) Circular motion of a conductor in a constant magnetic field.

(b) The induced voltage has the shape of a sine wave.

Circular Motion of a Field Near a Stationary Conductor

As in the case of a moving conductor, a magnetic field can be rotated near a conductor to produce a voltage with the shape of a sine wave, as illustrated in Figure 8–14. A permanent magnet that is rotated about its axis will sweep magnetic field lines through a stationary conductor that is perpendicular to the flux. In this case, the conductor is a single loop of a coil that is within an iron frame to concentrate the magnetic field lines from the permanent magnet. The changing magnetic field causes a sinusoidal voltage to be induced in the conductor.

FIGURE 8–14

A rotating magnet will generate a sinusoidal voltage in a conductor. The conductor shown is a single loop of a coil that is perpendicular to the magnet.

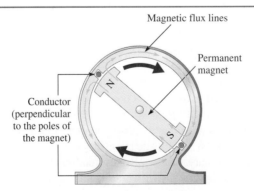
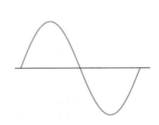

(a) Rotating magnet sweeps by a conductor

(b) Voltage waveform induced in the conductor

MOTORS AND GENERATORS

Alternators

An **alternator** is an electrical generator that has ac as its output. As you know, a voltage is generated when there is relative motion between a conductor and a magnetic field. Alternators take advantage of this principle to convert mechanical energy to electrical energy. The *stat*ionary portion of the alternator is called the **stator**; the *rot*ating portion is called the **rotor**. Parts of the stator include a frame, poles, field windings, and other nonmoving parts. The rotor is spun by energy supplied from an external source, such as a diesel engine, called a **prime mover**. All generators are really energy conversion devices; they simply convert the energy of a prime mover into electrical energy.

Figure 8–15 shows a greatly simplified drawing of a stationary field alternator that shows the key electrical features; the mechanical support structure is not shown. The simplified drawing shows two poles of a magnet—a north and a south pole. Thus, the configuration shown is called a two-pole alternator. A coil of wire is between the two poles (represented by a single loop), and the coil is turned is turned by applying a force to it. Recall that a single conductor rotating in a magnetic field will produce a sinusoidal voltage wave across the ends. By forming the wire into a one-turn coil, the voltage is twice that of a single conductor because twice as much wire is exposed to the magnetic field. To increase the exposed length even more, the single loop is formed into a multiturn coil that rotates in the magnetic field. The conductive coil that intercepts the magnetic field is called the **armature coil**. The **armature** is defined as that part of an alternator in which a voltage is induced due to the relative motion between a conductor and a magnetic field. For a stationary field alternator, the armature is the rotor; for a rotating field alternator, the armature is the stator.

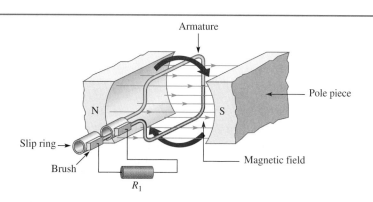

FIGURE 8–15

Simplified alternator. The armature of an actual alternator has many turns on an iron core.

For the stationary field alternator, the ends of the rotating coil will have an induced voltage across them as the armature spins. Each end of the rotating coil is connected to a solid circular ring called a **slip ring**. The slip rings rotate with the armature and are part of the electrical path for the rotor. Stationary conductors called **brushes** make contact with the moving slip rings. The brushes are constructed from a conductive material (typically carbon or carbon-graphite) that connect the rotating loop to the external circuit. The combination of the slip rings and brushes converts the rotating motion of the armature to a stationary source of ac.

Very small alternators, such as used to power the display on an exercise machine, use a permanent magnet to obtain a field for the rotating armature coil. For purposes of simplification, a permanent magnet is shown in Figure 8–15, but electromagnets are more frequently used in larger alternators because they can produce a higher flux density. Recall that an electromagnet is formed by a coil with a direct current in it. The coils that are formed for providing the magnetic field are appropriately named **field coils**. Alternators may use an external source to provide the necessary dc to the field coils or may use residual magnetism in the poles for the initial magnetic field. Then, as the current from the generator increases, a portion of it is converted to dc and connected through the field coils to build the magnetic field to its full value.

CHAPTER 8

Magnetic Structure

Recall from Chapter 7 that an air gap in a magnetic path has high reluctance. In practical generators and motors, it is necessary to reduce the reluctance as much as possible in order to have a strong magnetic field. To accomplish this, the rotor includes a magnetic material, which is part of the magnetic path, as shown in Figure 8–16 for a two-pole stationary field generator or motor. The field shown is that produced by the field coils with no armature current. The only air gap is between the rotating armature and the field poles, necessary for mechanical reasons. The spacing between the armature and field poles is kept to a minimum to minimize the reluctance.

FIGURE 8–16

Magnetic path for a two-pole alternator produced by the field poles.

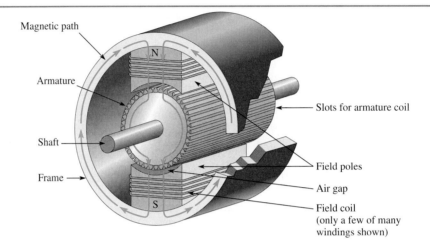

As mentioned, when electromagnets are used for the magnetic field, the field is produced by the field coils, which are wrapped around the poles. These create the magnetic field for the armature coil. The armature is also slotted to allow room for the armature coil windings. Notice that the frame is part of the magnetic circuit, so it is constructed using a ferromagnetic material. This is designed to produce a low reluctance path, which is necessary to have a high flux. Some parts of an alternator such as bearings, end caps, and brushes are omitted from Figure 8–16 for clarity.

A four-pole alternator is similar in design to a two-pole alternator, but the locations of the north and south poles are no longer directly across from each other; rather they are at 90° to each other. Figure 8–17 illustrates a four-pole alternator and its magnetic path.

FIGURE 8–17

Magnetic path for a four-pole alternator or motor produced by the field poles. Two of the four paths are shown.

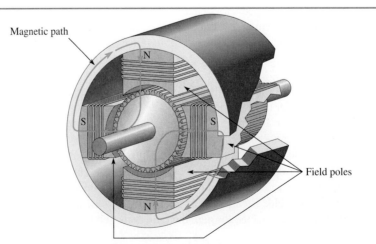

Alternator Frequency

In a two-pole alternator, one cycle is generated for each revolution. With a four-pole alternator, each time the armature makes one revolution it crosses two complete north-south magnetic poles. Thus, for a four-pole alternator, there are two electrical cycles for each revolution. The number of poles and the speed of the rotor determine the output frequency in accordance with the following equation:

$$f = \frac{Ns}{120} \qquad (8\text{--}5)$$

where f is frequency in hertz, N is number of poles, and s is speed in rpm of the rotor.

The number of poles in an alternator can vary significantly. Two-pole alternators are common with very high-speed alternators such as those driven with a turbine. Slower-speed alternators typically have four or more poles — in some cases as many as 100.

EXAMPLE 8–5

Problem
Assume an alternator rotates at 1800 rpm and has four poles. What is the output frequency?

Solution
Substitute into Equation 8–5.

$$f = \frac{Ns}{120} = \frac{(4)(1800 \text{ rpm})}{120} = \mathbf{60 \text{ Hz}}$$

Question
If the frequency drops to 50 Hz, what is the speed of the rotor?

Armature Reaction

As the rotating armature in a stationary field generator turns, a current that is determined by Ohm's law is delivered to the load. This armature current has a magnetic field of its own that interacts with the magnetic field set up by the field poles. The magnetic field set up by the armature is at a 90° angle from that produced by the field poles. The interaction of the two magnetic fields produces a distorted magnetic field in the magnetic assembly, as indicated in Figure 8–18. This distorted field tends to be rotated from its original position. The effect of an additional field created by the armature current is called **armature reaction**, and it creates problems for the designer of generators and motors, such as sparking at the brushes.

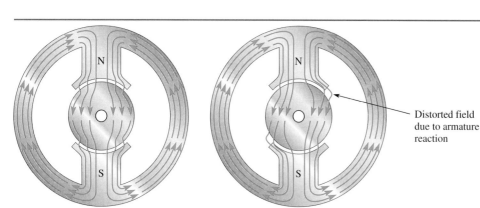

FIGURE 8–18

Current in the armature distorts the magnetic field in a stationary field generator or motor.

(a) Field lines with no armature current

(b) Field lines with armature current

Armature reaction can be minimized by several means. One common method is to add another set of small magnetic poles, called *interpoles,* to the magnetic structure perpendicular to the main field poles. They are wired in series with the armature, but in a direction to oppose the armature contribution to the field. Thus, as the armature field increases, the interpole field increases to counteract it.

Basic DC Generator

The inherent waveform from a spinning loop of wire is an alternating current. A dc generator must provide current in one direction only. The process of converting ac to dc is called **rectifying**. DC generators still generate an ac waveform in the armature, which is converted to pulsating dc by mechanical means. Smoothing the resulting waveform can be done with additional components.

Figure 8–19 is a simplified drawing of a dc generator. The mechanical conversion from ac to dc is done by dividing the slip rings into two parts (for a two-pole generator). This "segmented" slip ring is called a **commutator**. In the basic two-pole arrangement, each end of the rotating loop is connected to one-half of the commutator. The commutator reverses the connection to the external circuit just as the induced voltage on the loop changes polarity. As the loop is rotated in the magnetic field, the split commutator ring also rotates. Each half of the split ring rubs against brushes as in the ac generator. The loop changes polarity just as the split rings make contact with the opposite set of brushes. Thus, the output is pulsating dc called a *rectified sine wave.* Essentially, the only difference between an alternator and a dc generator is that the slip rings present in the alternator have been replaced with the commutator in a dc generator.

FIGURE 8–19

Simplified dc generator.

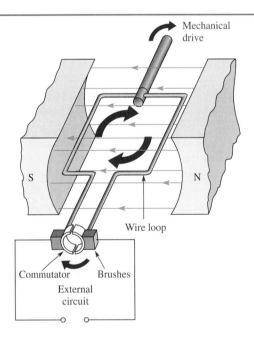

The process of generating a rectified sine wave is shown graphically in Figure 8–20. As the loop rotates through the magnetic field, it cuts through the flux lines at varying angles. At position *A* in its rotation, the loop of wire is effectively moving parallel with the magnetic field. Therefore, at this instant, the rate at which it is cutting through the magnetic flux lines is zero. As the loop moves from position *A* to position *B,* it cuts through the flux lines at an increasing

MOTORS AND GENERATORS

As the single loop rotates in the magnetic field, a pulsating voltage is generated.

FIGURE 8–20

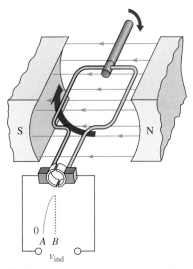

Position B: Loop is moving perpendicular to flux lines, and voltage is maximum.

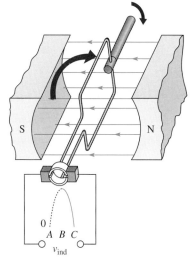

Position C: Loop is moving parallel with flux lines, and voltage is zero.

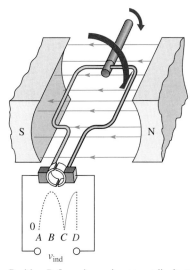

Position D: Loop is moving perpendicular to flux lines, and voltage is maximum.

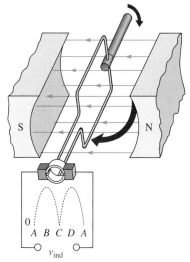

Position A: Loop is moving parallel with flux lines, and voltage is zero.

rate. At position B, it is moving effectively perpendicular to the magnetic field and thus is cutting through a maximum number of lines. As the loop rotates from position B to position C, the rate at which it cuts the flux lines decreases to minimum (zero) at C.

At position C, the polarity of the induced voltage to the output circuit is switched by the split ring commutator because the ends of the loop will now make contact with the opposite set of brushes. This causes the output to repeat the pattern that was observed in the first half of the revolution. As the loop rotates from position C to position D, the rate at which the loop cuts the flux lines increases to a maximum at D and then back to a minimum again at A. Thus, a second pulse is observed at the output. For each complete revolution of the loop, two output pulses are observed, as illustrated in Figure 8–21.

CHAPTER 8

FIGURE 8–21

Each rotation of the loop produces two pulses. Three rotations are shown.

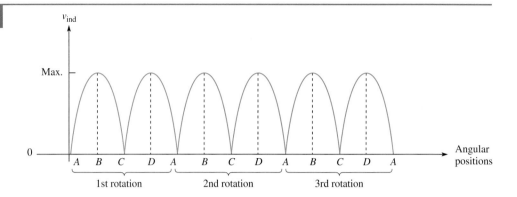

Three-Phase AC

The ac discussed to this point is single phase (abbreviated 1ϕ), meaning there is one voltage with respect to ground that varies sinusoidally. More useful in industrial applications is **three-phase** power (abbreviated 3ϕ), in which three wires carry voltages that are out of phase with each other by one-third of a cycle. The power company uses four conductors to transmit three voltages plus ground. Three-phase ac is illustrated in Figure 8–22. Phase-1 is a sinusoidal voltage that can be the reference for starting the other two phases. Phase-2 is shifted by 120° from phase-1 and phase-3 is shifted 240° from the first. Single-phase power is reasonable for household and other small power applications but three-phase power is much more efficient than single-phase power for driving heavy electrical loads.

FIGURE 8–22

Three-phase ac.

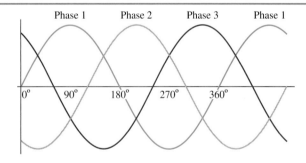

Three-Phase Alternators

Alternators that produce three-phase ac are of the rotating field type because they offer advantages over stationary field types. The principal advantage is that the slip rings handle a much lower current at a lower voltage for a given output than the stationary field type. This extends the life of the slip rings and brushes. Another advantage is that the load can be connected directly to the output. Three phases are also used in automobile alternators because it is easier to convert to regulated dc, necessary for operating automotive electrical systems. The conversion to dc is accomplished with solid-state diodes, contained within the same housing as the rest of the alternator.

The stator of a three-phase alternator contains three poles each 120° apart. Each has a separate but identical winding—in this case, there are three armature windings. The three armature windings on the stator develop voltages that are independent of each other but are 120° out of phase. Within the stator is a rotor that is composed of one or more field magnets and the associated field windings*. The field coils require dc, which

---

* Some specialized designs use a permanent magnet for the rotating field, eliminating the need for slip rings, brushes, and a field winding. The disadvantage is that they are more difficult to regulate.

MOTORS AND GENERATORS

is generally supplied from an external source called an **exciter**. It is brought to the rotor via slip rings and brushes. The three output voltages of the alternator are controlled by the exciter current. In an automobile, this is the job of the voltage regulator. The exciter can be an external dc supply or it can be a dc generator mounted on the same shaft as the alternator.

The armature coils on a three-phase alternator are connected in either of two basic configurations known as either a wye (think of the letter Y) or a delta (like the Greek letter Δ). The wye configuration is more popular and is illustrated in Figure 8–23(a). It uses a single lead from each coil as a common lead. The delta connection has all three coils connected in series and is shown in Figure 8–23(b). The resulting connection looks like a triangle, hence the name delta (from the Greek letter Δ). Although the normal configuration of the delta connection uses only three wires, a variation of the delta connection, called a four-wire delta, uses a center tap on one coil as ground or neutral.

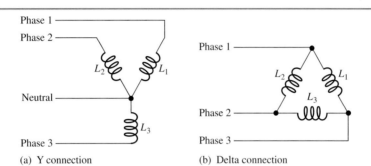

FIGURE 8–23

Delta and wye connections for a three-phase alternator.

Review Questions

11. What are the two mechanical parts of all alternators and generators called?
12. What class of generators has a moving armature?
13. What is the principal difference between an ac and a dc generator?
14. What is the advantage of a three-phase alternator in an automobile?
15. How is the ac from an alternator turned into dc in an automobile?

DC MOTORS 8–4

Motors convert electrical energy to mechanical energy by taking advantage of the force produced on a conductor that is carrying a current in the presence of a magnetic field. A dc motor operates from a dc source and may use either an electromagnet or a permanent magnet to supply the field.

In this section, you will learn about dc motors and how they convert electrical energy to mechanical motion.

Basic Operation

Motor action is the result of the interaction of magnetic fields. In a dc motor, the rotor field interacts with the magnetic field set up by current in the stator windings. The rotor in all dc motors contains the armature winding, which sets up a magnetic field. The rotor moves because of the attractive force between opposite poles and the repulsive force between like poles, as illustrated in the simplified diagram of Figure 8–24(a). The rotor moves because

FIGURE 8–24

Simplified diagram of armature rotation in a dc motor.

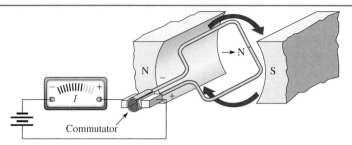

(a) The north pole of the rotor is attracted to the south pole of the field and vice versa.

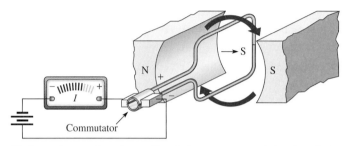

(b) Just as the opposite poles are aligned, the commutator switches the polarity, and unlike poles are next to each other so they repel.

of the attraction of its north pole with the south pole of the stator (and vice-versa). As the two poles near each other, the polarity of the rotor current is suddenly switched by the commutator, reversing the magnetic poles of the rotor. As in the case of the dc generator, the commutator serves as a mechanical switch to reverse the current in the armature just as the unlike poles are near each other. This is illustrated in Figure 8–24(b). Like poles are now near each other, so they repel, continuing the rotation of the rotor.

Back EMF

When a dc motor is first started, a magnetic field is present from the field windings. Armature current develops another magnetic field that interacts with the one from the field windings and starts the motor turning. The armature windings are now spinning in the presence of the magnetic field, so generator action occurs. In effect, the spinning armature has a voltage developed across it that opposes the original applied voltage. This self-generated voltage is called **back emf** (electromotive force). The term *emf* was once common for voltage but is not favored because voltage is not a "force" in the physics sense, but back emf is still applied to the self-generated voltage in motors. Back emf, also called *counter emf*, serves to significantly reduce the armature current when the motor is turning at constant speed.

Motor Ratings

Some motors are rated by the torque they can provide, others are rated by the power they produce. Torque and power are important parameters for any motor. Although torque and power are different physical parameters, if one is known, the other can be obtained.

The concept of torque was introduced in Section 1–3. Recall that torque tends to rotate an object. In a dc motor, the torque is proportional to the amount of flux and to the armature current. Torque in a dc motor can be calculated from the equation

$$T = K\phi I_A \qquad (8\text{–}6)$$

where T is torque in N-m, K is a constant that depends on physical parameters of the motor, ϕ is magnetic flux in Wb, and I_A is armature current in A.

Torque is easy to measure for a motor with a setup known as a **prony brake**, named after Gaspard de Prony, its inventor. In a typical arrangement, the motor turns a drum, which is connected to a friction belt, as shown in Figure 8–25. The belt acts as a load on the motor. As the motor turns against the brake, it tends to rotate the arm in the clockwise direction (in the setup illustrated) about the pivot. The scale and arm restrain the clockwise motion with an equal but opposite counterclockwise torque. Thus,

$$T = Fl = fr$$

where T is torque developed by motor in N-m, F is force on the scale in N, l is length of moment arm in m, f is force exerted by motor in N, and r is radius of drum in m. By measuring the torque, you can calculate power.

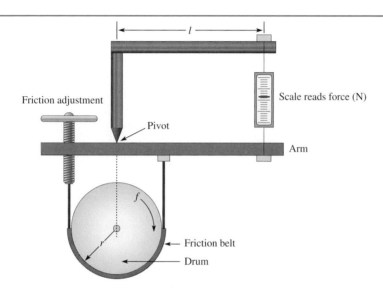

FIGURE 8–25
Prony brake arrangement for measuring motor torque.

Recall that power is defined as the rate of doing work. To calculate power from torque, you must know the speed of the motor in rpm for the torque that you measured. The equation to determine the power, given the torque at a certain speed, is

$$P = 0.105Ts \qquad (8\text{--}7)$$

where P is power in W, T is torque in N-m, and s is speed of motor in rpm.

EXAMPLE 8–6

Problem
What is the power developed by a motor that turns at 350 rpm when the torque is 3.6 N-m?

Solution
Substitute into Equation 8–7.

$$P = 0.105Ts = 0.105(3.6 \text{ N-m})(350 \text{ rpm}) = \mathbf{132\ W}$$

Question
There are 746 W in one hp. What is the hp of this motor under these conditions?

One characteristic of dc motors is that if they are allowed to run without a load, the torque can cause the motor to "run away" to a speed beyond the manufacturer's rating. Therefore, dc motors should always be operated with a load to prevent self-destruction.

Series DC Motors

The series dc motor has the field coil windings and the armature coil windings in series. A schematic of this arrangement is shown in Figure 8–26. The internal resistance is generally small and consists of field coil resistance, armature winding resistance, and brush resistance. As in the case of generators, dc motors may also contain an interpole winding, as shown, and current limiting for speed control. In a series dc motor, the armature current, field current, and line current are all the same.

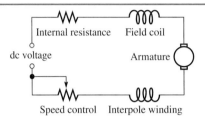

FIGURE 8–26

Simplified schematic of a series dc motor.

As you know, magnetic flux is proportional to the current in a coil. The magnetic flux created by the field windings is proportional to armature current because of the series connection. Thus, when the motor is loaded, armature current rises, and the magnetic flux rises too. Recall that Equation 8–6 showed that the torque in a dc motor is proportional to both the armature current and the magnetic flux. Thus, the series-wound motor will have a very high starting torque when the current is high because flux and armature current are high. For this reason, series dc motors are used when high starting torques are required (such as a starter motor in a car).

The plot of the torque and motor speed for a series dc motor is shown in Figure 8–27. The starting torque is at its maximum value. At low speeds, the torque is still very high, but drops off dramatically as the speed increases. As you can see in Figure 8–27, the speed can be very high if the torque is low; for this reason, the series-wound dc motor always is operated with a load.

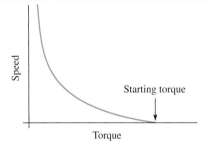

FIGURE 8–27

Torque-speed characteristic for a typical series dc motor.

Shunt DC Motors

A shunt dc motor has the field coil in parallel with the armature, as shown in the equivalent circuit in Figure 8–28. In the shunt motor, the field coil is supplied by a constant voltage source, so the magnetic field set up by the field coils is constant.* The armature resistance and the back emf produced by generator action in the armature determine the armature current.

The torque-speed characteristic for a shunt dc motor is quite different than for a series dc motor. When a load is applied, the shunt motor will slow down, causing the back emf to

---

* The exception is if resistive speed control is added to the field windings.

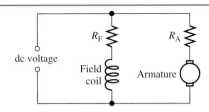

FIGURE 8–28

Simplified schematic of a shunt dc motor.

be reduced and the armature current to increase. The increase in armature current tends to compensate for the added load by increasing the torque of the motor. Although the motor has slowed because of the additional load, the torque-speed characteristic is nearly a straight line for the shunt dc motor as shown in Figure 8–29. At full load, the shunt dc motor still has high torque.

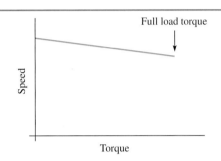

FIGURE 8–29

Torque-speed characteristic for a typical series dc motor.

Compound DC Motors

A third type of dc motor is the compound motor. The coils in a compound motor are wound in a series-parallel combination. The compound motor combines the strengths of the series and shunt motors.

Review Questions

16. What is the purpose of the commutator in a dc motor?
17. What creates back emf?
18. How does back emf affect armature current as the motor comes up to speed?
19. What quantities are necessary to measure the power of a motor?
20. What type of dc motor has the highest starting torque?

AC MOTORS 8–5

In ac motors, the stator field is not stationary but rotates because of the nature of the current in the windings.

In this section, you will compare synchronous motors with induction motors and be able to explain how each converts electrical energy to mechanical motion.

AC Motor Classification

AC motors are generally classified as synchronous motors or induction motors. These classifications are related to differences in the rotor. The stator windings are essentially identical in both types. A **synchronous motor** is one with a rotor that acts as a rotating magnet that moves in sync (at the same rate) with the rotating magnetic field of the stator. An **induction motor** is one that achieves excitation to the rotor by transformer action (or induction). A third category, the *universal motor*, is sometimes included as an ac motor but is basically a specially designed dc motor, complete with commutator and brushes, that can run on either ac or dc. Both synchronous and induction motors are available as either single-phase or three-phase motors. Smaller (fractional horsepower) motors and motors for household use (fans, refrigerators, washing machines) are single-phase and usually are induction motors. Larger motors for industrial applications are usually three-phase and may be either induction or synchronous motors.

Rotating Stator Field

Both synchronous and induction motors have a similar arrangement for the stator winding, which allows the magnetic field of the stator to rotate. The rotating stator field is equivalent to moving a magnet in a circle except that the rotating field is produced electrically, with no moving parts.

How can the magnetic field in the stator rotate if the stator itself does not move? The rotating field is created by the changing ac itself. Let's look at a rotating field with a three-phase stator, as shown in Figure 8–30. Notice that one of the three phases "dominates" at different times. When phase 1 is at 90°, the current in the phase 1 winding is at a maximum and current in the other windings is smaller. Therefore, the stator magnetic field will be oriented toward the phase-1 stator winding. As the phase-1 current declines, the phase-2 current increases, and the field rotates toward the phase-2 winding. The magnetic field will be oriented toward the phase-2 winding when current in it is a maximum. As the phase-2 current declines, the phase-3 current increases, and the field rotates toward the phase-3 winding. The process repeats as the field returns to the phase-1 winding. Thus, the field rotates at a rate determined by the frequency of the applied voltage. With a more detailed analysis, it can be shown that the magnitude of the field is unchanged; only the direction of the field changes.

FIGURE 8–30

Generation of the rotating field in a stator. The application of the three phases to the stator windings produces a net field as shown by the rotating arrow.

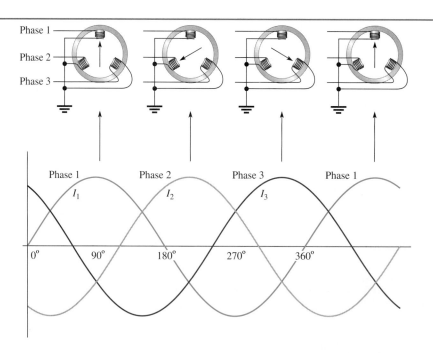

As the stator field moves, the rotor moves in sync with it in a synchronous motor but lags behind in an induction motor. The rate the stator field moves is called the *synchronous speed* of the motor.

MOTORS AND GENERATORS

Induction Motors

The induction motor is the most common type of small ac motor because of its high reliability. The theory of operation is essentially the same for both single-phase and three-phase motors. Both types use the rotating field described previously, but the single-phase motor requires a starting winding or other method to produce torque for starting the motor, whereas the three-phase motor is self-starting. When a starting winding is employed in a single-phase motor, it is removed from the circuit by a mechanical centrifugal switch as the motor speeds up.

The major advantage to an induction motor is that it does not have brushes or slip rings to wear. Essentially, the rotor current is "induced" by transformer action, therefore, the name *induction motor*. Recall from the discussion of transformers in Section 7–4 that a changing magnetic field can induce a voltage across a coil that, in turn, causes a current in the coil. The induced current creates the rotor's magnetic field.

The core of the induction motor's rotor consists of an aluminum frame that forms the conductors for the circulating current in the rotor. (Some larger induction motors use copper bars.) The aluminum frame is similar in appearance to the exercise wheel for pet squirrels (common in the early 20th century), so it is aptly called a **squirrel cage**, illustrated in Figure 8–31. The aluminum squirrel cage itself is the *electrical* path; it is embedded within a ferromagnetic material to provide a low reluctance *magnetic* path through the rotor. In addition, the rotor has cooling fins that may be molded into the same piece of aluminum as the squirrel cage. The entire assembly must be balanced so that it spins easily and without vibrating.

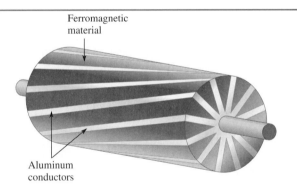

FIGURE 8–31

Diagram of a squirrel-cage rotor.

Operation of an Induction Motor

When the magnetic field from the stator moves across the squirrel cage of the inductor, a current is generated in the squirrel cage. This current creates a magnetic field that reacts with the moving field of the stator, causing the rotor to start turning. The rotor will try to "catch-up" with the moving field, but cannot, in a condition known as slip. **Slip** is defined as the difference between the synchronous speed of the stator and the rotor speed. The rotor can never reach the synchronous speed of the stator field because, if it did, it would not cut any field lines and the torque would drop to zero. Without torque, the rotor could not turn itself.

Initially, before the rotor starts moving, there is no back emf, so the stator current is high. As the rotor speeds up, it generates a back emf that opposes the stator current. As the motor speeds up, the torque produced balances the load and the current is just enough to keep the rotor turning. The running current is significantly lower than the initial start-up current because of the back emf. If the load on the motor is then increased, the motor will slow down and generate less back emf. This increases the current to the motor and increases the torque it can apply to the load. Thus, an induction motor can operate over a range of speeds and torque. Maximum torque occurs when the rotor is spinning at about 75% of the synchronous speed.

CHAPTER 8

Synchronous Motors

A synchronous motor gets its name from the fact that it runs at the synchronous speed of the rotating stator field, regardless of fluctuations in the line voltage. Recall that an induction motor develops no torque if it runs at the synchronous speed, so it must run slower than the synchronous speed, depending on the load. The synchronous motor will run at the synchronous speed and still develop the required torque for different loads. The only way to change the speed of a synchronous motor is to change the frequency*.

The fact that synchronous motors maintain a constant speed for all load conditions is a major advantage in certain industrial operations and in applications where clock or timing requirements are involved (such as a telescope drive motor or a chart recorder). In fact, the first application of synchronous motors was in electric clocks (in 1917).

Another important advantage to large synchronous motors is their efficiency. Although their original cost is higher than a comparable induction motor, the savings in power will often pay for the cost difference in a few years.

Operation of a Synchronous Motor

Essentially, the rotating stator field of the synchronous motor is identical to that of an induction motor. The primary difference in the two motors is in the rotor. Whereas the induction motor has a rotor that is electrically isolated from a supply, the synchronous motor uses a magnet to follow the rotating stator field. Small synchronous motors use a permanent magnet for the rotor; larger motors use an electromagnet. When an electromagnet is used, dc is supplied from an external source via slip rings as in the case of the alternator.

In larger synchronous motors, the rotor is designed as a series of projecting arms called *salient poles*. Each salient pole has a coil wrapped around it, forming an electromagnet, as illustrated in Figure 8–32. The rotor's north pole is never quite in line with the stator field's rotating south pole but lags behind by a small amount that depends on the load. This small lag creates the torque of a synchronous motor.

FIGURE 8–32

Typical rotor for a synchronous motor.

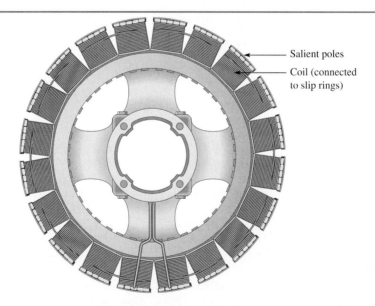

A provision must be made to start a synchronous motor. One method is to include squirrel-cage conductors in the rotor that allow it to start as an induction motor until the rotor turns at the synchronous speed. Once the motor is up to speed, no voltage is induced in the squirrel cage and the motor runs strictly as a synchronous motor.

---

* Power line frequencies are not the same in all parts of the world, so in critical applications, systems are designed to take into account the motor speed.

Review Questions

21. What is the main difference between an induction motor and a synchronous motor?
22. What happens to the magnitude of the rotating stator field as it moves?
23. What is the purpose of a squirrel cage?
24. With reference to motors, what does the term *slip* mean?
25. What are two advantages to large synchronous motors?

CHAPTER REVIEW

Key Terms

Alternator An ac generator.

Armature That part of an alternator in which a voltage is induced due to the relative motion between a conductor and a magnetic field.

Armature reaction The distortion of the magnetic field from the field coils due to an additional field created by the armature current.

Brushes Conductive contacts usually made from carbon or carbon-graphite that make contact with the moving slip rings and provide an electrical path from a source to the rotor of a generator or motor.

Commutator A segmented ring that is used in dc generators and motors to reverse the polarity of the rotor.

Faraday's law A law stating that the voltage that is induced across a wire that is moving relative to a magnetic field is determined by the strength of the field, length of the conductor, speed of the conductor, and the angle it makes with respect to the magnetic field. For a coil, Faraday's law gives the voltage as the product of the number of turns and the rate of change of magnetic flux.

Generator A device that converts mechanical work to electricity.

Induction motor An ac motor that achieves excitation to the rotor by transformer action.

Rotor The moving part of a motor or generator.

Slip The difference between the synchronous speed of the stator and the rotor speed in an induction motor.

Slip ring A solid circular ring connected to the rotating coil in a generator or motor that is part of the electrical path for the rotor.

Squirrel cage An aluminum frame within the rotor of an induction motor that forms the electrical conductors for the circulating rotor current.

Stator The stationary part of a motor or generator.

Synchronous motor A motor in which the rotor moves at the same rate as the rotating magnetic field of the stator.

Important Facts

❑ A voltage is induced across a conductor when it moves in a manner to cut magnetic field lines.

- A voltage is induced in a coil that is moving in a magnetic field. The induced voltage is proportional to the rate of change of the magnetic flux with respect to the coil and the number of turns in the coil.
- The magnetic field that surrounds a current-carrying conductor has the shape of concentric circles.
- The magnetic field around a current-carrying coil is shaped like a bar magnet, with a north end and a south end along the axis of the coil.
- The two main parts to an alternator or motor are the stator and the rotor.
- Slip rings and brushes are used to connect a rotating coil to a stationary circuit.
- The field coils are responsible for generating a magnetic field.
- The armature intercepts the flux from the magnetic field.
- Armature reaction distorts the magnetic field from the field coils.
- A rectified sine wave is a pulsating dc waveform that appears sinusoidal in shape except the negative portion is inverted, forming two duplicate positive portions.
- Three-phase ac is more efficient for high-current industrial motors.
- Motors convert electrical energy to mechanical motion by the interaction of magnetic fields.
- Motors are rated based on the torque and the power they can produce.
- A series-wound dc motor has the field coil, armature coil, and interpole winding in series.
- A shunt-wound dc motor has the field coil and armature coil in parallel.
- AC motors can be either the induction type or the synchronous type.
- Induction motors are widely used in applications that require fractional horsepower motors such as small appliances.
- Synchronous motors are very efficient and run at constant speed.

Formulas

Induced voltage in a moving conductor (Faraday's law):

$$v_{ind} = Bl \sin \theta \qquad (8\text{–}1)$$

Induced voltage in a moving conductor that is perpendicular to the magnetic field:

$$v_{ind} = B_{\perp}lv \qquad (8\text{–}2)$$

Force on a current-carrying conductor in a magnetic field:

$$F = BIl \sin \theta \qquad (8\text{–}3)$$

Force on a current-carrying conductor that is perpendicular to the magnetic field:

$$F = B_{\perp}Il \qquad (8\text{–}4)$$

Alternator frequency:

$$f = \frac{Ns}{120} \qquad (8\text{–}5)$$

DC motor torque:

$$T = K\phi I_A \qquad (8\text{–}6)$$

DC motor power:

$$P = 0.105Ts \quad (8\text{–}7)$$

Chapter Checkup

Answers are at the end of the chapter.

1. An induced voltage occurs whenever a conductor
   (a) is moved in a magnetic field
   (b) has a current in it
   (c) is connected to a voltage source
   (d) cuts magnetic field lines

2. A factor that determines the amount of voltage induced across a coil is the
   (a) rate of change of the magnetic flux
   (b) number of turns
   (c) both of the above
   (d) none of the above

3. The force on a conductor in a magnetic field is *not* dependent on
   (a) the magnetic field strength
   (b) current in the conductor
   (c) diameter of the conductor
   (d) length of the conductor

4. The force on a conductor in a magnetic field is stronger if the conductor is moved
   (a) perpendicular to the magnetic field
   (b) parallel to the magnetic field
   (c) at a 45° angle to the magnetic field
   (d) none of the above

5. The armature of a stationary field dc generator generates
   (a) steady dc   (b) pulsating dc
   (c) ac   (d) none of these answers

6. A motor that can operate on either dc or ac is a
   (a) synchronous motor   (b) universal motor
   (c) series-wound motor   (d) induction motor

7. Slip rings and brushes are not used in
   (a) dc generators   (b) alternators
   (c) synchronous motors   (d) induction motors

8. The frame for a motor or generator serves as part of the
   (a) electrical path   (b) magnetic path
   (c) both of the above   (d) none of the above

9. In a four-pole alternator, the number of full electrical cycles for each revolution of the rotor is
   (a) one   (b) two
   (c) four   (d) eight

10. You would expect to find a commutator in a
    (a) dc generator   (b) alternator
    (c) both of the above   (d) none of the above

11. Each phase of 3-phase ac is separated by
    (a) 60° (b) 90°
    (c) 120° (d) 180°

12. The most common motor for a household washing machine is a
    (a) single-phase induction motor
    (b) three-phase induction motor
    (c) single-phase synchronous motor
    (d) three-phase synchronous motor

13. An exciter supplies the rotor of
    (a) an induction motor with ac
    (b) an induction motor with dc
    (c) an alternator with ac
    (d) an alternator with dc

14. An example of a motor with high starting torque is a
    (a) single-phase induction motor
    (b) synchronous motor
    (c) series-wound dc motor
    (d) shunt-wound dc motor

15. A motor that runs at the same speed as its rotating magnetic field is a
    (a) single-phase induction motor
    (b) three-phase induction motor
    (c) synchronous motor
    (d) dc motor

16. The difference in the synchronous speed and the rotor speed of a motor is called
    (a) differential speed (b) loading
    (c) lag (d) slip

Questions

Answers to odd-numbered questions are at the end of the book.

1. Can a conductor be moved in a magnetic field *without* inducing a voltage? Explain your answer.
2. How can the right hand can be used to show the direction of conventional current in a conductor when it is moved in a magnetic field?
3. What happens to the induced voltage in a conductor moving in a magnetic field if the conductor is moved faster?
4. What happens to the induced voltage in a conductor moving in a magnetic field if the direction the conductor is moving is reversed?
5. What two factors determine the voltage induced in a coil when it is cutting magnetic field lines?
6. What happens to the magnetic field between two parallel current-carrying conductors if the current is in the same direction?
7. What happens to the magnetic field between two parallel current-carrying conductors if the current is in the opposite direction?
8. Why do solenoids and relays use a ferromagnetic core?

9. What is the difference between a stationary field generator and a rotating field generator?
10. What is a prime mover?
11. In a rotating field alternator, what part is the armature?
12. What is the purpose of brushes and slip rings?
13. What is the advantage of a wound rotor over a permanent magnet for an alternator?
14. What is the purpose of field coils?
15. What is the purpose of an air gap in a motor or generator?
16. Why should the air gap in a motor or generator be as small as practical?
17. In a four-pole generator or motor, where are the two identical poles located with respect to each other?
18. How does the armature reaction tend to rotate the magnetic field from its original position?
19. How is a rotating stator field created?
20. What is meant by a three-phase connection that is described as a wye connection?
21. What is meant by a three-phase connection that is described as a delta connection?
22. Why is the initial starting current in a motor higher than the running current?
23. What is a prony brake? What does it measure?
24. Why should a dc motor always be run with a load?
25. What type of dc motor has high torque at full load?
26. What is a universal motor?
27. What type of motor never uses slip rings or brushes?
28. What is meant by the synchronous speed of an ac motor?
29. Why can't an induction motor turn at its synchronous speed?
30. In what type of motor would you expect to find a centrifugal switch?
31. If a synchronous motor has a squirrel cage, what is its purpose?
32. What is a salient pole?

Basic Problems

PROBLEMS

Answers to odd-numbered problems are at the end of the book.

1. A 10 cm long conductor is moving at a velocity of 0.9 m/s through a magnetic flux density of 1.5 T. The motion is perpendicular to the flux. What voltage is induced across the conductor?
2. Repeat Problem 1 for the case where the conductor is moving at an angle of 30° with respect to the flux.
3. A wire is moved with a constant 1.5 m/s at an angle of 75° to a magnetic field. The poles of the magnet are 6 cm across. If a voltage of 16 mV is induced across the inductor, what is the flux density of the magnetic field?
4. Repeat Problem 3 for the case where the induced voltage is 120 mV.
5. A 2.0 cm section of wire is placed between the poles of a magnet at a right angle to the poles. A force of 0.19 N is measured on the wire with a current of 15 A. From this data, calculate the flux density in teslas.
6. Assume that the north pole of the magnet in Problem 5 is on your left and the conventional current is directed away from you. What direction is the force on the wire?

7. A 4.0 cm conductor is placed at right angles between the poles of a magnet with a flux density of 0.2 T. A 1.0 Ω limiting resistor is placed in series with a 12 V battery. Assume the resistance of the wire is negligible. What is the force on the wire?

8. What is the frequency of a four-pole alternator that is turning at 2000 rpm?

9. The frequency of a four-pole alternator is measured as 200 Hz. What is the speed of the rotor in rpm?

10. What is the power developed by a motor that turns at 800 rpm when the torque is 3.2 N-m?

11. What is the horsepower of the motor in Problem 10?

Basic-Plus Problems

12. A 20 cm long conductor is moving upward between the poles of a magnet as shown in Figure 8–33. The pole faces are 8.5 cm on each side and the flux is 1.24 mWb. The motion produces an induced voltage across the conductor of 44 mV. What is the speed of the conductor?

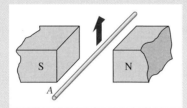

FIGURE 8–33

13. (a) For the conductor shown in Figure 8–33, what is the polarity of the end marked with the letter A? (b) Assuming a complete path is provided and current is in the direction you indicated, what direction is the induced force on the conductor?

14. Assume that the pole faces in Figure 8–33 are 8.0 cm on a side and a 20 cm conductor is moved upward at 4.0 m/s. (a) What must be the flux density to induce a voltage of 60 mV across the conductor? (b) What is the flux per pole for this case?

15. An alternator generates four pulses of dc every time it makes one revolution. How many poles does it have? Explain your answer.

16. If the alternator in Problem 15 is spinning at 2250 rpm, how many pulses will be observed in one second?

Example Questions

8–1: 150 mV

8–2: 1.0 T

8–3: 1.08 N

8–4: 11.25 N

8–5: 1500 rpm

8–6: 0.18 hp

Review Questions

1. Two requirements for inducing a voltage are (a) relative motion of a conductor and a magnetic field and (b) motion must be such that magnetic field lines are "cut".
2. Four factors that determine the amount of induced voltage:
   - ❏ The strength of the magnetic field
   - ❏ The length of the conductor exposed to the magnetic field
   - ❏ The speed that the conductor is moving relative to the magnetic field
   - ❏ The angle the conductor is moving relative to the magnetic field
3. A voltage can be induced in a stationary conductor if a magnetic field moves such that lines are "cut' by the conductor.
4. The polarity is reversed.
5. Zero volts
6. The magnetic field lines form concentric circles.
7. The current in the conductor
8. The force on each conductor is toward the other (attractive forces).
9. The poles will be reversed.
10. The current, the number of turns, and the core material
11. The rotor and stator
12. A stationary field generator
13. An ac generator (alternator) uses slip rings and brushes.
14. The output is easier to convert to regulated dc.
15. With solid-state diodes
16. The commutator reverses the current in the armature when unlike poles are near each other in the motor.
17. Back emf is a voltage developed in a motor because of generator action in the armature. It opposes the original supply current to the armature.
18. As a motor speeds up, the armature current is reduced.
19. The speed of the motor for a given torque
20. A series-wound motor
21. The difference is the rotors. In an induction motor, the rotor obtains current by transformer action; in a synchronous motor, the rotor is a permanent magnet or an electromagnet that is supplied current from an external source through slip rings and brushes.
22. The magnitude is constant.
23. The squirrel cage is composed of the electrical conductors that generate current in the rotor.
24. Slip is the difference between the synchronous speed of the stator and the rotor speed.
25. Large synchronous motors run at constant speed and they are efficient.

Chapter Checkup

1. (d)	2. (c)	3. (c)	4. (a)	5. (c)
6. (b)	7. (d)	8. (b)	9. (b)	10. (a)
11. (c)	12. (a)	13. (d)	14. (c)	15. (c)
16. (d)				

CHAPTER 9

CHAPTER OUTLINE

9–1 The Sine Wave
9–2 Phasor Representation of a Sine Wave
9–3 Nonsinusoidal Waveforms
9–4 Function Generators
9–5 Oscilloscopes

ALTERNATING CURRENT

INTRODUCTION

The sine wave is the fundamental ac waveform because all periodic waves can be constructed from a set of sine waves of the appropriate amplitude and frequency. In this chapter, the sine wave and a method to represent sine waves called phasors are introduced. In the laboratory, sine waves and other waveforms can be produced by an instrument called the function generator. Sine waves and other waveforms are measured by oscilloscopes. The operation of both function generators and oscilloscopes are described.

Study aids for this chapter are available at

http://www.prenhall.com/SOE

KEY OBJECTIVES

A section number is given for each objective. After completing this chapter, you should be able to

9–1 Describe the parameters used to characterize a sinusoidal wave

9–2 Explain how phasors can be used to represent one or more sine waves

9–3 Describe the key parameters used to characterize pulses, triangular waves, and sawtooth waves

9–4 Describe the principal features and controls for a function generator

9–5 Describe the four major sections in a block diagram of an oscilloscope and the controls associated with each section

LABORATORY EXPERIMENTS DIRECTORY

The following exercises are for this chapter.

◆ **Experiment 16**
The Oscilloscope

◆ **Experiment 17**
Sine Wave Measurements

KEY TERMS

- Radian
- Oscillator
- Peak value
- Peak-to-peak value
- Phase shift
- Phasor
- Rise time
- Fall time
- Duty cycle
- Function generator
- Oscilloscope

ON THE JOB...

(Getty Images)

Social skills include your ability to properly take instruction and directions from your supervisor and to get along with the people with whom you work. It is important to foster a good working environment for all. In particular, if you are dealing with customers, you must be able to work with them in a courteous and professional manner. If a situation looks like it is getting out of hand, ask your supervisor for help. The impression that you make on customers will affect current and future business for your company.

291

CHAPTER 9

Sci Hi SCIENCE HIGHLIGHT

You are familiar with many types of waves such as sound waves, water waves, light waves, earthquake waves, and vibrations along a stretched slinky toy. Waves such as sound waves and water waves require a physical material (air or water) in which to travel. Light waves, radio waves, and x-rays are part of the electromagnetic spectrum and can travel through a vacuum.

One of the early great debates of science centered on the nature of light. Isaac Newton, the father of modern physics, announced that light was corpuscles of matter that moved through space in a straight-line path. Christian Huygens, a contemporary of Newton, believed light consisted of wave motion. Unfortunately, neither theory could fully explain the properties of light.

The nature of light was not solved during Newton's or Huygens' time. In the nineteenth century, James Clerk Maxwell, a brilliant Scottish mathematician and physicist, pictured light as being due to changing electric and magnetic fields that are at right angles to each other. Maxwell was able to show that a changing magnetic field should induce a changing electric field and a changing electric field should induce a changing magnetic field. He found that the pattern of the changing electric and magnetic fields would move through space with the speed of light. Maxwell's equations apply to all of the electromagnetic spectrum, not just light, and today they form the basis for all theoretical work in electromagnetism.

9–1 THE SINE WAVE

The sine wave is the fundamental waveform of ac circuits. It is also fundamental to many other basic electronic circuits, including those found in communication systems.

In this section, you will learn the parameters used to characterize a sinusoidal wave.

The Sine Function

Recall that *alternating current* was defined in Section 3–2 as current that changes direction or "alternates" back and forth. Alternating current has the same shape as the trigonometric sine (abbreviated sin) function. The sine function is defined from a right triangle (one with a 90° angle), as illustrated in Figure 9–1. For the reference angle labeled θ, the sine is the ratio of the opposite side to the hypotenuse, so it is a dimensionless number.

The angular measurement of the sine function is based on one revolution or 360°, as shown in the plot in Figure 9–2. As the angle θ increases from 0° to 360°, the sine of θ varies from zero to a positive peak of $+1$ (at 90°), back through zero (at 180°) to a negative peak of -1 (at 270°), and finally back to zero as shown. At 90° and 270°, the length of the opposite side of the right triangle is equal to the length of the hypotenuse; thus, the sine function is 1 and -1, respectively.

FIGURE 9–1

The definition of the sine function is based on a right triangle.

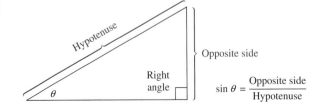

ALTERNATING CURRENT

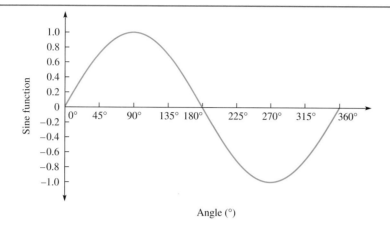

FIGURE 9-2

A plot of the sine function as the angle is varied from zero to 360°.

Another important angular measurement of sine waves is the radian, which is based on the radius of a circle. The **radian** (abbreviated rad) is an angle whose arc is equal to the radius of the circle, as illustrated in Figure 9-3.

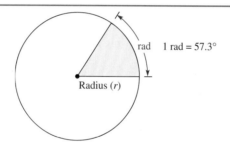

FIGURE 9-3

The radian (rad) is an angle that is formed when the arc is equal to the radius of a circle.

From geometry, the circumference of a circle is 2π times the radius. With radian measurement, one revolution (360°) contains 2π radians. The mathematical sine function is plotted with radian angular measurement in Figure 9-4. Because the sine function can be plotted with radians or degrees, you need to be sure which of the two angular measurements is used in a given situation. For electronics, only the degree and radian are important. (The grad is another angular measurement used in surveying work that divides the circle into 400 parts.) All scientific calculators allow you to choose degrees, radians, or grads as an angular unit of measurement.

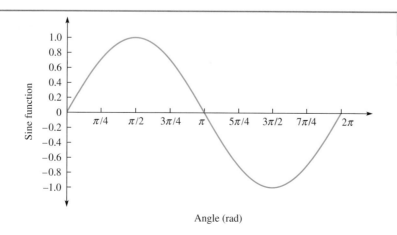

FIGURE 9-4

A plot of the sine function as the angle is varied from zero to 2π radians.

293

CHAPTER 9

The relationship between degrees and radians is simple if you keep in mind that there are 360° in a circle and 2π radians in a circle. You can find the number of degrees in a radian by dividing 360° by 2π radians.

$$\frac{360°}{2\pi \text{ rad}} = 57.3°/\text{rad}$$

You can use this ratio to convert from degrees to radians.

$$\text{rad} = \left(\frac{2\pi \text{ rad}}{360°}\right) \times \text{degrees} \tag{9–1}$$

To convert from radians to degrees, the formula is

$$\text{deg} = \left(\frac{360°}{2\pi \text{ rad}}\right) \times \text{rad} \tag{9–2}$$

EXAMPLE 9–1

Problem
Convert the following number of degrees to radians.

(a) 50°   (b) 120°   (c) 290°

Solution
Apply Equation 9–1.

(a) $\text{rad} = \left(\dfrac{2\pi \text{ rad}}{360°}\right) \times \text{degrees} = \left(\dfrac{2\pi \text{ rad}}{360°}\right) 50° = \mathbf{0.873 \text{ rad}}$

(b) $\text{rad} = \left(\dfrac{2\pi \text{ rad}}{360°}\right) 120° = \mathbf{2.094 \text{ rad}}$

(c) $\text{rad} = \left(\dfrac{2\pi \text{ rad}}{360°}\right) 290° = \mathbf{5.061 \text{ rad}}$

For the TI-36X calculator, you can make the conversion using the 3rd function DRG key. First, check that you are in degree mode—the letters DEG appear in small print in the display if you are. (If necessary to change the display, press [3rd] [DRG/HYP].) Then enter the number of degrees and press [3rd] [DRG/HYP] again for the conversion to rad.

Question*
How many rads are in 100°?

Answers are at the end of the chapter.

EXAMPLE 9–2

Problem
Convert the following number of radians to degrees.

(a) $\pi/3$ rad (b) 1.2 rad (c) 3.9 rad

Solution
Apply Equation 9–2.

(a) $\deg = \left(\dfrac{360°}{2\pi \text{ rad}}\right) \times \text{rad} = \left(\dfrac{360°}{2\pi \text{ rad}}\right)\dfrac{\pi}{3}\text{ rad} = \mathbf{60°}$

(b) $\deg = \left(\dfrac{360°}{2\pi \text{ rad}}\right)1.2 \text{ rad} = \mathbf{68.8°}$

(c) $\deg = \left(\dfrac{360°}{2\pi \text{ rad}}\right)3.9 \text{ rad} = \mathbf{223°}$

Question
How many degrees are in 1.5 radians?

Electrical Waves

Any wave that has the same shape as the sine function is said to be **sinusoidal**. Thus, a voltage or current waveform with the same shape of the sine function is a sinusoidal wave; however, it is common practice to call a voltage or current waveform with this shape a sine wave. The terms *sine wave* and *sinusoidal wave* are often used interchangeably. Figure 9–5 shows a voltage sine wave with a peak voltage of 40 V. For an electrical wave, voltage (or current) is usually displayed on the vertical (y) axis. Either time (t) or the angle (θ) can be shown on the horizontal (x) axis. In this case, time is shown as the x variable.

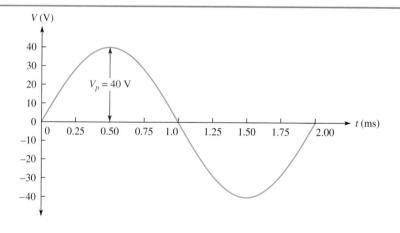

FIGURE 9–5

A voltage sine wave. The peak voltage, V_p, is the amplitude of the wave and is equal to 40 V in this case.

Electrical sine waves are produced by two types of sources: rotating electrical machines (alternators) or electronic oscillators. A sine wave is the natural output when a loop of wire is spinning in a magnetic field as you saw in Section 8–3. An **oscillator** is an electronic circuit that also produces this natural frequency but can oscillate much faster than the mechanical rotation of an alternator. Electronic oscillators are found in many different circuits, including instruments known as function generators.

CHAPTER 9

As you have seen, a sine wave changes polarity at its zero value; that is, it alternates between positive and negative values. When a sinusoidal voltage source (V_s) is applied to a resistive circuit, an alternating sinusoidal current results, as indicated in Figure 9–6. The shape of the current waveform is the same as the shape of the voltage waveform throughout. In a resistive circuit, the voltage and current waveforms change "in sync" with each other.

FIGURE 9–6

A sinusoidal voltage applied to a resistive circuit produces a sinusoidal current.

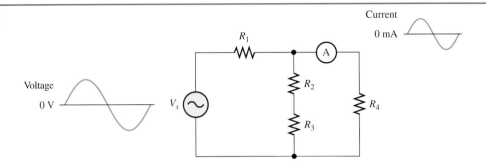

Electrical Wave Parameters

Sinusoidal waves used in electronics are characterized by certain parameters. These include peak value, peak-to-peak value, rms value, average value, frequency, and period.

Peak Value

The voltage sine wave illustrated in Figure 9–5 has a maximum value of +40 V, measured from the center of the wave. This is called the **peak value** of the waveform and is a fixed value for a given wave. The peak value is designated as V_p for a voltage wave or I_p for a current wave. The peak value is also called the **amplitude** of the wave. The voltage at any instant in time changes, but the peak value or amplitude is constant.

Peak-to-Peak Value

The magnitude of the peak positive value is equal to the magnitude of the peak negative value. Thus, the **peak-to-peak value** is twice the peak value, as indicated in the Figure 9–7. Peak-to-peak values are represented by V_{pp} or I_{pp}. The relationship between the peak and the peak-to-peak value is expressed in the following equations for voltage and current.

$$V_{pp} = 2V_p \qquad (9\text{–}3)$$

$$I_{pp} = 2I_p \qquad (9\text{–}4)$$

SAFETY NOTE

You should never wear metallic jewelry when you work on circuits. If it accidentally contacts a voltage source, the jewelry can heat up instantly, resulting in serious burns. Always remove watches and rings before working on a circuit.

FIGURE 9–7

The peak-to-peak value is twice the peak value for either voltage or current.

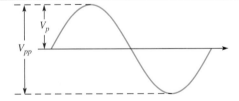

RMS Value

A more common way to express the voltage of a sine wave is the rms (root mean square) value. Recall from Section 3–2 that the rms value allows you to compare the power delivered by an ac source directly with that of a dc source. As an example, consider the circuit in Figure 9–8. The dc source is 120 V. The ac source is specified as 120 Vrms and is therefore equivalent to the dc source. Because the sources have identical voltages, the bulb will

ALTERNATING CURRENT

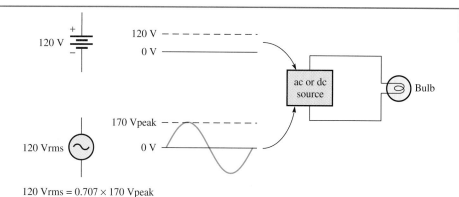

FIGURE 9–8
The dc and the ac source deliver the same power to the bulb.

dissipate the same power with either source in the box. Because an rms voltage is equivalent to the same dc voltage, it is sometimes called *effective*.

The rms value of a sine wave is related to the peak value by the following equations for voltage and current:

$$V_{rms} = 0.707 V_p \quad (9\text{–}5)$$

$$I_{rms} = 0.707 I_p \quad (9\text{–}6)$$

Average Value
Another way that is used occasionally to express the value of a sine wave is the **average value**. The true average of any sine wave is 0 V because the positive and negative values cancel. The term *average* is applied to sinusoidal waves assuming all values are positive. Taking the unsigned average of many equally spaced points on a voltage or current waveform gives these relationships:

$$V_{avg} = 0.637 V_p$$

$$I_{avg} = 0.637 I_p$$

HISTORICAL NOTE
The unit of frequency, the hertz, is named in honor of Heinrich Rudolf Hertz (1857–1894). Hertz, a German physicist, was the first to broadcast and receive electromagnetic (radio) waves. He produced electromagnetic waves in the laboratory and measured their parameters. Hertz also proved that the nature of the reflection and refraction of electromagnetic waves was the same as that of light.

EXAMPLE 9–3

Problem
Calculate the rms and peak-to-peak values of the voltage waveform shown in Figure 9–9.

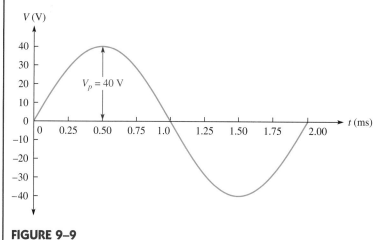

FIGURE 9–9

297

CHAPTER 9

Solution
From Equation 9–5,
$$V_{rms} = 0.707 V_p$$
The peak voltage is $V_p = 40$ V; therefore,
$$V_{rms} = 0.707(40 \text{ V}) = \textbf{28.3 V}$$
$$V_{pp} = 2V_p = 2(40 \text{ V}) = \textbf{80 V}$$

Question
What is the peak-to-peak value for a 120 Vrms voltage?

Frequency and Period

The frequency of a repeating wave was defined in Section 3–2 as the number of complete cycles that occur in one second. The measurement unit for frequency is the hertz. Another important term is the period, T, which was defined as the time required for a given wave to complete one full cycle.

Frequency and period are reciprocals of each other. If you know the frequency, you can calculate the period by using the reciprocal relationship between the two quantities. Likewise, if you know the period, you can calculate the frequency by taking the reciprocal. In either case, you simply take the reciprocal of the known quantity to find the unknown quantity.

EXAMPLE 9–4

Problem
(a) Calculate the frequency of the sine wave in Figure 9–10.

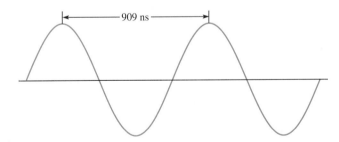

FIGURE 9–10

(b) Calculate the frequency of the sine wave in Figure 9–11. The time is shown at an arbitrary level along the curve but will be the same time anywhere it is measured, as long as the level and slope are the same for both points.

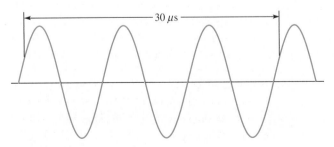

FIGURE 9–11

(c) Calculate the period of a 60 Hz power line.

(d) How many cycles of a 10 kHz wave will occur in 20 ms?

Solution

(a) The period is 909 ns. Calculate the frequency by taking the reciprocal of the period.

$$f = \frac{1}{T} = \frac{1}{909 \text{ ns}} = \textbf{1100 kHz}$$

This is a frequency in the am radio band.

(b) Three cycles occur in 30 μs. Therefore, the period is

$$T = \frac{30 \text{ }\mu\text{s}}{3} = 10 \text{ }\mu\text{s}$$

The frequency is

$$f = \frac{1}{T} = \frac{1}{10 \text{ }\mu\text{s}} = \textbf{100 kHz}$$

(c) The frequency is 60 Hz. Therefore, the period is

$$T = \frac{1}{f} = \frac{1}{60 \text{ Hz}} = \textbf{16.7 ms}$$

(d) First, calculate the period of 10 kHz.

$$T = \frac{1}{f} = \frac{1}{10 \text{ kHz}} = 100 \text{ }\mu\text{s}$$

20 ms = 20,000 μs. The number of cycles in 20,000 μs is

$$\frac{20,000 \text{ }\mu\text{s}}{100 \text{ }\mu\text{s/cycle}} = \textbf{200 cycles}$$

Question

What is the frequency of a wave that has a period of 500 μs?

Equation for Electrical Sinusoidal Waves

As you know, a sinusoidal wave is generated when a conductor rotates in a magnetic field. The equation for the instantaneous value of the sinusoidal voltage that is generated in the conductor is

$$v = V_p \sin \theta \qquad (9\text{--}7)$$

where v is voltage at any instant in time, V_p is peak voltage, and θ is angle between the conductor and the magnetic field. Equation 9–7 is useful if you need to find the instantaneous voltage at some point on a voltage waveform that starts at zero degrees.

CHAPTER 9

EXAMPLE 9–5

Table 9–1

θ (°)	sin θ (°)	v = 12 sin θ (V)
0	0	0
10	0.173648	2.083778
20	0.34202	4.104242
30	0.5	6
40	0.642788	7.713451
50	0.766044	9.192533
60	0.866025	10.3923
70	0.939693	11.27631
80	0.984808	11.81769
90	1	12
100	0.984808	11.81769
110	0.939693	11.27631
120	0.866025	10.3923
130	0.766044	9.192533
140	0.642788	7.713451
150	0.5	6
160	0.34202	4.104242
170	0.173648	2.083778
180*	1.23E-16	1.47E-15

* The sin θ at 180° is actually 0. Because of the algorithm used to calculate the function, Excel gets a value of 1.23×10^{216}, which is nearly 0.

Problem
Calculate the instantaneous voltage every 10° from 0° to 180° for a sinusoidal voltage waveform with a peak voltage of 12 V.

Solution
Apply Equation 9–7 for each data point. Table 9–1 is set up for the calculation. If you are familiar with spreadsheet computer programs such as Excel, this can easily be set up to calculate the answers and even plot the result. The values in Table 9–1 were calculated with Excel, but you can easily do the calculations with a calculator. The calculator solution using the TI-36X for 10^0 is

The data are plotted in Figure 9–12.

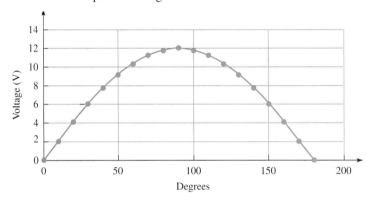

FIGURE 9–12

Question
What are the instantaneous voltages from 180° to 360° every 10 degrees?

Review Questions

Answers are at the end of the chapter.

1. What is a radian?
2. How many degrees are in π/2 radians?
3. What is the rms value of a sine wave with a peak voltage of 20 V?
4. If 5 complete cycles of a sine wave occur in 10 ms, what is the frequency?
5. What is the period of a 20 MHz sine wave?

9–2 PHASOR REPRESENTATION OF A SINE WAVE

When sinusoidal waves of the same frequency are compared, it is necessary to specify the amplitude and the phase relationship between them. Phasors, which indicate the amplitude and the phase relationship, simplify the description of sinusoidal waveforms.

In this section, you will learn how phasors can be used to represent one or more sine waves.

The **phase** of a sinusoidal wave is an angular measurement that specifies the position of the wave relative to a reference. In an actual circuit, one waveform is selected as the reference. Typically, this will be the source voltage, such as the output of a function generator. Figure 9–13 shows one cycle of a sinusoidal wave to be used as the reference. The reference waveform has a positive-going crossing of the horizontal axis (zero crossing) at 0° (0 rad) and a positive peak at 90° ($\pi/2$ rad). The negative-going zero crossing is at 180° (π rad), and the negative peak is 270° ($3\pi/2$ rad). The cycle is completed at 360° (2π rad).

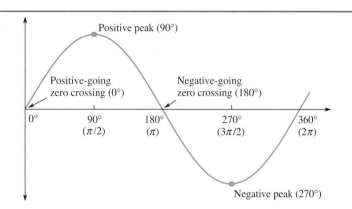

FIGURE 9–13

Phase reference.

When another sinusoidal wave of the same frequency is shifted left or right with respect to a reference waveform, there is a phase shift. **Phase shift** is the angular difference between two waves of the same frequency and can be measured in degrees or radians. Although the two waves must be of the same frequency, a phase shift measurement on two waves of different amplitudes can be made at the center of the waves.

Leading and Lagging Waves

Figure 9–14 illustrates phase shifts of a sine wave (*B*) compared to the reference sine wave (*A*). In part (a), sine wave *B* is shifted to the right by 90° ($\pi/2$ rad). Thus, there is a phase shift of +90° of sine wave *B* with respect to sine wave *A*. In terms of time, the positive zero crossing of sine wave *B* occurs later than the positive zero crossing of sine wave *A* because time increases to the right along the horizontal axis. In this case, sine wave *B* is said to **lag** sine wave *A* by 90° or $\pi/2$ radians.

Illustration of phase shift. **FIGURE 9–14**

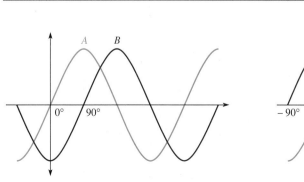
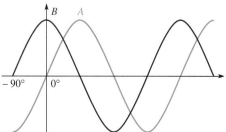

(a) *B* lags *A* by 90°.

(b) *B* leads *A* by 90°.

CHAPTER 9

In Figure 9–14(b), sine wave B is shown shifted left by 90°. In this case, the zero crossing of sine wave B occurs earlier in time than that of sine wave A (the reference). Sine wave B is said to **lead** sine wave A by 90°. In both cases, there is a 90° angle between the two waveforms.

The determination of a leading or lagging sine wave is strictly determined by the reference sine wave, so it is important that the reference is specified or understood. If you say, for example, that sine wave B is *lagging* sine wave A, then you can infer that sine wave A is *leading* sine wave B. The two ideas are equivalent but express a different reference.

EXAMPLE 9–6

Problem
What are the phase angles between the two sine waves in Figure 9–15(a) and 9–15(b)?

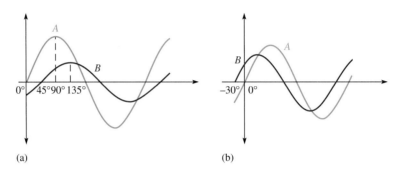

FIGURE 9–15

Solution
In Figure 9–15(a), the positive zero crossing of sine wave A is at 0°, so it is the reference. The corresponding positive zero crossing of sine wave B occurs **45°** later with respect to A. Thus, you can say that there is a 45° phase angle between the two waveforms with sine wave B lagging sine wave A.

In Figure 9–15(b), the positive zero crossing of sine wave B is at −30°, and the corresponding positive zero crossing of sine wave A is at 0°. There is a **30°** phase angle between the two waveforms with sine wave B leading sine wave A.

Question
How can you express the leading and lagging relationship between the waves if sine wave B is the reference in both cases?

Certain quantities called vectors require two numbers for a complete description. One of the numbers specifies the magnitude (how much) and the other specifies the direction. In mathematics, such numbers are called *complex quantities,* meaning they require two numbers to describe them. (This topic is covered further in Section 12–1.) Frequently, vectors are described using an arrow. The length of the arrow represents the magnitude, and the arrow points in the direction of the vector. In physics, vectors include force, velocity, and acceleration as well as many other quantities. Figure 9–16 shows an example of velocity vectors for a diver. The length of the arrow represents the diver's speed (magnitude), and the arrow shows his direction. Combined, magnitude and direction form the velocity vector.

ALTERNATING CURRENT

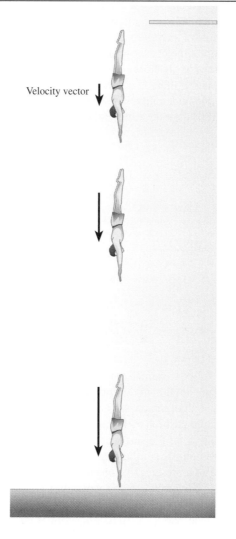

FIGURE 9–16

Velocity vectors for a diver. As the diver falls, his speed is represented by a longer arrow. The direction component of motion is indicated by the arrow.

Just as a vector is used to describe the velocity of a diver, a vector can describe a sinusoidal wave. A sinusoidal wave can be thought of in terms of a rotating vector, which can simplify the description of complex circuit behavior, as you will see later. The two quantities that are used to describe the sine wave are the amplitude (magnitude) and the phase (direction). To show that direction is indicated by phase, the rotating vector is given the special name **phasor**. A phasor is represented graphically as an arrow that rotates around a fixed point. A "snapshot" in time of a phasor when it rotates counterclockwise by 45° from the starting point is shown in Figure 9–17. The projection of the tip of the phasor represents one point on the sinusoidal graph.

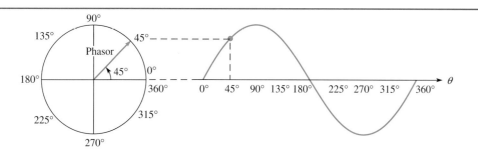

FIGURE 9–17

As the phasor rotates, the projection produces a sinusoidal wave.

303

CHAPTER 9

Figure 9–18 illustrates a reference phasor rotating counterclockwise through a complete revolution of 360°. It is the reference because it starts at an angle of 0°, located on the right side of the reference circle. Each point represents a different time starting from zero. If the tip of the phasor is projected over to a graph with phase angles running along the horizontal axis, a sine wave is "traced out," as shown in the figure. At each angular position of the phasor, there is a corresponding magnitude value. As you can see, at 90° and at 270° the amplitude of the sine wave is maximum and equal to the length of the phasor. At 0° and at 180° the sine wave is equal to zero because the phasor lies horizontally at these points.

FIGURE 9–18

Sine wave represented by a rotating phasor arrow.

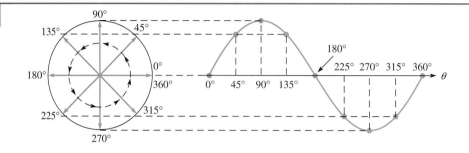

Let's examine the "snapshot" of the reference phasor at 45° in more detail. Usually, phasors are drawn to represent voltage, so it will be labeled as V_p, which represents both the length of the phasor and the peak voltage. Figure 9–19 shows the voltage phasor at an angular position of 45° and the corresponding point on the sine wave. The instantaneous voltage, v, of the sine wave at this point is related to both the position (angle) and the length (amplitude) of the phasor. The vertical distance from the phasor tip down to the horizontal axis represents the instantaneous voltage at that point.

Notice that when a vertical line is drawn from the phasor tip down to the horizontal axis, a right triangle is formed, as shown in Figure 9–19. The phasor is the hypotenuse of the triangle, and the vertical projection is the opposite side. The vertical projection represents the instantaneous voltage and its value can be found from Equation 9–7, given earlier.

FIGURE 9–19

Relationship of a sine wave to a right triangle. The amplitude of the sine wave at 45° is given by the equation $v = V_p \sin \theta$.

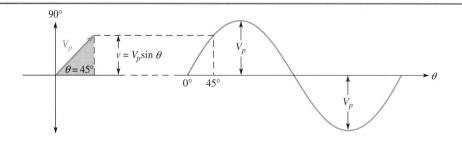

EXAMPLE 9–7

Problem
What is the instantaneous voltage for a sine wave that has a peak voltage of 40 V at each of the following angles? Plot the points on a sine wave.

(a) 45°   (b) 75°   (c) 130°   (d) 205°

ALTERNATING CURRENT

Solution

Apply Equation 9–7 for each angle.

(a) $v = V_p \sin\theta = 40\text{ V}\sin 45° = 40\text{ V}(0.707) = \mathbf{28.3\text{ V}}$

(b) $v = 40\text{ V}\sin 75° = 40\text{ V}(0.966) = \mathbf{38.6\text{ V}}$

(c) $v = 40\text{ V}\sin 130° = 40\text{ V}(0.766) = \mathbf{30.6\text{ V}}$

(d) $v = 40\text{ V}\sin 205° = 40\text{ V}(-0.423) = \mathbf{-16.9\text{ V}}$

These points are plotted in Figure 9–20. Each point falls on the curve for the sine wave.

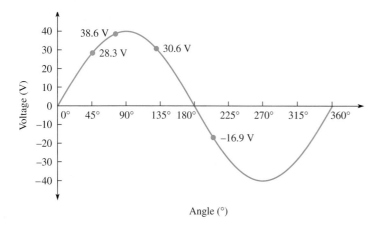

FIGURE 9–20

Question
What is the voltage at an angle of 330°?

Expressions for Phase-Shifted Sine Waves

A sine wave can be shifted so that it is leading or lagging the reference wave. When it is shifted to the right less than one-half cycle, it is said to be lagging as discussed previously. When it is shifted to the left of the reference wave by less than one-half cycle, it is said to be leading. The equation for a sine wave that starts before or after the reference is modified to include a phase shift term.

For a voltage waveform that is lagging the reference wave, the equation is

$$v = V_p\sin(\theta - \phi) \tag{9-8}$$

where v is the instantaneous voltage, V_p is peak voltage, θ is angle measured from the positive slope zero crossing of the wave, and ϕ is phase angle measured from the reference wave.

With a lagging waveform, the sign of the phase shift (in Equation 9–8) is negative. Figure 9–21 illustrates a lagging wave (in red). In the case illustrated, the peak voltage of the reference is 40 V peak and the peak voltage of the shifted sine wave is 30 V. It is common that both waveforms will have different amplitudes as shown. When this is the case, it is necessary to measure the phase shift along the *x*-axis to avoid an error or to adjust the waves to *appear* to have the same amplitude (discussed in Section 9–5).

305

CHAPTER 9

FIGURE 9–21

A voltage waveform that lags the reference. More than a full cycle is shown.

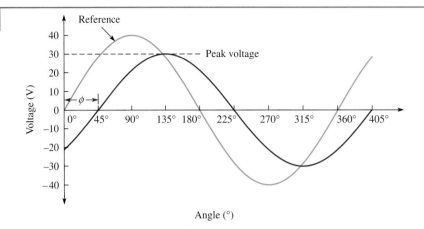

EXAMPLE 9–8

Problem

(a) What is the equation for the lagging sine wave shown in Figure 9–21?

(b) What is the instantaneous voltage of the lagging waveform at 90°?

Solution

(a) The peak voltage (amplitude) of the wave is 30 V. It lags the reference by 45°. Substitute these values into Equation 9–8.

$$v = V_p\sin(\theta - \phi) = \mathbf{30\ V\ sin(\theta - 45°)}$$

(b) $v = 30\ V\ \sin(90° - 45°) = 30 \sin 45° = \mathbf{21.2\ V}$

Question
What is the equation for the reference wave?

For a voltage waveform that is leading the reference wave, the equation is

$$v = V_p\sin(\theta + \phi) \qquad (9\text{–}9)$$

With a leading waveform, the sign of the phase shift (in Equation 9–9) is positive. Figure 9–22 illustrates a leading wave. Notice that the reference defines the location of 0°; therefore, the wave leads the reference by 45°. Again, the two waves do not have the same amplitude, so the phase shift should be measured on the *x*-axis.

FIGURE 9–22

A voltage waveform that leads the reference.

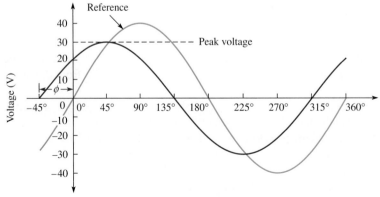

306

As a memory aid for the sign of the phase shift, notice that if the shifted waveform is above the *x*-axis at 0°, the sign of the phase shift term is positive and the wave is said to be leading. If the waveform is below the *x*-axis at 0°, the sign of the phase shift term is negative and the wave is said to be lagging.

Phasor Graphs With Two Sine Waves

As you have seen, the generation of a sine wave can be represented by the projection of the tips of the rotating phasors. Frequently, it simplifies the analysis of circuits to work directly with the phasors that represent the sine waves, rather than the sine waves themselves. A "snapshot" in time of the rotating phasors shows the same information as the sine wave plots shown previously. Figure 9–23 shows the phasors for the two sine waves shown in Figure 9–21. The phasor diagram shown here is a "snapshot" at 0°; it can be rotated to another position if desired. The blue reference phasor is drawn in the horizontal position because at 0°, the instantaneous voltage is zero. The red phasor lags the reference phasor by 45°, which corresponds to these phase difference shown by the waveforms.

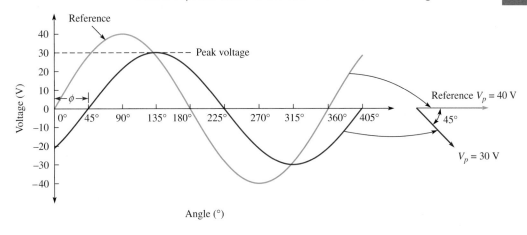

FIGURE 9–23

Phasor representation of the two sine waves shown in Figure 9–21.

Figure 9–24 shows the phasors for the two waves shown in Figure 9–22. The phasor diagram is again shown at 0°, so the blue reference phasor is drawn in the horizontal position. The red phasor leads the reference phasor by 45°, which corresponds to the phase difference shown by the waveforms.

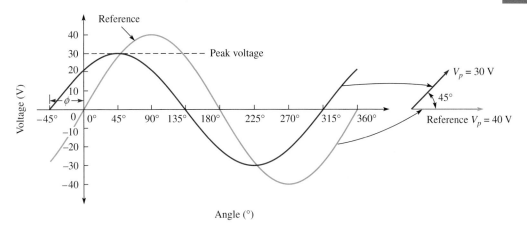

FIGURE 9–24

Phasor representation of the two sine waves shown in Figure 9–22.

CHAPTER 9

Review Questions

6. Why is it necessary to have a reference wave specified when referring to phase shift?
7. How does a phasor differ from an ordinary vector?
8. When a sine wave reaches the negative peak, what is the orientation of the associated phasor?
9. What is the equation for a sinusoidal voltage wave with a peak voltage of 20 V that lags the reference by 90°?
10. What is the instantaneous voltage at 130° for the sine wave in Question 9?

9–3 NONSINUSOIDAL WAVEFORMS

The pulse waveform is essential in digital circuits and is used in testing both digital and analog circuits. Two other useful nonsinusoidal waveforms are the triangle wave and the sawtooth wave.

In this section, you will learn how pulses, triangle waves, and sawtooth waves are specified.

Pulse Waveforms

A **pulse** can be described as a very rapid transition (leading edge) from one voltage or current level (baseline) to a constant level and then, after an interval of time, a very rapid transition (trailing edge) back to the original baseline level. The transitions in level are called *steps*. An ideal pulse consists of two equal but opposite-going instantaneous steps. When the leading or trailing edge is positive-going, it is called a *rising edge*. When the leading or trailing edge is negative-going, it is called a *falling edge*. For a pulse, the *amplitude* is its height measured from the baseline.

Figure 9–25(a) shows an ideal positive-going pulse. The time interval between the positive transition and the negative transition is called the pulse width. Figure 9–25(b) shows an ideal negative-going pulse.

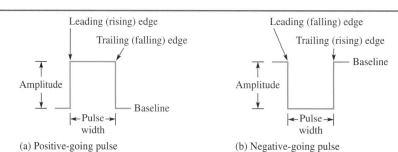

FIGURE 9–25

Ideal pulses.

In many applications, analysis is simplified by treating all pulses as ideal (composed of instantaneous steps and perfectly rectangular in shape). Actual pulses, however, are never ideal. All pulses possess certain characteristics that cause them to be different from the ideal.

In practice, pulses cannot change from one level to another instantaneously. Time is always required for a transition (step), as illustrated in Figure 9–26(a). During the rising edge, there is a time that the pulse is going from its lower value to its higher value. This interval is called the rise time, t_r. **Rise time** is the time required for the pulse to go from 10% of its amplitude to 90% of its amplitude.

ALTERNATING CURRENT

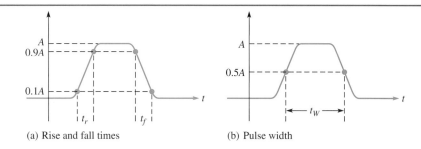

FIGURE 9–26

Nonideal pulses. A represents the amplitude.

The interval of time during the falling edge in which the pulse is going from its higher value to its lower value is called the fall time, t_f. **Fall time** is the time required for the pulse to go from 90% of its amplitude to 10% of its amplitude.

Pulse width, t_W, also requires a precise definition for the nonideal pulse because rising and falling edges are not vertical. **Pulse width** is the time between the point on the rising edge where the value is 50% of the amplitude to the point on the falling edge where the value is 50% of the amplitude. The definition of pulse width is illustrated in Figure 9–26(b).

Repetitive Pulses

Any waveform that repeats itself at fixed intervals is periodic. Some examples of periodic pulse waveforms are shown in Figure 9–27. Notice that, in each case, the pulses repeat at regular intervals. The rate at which the pulses repeat is the pulse repetition frequency. The frequency can be expressed in hertz or in pulses per second. The time from one pulse to the corresponding point on the next pulse is the period, T. The relationship between frequency and period is the same as with the sine wave, $f = 1/T$.

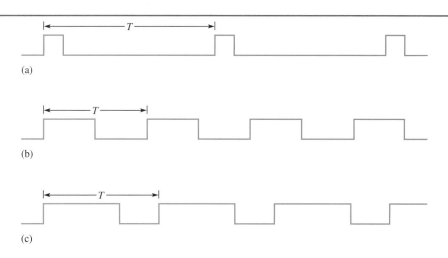

FIGURE 9–27

Examples of repetitive pulse waveforms.

An important characteristic of repetitive pulse waveforms is the duty cycle. The **duty cycle** is the ratio of the pulse width (t_W) to the period (T) and is usually expressed as a percentage.

$$\text{percent duty cycle} = \left(\frac{t_W}{T}\right)100\% \qquad (9\text{–}10)$$

CHAPTER 9

EXAMPLE 9–9

Problem
Determine the period, frequency, and duty cycle for the ideal pulse waveform in Figure 9–28.

FIGURE 9–28

Solution
Measure the period from any two corresponding points on a wave. The period is determined by observing the time from the rise of the first pulse to the rise of the second pulse.

$$T = 5\ \mu s$$

The frequency is

$$f = \frac{1}{T} = \frac{1}{5\ \mu s} = 200\ \text{kHz}$$

Determine the percent duty cycle from the pulse width and the period. In Figure 9–28, the pulse width (time from the rise to the fall) is 1 μs. Substitute into Equation 9–10.

$$\text{percent duty cycle} = \left(\frac{t_W}{T}\right)100\% = \left(\frac{1\ \mu s}{5\ \mu s}\right)100\% = 20\%$$

Question
If $f = 400$ kΩ and the duty cycle is unchanged, what happens to the pulse width?

Square Waves
A square wave is a pulse waveform with a duty cycle of 50%. Thus, the pulse width is equal to one-half of the period. A square wave is shown in Figure 9–29.

FIGURE 9–29

Square wave.

Triangular and Sawtooth Waveforms

Triangular and sawtooth waveforms are formed by voltage or current ramps. A ramp is a linear increase or decrease in the voltage or current. Figure 9–30 shows a positive and a negative ramp.

FIGURE 9–30

Ramps.

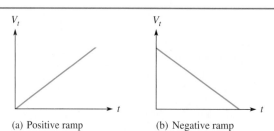

(a) Positive ramp   (b) Negative ramp

Triangular Waveforms

Figure 9–31 shows that a triangular waveform is composed of both a positive-going and a negative-going ramp having equal slopes. This implies that the time for the positive-going ramp and the negative-going ramp are equal. Like any waveform, the period is measured from one point on the wave to the corresponding point on the next wave. For the triangle waveform shown, the peaks are convenient points to use, as illustrated. In this case, the waveform has both positive and negative excursions, so it is alternating.

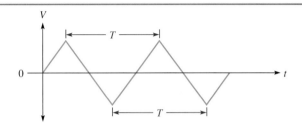

FIGURE 9–31

Alternating triangular waveform.

Sawtooth Waveforms

The sawtooth waveform is actually a special case of the triangular wave. Like the triangular wave, the sawtooth consists of two ramps, but one of the ramps is much longer than the other. Sawtooth waveforms are used in many electronic systems. For example, the electron beam that sweeps across the screen of your TV receiver, creating the picture, is controlled by sawtooth voltages and currents. One sawtooth wave produces the horizontal beam movement, and the other produces the vertical beam movement. A sawtooth voltage is sometimes called a *sweep voltage*.

Figure 9–32 is an example of a sawtooth wave, which consists of a positive-going ramp of relatively long duration, followed by a negative-going ramp of relatively short duration. Generally, the time for each ramp is specified for a sawtooth waveform. Again, the period is the time between two corresponding points on a waveform.

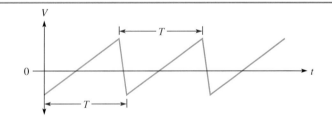

FIGURE 9–32

Alternating sawtooth waveform.

Review Questions

11. Where on a pulse waveform is the pulse width defined?
12. How is rise time defined for a pulse?
13. What is percent duty cycle?
14. What is the special name for a pulse waveform with a duty cycle that is 50%?
15. What is the difference between a triangle waveform and a sawtooth waveform?

CHAPTER 9

9-4 FUNCTION GENERATORS

Repetitive waves such as sine waves, pulses, and triangle waveforms are produced by circuits called oscillators. In the laboratory, an instrument called the function generator has an internal oscillator that allows you to produce a variety of waveforms for test purposes.

In this section, you will learn the principal features and controls for a function generator.

Nearly all electronic systems require an oscillator within the system. Oscillators are also necessary in the laboratory for testing the response of other circuits.

All oscillators convert energy from the dc power supply into a periodic waveform. There is no need for an input signal. Oscillations are used for various timing, control, or signal generation applications. A common type of oscillator returns a fraction of the output back to the input, where it is amplified. This type of oscillator is called a *feedback* oscillator. Figure 9–33 shows the general idea.

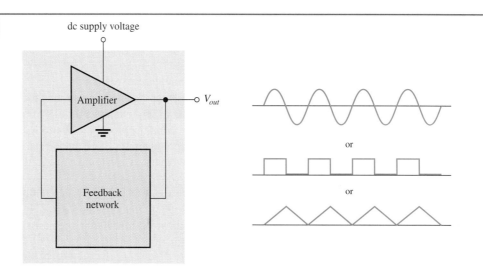

FIGURE 9–33

Basic feedback oscillator.

Laboratory Signal Sources

Certain types of laboratory instruments called *signal sources* are used to drive a circuit under test in order to measure the test circuit's response. These instruments produce waveforms within a specified range of frequencies. Generally, the output frequency and amplitude are calibrated and can be varied over the specified range. Some instruments produce only sine waves; others produce complex waveforms such as those used in a television receiver. Technicians choose the instrument to use depending on the circuit or system they are testing.

Signal sources are generally categorized by the frequencies they can produce. Broadly speaking, these frequencies are divided into three categories: low frequencies that range from less than 1 Hz to 1 MHz, radio frequencies that overlap this range from about 10 kHz to 1 GHz, and microwave frequencies that are higher than 1 GHz. Some instruments have greater ranges than these.

The most versatile and common instrument for producing a variety of low-frequency waveforms is the **function generator**. Function generators use an oscillator to provide a selection of sine, square, and triangle waveforms in a single instrument. Most also produce pulses for digital work. Because function generators are the most widely used instrument for generating sine waves in the laboratory, it will be the focus for the remainder of this section. A typical function generator is shown in Figure 9–34.

ALTERNATING CURRENT

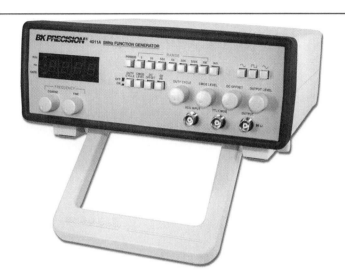

FIGURE 9–34

Precision function generator. (Courtesy of B + K Precision)

Function Generator Basic Controls

For the basic function generator, controls are divided into three groups: function selection, frequency controls, and amplitude controls. Most function generators have additional controls, but all will have these basic three.

- *Function selection* Function generators are designed to produce several different types of waves including sine waves. They all produce sine, square, and triangle waves and may include pulses and ramp (sawtooth) waveforms. Typical waves produced by function generators were introduced in Section 9–3 and are illustrated in Figure 9–35.

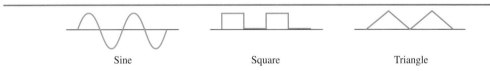

FIGURE 9–35

Typical function generator waveforms.

- *Frequency range* The frequency range is the span of frequencies over which the instrument performance is specified. For many function generators, the frequency is selected by first selecting a range switch, which chooses a set of frequencies, and then dialing in the exact frequency with a dial control. Other function generators use a digital keypad to select the frequency. The accuracy of frequency settings depends on the internal oscillator as well as on the type of dial controls (mechanical or digital). Typically, function generators can produce frequencies from less than 1 Hz to 1 MHz or more; but some generators can cover a much larger range.
- *Output amplitude* Typically, function generators produce low-voltage outputs that range from a few hundred millivolts to several volts, depending on the generator. Some generators allow you to set very precise outputs with digital controls or set very small voltages by reducing the output a certain amount with an attenuator. An **attenuator** is a calibrated resistive divider that reduces a waveform by a known amount. Generally, the output attenuator is selected in fixed increments.

Other Function Generator Controls

In addition to the basic controls, nearly all function generators allow you to add or subtract a dc component to the output signal. The control is usually labeled **dc offset**. The function of the dc offset control is illustrated in Figure 9–36.

FIGURE 9–36

Adding a dc component to a sine wave from a function generator using the dc offset control.

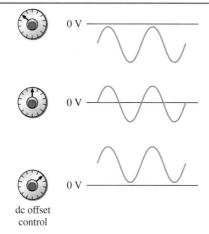

Another common control is the **symmetry** (or *duty cycle*) control. This control allows you to adjust the rise and fall time of sine and sawtooth waves and the duty cycle of pulses. Figure 9–37 reviews the definition of duty cycle and illustrates how the symmetry control affects the duty cycle.

FIGURE 9–37

The symmetry control varies the duty cycle of the pulse output.

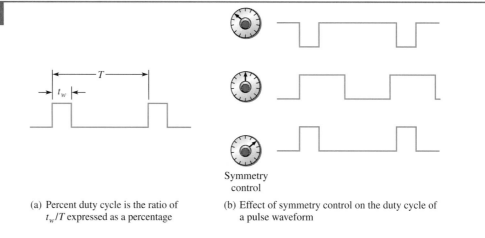

(a) Percent duty cycle is the ratio of t_w/T expressed as a percentage

(b) Effect of symmetry control on the duty cycle of a pulse waveform

You may also see a control labeled *sweep*. This control has the effect of changing the output frequency in a regular manner so that it repeatedly sweeps through a specified range of frequencies. The frequency response can be directly plotted on an oscilloscope. The sweep control is useful for testing the frequency response of filters or other frequency-dependent circuits. The frequency response of filters is discussed in Sections 12–6 and 13–5.

Output Resistance

The *output resistance* is an important specification for a function generator. From the perspective of the output terminals, the generator can be thought of as a Thevenin circuit. Recall (Section 6–3) that a Thevenin circuit consists of a voltage source (V_{TH}) in series with a resistance (R_{TH}). This resistance is the equivalent internal resistance that is in series with the output. Normally, the output resistance is marked on the output terminals. Typical values are 50 Ω, 75 Ω, or 600 Ω.

It is important to know the output resistance of the generator because whenever you connect a load to a generator that has a finite output resistance, the output voltage will change due to a loading effect. Loading occurs because current in the output resistance drops some voltage, reducing the output as shown in Example 9–10.

EXAMPLE 9–10

Problem
A function generator with an output resistance of 600 Ω is set to 5.0 Vpp and then is connected to a 1.0 kΩ resistor. What is the voltage across the resistor after the generator is connected?

Solution
The function generator looks like a voltage source in series with its output resistance, as shown in Figure 9–38. Calculate the loaded voltage from the voltage-divider theorem.

$$V_{out} = \left(\frac{R_L}{R_L + R_{TH}}\right) V_{in} = \left(\frac{1000\ \Omega}{1000\ \Omega + 600\ \Omega}\right) 5.0\ \text{Vpp} = \mathbf{3.13\ Vpp}$$

Question
What is the effect of loading if a function generator with a 50 Ω output resistance were used instead?

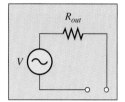

FIGURE 9–38

Applications

The basic waveforms (sine, square, and triangle) are used in many tests of electronic circuits and equipment. One of the common applications of a function generator is to inject a sine wave at some point in a circuit to test the circuit's response to various frequencies. The output is normally observed on an oscilloscope; it is relatively easy to ascertain if the circuit is operating properly by checking the output signal at various frequencies. Other parameters, such as distortion or gain, can be measured by comparing the input from the generator with the output from the circuit under test. You will gain experience in these tests as you use a function generator in the laboratory.

Review Questions

16. What is an oscillator?
17. What is a function generator?
18. What are the three basic controls on all function generators?
19. What is the purpose of the symmetry control on a function generator?
20. What is meant by the output resistance of a function generator?

9–5 OSCILLOSCOPES

The oscilloscope ("scope" for short) is an important general-purpose measuring instrument that allows you to visualize circuit operation.

In this section, you will learn the four major sections in a block diagram of an oscilloscope and the controls associated with each section.

Analog and Digital Oscilloscopes

The **oscilloscope** is a versatile instrument that shows a graph of the voltage in a circuit as a function of time. Figure 9–39 shows an analog oscilloscope. There are two basic types of oscilloscope—analog and digital. Both types perform the same major function; that is,

CHAPTER 9

FIGURE 9-39

Analog oscilloscope. (Courtesy of B+K Precision)

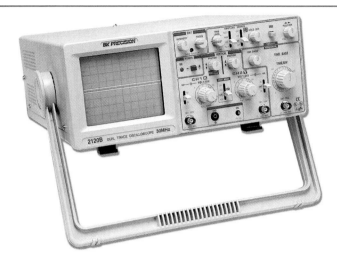

HISTORICAL NOTE

The first true digital oscilloscope was inspired by a high school science project in 1970. Student Joel Shackleford corroborated with his stepfather, Robert Schumann, to work on the project. Schumann set to work on the project at his company, Nicolet Instruments. After months of development, the first commercial digital oscilloscope, the 1090A Explorer, was marketed by Nicolet in 1972. Nicknamed "Schumann's folly" by company executives, one executive bet Schumann that the company would not sell 100 digital scopes. The loser was to roll a peanut with his nose for one mile down State Street in Madison, Wisconsin. Although Schumann won the bet, it was never paid off.

graphing the voltage on the vertical (y-axis) versus time on the horizontal (x-axis). Analog scopes plot the waveform instantaneously on a cathode ray tube (CRT). Digital oscilloscopes are rapidly replacing analog scopes because of their ability to store waveforms, because of measurement automation, and because of many other features such as connections for computers. No matter which type you use, the basic controls and functions are the same; however, a digital scope has more features and certain capabilities not available on analog scopes.

Both types of scopes have a grid on the display called the *graticule,* which is composed of solid lines that form vertical and horizontal divisions. Small "tic" marks subdivide the major divisions. The grid is important because the major controls for time and voltage are calibrated in time per division and volts per division. Typically, the display area has 10 divisions in the horizontal direction and 8 divisions in the vertical direction, as illustrated in Figure 9–40. These blocks are generally 1 cm by 1 cm.

FIGURE 9-40

Typical oscilloscope display area.

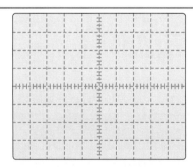

Oscilloscope Block Diagram

All general-purpose oscilloscopes, analog or digital, contain four functional blocks (or sections), as illustrated in Figure 9–41. Although there are many details shown within the blocks and they vary with different scopes, you should focus on the four main blocks. Shown within the four major blocks on the figure are the most important controls.

- *Vertical section*   Each of two input channels is connected to the vertical section. The input signals are amplified (or attenuated) as necessary to provide the proper scaling of the input

Block diagram of a basic oscilloscope. There are some differences in how the signal is processed between analog and digital scopes, but the four essential blocks are the same on both.

FIGURE 9–41

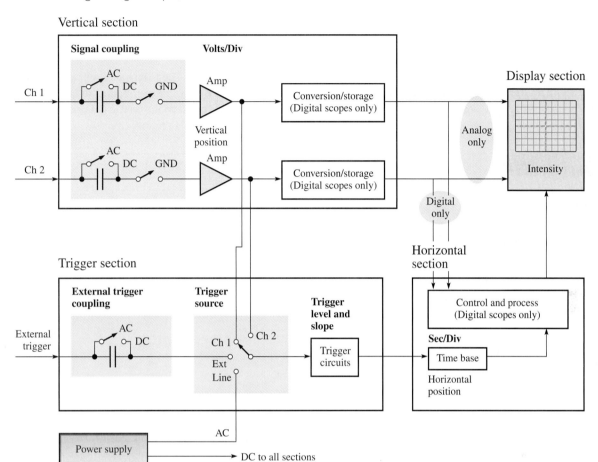

signal for display or further processing. This is done with the Volts/Div control; there is one for each channel. The vertical section controls allow you to adjust the amplitude of the displayed signal and control how the signal is coupled to the scope.

- *Trigger section* The trigger section "picks off" a sample of the input signal from one of the channels or another source and synchronizes the horizontal display. This action causes the signal to appear to stop, allowing you to examine the signal. The trigger circuits determine when the signal will start on the horizontal axis. You can choose one of several trigger sources with the Source control to generate the synchronizing trigger. The trigger controls include a Level and a Slope control that position the exact "pick-off" point on the signal.
- *Horizontal section* The horizontal axis is calibrated in units of time, and the horizontal section controls how much time elapses for the display. Because the beam moves across the display in a motion reminiscent of using a broom, the controls in this section are also referred to as the *sweep* controls. The most important control in the horizontal section is the Sec/Div control.
- *Display section* The display section contains the screen and controls for proper intensity and focus (analog scopes). There may be other features such as a probe compensation jack for checking calibration of the probes.

CHAPTER 9

Basic Scope Controls

Differences in various oscilloscopes are dependent on the type of scope (digital or analog) and the manufacturer. Despite these differences, certain controls are common to all scopes. The numbers associated with various controls will be printed on the front panel or they will be displayed on the screen. For digital scopes, and advanced analog scopes, these controls may be set by menus, like those common in computers.

Figure 9–42 shows a generic oscilloscope front panel. The front panel will always have controls for a common section grouped together. Look for these groupings to help you understand a control. In our generic scope, each of the four sections (vertical, horizontal, trigger, and display) is identified with the controls used in that section.

FIGURE 9–42 Front panel of a generic oscilloscope. The controls shown are for both digital and analog scopes.

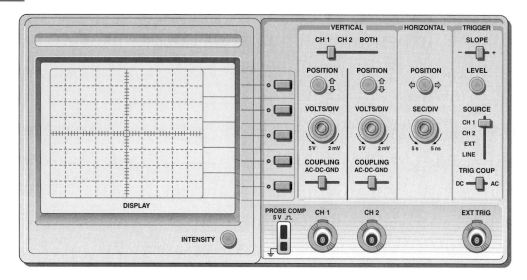

The *vertical* controls are shown to the right of the screen. At the top is a control that allows you to select which of the two channels, or both, is selected for display. Each of the other controls in the vertical section is duplicated for the two channels. The Position control allows you to move the selected trace up or down. You can use it in conjunction with the Coupling in the Gnd position to set 0 V to a convenient level on the display. The Volts/Div control sets the scale for the y-axis. Along with the course setting, a fine adjust is available. On our generic scope, the concentric knob on the Volts/Div control is for fine adjusting the vertical scale. The Coupling control is the last control in the vertical section. It has three positions: AC, DC, and Gnd (ground). Normally, the scope should be coupled with dc, unless you need to block dc with a capacitor (discussed in the next chapter). The Gnd position does not actually ground the input—instead it opens the input; it allows you to see the ground reference position on the display when the input is disconnected from the scope.

In addition to the controls mentioned, the vertical section has jacks for the input signals. The input signals should always be connected to the scope through a probe. Probes are provided with the oscilloscope by the manufacturer and are part of the measuring system.

To the right of the vertical controls are the *horizontal* controls. The top control is the horizontal position control, used for fine adjusting the displayed signal along the x-axis. The most important horizontal control is the Sec/Div control. This control determines how much time is displayed on the horizontal axis. It too has a course and a fine adjust knob; in our generic scope, it is shown as a concentric knob. The standard scope will have 10 horizontal divisions across the screen. Thus, if the Sec/Div control is set to 5.0 μs/div, the total time on the screen is 50 μs (5.0 μs/div × 10 div = 50 μs). The proper setting of this control depends on the speed of the signal and the particular part of the signal that you want to observe. If you

ALTERNATING CURRENT

are making a time measurement, you need to be sure the controls are calibrated to read the correct Sec/Div, as some scopes allow you to have controls in an uncalibrated position.

On the far right are the *trigger* controls. Recall that the trigger controls determine the pick-off point on the signal that synchronizes the display. The trigger Slope control determines if the signal is rising (+) or falling (−) when the trigger occurs. The Level control determines the voltage level (above or below ground) when the trigger occurs. Together these important controls will help you obtain a stable display.

Under the trigger Slope and Level controls is the Source selector switch. This switch is used to determine whether the trigger is derived from channel 1 or channel 2, or another source. Usually, one of the channels is selected for the source of triggers. For complicated or specialized signals such as encountered in TV or radar, an external trigger can be selected. In this case, a trigger signal must be connected to the Ext Trig input. The Line position synchronizes the horizontal display with the ac power applied to the scope. This is useful if you are looking at a signal that is timed to the power line (or if you want to confirm power line interference).

Oscilloscope Probes

Signals should *always* be coupled into an oscilloscope through a probe. A probe is the interface between the scope and the circuit under test, and it performs two key functions. First, it reduces the loading of the scope on the circuit so that the circuit is effectively unchanged by the presence of the instrument. Secondly, it corrects the imperfect input of the scope so that a true picture of the waveforms in the circuit under test can be observed.

Various types of probes are provided by manufacturers, but the most common type is a 10:1 attenuating probe that comes with most general-purpose oscilloscopes. These probes have a short ground lead that should be connected to a nearby circuit ground point to avoid oscillation and power line interference. The probe tip is touched to the test circuit and passes the signal through a flexible, shielded cable to the oscilloscope. The shielding has the added benefit of protecting the signal from external noise pickup.

Begin any session with the oscilloscope by checking the probe compensation on each channel. The procedure is illustrated for a typical digital scope in Figure 9–43. Adjust the

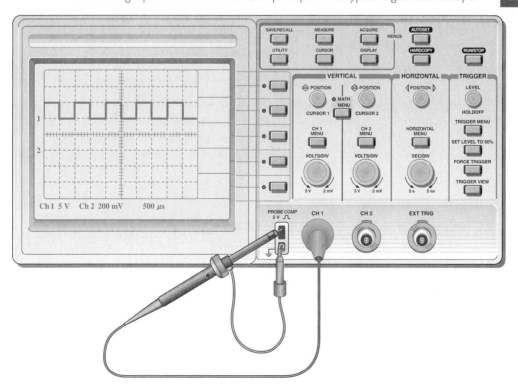

Connecting a probe to the Probe Comp output on a typical digital oscilloscope.   **FIGURE 9–43**

CHAPTER 9

probe (if necessary) for a flat-topped square wave while observing the scope's calibrator output. A small adjustment screw is usually on the probe. Figure 9–44(a) illustrates the correct waveform you should observe when you compensate a probe, and Figure 9–44(b) shows the adjustment. A properly compensated probe assures that the scope and probe are matched, and the waveform will not be distorted.

FIGURE 9–44

Probe compensation waveforms and typical adjustment.

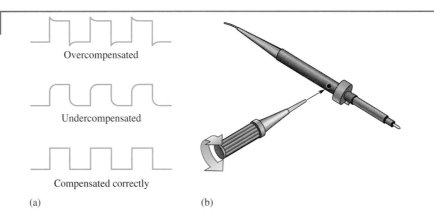

Oscilloscope Measurements

An oscilloscope is one of the most versatile instruments for making measurements of various parameters on signals. Many scopes use automated methods to make standard measurements such as the frequency or period of a signal. Results can be displayed directly on the display with little input from the operator. Other measurements require a manual setting of the controls and skilled observation of the waveform. You can develop an intuitive understanding of oscilloscope measurements by reviewing how these manual measurements are made.

Generally, most oscilloscope measurements are either a voltage measurement or a time measurement. The y-axis is calibrated in Volts/Div, and the x-axis is calibrated in Sec/Div; therefore, for voltage and time measurements, you will depend on these calibrated values when making measurements on a waveform. Some scopes can be taken out of calibration with the vernier controls, so you need to check these if you are depending on the dial settings. The following three examples illustrate basic measurements with specific suggestions for making the measurement.

EXAMPLE 9–11

Problem

The pulse width of a pulse is the time between the rise and the fall measured at the 50% level, as discussed earlier. For a pulse width measurement, the scope is set up to display the rise and fall over a large area of the display and the horizontal calibration is checked. The Sec/Div control is set to 1.0 ms/div, and a pulse is displayed as shown in Figure 9–45. What is the pulse width?

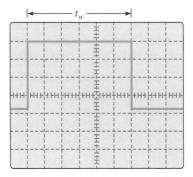

FIGURE 9–45

Solution
The rise time and fall time are much faster than the pulse width, so finding the exact center of the pulse is not necessary in this case.

Find the pulse width by multiplying the time per division on the Sec/Div control by the number of divisions observed on the display for the pulse. Counting from the rise to the fall shows 6.0 divisions for the pulse width.

$$t_W = (\text{sec/div})(\text{number of divisions}) = (1.0 \text{ ms/div})(6.0 \text{ div}) = \mathbf{6.0 \text{ ms}}$$

Notice that the divisions (div) cancel, leaving the time in ms.

Question
If the Sec/Div control is changed to 2 ms/div, how will the display change for this pulse?

EXAMPLE 9–12

Problem
An oscilloscope display is shown in Figure 9–46. The Sec/Div control is set to 20 μs/div and the Volts/Div is set to 2.0 V/div for the displayed channel. What are the period, the frequency, and the peak-to-peak voltage?

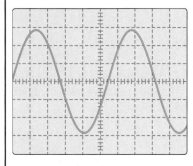

FIGURE 9–46

Solution
Find the period by multiplying the time per division on the Sec/Div control by the number of divisions observed on the display for one complete cycle. You can measure the number of divisions for one cycle along any horizontal line. It is observed to be 5.5 divisions.

$$T = 20 \text{ }\mu\text{s/div} \times 5.5 \text{ div} = \mathbf{110 \text{ }\mu\text{s}}$$

The frequency is the reciprocal of the period.

$$f = \frac{1}{T} = \frac{1}{110 \text{ }\mu\text{s}} = \mathbf{9.09 \text{ kHz}}$$

Find the peak-to-peak voltage by multiplying the Volts/Div setting by the number of vertical divisions between the minimum and maximum excursion of the wave. The number of divisions is counted and noted to be 5.8 div.

$$V_{pp} = (2.0 \text{ V/div})(5.8 \text{ div}) = \mathbf{11.6 \text{ Vpp}}$$

Notice that the "div" again cancels, leaving the voltage.

Question
What is the rms voltage for the wave?

EXAMPLE 9–13

Problem
What is the duty cycle for the pulse shown in Figure 9–47? The Sec/Div control is set to 10 μs/div.

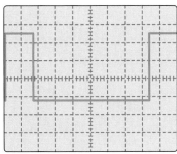

10 μs/div

FIGURE 9–47

Solution
Measure the period and the pulse width. Find the period by multiplying the time per division by the number of divisions between the leading edge of one pulse and the leading edge of the next pulse. The period is

$$T = 10 \; \mu\text{s/div} \times 8.3 \; \text{div} = 83 \; \mu\text{s}$$

To measure the pulse width accurately, change the Sec/Div control to 2.0 μs/div. This spreads the waveform across the screen as shown in Figure 9–48. You may want to adjust the pulse so that the center (50%) level can be read easily as shown.

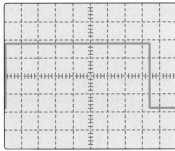

2.0 μs/div

FIGURE 9–48

The pulse width is

$$t_W = 8.3 \; \text{div} \times 2.0 \; \mu\text{s/div} = 16.6 \; \mu\text{s}$$

Substitute the measured pulse width and period into Equation 9–10 to get the percent duty cycle.

$$\text{percent duty cycle} = \left(\frac{t_W}{T}\right)100\% = \left(\frac{16.6 \; \mu\text{s}}{83 \; \mu\text{s}}\right)100\% = \mathbf{20\%}$$

Question
Assume that the Volts/Div control is set to 2.0 V/div. What is the amplitude of the pulse?

Phase Shift Measurements

Figure 9–49(a) shows two waves of the same frequency but different amplitudes. To measure the phase shift between the two waves, compare the time shift between two consistent points on the waves to the period. This is the fraction of 360° that one wave is shifted with respect to the other. The best way to make an accurate measurement of the time difference between the two waves is to first adjust the Volts/Div and the vernier on the Volts/Div control on one channel until the two waves *appear* to have the same amplitude and they are centered on the vertical axis. To assure the waves are centered vertically, you can first ground both signals, and superimpose the horizontal lines of both channels using the vertical position controls. Then ac couple both channels. The waveforms are adjusted to appear as illustrated in Figure 9–49(b).

When the waves appear to be the same, a time difference measurement will be more accurate because any misalignment will tend to be negated. To measure the phase shift, note that one full cycle corresponds to 360°. In Figure 9–49, a full cycle is 8.5 divisions or 360°. The waves are separated by 0.7 divisions, which represent

$$\frac{0.7 \text{ div}}{8.5 \text{ div}} \times 360° = 30.4°$$

The method given here works equally well for analog and digital oscilloscopes.

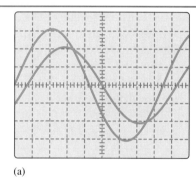
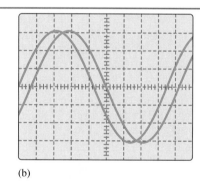

FIGURE 9–49

Phase shift measurement.

(a) (b)

Features of Digital Oscilloscopes

The digital oscilloscope uses a fast analog-to-digital converter (ADC) in the vertical section of each channel to convert the input voltage into a series of numbers. The digitizer samples the input at a uniform rate called the sample rate; the optimum sample rate depends on the speed of the signal. The data are plotted on the display area of the scope; the appearance is that of a continuous trace that represents the original signal.

Although the process of obtaining the data and plotting it are very different from an analog scope, the result is the same as an analog scope, where the signal is sent directly to the display section. The major advantages to digitizing the data are accuracy, waveform analysis, and for viewing hard to see events (such as one-time signals). For these reasons, digital oscilloscopes are generally preferred.

Important parameters with DSOs (digital storage oscilloscopes) include the resolution, maximum digitizing rate, and the size of the acquisition memory as well as the available analysis options. The resolution is determined by the number of bits digitized by the ADC. A low-resolution DSO may use only six bits (one part in 64). A typical DSO may use 8 bits, with each channel sampled simultaneously. High-end DSOs may use 12 bits. The maximum digitizing rate is important to capture rapidly changing signals; typically, the maximum rate is 1 Gs/s (gigasample/second). The size of the memory determines the length of time the sample can be taken; it is also important in certain waveform measurement functions.

CHAPTER 9

Triggering

One useful feature of digital storage oscilloscopes is their ability to capture waveforms either *before* or *after* the trigger event. Any segment of the waveform can be captured for analysis. **Pretrigger capture** refers to acquisition of data that occurs before a trigger event. This is possible because the data is digitized continuously, and a trigger event can be selected to stop the data collection at some point within the sample window. This is particularly useful if a circuit has an occasional problem such as a random noise spike or "glitch." With pretrigger capture, the scope can be triggered on the fault condition, and the signals that immediately preceded the fault condition can be observed. By employing pretrigger capture, one can analyze trouble leading to the fault. A similar application of pretrigger capture is in certain types of failure analysis studies where the events leading to the failure are of interest to an investigator but the failure itself causes the scope triggering.

Automated Measurements

One of the most important features on a digital scope is its ability to analyze the waveform and show certain important parameters such as frequency or rms voltage. In addition, digital scopes can measure and display the time or voltage between a set of two user-controlled markers (called *cursors*). In this way, the user can read the time or voltage directly between the cursors and avoid the possibility of misreading a dial or making a calculational mistake. Results are generally more accurate than reading the number of divisions and multiplying by a dial setting. The voltage of a given wave can be displayed in various forms also. For example, a sine wave can be analyzed for its peak-to-peak or its rms voltage without the user having to read the settings of the controls.

Because of the large number of functions that can be accomplished by even basic DSOs, manufacturers have largely replaced the large number of controls with menu options, similar to computer menus. The settings of controls can be easily viewed on the display and options can be selected, depending on the control. CRTs have been replaced by liquid crystal displays, similar to those on laptop computers. As an example, the display for a typical digital storage oscilloscope is Figure 9–50. Although this is a basic display, the information available to the user right at the display is impressive.

FIGURE 9–50

The display area for a typical digital oscilloscope.

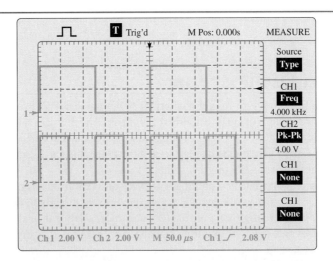

The user can readily see the waveform and the settings of all important controls such as the time/division and the volts/division. The display also shows exactly where on the waveform triggering occurs. In addition, specific measured quantities can be read out on the display, such as the frequency or amplitude of a waveform.

Two Specific DSOs

A photograph of a TDS1012 and a TDS2024 is shown in Figure 9–51. Operation is similar to that of an analog scope except more of the functions are menu controlled and a number of

ALTERNATING CURRENT

common measurements are done automatically by the scope. On these oscilloscopes, multiple menus are accessed to select various controls and options. For example, the Measure function brings up a menu from which the user can select various automated measurements including voltage, frequency, period, rise and fall time of pulses, pulse width, and averaging.

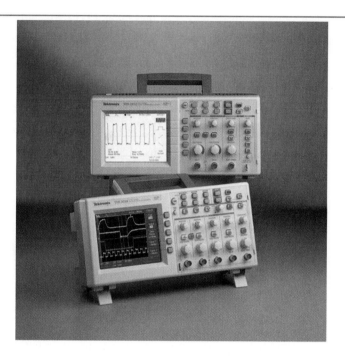

FIGURE 9-51
The TDS1012 and the TDS2024. (Used with permission from Tektronix, Inc.)

Review Questions

21. What are the four major sections in an oscilloscope?
22. What is the purpose of the Level and Slope controls on an oscilloscope?
23. What is the purpose of the Sec/Div control on an oscilloscope?
24. What are two reasons for using a scope probe to connect a signal to an oscilloscope?
25. What is meant by pretrigger capture?

CHAPTER REVIEW

Key Terms

Duty cycle The ratio of the pulse width to the period for a pulse waveform. It is often expressed as a percentage.

Fall time The time required for a pulse to go from 90% of its amplitude to 10% of its amplitude.

Function generator A versatile instrument for producing a variety of low-frequency waveforms.

325

Oscillator An electronic circuit that produces a repetitive waveform.

Oscilloscope A versatile instrument that shows a graph of the voltage in a circuit as a function of time.

Peak-to-peak value The magnitude of a sine wave measured from the maximum negative level to the maximum positive level.

Peak value The magnitude of a sinusoidal wave measured from the center to its maximum positive level.

Phase shift The angular difference between two waves of the same frequency.

Phasor A rotating vector.

Radian An angular measurement based on a circle. One radian is the angle formed when the arc of a circle is equal to its radius; it is equal to 57.3°.

Rise time The time required for a pulse to go from 10% of its amplitude to 90% of its amplitude.

Important Facts

- A sinusoidal waveform has the same shape as the trigonometric sine function.
- Three angle measurements are degrees, radians, and grads.
- The phase of a sinusoidal wave is an angular measurement that specifies the position of the wave relative to a reference.
- A phasor is a rotating vector. The projection of its rotation produces a sinusoidal wave.
- A pulse waveform is one that changes from one voltage or current level (baseline) to an amplitude level, and then, after an interval of time, returns to the original level.
- A square wave is a pulse waveform with a duty cycle of 50%.
- Triangular and sawtooth waveforms are formed by voltage or current ramps.
- Function generator controls include frequency, amplitude, dc offset, and symmetry.
- A function generator with a sweep control can provide the signal to show the response of a frequency-dependent circuit.
- Oscilloscopes have four functional blocks: vertical, horizontal, trigger, and display.
- Some important oscilloscope controls are *vertical:* volts/div, input coupling, and vertical position; *horizontal:* sec/div and horizontal position; *trigger:* source, level and slope controls.

Formulas

Conversion from degrees to radians:

$$\text{rad} = \left(\frac{2\pi \text{ rad}}{360°}\right) \times \text{degrees} \tag{9-1}$$

Conversion from radians to degrees:

$$\text{deg} = \left(\frac{360°}{2\pi \text{ rad}}\right) \times \text{rad} \tag{9-2}$$

Conversion from peak to peak-to-peak:

$$V_{pp} = 2V_p \tag{9-3}$$
$$I_{pp} = 2I_p \tag{9-4}$$

Conversion from peak to rms:

$$V_{rms} = 0.707 V_p \tag{9-5}$$

$$I_{rms} = 0.707 I_p \tag{9-6}$$

Instantaneous sinusoidal voltage:

$$v = V_p \sin\theta \tag{9-7}$$

Voltage waveform that is lagging a reference wave by an angle of ϕ:

$$v = V_p \sin(\theta - \phi) \tag{9-8}$$

Voltage waveform that is leading a reference wave by an angle of ϕ:

$$v = V_p \sin(\theta + \phi) \tag{9-9}$$

Percent duty cycle:

$$\text{Percent duty cycle} = \left(\frac{t_W}{T}\right) 100\% \tag{9-10}$$

Chapter Checkup

Answers are at the end of the chapter.

1. The largest value of the sine function is
   (a) +0.5
   (b) +1
   (c) +10
   (d) infinity

2. At 180°, the value of the sine function is
   (a) −1
   (b) 0
   (c) +1
   (d) infinite

3. The number of radians in a circle is
   (a) 1
   (b) 2
   (c) 4
   (d) more than 6

4. If the period of a sine wave is 10 ms, the frequency is
   (a) 100 Hz
   (b) 1 kHz
   (c) 10 kHz
   (d) none of these answers

5. If a sine wave has a peak voltage of 100 V, the rms voltage is
   (a) 50 V
   (b) 141 V
   (c) 200 V
   (d) none of these answers

6. To say one wave is phase shifted from another, the two waves must have
   (a) different frequencies
   (b) different amplitudes
   (c) the same frequency
   (d) the same amplitude

7. A phasor is a
   (a) constant voltage
   (b) pulse
   (c) phase shift
   (d) rotating vector

8. The period of a wave that completes twenty cycles in 100 μs is
   (a) 5 μs
   (b) 20 μs
   (c) 100 μs
   (d) 2 ms

9. The rise time of a pulse is normally measured between
   (a) the 0% and 50% levels
   (b) the 50% and 100% levels
   (c) the 10% and 90% levels
   (d) the 30% and 70% levels

10. Duty cycle is a ratio of the pulse width to the
    (a) amplitude (b) period
    (c) frequency (d) rise time

11. The duty cycle of a square wave
    (a) varies with frequency (b) varies with pulse width
    (c) both (a) and (b) (d) is 50%

12. For some applications such as moving a beam in an oscilloscope, a sawtooth waveform is referred to as a
    (a) triangle (b) sweep
    (c) ramp (d) stairstep

13. A control on a function generator that allows you to change the duty cycle of a pulse waveform is
    (a) dc offset (b) amplitude
    (c) symmetry (d) frequency

14. The oscilloscope controls that position the exact point on a signal for generating a trigger are
    (a) Slope and Source (b) Level and Slope
    (c) Sec/Div and Source (d) Volts/Div and Sec/Div

15. If an oscilloscope coupling control is set to the Ground position, the input is
    (a) disconnected
    (b) connected to a high resistance
    (c) shorted to ground
    (d) connected to a low resistance

16. If the Sec/Div control of an oscilloscope is set to 2 μs/div and a pulse is observed to rise and fall in 8 divisions, the pulse width is
    (a) 0.25 μs (b) 2 μs
    (c) 8 μs (d) 16 μs

Questions

Answers to odd-numbered questions are at the end of the book.

1. Why are periodic sine waves considered the fundamental wave in electronics?
2. At what two angles does the sine function have a value of zero?
3. At what angle does the sine function have a value of $+1$?
4. How many radians are in one complete cycle?
5. What are two ways to produce electrical sine waves?
6. What is meant by the amplitude of a sinusoidal wave?
7. How are frequency and period related?
8. What are the two quantities required to specify a phasor?

9. How is a phasor represented?
10. In generating a sine wave, which way does a phasor rotate: clockwise or counterclockwise?
11. With respect to a phase shift, what is meant by a reference waveform?
12. Does the term *phase shift* have meaning if two waves of the same frequency have different amplitudes?
13. Does the term *phase shift* have meaning if two waves of the same amplitude have different frequencies?
14. What is meant when a sinusoidal wave is said to be lagging?
15. What is the definition of a pulse?
16. What is meant by an ideal pulse?
17. How is the *amplitude* of a pulse defined?
18. What is the percent duty cycle for a square wave?
19. What type of waveform is a sweep voltage?
20. What type of output is produced by all oscillators?
21. What are microwaves?
22. What are common waveforms from function generators?
23. What is the purpose of the dc offset control on a function generator?
24. Assume a 50 Ω load is connected across the output of a function generator. Will the output be changed more by a generator with a 50 Ω output resistance or one with a 600 Ω output resistance? Explain your answer.
25. What are the two basic types of oscilloscopes?
26. What is the purpose of the trigger Source control on an oscilloscope?
27. When would you use Line triggering on an oscilloscope?
28. Why is it a good idea to always connect a signal to the oscilloscope through a probe?
29. How is a scope probe compensated?
30. What are cursors on an oscilloscope?
31. What are three common automated measurements on a digital oscilloscope?

Basic Problems

PROBLEMS

Answers to odd-numbered problems are at the end of the book.

1. Convert each of the following numbers of degrees to radians:
   (a) 30° (b) 90°
   (c) 150° (d) 250°
   (e) 325°

2. (a) How many degrees are in a half-cycle?
   (b) How many radians are in a half-cycle?

3. How many degrees are represented in each of the following numbers of radians?
   (a) $\pi/4$ rad (b) $2\pi/3$ rad
   (c) 1.5 rad   (d) 2.5 rad
   (e) 2π rad

4. If a sinusoidal current waveform has an amplitude of 2.0 mA, what is the peak-to-peak current?

5. If a sinusoidal voltage waveform has an amplitude of 21 V, what is the peak-to-peak voltage?

6. Draw one cycle of a sinusoidal wave with an amplitude of 20 V and a period of 100 μs. Label the y-axis with voltage and the x-axis with time. Show on your drawing the time when the voltage is a maximum.

7. Assume a small heater can run on ac or dc. If it is connected to a 12 Vdc source, it delivers 200 W. What power will it deliver if it is connected to a 12 Vrms ac source?

8. A sinusoidal wave has a peak-to-peak voltage of 68 Vpp. What is the rms voltage for this source?

9. What is the frequency for a wave that has a period of 250 μs?

10. A sinusoidal wave is displayed on an oscilloscope as shown in Figure 9–52. Assume the Volts/Div control is set to 10 V/div. What are the peak voltage, the peak-to-peak voltage, and the rms voltage of the wave?

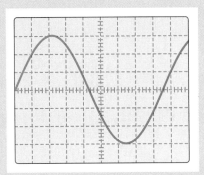

FIGURE 9–52

11. A periodic wave shows three complete cycles in 1 ms.

    (a) What is the period?

    (b) What is the frequency?

12. Write the equation for a reference sinusoidal voltage waveform with an amplitude of 17 V.

13. What is the rms voltage of the waveform in Question 12?

14. Write the equation for a sinusoidal voltage with an amplitude of 20 V that is lagging the reference by 40°.

15. What is the instantaneous voltage for a sine wave that has a peak voltage of 10 V at each of the following angles?

    (a) 25°   (b) 105°
    (c) 225°   (d) 330°

16. What is the rms voltage for a voltage waveform with the following equation: $v = 14.1$ V $(\sin \theta)$?

17. Assume a voltage waveform is expressed by the equation $v = 35$ V $\sin(\theta - 22°)$.

    (a) What is the amplitude?

    (b) What is the phase shift?

    (c) Is the wave leading or lagging the reference?

18. Draw a pulse wave with a duty cycle of 20% and a period of 200 μs. Label the time on your drawing.

19. A function generator with an output resistance of 600 Ω is set to 12 Vpp and then is connected to a 600 Ω resistor. What is the voltage across the resistor after the generator is connected?

20. For the display shown in Figure 9–53, assume the Volts/Div control is set to 5 V/div and the Sev/Div control is set to 2 μs/div.

    (a) What is the rms voltage?

    (b) What is the frequency?

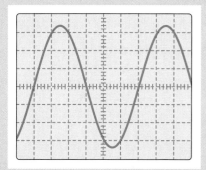

FIGURE 9–53

21. For the display shown in Figure 9–53, assume the Volts/Div control is set to 50 mV/div and the Sev/Div control is set to 10 μs/div.

    (a) What is the rms voltage?

    (b) What is the frequency?

Basic-Plus Problems

22. How many cycles of a 20 kHz wave will occur in 250 μs?

23. A pulse wave has a 25% duty cycle and a frequency of 40 kHz.

    (a) What is the period?

    (b) What is the pulse width?

24. A 100 Ω resistor is connected across the output of a function generator. If the output voltage drops to 2/3 of the level before the resistor was connected, what is the output resistance of the function generator?

25. What setting of the Sec/Div control will allow one full cycle of a 200 Hz signal to be displayed on an oscilloscope? Assume there are 10 divisions on the horizontal axis.

26. What setting of the Volts/Div control will show the maximum possible signal that is still within the displayed area for a 6 Vrms sinusoidal wave? Assume there are 8 divisions on the vertical axis.

Example Questions

ANSWERS

9–1: 1.75 rad

9–2: 86°

9–3: 339 Vpp

9–4: 2 kHz

9–5: The results for the next 180° are summarized in Table 9–2 (from an Excel worksheet). Notice that the voltages are identical to Table 9–1, except for the sign. The values at 180° and 360° have a very tiny error due to the algorithm.

Table 9–2

θ (°)	sin θ (°)	v = 12 sin θ (V)
180	1.23E-16	1.47E-15
190	−0.17365	−2.08378
200	−0.34202	−4.10424
210	−0.5	−6
220	−0.64279	−7.71345
230	−0.76604	−9.19253
240	−0.86603	−10.3923
250	−0.93969	−11.2763
260	−0.98481	−11.8177
270	−1	−12
280	−0.98481	−11.8177
290	−0.93969	−11.2763
300	−0.86603	−10.3923
310	−0.76604	−9.19253
320	−0.64279	−7.71345
330	−0.5	−6
340	−0.34202	−4.10424
350	−0.17365	−2.08378
360	−2.5E-16	−2.9E-15

9–6: If sine wave B is the reference, then in Figure 9–15(a), sine wave A leads by 45°. In Figure 9–15(b), sine wave A lags by 30°.

9–7: −20 V

9–8: $v = 40$ V sin θ

9–9: The pulse width will be smaller.

9–10: The output voltage will be 4.76 Vpp.

9–11: The pulse will show three divisions.

9–12: 4.1 Vrms

9–13: 7.0 V

Review Questions

1. One radian is the angle formed when the arc of a circle is equal to its radius; it is equal to 57.3°.
2. 90°
3. 14.1 V
4. 500 Hz
5. 50 ns
6. Phase shift is the angular difference between *two* waves. It is meaningless to talk about a shift unless one wave is compared to another, the reference.

7. A normal vector has magnitude and direction. A phasor is a rotating vector, so the direction component is constantly changing.

8. Down, along the negative y-axis.

9. $v = 20 \sin(\theta - 90°)$

10. $v = 12.9$ V

11. At the 50% level

12. The time between the 10% level and the 90% level of a rising edge of a pulse

13. It is the ratio of the pulse width to the period, expressed as a percentage.

14. A square wave

15. A triangle waveform has a positive and a negative ramp of equal duration and slope. A sawtooth waveform has a positive ramp that is relatively long and a negative ramp that is of short duration.

16. An electronic circuit that produces a repetitive waveform.

17. A laboratory instrument for producing a variety of low-frequency waveforms including sine, square, and triangle waveforms. Some also produce pulses.

18. Function selection, frequency and amplitude

19. The symmetry control allows you to adjust the rise and fall time of sine and sawtooth waves and the duty cycle of pulses.

20. The output resistance is the Thevenin resistance presented by the generator "looking back" from the output.

21. Vertical, horizontal, trigger, and display

22. The Level and Slope controls determine when the trigger occurs.

23. The Sec/Div control determines the amount of time displayed on the horizontal axis.

24. A scope probe reduces the loading of the scope on the circuit under test, and it corrects the input of the scope to provide a true picture of the waveform in the circuit under test.

25. Pretrigger capture is available on digital scopes. It refers to acquisition of data that occurs before a trigger event.

Chapter Checkup

1. (b)	2. (b)	3. (d)	4. (a)	5. (d)
6. (c)	7. (d)	8. (a)	9. (c)	10. (b)
11. (d)	12. (b)	13. (c)	14. (b)	15. (a)
16. (d)				

CAPACITORS

INTRODUCTION

Capacitors are fundamental passive components used in many circuit applications. Unlike the resistor, the ideal capacitor does not dissipate energy in a circuit but can store energy and later return it to the circuit. When a sine wave is applied to a capacitor, the capacitor produces an opposition to current called *capacitive reactance*. This opposition shifts the phase of current and voltage in a circuit. This makes the response of capacitors to ac signals different from the response of resistors.

In this chapter, you will learn about capacitors and their construction. You will learn the effects of connecting them in series and in parallel and learn how they respond in both dc and ac circuits. The chapter concludes with a survey of some important applications of capacitors.

CHAPTER OUTLINE

10–1 Capacitance

10–2 Types of Capacitors

10–3 Series and Parallel Capacitors

10–4 Capacitors in DC Circuits

10–5 Capacitors in AC Circuits

10–6 Applications of Capacitors

Study aids for this chapter are available at

http://www.prenhall.com/SOE

KEY OBJECTIVES

A section number is given for each objective. After completing this chapter, you should be able to

10–1 Describe the relationship between charge, voltage, and capacitance for a basic capacitor

10–2 Compare the characteristics of various types of capacitors, including polarized and nonpolarized ones

10–3 Calculate the total capacitance of series and parallel capacitors

10–4 Use the universal charging and discharging curves to predict the instantaneous current and voltage for a capacitor in a dc circuit

10–5 Calculate the capacitive reactance for a capacitor in an ac circuit

10–6 List several important applications of capacitors

COMPUTER SIMULATIONS DIRECTORY

The following figures have Multisim circuit files associated with them.

◆ Figure 10–21 Page 353

◆ Figure 10–28 Page 358

LABORATORY EXPERIMENTS DIRECTORY

The following exercises are for this chapter.

◆ **Experiment 18** Capacitors

◆ **Experiment 19** Capacitive Reactance

KEY TERMS

- Capacitor
- Dielectric
- Capacitance
- Farad
- Relative permittivity
- Dielectric constant
- Working voltage
- Electrolyte
- Time constant
- Capacitive reactance

ON THE JOB...

(Getty Images)

All valuable employees are reliable. It is important to arrive at your workplace on time and be ready to work. If there is a valid reason you can't be at work or will be late, call your supervisor and explain the situation. Most supervisors will appreciate knowing about a problem before you are absent or late rather than afterward.

CHAPTER 10

Sci Hi SCIENCE HIGHLIGHT

A very specialized circuit called a CCD (for Charge-Coupled Device) forms the heart of most digital cameras. CCDs have enabled great advances in astronomical imaging because of their high sensitivity to light. The CCD chip takes the place of film.

Most CCDs are arranged in an x-y array of tiny oxide-film photocapacitors, which are called "wells." The wells are able to store charge and move it when directed. Each well represents one pixel (picture element) in the digital image, and the entire image is captured on the CCD at once. Electrons that are released when covalent bonds in a silicon layer are broken by photons of light striking the layer charge the individual wells. The charge is greater where more light strikes the silicon.

After exposure, the charge in each photocapacitor is transferred from row to row and read out by special registers connected to the chip. CCDs can transfer charge from one photocapacitor to the next. As the charge in one row is moved vertically, the next row of charge (which is "coupled" to it) shifts into the photocapacitor row that was just vacated.

The next step in the process is to convert the charge to a voltage. This step is part of the processing of the image within the CCD; after leaving the CCD, the image is processed further for display.

The CCD itself is a monochromatic device. There are various methods of obtaining a color image. One is to expose the scene through three successive exposures while switching in optical filters that have the desired color characteristics.

10-1 CAPACITANCE

When unlike charges are on two parallel conductive plates with a separation between them, an electric field exists between the plates and energy is stored. This electric field between the plates can be derived from Coulomb's law.

In this section, you will learn the relationship between charge, voltage, and capacitance for a basic capacitor.

The Capacitor

A **capacitor** is an electronic component that has the ability to store an electric charge. In its simplest form, a capacitor consists of two conductive plates separated by an insulator, called the **dielectric**, as indicated in Figure 10–1. The plates can acquire a charge by connecting them momentarily across a voltage source such as a battery. One of the plates will acquire an excess of electrons and thus be negatively charged; the other plate will have a deficiency of electrons and thus a positive charge. Because the plates have an opposite charge, they are attracted to each other. An electric field is present in the insulating region between the plates. The ability to store a charge by means of an electrostatic field is called **capacitance**.

FIGURE 10–1

A basic capacitor.

CAPACITORS

Unit of Capacitance

Capacitance is a measure of how much charge a capacitor can hold with a given voltage. The formula for capacitance is

$$C = \frac{Q}{V} \qquad (10\text{--}1)$$

where C is capacitance in farads, Q is charge in coulombs (on one of the plates), and V is voltage in volts.

The farad, in honor of Michael Faraday, is the unit of capacitance. One **farad (F)** is the amount of capacitance when one coulomb of charge is stored with a potential difference of one volt across the plates. Think of the word *capacitance* in terms of the root word *capacity*. A capacitor has a certain capacity to hold charge, which is measured in farads. A farad is a very large unit of capacitance, so for most applications the microfarad (10^{-6} farad) or the picofarad (10^{-12} farad) are more practical units. (A 1 F capacitor constructed in air with square plates 1 mm apart would be over 6 miles on a side!) Generally, capacitors are marked with either microfarads or picofarads as the preferred metric prefixes.

EXAMPLE 10–1

Problem
Convert each of the following values to microfarads (μF):

(a) 470 nF (b) 10,000 pF
(c) 0.0027 F (d) 0.15 nF

Solution
(a) $470 \text{ nF} \times 10^{-3} \text{ } \mu\text{F/nF} = \mathbf{0.47 \text{ } \mu F}$
(b) $10,000 \text{ pF} \times 10^{-6} \text{ pF/}\mu\text{F} = \mathbf{0.01 \text{ } \mu F}$
(c) $0.0027 \text{ F} \times 10^{6} \text{ } \mu\text{F/F} = \mathbf{2700 \text{ } \mu F}$
(d) $0.15 \text{ nF} \times 10^{-3} \text{ } \mu\text{F/nF} = \mathbf{0.000\ 15 \text{ } \mu F}$

Question*
How many nanofarads is a 100 μF capacitor?

The amount of charge that a capacitor can hold is determined by physical properties of the capacitor itself, which determine its capacitance, and the applied voltage. A larger voltage will allow more charge to be stored on a given capacitor. By rearranging the terms in Equation 10–1, charge can be expressed as

$$Q = CV \qquad (10\text{--}2)$$

Charge in a capacitor can be thought of as the quantity of air in a pressurized cylinder. The quantity of air (analogous to charge) that can be put in a cylinder depends on the capacity of the bottle (analogous to capacitance) and the pressure (analogous to voltage). Thus, to put more air in a cylinder you could obtain a larger cylinder or increase the pressure within the cylinder. Likewise, to store more charge on a capacitor, obtain one with more capacitance or increase the voltage.

---

*Answers are at the end of the chapter.

CHAPTER 10

EXAMPLE 10–2

Problem
A capacitor stores 81 µC (microcoulombs) when 3 V are applied across its plates. What is its capacitance?

Solution
Apply Equation 10–1.

$$C = \frac{Q}{V} = \frac{81 \times 10^{-6}}{3 \text{ V}} = 27 \text{ µF}$$

Question
If the voltage is doubled on the capacitor, what will happen to the charge?

How a Capacitor Stores Charge

In the neutral state, both plates of a capacitor have an equal number of free electrons, as indicated in Figure 10–2(a). When the capacitor is connected to a voltage source through a resistor, as shown in part (b), electrons (negative charge) are removed from plate A and an equal number are deposited on plate B. As plate A loses electrons and plate B gains elec-

FIGURE 10–2

Illustration of a capacitor storing charge.

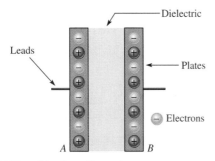

(a) Neutral (uncharged) capacitor (same charge on both plates)

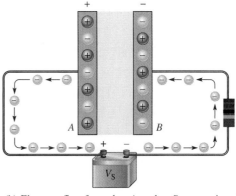

(b) Electrons flow from plate A to plate B as capacitor charges.

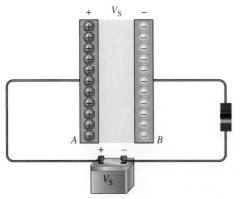

(c) Capacitor charged to V_S. No more electrons flow.

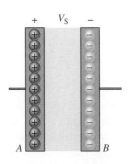

(d) Capacitor retains charge when disconnected from source.

trons, plate A becomes positive with respect to plate B. During this charging process, electrons flow only through the connecting leads and the source. No electrons move through the dielectric of the capacitor because it is an insulator. The movement of electrons ceases when the voltage across the capacitor equals the source voltage, as indicated in Figure 10–2(c). If the capacitor is disconnected from the source, it retains the charge for a period of time (the length of time depends on the type of capacitor and any tiny "leakage" current). It will still have the voltage across it, as shown in Figure 10–2(d). Actually, a large charged capacitor can act as a temporary battery and supply current for a short time in certain applications.

Factors Affecting Capacitance

The amount of capacitance for a capacitor is determined by three factors: the area of the plates, the separation of the plates, and the insulating material (dielectric). The plate area affects capacitance by providing a surface for charge to accumulate. A larger surface means more charge can be stored; and therefore the capacitance is greater.

The separation of the plates is a second factor that determines the capacitance. If the plates of a capacitor are brought closer together (but not touching), the voltage across the plates will drop. This effect occurs because opposite electrical charges have a greater attraction to each other, and there is less energy per charge (recall the definition of voltage). This is equivalent to an increase in capacitance, as shown in Equation 10–1.

The third factor that affects the amount of capacitance is the dielectric material between the plates. The electrons in a dielectric are not free to migrate, but the electrons and protons in each atom shift from their normal positions and create tiny dipoles (pairs of positive and negative charges). The dipoles tend to align themselves with the original electric field set up between the plates, with their positive ends aligned toward the negative plate and their negative ends aligned toward the positive plate, as illustrated in Figure 10–3. The dielectric has its own electric field that opposes the original one set up between the plates. The superposition of the two fields reduces the original field that would be present without the dielectric. As a result, more charge for a given voltage can be placed on the capacitor, thus increasing the capacitance.

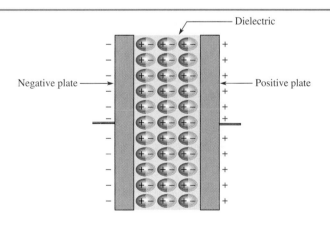

FIGURE 10–3

The dielectric material in a capacitor develops an electric field that opposes the original field and thus increases the capacitance.

The three factors that affect capacitance are related by the formula

$$C = 8.85 \times 10^{-12} \text{ F/m} \left(\frac{\varepsilon_r A}{d} \right) \tag{10-3}$$

where C is capacitance in farads, ε_r is the dielectric constant relative to air, A is area of one of the plates in meters squared, and d is separation of the plates in meters.

CHAPTER 10

The constant, 8.85×10^{-12} F/m, is called the *permittivity* of air because it is a measure of how easily the air dielectric will "permit" the establishment of flux lines. The second constant in the equation, ε_r, is the **relative permittivity**, also called the **dielectric constant**. The relative permittivity compares the effectiveness of a given dielectric to that of air. It is a pure number, meaning it has no units associated with it. Some common dielectrics and dielectric constants are listed in Table 10–1.

TABLE 10–1

Material	Typical ε_r values
Air (vacuum)	1.0
Teflon	2.0
Paper	2.5
Oil	4.0
Mica	5.0
Glass	7.5
Aluminum oxide	10
Ceramic	1200

EXAMPLE 10–3

Problem
Two rectangular sheets of copper are 10 cm on a side and are separated by a layer of Teflon that is 0.2 mm thick. Determine the capacitance.

Solution
The area of a plate is 0.1 m $\times$ 0.1 m = 0.01 m². The separation in meters is 2×10^{-4} m. From Table 10–1, the dielectric constant of Teflon is 2.0. Substituting into Equation 10–3,

$$C = 8.85 \times 10^{-12} \text{ F/m} \left(\frac{\varepsilon_r A}{d} \right) = 8.85 \times 10^{-12} \text{ F/m} \left(\frac{(2.0)(0.01 \text{ m}^2)}{2 \times 10^{-4} \text{ m}} \right) = \mathbf{885 \text{ pF}}$$

Question
What is the capacitance if the Teflon dielectric is replaced with a 0.4 mm paper dielectric?

Capacitive Voltage Limit

There is a limit to how much voltage can be across a given capacitor. If the voltage is too large, electrons in the dielectric break loose and temporarily make the dielectric conductive. This condition is known as *breakdown*. The breakdown causes a spark between the plates, discharging the capacitor. If the dielectric is oil or air, the failure will only be temporary; but in solid dielectrics, a hole may be permanently burnt through the material, reducing the voltage it can handle permanently.

Manufacturers of capacitors rate them on the amount of voltage they can withstand before breakdown. This maximum voltage is the dc voltage in the circuit, so it is given as the **working voltage** (WV or WVDC) and is generally printed on the body of the capacitor. (Some capacitors can exceed the working voltage for a very short time; this is referred to as the surge voltage.) It is important to take into account the maximum voltage as well as the capacitance when selecting a capacitor for an application.

Review Questions

Answers are at the end of the chapter.

1. What is capacitance?
2. (a) How many microfarads in one farad?
   (b) How many picofarads are there in one nanofarad?
   (c) How many picofarads are there in one microfarad?
3. What is the capacitance of a two-plate capacitor that stores a charge of 10.8×10^{-3} C when the potential difference across the plates is 40 V?
4. (a) When the plate area of a capacitor is increased, does the capacitance increase or decrease?
   (b) When the distance between the plates is increased, does the capacitance increase or decrease?
5. A ceramic capacitor ($\varepsilon_r = 1200$) has rectangular plates that are 2 cm × 3 cm. The thickness of the dielectric is 1.0 mm. What is the capacitance?

TYPES OF CAPACITORS 10–2

Capacitors normally are classified according to the type of dielectric material and whether they are polarized or nonpolarized. It is important to place polarized capacitors in the correct orientation in a circuit.

In this section, you will learn the characteristics of various types of capacitors, including polarized and nonpolarized ones.

Fixed Capacitors

Capacitors are named based on the type of dielectric material. The most common dielectric materials are mica, ceramic, plastic film, and electrolytic materials. Most capacitors are fixed, meaning their value cannot be changed. Fixed capacitors are available in a large range of sizes, from a one picofarad (a trillionth of a farad) to greater than one farad.

Mica Capacitors

The basic construction of the stacked-foil type is shown in Figure 10–4. It consists of alternate layers of metal foil and thin sheets of mica. The metal foil forms the plate, with alternate foil sheets connected together to increase the plate area and thus to increase the capacitance. The entire assembly is encapsulated in an insulating material as shown.

FIGURE 10–4

Construction of a typical mica capacitor.

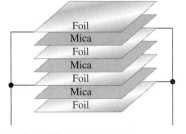

(a) Stacked layer arrangement

(b) Layers are pressed together and encapsulated.

Mica capacitors are available with capacitance values ranging from 1 pF to 0.1 μF. The major advantage of mica capacitors is that the working voltage range is high—typically from 100 V to over 2500 V.

Ceramic Capacitors

Ceramic dielectrics provide very high dielectric constants (1200 typical). As a result, comparatively high capacitance values can be achieved in a small physical size. This is a major advantage for printed circuit boards, where size is critical. The construction of a small ceramic disk capacitor is shown in Figure 10–5. Ceramic capacitors are available in other forms as well.

FIGURE 10–5 Construction of a typical ceramic disk capacitor.

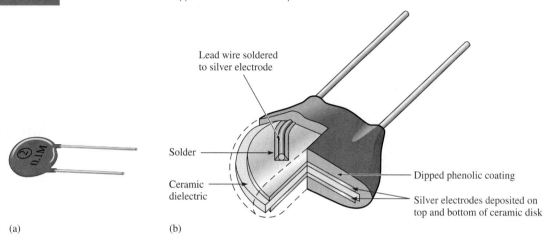

In addition to small physical size, ceramic capacitors have high working voltage ratings. They are also very stable, meaning they do not change values with changes in temperature. They are available in capacitance values ranging from 1 pF to 2.2 μF.

Plastic-Film Capacitors

There are several types of plastic-film capacitors. Figure 10–6 shows one common construction method used in many plastic-film capacitors. A thin strip of plastic-film dielectric

FIGURE 10–6

Construction of a typical plastic-film capacitor.

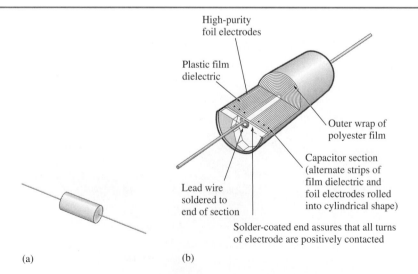

is sandwiched between two thin metal strips that act as plates. One lead is connected to the inner plate and one lead is connected to the outer plate as indicated. The strips are then rolled in a spiral configuration and encapsulated in a molded case. Thus, a large plate area can be packaged in a relatively small physical size, thereby achieving large capacitance values. Typically, these types have capacitance values from 1000 pF to about 1 μF.

Although plastic-film capacitors are not polarized, they are usually marked with a band on one end that indicates the outside foil. The banded end is connected to ground to help shield the other plate. This is useful in high-frequency circuits to avoid noise.

Electrolytic Capacitors

The electrolytic capacitor is different from other types of capacitors in that one of its conducting surfaces is a metallic plate; the other conducting surface is a conducting paste or electrolyte, which is applied to a conductive material such as plastic film. An **electrolyte** is a conductive material in which charge movement is by ions, rather than electrons. The electrolyte in combination with the plastic film forms the negative plate. The dielectric is an extremely thin layer of oxide grown on the metallic plate, which is either aluminum or tantalum. Because of the process used for forming the dielectric, the metallic plate must always be connected so that it is always positive with respect to the electrolyte.

The polarity of an electrolytic capacitor is always indicated by a plus or minus sign or some other obvious marking. The capacitor must be connected in a dc circuit where the voltage across the capacitor does not change polarity, regardless of any ac present. Reversal of the polarity of the voltage will cause the dielectric to act as a conductor and usually will result in complete destruction of the capacitor.

The vast majority of electrolytic capacitors are polarized; however, a certain group are not. These are so-called nonpolarized electrolytic capacitors, which are really two normal electrolytic capacitors connected internally in series with negative plates connected together. These capacitors are not widely used due to cost, but care must be taken to replace a nonpolarized electrolytic with another nonpolarized one, should the need arise.

The capacitance of an electrolytic capacitor is determined by the same factors that apply to any other capacitor. The extremely thin dielectric coating (10^{-5} cm) allows electrolytic capacitors to have much larger values for their size than conventional capacitors. For this reason, electrolytic capacitors are generally preferred for applications that require capacitance values over 1 μF and are readily available in sizes of up to 10,000 μF and even larger.

Although they are widely used because of the large capacitance values available, electrolytic capacitors have several disadvantages. They have relatively low breakdown voltages, which must be taken into account before using them in a circuit. The breakdown voltage is widely variable, from a few volts to several hundred volts, so it is important to check the voltage rating before using an electrolytic capacitor. In addition to low breakdown voltage, many electrolytic capacitors tend to have higher amounts of leakage current, meaning that a small steady current is present across the dielectric. (Low-leakage types are available in applications where leakage is a problem.) In addition, electrolytic capacitors are not precision components. Their tolerance is generally 20% or more of the value marked on the case.

There are two types of electrolytic capacitors—aluminum and tantalum. Aluminum electrolytic capacitors are the most common type and cost less than tantalum. The two plates of an electrolytic capacitor are separated by an extremely thin layer of an oxide that forms on the surface of the metal plate. Figure 10–7(a) illustrates the basic construction of a typical aluminum electrolytic capacitor with axial leads. Other electrolytic capacitors with radial leads are shown in Figure 10–7(b). The symbol for a polarized capacitor (most electrolytic capacitors are polarized) is shown in part (c). The curved plate on the symbol is connected to the more negative potential.

Tantalum electrolytic capacitors can be in either a tubular configuration as shown in Figure 10–7 or "teardrop" shape as shown in Figure 10–8. In the teardrop configuration, the positive plate is actually a pellet of tantalum powder rather than a sheet of foil. Tantalum pentoxide forms the dielectric, and manganese dioxide forms the negative plate.

SAFETY NOTE

It is important to observe the polarity of electrolytic capacitors when installing one in a circuit. An incorrectly installed capacitor can explode due to heating. Always check the polarity of the capacitor and the circuit before installing one.

CHAPTER 10

FIGURE 10–7
Electrolytic capacitors.

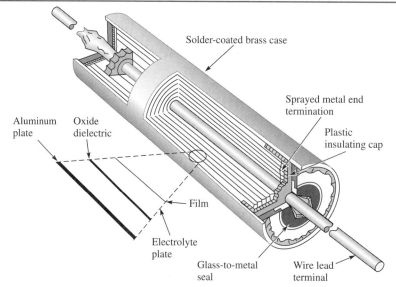

(a) Construction view of an axial-lead electrolytic capacitor

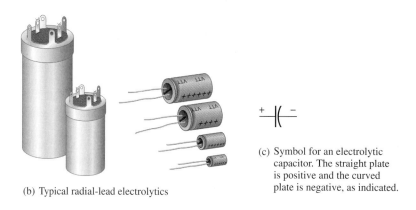

(b) Typical radial-lead electrolytics

(c) Symbol for an electrolytic capacitor. The straight plate is positive and the curved plate is negative, as indicated.

FIGURE 10–8
Construction of a "teardrop" tantalum electrolytic capacitor.

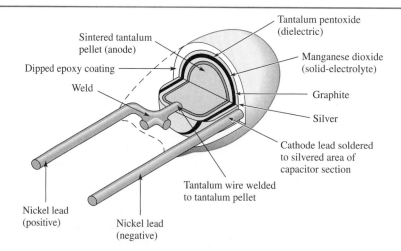

Variable Capacitors

Variable capacitors are used in a circuit when there is a need to adjust the capacitance value either manually or automatically. They function by changing the plate area or the spacing between the plates and can change the capacitance over a 10:1 range (typically). Variable capacitors are available only in small sizes, typically from 10 pF to about 1000 pF. The schematic symbol for a variable capacitor is shown in Figure 10–9. The arrow drawn through the symbol indicates that it is variable.

Small adjustable capacitors for very fine adjustments in a circuit are called **trimmer capacitors**. The value of these capacitors is usually less than 100 pF. Ceramic or mica is a common dielectric, and the capacitance usually is changed by adjusting the plate separation with a slotted screw adjustment. Figure 10–10 shows typical devices.

FIGURE 10–9

Schematic symbol for a variable capacitor.

Capacitor Labeling

Capacitor values are indicated on the body of a capacitor by typographical labels. Typographical labels consist of letters and numbers that indicate various parameters such as capacitance, voltage rating, and tolerance.

Many capacitors do not show the units for capacitance. In these cases, the units will be either microfarads (μF) or picofarads (pF) and are implied by the value indicated and the size of the capacitor itself. For example, a small ceramic capacitor marked .001 or .01 has units of microfarads because picofarad values that small are not available. As another example, a small ceramic capacitor labeled 50 or 330 has units of picofarads because microfarad units that large normally are not available in this type. In some cases, a 3-digit designation is used. The first two digits are the first two digits of the capacitance value. The third digit is the number of zeros after the second digit. For example, 103 means 10,000 pF. In some instances, the units are labeled as PF or μF; sometimes the microfarad unit is labeled as MF or MFD.

If the voltage rating is shown, it will be given with the abbreviation WV or WVDC (for "working voltage"). When it is omitted, the voltage rating can be determined from information supplied by the manufacturer.

FIGURE 10–10

Trimmer capacitors.

Review Questions

6. What are advantages to ceramic capacitors?
7. What is an electrolyte?
8. What type of dielectric is in an electrolytic capacitor?
9. What precautions are necessary when one connects a polarized capacitor in a circuit?
10. What two metric prefixes are normally used in identifying capacitance values?

SERIES AND PARALLEL CAPACITORS 10–3

When capacitors are connected in parallel, the total capacitance increases. When capacitors are connected in series, the total capacitance is less than the smallest capacitor.

In this section, you will learn to calculate the total capacitance of series and parallel capacitors.

Parallel Capacitors

When capacitors are connected in parallel, the effective plate area increases and more charge can be stored for a given voltage. As a result, the total capacitance increases, as illustrated in Figure 10–11. The total capacitance of two capacitors in parallel is the sum of the individual capacitances.

$$C_T = C_1 + C_2 \qquad (10\text{–}4)$$

FIGURE 10–11

Parallel capacitors give an effective plate area that is the sum of the plate areas of each.

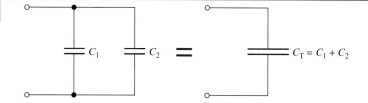

The same idea applies to any number of capacitors in parallel. The total capacitance is just the sum of the individual capacitors. The expanded formula is as follows where the subscript n is any number:

$$C_T = C_1 + C_2 + C_3 + \cdots + C_n \qquad (10\text{--}5)$$

The rule for finding the total capacitance of parallel capacitors has the same form as the rule for finding the total resistance of series resistors.

EXAMPLE 10–4

Problem
A 27,000 pF capacitor is in parallel with a 0.01 μF capacitor. What is the total capacitance?

Solution
First, convert the units to be consistent.

$$27{,}000 \text{ pF} \times 10^{-6} \text{ }\mu\text{F/pF} = 0.027 \text{ }\mu\text{F}$$

Substituting into Equation 10–4,

$$C_T = C_1 + C_2 = 0.027 \text{ }\mu\text{F} + 0.01 \text{ }\mu\text{F} = \mathbf{0.037 \text{ }\mu\text{F}}$$

Question
What is the total capacitance in pF?

Series Capacitors

When capacitors are connected in series, the total capacitance is less than the smallest capacitor. Figure 10–12(a) illustrates two unequal capacitors connected in series. This is equivalent to joining the two capacitors with a common plate between them as shown in Figure 10–12(b). The common plate does not affect the capacitance, so it is removed in Figure 10–12(c). The equivalent capacitor in part (c) has one plate from the smaller capacitor and greater plate separation than either of the original capacitors, implying less total capacitance than the smaller capacitor in the original series connection.

FIGURE 10–12

The equivalent series capacitance of two series capacitors is smaller than the smallest individual capacitor.

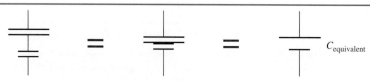

(a) Two unequal capacitors connected in series

(b) The common plates are touching. Electrically this is equivalent to (a).

(c) The common plates are removed because they are unnecessary.

CAPACITORS

Another point to consider for series capacitors is that the total voltage is split between the capacitors, and thus each capacitor holds less charge than if it were connected directly to the source. For two capacitors, the relationship between the total capacitance and the individual capacitances is expressed as

$$\frac{1}{C_T} = \frac{1}{C_1} + \frac{1}{C_2} \quad (10\text{–}6)$$

Taking the reciprocal of both sides of Equation 10–6 gives the formula for the total capacitance of two capacitors in series.

$$C_T = \frac{1}{\frac{1}{C_1} + \frac{1}{C_2}} \quad (10\text{–}7)$$

Equation 10–7 can be manipulated into the product-over-sum form that you are familiar with from the parallel resistor rule. Remember that the rule for two series capacitors has the same form as the two parallel resistor rule.

$$C_T = \frac{C_1 C_2}{C_1 + C_2} \quad (10\text{–}8)$$

Equation 10–8 is only valid for the case of two capacitors in series. It is unusual to have a circuit application with more than two capacitors in series. However, if you should encounter this case, Equation 10–6 can be extended to include any number of series capacitors (2 through n) as follows:

$$\frac{1}{C_T} = \frac{1}{C_1} + \frac{1}{C_2} + \frac{1}{C_3} + \cdots + \frac{1}{C_n} \quad (10\text{–}9)$$

Solving for C_T, the total capacitance for series capacitors is

$$C_T = \frac{1}{\frac{1}{C_1} + \frac{1}{C_2} + \frac{1}{C_3} + \cdots + \frac{1}{C_n}} \quad (10\text{–}10)$$

The rule for finding the total capacitance of series capacitors has the same form as the rule for finding the total resistance of parallel resistors.

EXAMPLE 10–5

Problem
A 3300 pF capacitor is in series with a 0.001 μF capacitor. What is the total capacitance?

Solution
Convert the units to be consistent. (One option is to convert the 0.001 μF capacitor.)

$$0.001 \ \mu F \times 10^6 \ pF/\mu F = 1000 \ pF$$

Substituting into Equation 10–8,

$$C_T = \frac{C_1 C_2}{C_1 + C_2} = \frac{(3300 \ pF)(1000 \ pF)}{3300 \ pF + 1000 \ pF} = \mathbf{767 \ pF}$$

CHAPTER 10

For the TI-36X calculator, this can be entered as

[3] [3] [0] [0] [×] [1] [0] [0] [0] [÷] [(] [3] [3] [0] [0] [+] [1] [0] [0] [0] [)] [=]

In this sequence, the capacitor values are entered in units of pF and the result is read in pF. You may prefer to enter the capacitor values in units of F. The key sequence for this case is

[3] [.] [3] [EE] [+/−] [9] [×] [1] [EE] [+/−] [9] [÷] [(] [3] [.] [3] [EE] [+/−] [9] [+] [1] [EE] [+/−] [9] [)] [=]

The display will show the result as 767^{-12}, which is read as 767×10^{-12} F, or 767 pF.

Question
What is the total capacitance in μF?

Capacitor Voltages

Parallel Capacitors

As you know, all components that are connected in parallel with a voltage source will have the same voltage across them, which is the same as the source voltage. In parallel circuits, the voltage is the same across all components, as expressed by the relation

$$V_S = V_1 = V_2 = V_n$$

where n represents the last component. Although the voltage will be the same across each capacitor, the charge depends on the size of the capacitor. In parallel, a larger capacitor can store more charge than a smaller one.

Series Capacitors

When capacitors are connected in series to a voltage source, the charge (Q) is the same across each capacitor; and the voltages across each capacitor must sum to the source voltage by Kirchhoff's voltage law.

$$Q_S = Q_1 = Q_2 = \cdots = Q_n$$

Substituting $Q = CV$ for the series case gives

$$C_T V_S = C_1 V_1 = C_2 V_2 = \cdots = C_n V_n$$

The voltage, V_x, across any series capacitor, C_x, can be expressed as

$$V_x = \left(\frac{C_T}{C_x}\right) V_S \tag{10–11}$$

This equation is referred to as the capacitive voltage-divider rule. Notice that the quantity in parentheses is always a fraction less than 1 because C_T is smaller than the smallest capacitor. Compare this equation to the resistive voltage-divider rule given as Equation 5–5. Equation 10–11 allows you to solve for the voltage across any capacitor in a series of capacitors. The largest capacitor will have the smallest voltage across it. Likewise, the smallest capacitor will have the largest voltage across it.

EXAMPLE 10–6

Problem
A 220 μF capacitor is in series with a 100 μF capacitor and connected across a 10 V source as shown in Figure 10–13. What is the voltage across the 100 μF capacitor?

Solution
Use Equation 10–8 to determine the total capacitance.

$$C_T = \frac{C_1 C_2}{C_1 + C_2} = \frac{(220\ \mu F)(100\ \mu F)}{220\ \mu F + 100\ \mu F} = 68.8\ \mu F$$

Use Equation 10–11 to calculate the voltage across the 100 μF capacitor.

$$V_2 = \left(\frac{C_T}{C_2}\right) V_S = \left(\frac{68.8\ \mu F}{100\ \mu F}\right) 10\ V = \mathbf{6.88\ V}$$

Notice that, unlike resistors, the larger voltage is across the smaller capacitor.

Question
What voltage is across the 220 μF capacitor?

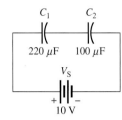

FIGURE 10–13

Review Questions

11. How is the total capacitance of parallel capacitors determined?
12. What is the total capacitance of five parallel 10,000 μF capacitors?
13. How is the total capacitance of series capacitors determined?
14. What is the total capacitance of five series 10,000 μF capacitors?
15. Assume two unequal capacitors are connected in series across a voltage source. Which will have the larger voltage across it?

10–4 CAPACITORS IN DC CIRCUITS

A capacitor will have a charging current when it is connected to a dc voltage source. The charging rate is dependent on the capacitance and the resistance in a circuit. The capacitor responds only to changes, so once it is charged, the capacitor acts as a block to dc.

In this section, you will learn to use the universal charging and discharging curves to predict the instantaneous current and voltage for a capacitor in a dc circuit.

Charging a Capacitor

When a capacitor is initially uncharged, there will be no net negative or positive charge on either plate. If it is placed in a circuit, as shown in Figure 10–14(a), the capacitor will remain uncharged until the switch is closed. When the switch closes, immediately electrons are drawn from the positive plate, through R, to the positive source terminal. Simultaneously, electrons move from the negative source terminal to the negative plate, causing the capacitor to acquire a charge, as shown in Figure 10–14(b). At first, there is a large current but it decays as the capacitor charges. Although current will be indicated in the meter, there is no current in the dielectric of the capacitor during this charging phase. It is not necessary for electrons to pass through the dielectric in order to have current in the circuit.

FIGURE 10–14   Charging a capacitor through a resistor.

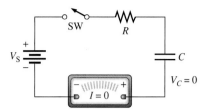
(a) Before the switch is closed, there is no current in the circuit and the capacitor is uncharged.

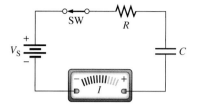
(b) Immediately after the switch is closed, current is indicated on the meter as the capacitor charges.

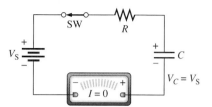
(c) Current decreases to zero when the capacitor is fully charged.

SAFETY NOTE

When a charged capacitor is disconnected from the source, it can remain charged for a long period of time, depending on its leakage current and any load connected to it. Electrolytic capacitors in particular can cause severe electrical shock because of this stored charge. You should never assume a circuit is safe just because it is disconnected from the source.

As this charging process continues, the voltage across the plates builds up until it is equal to the applied voltage, V_S, but opposite in polarity, as shown in Figure 10–14(c). Once the capacitor is fully charged, there is no current in the circuit and the capacitor retains the charge. No current is through the dielectric of the capacitor during charging or discharging because the dielectric is an insulating material. Current is from one plate to the other only through the external circuit.

Discharging a Capacitor

A charged capacitor can be discharged directly by connecting it across a conductor. Immediately, the excess of electrons on the negative plate is transferred through the conductor to the positive plate. A spark may accompany this transfer.

If the capacitor is discharged through a resistor, the process is slower. Figure 10–15 illustrates discharging a capacitor through a resistor. The direction of the current during discharge is opposite to that of the charging current. Before the switch is closed, the capacitor is charged to a voltage of V_C and no current is indicated on the meter, as shown in Figure 10–15(a). Immediately after the switch is closed, current from the capacitor is in the circuit (except in the dielectric of the capacitor), as indicated in Figure 10–15(b). This current decays to zero as the capacitor discharges. At this time, there is no more current in the circuit, and the capacitor is discharged, as shown in Figure 10–15(c).

FIGURE 10–15   Discharging a capacitor through a resistor.

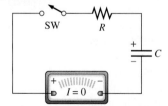
(a) The capacitor is initially charged; the meter polarity is reversed from the charging case.

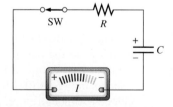
(b) Immediately after the switch is closed, current is indicated on the meter as the capacitor discharges.

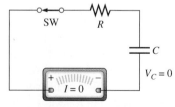
(c) Current decreases to zero when the capacitor is fully discharged.

Current and Voltage During Charging and Discharging

Let's look at the charging and discharging waveforms for a capacitor after the switch is closed. The time required to charge or discharge a capacitor is determined by the values of R and C, but the shape of the charging curve is the same as the shape of the discharging curve for all RC combinations. Figure 10–16(a) illustrates the shape of the voltage across the capacitor during the charging phase. Figure 10–16(b) shows the current in the circuit for the same charging time. Eventually, the voltage reaches a final voltage equal to the applied source voltage as the current decays to zero.

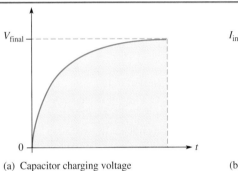
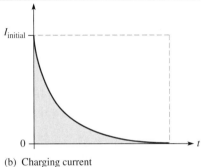

FIGURE 10–16

Charging waveforms.

(a) Capacitor charging voltage (b) Charging current

Figure 10–17(a) shows the shape of the voltage across the capacitor during the discharging phase. Eventually, the voltage across the capacitor is zero, matching the applied voltage of zero. The current that accompanies the discharging of the capacitor is shown in Figure 10–17(b). Although the discharging current is in the opposite direction as the charging current, the shape is the same.

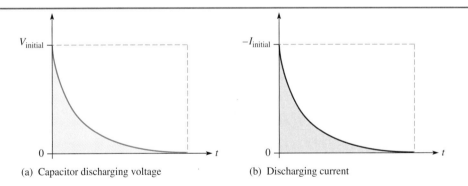

FIGURE 10–17

Discharging waveforms. Usually negative quantities are plotted on the negative y-axis but is done the opposite here to emphasize the similar shape of the charging and discharging current.

(a) Capacitor discharging voltage (b) Discharging current

For both the charging and discharging curves, the maximum current is in the circuit initially after the switch is closed and eventually the current reaches zero. The current in the charging and discharging cases differs only that it is in the opposite direction.

When the current reaches zero, the capacitor looks like an open circuit to dc. In both the charging and discharging phase, the capacitor responded only to a change in the applied circuit voltage. For steady-state dc, the capacitor appears to be open.

A capacitor is equivalent to an open circuit for constant dc.

The shape of the charging and discharging current also reveals another important point. At the instant the switch is closed, the current in both cases is a maximum. This current is limited *only* by the resistance in the circuit. During this brief time, the capacitor looked like a short.

A capacitor is equivalent to a short circuit for a sudden change in voltage.

CHAPTER 10

Square Wave Response of an RC Circuit

The charging voltage observed across the capacitor when a switch closes is the same response that will be observed when a square wave is substituted for the battery and switch. Figure 10–18 shows the circuit, and Figure 10–19 shows its response. The period of the square wave must be long enough in order to see the full response.

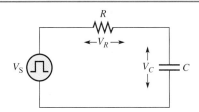

FIGURE 10–18

A series RC circuit driven by a square wave.

An important difference between the switch closure and the square wave generator is that the generator has its own internal resistance (the Thevenin resistance), which adds to the resistance of the circuit. Also another important difference is that a square wave generator simulates a closed switch when it is high but does not act like an open switch when the square wave is low. An open switch looks like infinite resistance, but the square wave when it is low represents a low resistance path across the circuit. Thus, when the square wave goes low, the capacitor discharges back through the generator.

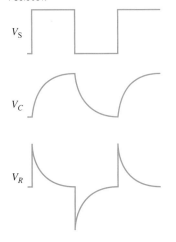

FIGURE 10–19

The response of the series RC circuit to a square wave. V_S is the source voltage, V_C is the voltage across the capacitor, and V_R is the voltage across the resistor.

The RC Time Constant

When a capacitor charges or discharges through a resistance, a certain time is required for the capacitor to fully charge or discharge. The voltage across the capacitor cannot change instantaneously because time is required to move charge. The rate the charge moves is determined by the time constant of the circuit. A **time constant** is a fixed time interval that determines the time response of an RC circuit or an RL circuit. (The RL time constant is discussed in Section 11–4.)

The time constant for a series RC circuit is a time interval determined by the product of the resistance and the capacitance. It establishes the time required for charging and discharging a capacitor after a transient event such as a switch closure or a pulse changing levels. The time constant is symbolized by the Greek letter tau (τ), and the formula is

$$\tau = RC \tag{10-12}$$

where τ is the time constant expressed in seconds when the resistance is in ohms and the capacitance is in farads.

EXAMPLE 10–7

Problem
A 47 μF capacitor is in series with a 100 kΩ resistor. What is the time constant?

Solution
Substituting into Equation 10–12,

$$\tau = RC = (100 \text{ k}\Omega)(47 \text{ }\mu\text{F}) = \mathbf{4.7 \text{ s}}$$

Question
What is the time constant for a 3300 pF capacitor in series with a 27 kΩ resistor?

352

The time constant is useful to determine the values to place on the generalized plots shown in Figures 10–16 and 10–17. An uncharged capacitor will charge to 63% of the final voltage in one time constant. A fully charged capacitor will discharge by 63%, implying that it will have 37% of its initial charge after one time constant.

The charging and discharging curves for current and voltage follow a precise mathematical form called an *exponential curve*. The charging curve for the voltage across a capacitor is called a *rising exponential*. As the voltage increases across the capacitor, the charging current decreases, so it is shown as a *falling exponential*. The discharging curves for current and voltage are both falling exponentials. These exponential curves require five time constants to change the voltage or current by 99%. This interval of five time constants is accepted as the time to fully charge or discharge a capacitor.

The charging and discharging curves are plotted in Figure 10–20 and represent the transient response of an *RC* circuit. (In chapter 11, you will see that the curves are also applied to inductive, or *RL* circuits, also.) The curves are called **universal curves** because the *x* and *y* axes are not scaled to specific values but are plotted as a percentage of the final value, either current or voltage. The universal curves can be used to find voltage or current in a circuit at any instant of time. The *x*-axis represents the number of time constants that have elapsed after the charging or discharging begins. If you know the shape of the curve (rising or falling) that you are interested in, you can use the universal curves to find a particular solution. Values are assigned for a particular case as will be illustrated in the following two examples.

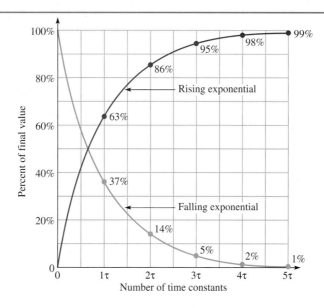

FIGURE 10–20

Universal time constant curves.

EXAMPLE 10–8

Problem
A 27 kΩ resistor is in series with an uncharged 33 µF capacitor and connected through a switch to a 10 V source, as shown in Figure 10–21. What is the approximate voltage on the capacitor 1.2 s after the switch is closed?

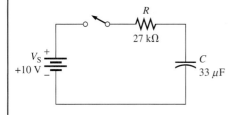

FIGURE 10–21

Solution

First, find the time constant.

$$\tau = RC = (27\ \text{k}\Omega)(33\ \mu\text{F}) = 891\ \text{ms}$$

Next, find how many time constants are represented by 1.2 s.

$$\text{Number of time constants} = \frac{1.2\ \text{s}}{0.891\ \text{s}} = 1.14$$

On the *x*-axis of the universal time-constant curve, find 1.14 time constants and note that the rising exponential is 68% of the final value. This reading of the curve is illustrated in Figure 10–22. The final value in this case is the source voltage of 10 V. Therefore,

$$v_C = 68\% \times 10\ \text{V} = \mathbf{6.8\ V}$$

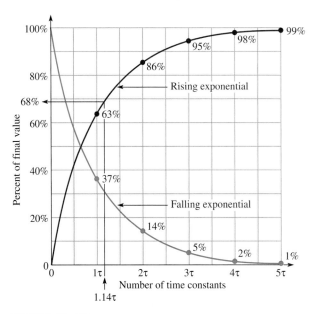

FIGURE 10–22

Question
What is the voltage on the capacitor at a time of 2.0 s after the switch is closed?

COMPUTER SIMULATION

Open the Multisim file F10-21AC on the website. Using the oscilloscope, observe the waveform on the resistor. The dc voltage source has been replaced with a square wave in the exercise.

EXAMPLE 10–9

Problem

Find the charging current the instant the switch is closed in Example 10–8 and at a time of 1.5 s.

CAPACITORS

Solution
The charging current the instant the switch is closed is determined by the voltage and resistance in the circuit. The capacitor at this instant looks like a short. Therefore,

$$i_{\text{instantaneous}} = \frac{V_S}{R} = \frac{10 \text{ V}}{27 \text{ k}\Omega} = 0.37 \text{ mA}$$

This is the maximum current and the current that will be in the circuit the instant the switch is closed.

To find the current at 1.5 s, determine the number of time constants that elapse. From Example 10–8, $\tau = 0.891$ s.

$$\text{Number of time constants} = \frac{1.5 \text{ s}}{0.891 \text{ s}} = 1.68$$

Reading the falling exponential on the universal time constant curve for 1.68 time constants indicates that the value has decayed to approximately 19% of the original value. This reading is illustrated in Figure 10–23. Therefore,

$$i_{\text{instantaneous}} = 19\% \times 0.37 \text{ mA} = \mathbf{0.70 \text{ mA}}$$

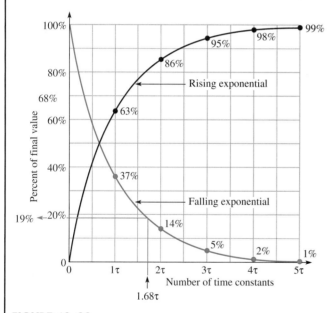

FIGURE 10–23

Question
What is the current 1 s after the switch is closed?

Review Questions

16. During the time that a capacitor is charging, how much current is in the dielectric?
17. What is the time constant for a 0.33 μF capacitor in series with a 470 kΩ resistor?
18. What are the universal time-constant curves?

CHAPTER 10

19. What limits the initial charging current in an *RC* circuit that is connected to a dc source?
20. If a capacitor is initially charged to 100 V and is discharged by switching it across a resistor, what voltage is on the capacitor after one time constant has elapsed?

10–5 CAPACITORS IN AC CIRCUITS

A capacitor blocks dc but allows a transient (changing) current to pass because of the changing voltage. A capacitor can pass ac for the same reason, but there is an opposition called reactance that is determined by frequency.

In this section, you will learn to calculate the capacitive reactance for a capacitor in an ac circuit.

Recall that in a dc circuit, a capacitor charges for approximately five time constants. After this time, there is no more current in the circuit and the capacitor appears to be an open. A capacitor only passes a changing voltage.

When a capacitor is connected to an ac voltage source, such as shown in Figure 10–24, the capacitor charges and discharges at a rate equal to the frequency of the source. Because ac is constantly changing, the charging and discharging continues; and the circuit has current in it as indicated on the ac ammeter.

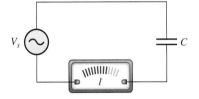

FIGURE 10–24

Current is in the circuit because the capacitor passes ac.

Capacitive Reactance

The current in the circuit in Figure 10–24 is limited by how fast the capacitor charges and discharges. Although the capacitor (ideally) does not have resistance and therefore does not dissipate any power from the source, it still limits the current that is in the circuit. This opposition is called **capacitive reactance**. The symbol for capacitive reactance is X_C, and its unit is the ohm (Ω). Although capacitive reactance is an opposition to current, there is an important difference between capacitive reactance and resistance. A capacitor (ideally) does not dissipate any energy as a resistor does; it only stores it during a portion of the ac cycle and returns it to the source during another portion of the cycle.

How Frequency Affects Capacitive Reactance

Figure 10–25 illustrates a function generator used as a source of constant amplitude ac voltage. If the frequency is increased, as in Figure 10–25(a), the current increases also. When the frequency increases, its rate of change also increases. When the frequency of the source is decreased, as in Figure 10–25(b), the current decreases. In this case, the rate of change decreases.

The rate of change depends on frequency, as illustrated in Figure 10–26. Notice that the higher-frequency wave has a steeper slope. This implies a greater rate of voltage change, so the amount of charge moving in the circuit is greater for this case. More charge moving in a given time interval implies a larger current.

CAPACITORS

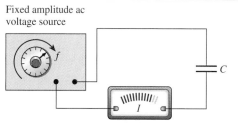

(a) Current increases (b) Current decreases

FIGURE 10–25

Current increases when the frequency is raised; current decreases when the frequency is lowered.

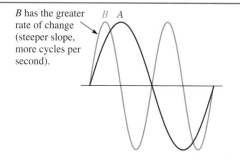

FIGURE 10–26

The rate of change of a sine wave increases when frequency increases.

An increase in the current for a fixed amount of voltage indicates that the opposition to the current has *decreased*. Conversely, a decrease in the current means that the opposition has *increased*. Therefore, the capacitive reactance varies inversely with frequency.

How Capacitance Affects Capacitive Reactance

Figure 10–27(a) shows that when a sinusoidal voltage with a fixed amplitude and fixed frequency is applied to a capacitor, there will be a certain amount of alternating current in the circuit. If the total capacitance is doubled by adding a second identical capacitor in parallel, as in Figure 10–27(b), twice as much charge will flow for the same frequency and applied voltage as before. Of course, this is equivalent to replacing the original capacitor with one twice as large. Thus, a larger capacitance implies a larger current or less opposition to current. Capacitive reactance is inversely proportional to capacitance.

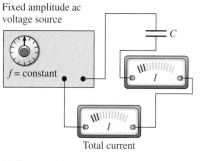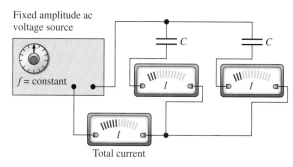

(a) Less capacitance (b) More capacitance

FIGURE 10–27

A greater capacitance increases the total current in the circuit, which represents less opposition.

Formula for Capacitive Reactance

Capacitive reactance is inversely proportional to both the frequency and the capacitance. The two ideas can be combined as a proportionality.

CHAPTER 10

X_C is proportional to $\dfrac{1}{fC}$

The constant of proportionality that relates X_C to $1/fC$ is $1/(2\pi)$. The 2π term comes from the fact that capacitive reactance is based on radian frequency rather than hertz. As you learned in Section 9–1, one revolution contains 2π radians; so the 2π term is introduced to convert f to hertz, which is generally more convenient. Therefore, the formula for capacitive reactance (X_C) is

$$X_C = \dfrac{1}{2\pi fC} \qquad (10\text{--}13)$$

where X_C is capacitive reactance in ohms, f is frequency in hertz, and C is capacitance in farads.

EXAMPLE 10–10

Problem
Find the capacitive reactance for a 0.047 µF capacitor if the frequency is 1.0 kHz.

Solution
Apply Equation 10–13.

$$X_C = \dfrac{1}{2\pi fC} = \dfrac{1}{2\pi(1.0\text{ kHz})(0.047\ \mu\text{F})} = 3.39\text{ k}\Omega$$

The TI-36X calculator sequence is

Question
What is the frequency that is required to make $X_C = 100\ \Omega$?

Ohm's Law in Capacitive AC Circuits

The reactance of a capacitor is analogous to the resistance of a resistor. In fact, both are expressed in ohms. Since both R and X_C are forms of opposition to current, Ohm's law applies to capacitive circuits as well as to resistive circuits. For a capacitive circuit, Ohm's law is stated as

$$I = \dfrac{V}{X_C} \qquad (10\text{--}14)$$

When applying Ohm's law in ac circuits, you must express both the current and the voltage in the same way, that is, rms, peak, or peak-to-peak. In general, rms is preferred because it can be used directly for power calculations whereas the others cannot.

EXAMPLE 10–11

Problem
Find the current for the circuit in Figure 10–28.

FIGURE 10–28

CAPACITORS

Solution
First, find the capacitive reactance.

$$X_C = \frac{1}{2\pi fC} = \frac{1}{2\pi(10 \text{ kHz})(6800 \text{ pF})} = 2.34 \text{ k}\Omega$$

Apply Ohm's law for a capacitive circuit.

$$I = \frac{V_s}{X_C} = \frac{5.0 \text{ V}}{2.34 \text{ k}\Omega} = \mathbf{2.14 \text{ mA}}$$

Question
What is the total current from the source if a second 6800 pF capacitor is placed in parallel with the first one?

COMPUTER SIMULATION

Open the Multisim file F10-28AC on the website. Verify the current in the circuit.

Capacitive Phase Relationship

A sinusoidal voltage is shown in Figure 10–29. Notice that the rate at which the voltage is changing varies along the sine wave curve, as indicated by the slope of the curve. (Slope is a measure of the "steepness" of a curve.) At the zero crossings, the curve is changing at a faster rate than anywhere else along the curve. At the peaks, the curve has a zero rate of change because it has just reached its maximum and is at the point of changing direction.

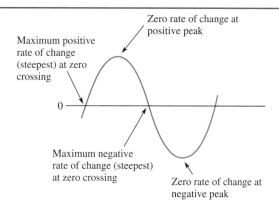

FIGURE 10–29

The rates of change of a sine wave.

The amount of charge stored by a capacitor determines the voltage across it. Therefore, the rate at which the charge is moved ($Q/t = I$) from one plate to the other determines the rate at which the voltage changes. When the current is changing at its maximum rate (at the zero crossings), the voltage is at its maximum value (peak). When the current is changing at its minimum rate (zero at the peaks), the voltage is at its minimum value (zero). This phase relationship is illustrated in Figure 10–30. As you can see, the current peaks occur a quarter of a cycle before the voltage peaks. Thus, the current leads the voltage by 90° in a capacitor.

CHAPTER 10

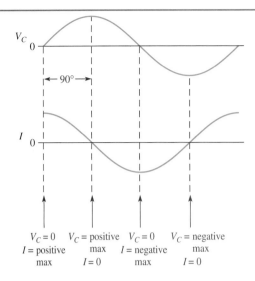

FIGURE 10–30

Current is always leading the capacitor voltage by 90°.

Review Questions

21. What factors determine the capacitive reactance of a capacitor?
22. What is the measurement unit for capacitive reactance?
23. How does capacitive reactance differ from resistance?
24. What is Ohm's law as applied to a capacitive circuit?
25. What is the phase relationship between current and voltage in a capacitor for an ac circuit?

10–6 APPLICATIONS OF CAPACITORS

Capacitors are widely used in electrical and electronic applications.

In this section, you will learn several applications of capacitors.

If you pick up any circuit board, open any power supply, or look inside any piece of electronic equipment, chances are you will find capacitors of one type or another. These components are used for a variety of reasons in both dc and ac applications.

Power Supply Filtering

A basic dc power supply consists of a circuit known as a rectifier followed by a filter. In general, a **rectifier** is a circuit that converts ac into pulsating dc. With respect to a power supply, a **filter** is a low-pass network that smooths out variations in dc. In a power supply, a rectifier converts the 110 V, 60 Hz utility line voltage into a pulsating dc voltage. Figure 10–31 shows a full-wave rectifier that reverses the polarity of the negative alteration of each cycle. The output is pulsating dc because it does not alternate polarity.

FIGURE 10–31

Full-wave rectification.

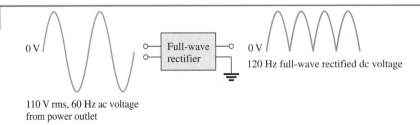

The rectified voltage must be converted to constant dc voltage because virtually all electronic circuits require constant voltage. As indicated in Figure 10–32, a capacitor is used as a filter to eliminate the fluctuations in the rectified voltage and provide a smooth constant-value dc to the load, which is the electronic circuit. It provides filtering action because of the capacitor's ability to store charge. The capacitor charges during the peak of the input and discharges at other times, keeping the output fairly steady. The amount of discharge is typically very small and is exaggerated in the figure for purposes of illustration. Because the output current is supplied by the capacitor after the peak has passed, the capacitor is usually a large-value electrolytic type.

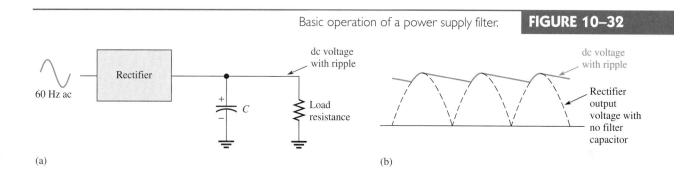

FIGURE 10–32 Basic operation of a power supply filter.

DC Blocking and AC Coupling

Capacitors are commonly used to block the constant dc voltage in one part of a circuit from getting to another part. For example, a capacitor is connected between two stages of an amplifier to prevent the dc voltage at the output of stage 1 from affecting the dc voltage at the input of stage 2, as illustrated in Figure 10–33. Assume that, for proper operation, the output of stage 1 has a zero dc voltage and the input to stage 2 has a 3 Vdc voltage. The capacitor prevents the 3 Vdc at stage 2 from getting to stage 1 and affecting its zero value, and vice versa.

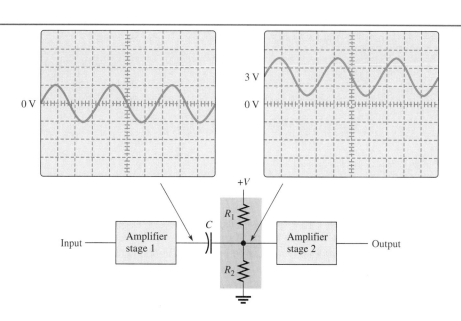

FIGURE 10–33 An application of a capacitor to block dc and pass ac in an amplifier.

CHAPTER 10

If a sinusoidal signal voltage is applied to the input to stage 1, the signal voltage is increased (amplified) and appears on the output of stage 1, as shown. The amplified signal voltage is then coupled through the capacitor to the input of stage 2 where it is superimposed on the 3 Vdc level and then again amplified by stage 2. In order for the signal voltage to be passed through the capacitor without being reduced, the capacitor must be large enough so that its reactance at the frequency of the signal voltage is negligible. In this application, the capacitor is known as a *coupling capacitor,* which ideally acts as an open to dc and as a short to ac.

Power Line Decoupling

Capacitors connected from the dc supply voltage line to ground are used on circuit boards to decouple unwanted voltage transients or spikes that occur on the dc voltage because of fast-switching digital circuits. A voltage transient contains high frequencies that may affect the operation of the circuits. These transients are shorted to ground through the very low reactance of the decoupling capacitors.

Bypassing

Another capacitor application is in bypassing an ac voltage around a resistor in a circuit without affecting the dc voltage across the resistor. In amplifier circuits, for example, dc voltages called bias voltages are required at various points. For the amplifier to operate properly, certain bias voltages must remain constant and, therefore, any ac voltages must be removed. A sufficiently large capacitor connected from a bias point to ground provides a low reactance path to ground for ac voltages, leaving the constant dc bias voltage at the given point. This bypass application is illustrated in Figure 10–34.

FIGURE 10–34

Example of a bypass capacitor. Point A is at ac ground due to the low reactance path through the capacitor. A fraction of the dc is at point A.

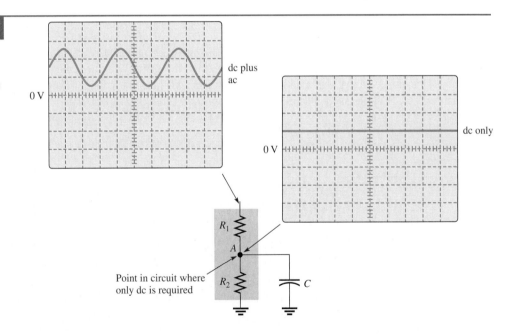

Timing Circuits

Another important area in which capacitors are used is in timing circuits that generate specified time delays or produce waveforms with specific characteristics. Recall that the time constant of a circuit with resistance and capacitance can be controlled by selecting appropriate values for R and C. The charging time of a capacitor can be used as a time delay in various types of circuits. The capacitor voltage increases (or decreases) in a controlled manner, so that the time delay is consistent. An example is the circuit that controls the indicators on your car where the light flashes on and off at regular intervals. This application is described further in Experiment 2 of the lab manual.

Review Questions

26. How is a rectified dc voltage smoothed by a filter capacitor?
27. Why are electrolytic capacitors used for power supply filtering?
28. What is the purpose of a coupling capacitor?
29. What is the purpose of a decoupling capacitor?
30. What characteristic of a capacitor is most important in time-delay applications?

CHAPTER REVIEW

Key Terms

Capacitance The ability to store a charge by means of an electrostatic field.

Capacitive reactance Opposition to ac by a capacitor.

Capacitor A fundamental component that has two conductive plates separated by an insulating region called the dielectric. A capacitor has the ability to store an electric charge.

Dielectric The insulating material that is between the conductive plates of a capacitor.

Electrolyte A conductive material in which charge movement is by ions. The electrolyte in combination with a plastic film forms the negative plate of an electrolytic capacitor.

Farad The unit of capacitance; one farad is the capacitance when a charge of one coulomb is stored with a potential difference of one volt across the plates.

Relative permittivity A pure number that compares the effectiveness of a dielectric to that of air; also known as *dielectric constant*.

Time constant A fixed time interval that determines the time response of an *RC* circuit or an *RL* circuit.

Working voltage The maximum sustained dc voltage a capacitor can withstand.

Important Facts

❑ Factors that affect capacitance are the area of the plates, the plate separation, and the type of dielectric.

❑ Fixed capacitors are available as mica, ceramic, plastic-film, and electrolytic types and have a large range of sizes.

❑ The metric prefixes micro (μ) and pico (p) are usually used to designate capacitor sizes.

❑ The total capacitance of parallel capacitors is the sum of the individual capacitors.

❑ The total capacitance of series capacitors is less than the smallest capacitor. It is the reciprocal of the sum of reciprocals of the individual capacitors.

❑ When a capacitor charges in a series *RC* circuit, the voltage across the capacitor increases exponentially and the charging current decreases exponentially.

- When a capacitor discharges in a series *RC* circuit, the voltage across the capacitor and the discharging current both decrease exponentially.
- The time constant of a series *RC* circuit is a time interval determined by the product of the resistance and the capacitance.
- Capacitive reactance varies inversely with frequency and capacitance.
- Capacitor applications include power supply filtering, blocking dc while allowing ac to pass, power line decoupling, and bypassing.

Formulas

Capacitance defined in terms of charge and voltage:

$$C = \frac{Q}{V} \tag{10-1}$$

Quantity of charge on a capacitor:

$$Q = CV \tag{10-2}$$

Capacitance in terms of physical parameters of a capacitor:

$$C = 8.85 \times 10^{-12} \text{ F/m} \left(\frac{\varepsilon_r A}{d}\right) \tag{10-3}$$

Total capacitance of two parallel capacitors:

$$C_T = C_1 + C_2 \tag{10-4}$$

Total capacitance of multiple parallel capacitors:

$$C_T = C_1 + C_2 + C_3 + \cdots + C_n \tag{10-5}$$

Reciprocal of total capacitance for two series capacitors:

$$\frac{1}{C_T} = \frac{1}{C_1} + \frac{1}{C_2} \tag{10-6}$$

Total capacitance of two series capacitors:

$$C_T = \frac{1}{\dfrac{1}{C_1} + \dfrac{1}{C_2}} \tag{10-7}$$

or

$$C_T = \frac{C_1 C_2}{C_1 + C_2} \tag{10-8}$$

Reciprocal of total capacitance for multiple series capacitors:

$$\frac{1}{C_T} = \frac{1}{C_1} + \frac{1}{C_2} + \frac{1}{C_3} + \cdots + \frac{1}{C_n} \tag{10-9}$$

Total capacitance of multiple series capacitors:

$$C_T = \frac{1}{\dfrac{1}{C_1} + \dfrac{1}{C_2} + \dfrac{1}{C_3} + \cdots + \dfrac{1}{C_n}} \tag{10-10}$$

Capacitive voltage-divider rule:

$$V_x = \left(\frac{C_T}{C_x}\right) V_s \tag{10-11}$$

Time constant in a series *RC* circuit:

$$\tau = RC \tag{10–12}$$

Capacitive reactance in an ac circuit:

$$X_C = \frac{1}{2\pi f C} \tag{10–13}$$

Ohm's law for a capacitive circuit:

$$I = \frac{V}{X_C} \tag{10–14}$$

Chapter Checkup

Answers are at the end of the chapter.

1. The unit of capacitance is the
   - (a) ohm
   - (b) weber
   - (c) volt
   - (d) farad

2. The unit of capacitive reactance is the
   - (a) ohm
   - (b) weber
   - (c) volt
   - (d) farad

3. The dielectric in an electrolytic capacitor is made of
   - (a) a thin oxide layer
   - (b) a metal plate
   - (c) an electrolyte
   - (d) mica

4. What type of capacitor is a trimmer?
   - (a) an electrolytic
   - (b) a fixed ceramic
   - (c) a small variable
   - (d) any of these answers

5. The maximum sustained voltage that a capacitor can withstand is called the
   - (a) surge voltage
   - (b) peak voltage
   - (c) working voltage
   - (d) breakdown voltage

6. The total capacitance of two series capacitors is equal to
   - (a) the sum
   - (b) the product over the sum
   - (c) the sum over the product
   - (d) the sum of the reciprocals

7. A 1000 pF capacitor is in parallel with a 0.001 μF capacitor. The total capacitance is
   - (a) 2000 pF
   - (b) 2 nF
   - (c) 0.002 μF
   - (d) Answers (a), (b), and (c) are correct.
   - (e) None of the answers are correct.

8. Two unequal capacitors are connected in parallel across an ac source. Compared to the larger capacitor, the voltage across the smaller capacitor is
   - (a) larger
   - (b) equal
   - (c) smaller

9. Refer to the circuit in Figure 10–35. The time constant is
   (a) 100 ns   (b) 100 μs
   (c) 73 μs   (d) 73 ms

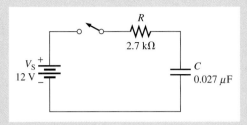

FIGURE 10–35

10. Refer to the circuit in Figure 10–35. *Immediately* after the switch is closed the voltage across the resistor will be
    (a) 0 V   (b) 6 V
    (c) 7.56 V   (d) 12 V

11. Refer to the circuit in Figure 10–35. After the switch is closed and five time constants have elapsed, the voltage across the capacitor will be
    (a) 0 V   (b) 6 V
    (c) 7.56 V   (d) 12 V

12. For constant dc, a capacitor is equivalent to
    (a) an open   (b) a resistor
    (c) a short

13. Which of the following does *not* affect capacitive reactance?
    (a) the applied voltage   (b) the frequency
    (c) the size of the capacitor

14. When ac is applied to a capacitor, the current leads the voltage by
    (a) 45°   (b) 90°
    (c) 120°   (d) answer depends on the capacitor

15. The type of capacitor normally used for a power supply filter is
    (a) variable   (b) electrolytic
    (c) ceramic   (d) mica

16. The purpose of a power line decoupling capacitor is to
    (a) block ac
    (b) pass dc
    (c) pass unwanted transients to ground
    (d) all of the above

Questions

Answers to odd-numbered questions are at the end of the book.

1. What is a dielectric?
2. What determines the amount of charge on a capacitor?

3. What two metric prefixes are most common for describing the size of a capacitor?
4. What are three factors that determine the capacitance of a capacitor?
5. If a capacitor is marked 104, what is its value?
6. In addition to the value of a capacitor, what other specification is generally given on a capacitor?
7. What are the units for permittivity?
8. What is meant by relative permittivity?
9. What happens when the dielectric of a capacitor breaks down?
10. What is the advantage of mica capacitors?
11. Of mica or ceramic, which type of capacitor has the highest dielectric constant?
12. What does the band on one end of a plastic-film capacitor indicate?
13. What are the two types of electrolytic capacitors?
14. What is the major advantage of electrolytic capacitors?
15. What are some disadvantages to electrolytic capacitors?
16. What does an arrow drawn through the symbol of a capacitor indicate?
17. A capacitor is marked 222. What is its value?
18. What do the letters WVDC mean on a capacitor?
19. What is the total capacitance of a 220 μF capacitor that is in series with a 1000 μF capacitor?
20. What is the total capacitance of a 220 μF capacitor that is in parallel with a 1000 μF capacitor?
21. What is the total capacitance of a 4700 pF capacitor that is in series with a 0.015 μF capacitor?
22. What is the total capacitance of a 4700 pF capacitor that is in parallel with a 0.015 μF capacitor?
23. What is the time constant for a 22 μF capacitor in series with a 4.7 kΩ resistor?
24. For a dc circuit with a series resistor and capacitor and switch, how many time constants are required for a capacitor to reach 82% of its final charge after the switch has been closed?
25. For a dc circuit with a series resistor and capacitor and switch, how many time constants are required for a capacitor to reach its full charge after the switch has been closed?
26. For a dc circuit with a series resistor and capacitor and switch, when is the current a maximum in the circuit?
27. What limits current in an ac circuit containing a single capacitor?
28. How does frequency affect capacitive reactance?
29. If a capacitor in an ac circuit is replaced with a larger one, what happens to the current?
30. What is Ohm's law for a capacitive circuit?
31. When does a sine wave have the highest rate of change?
32. What is the purpose of a power supply rectifier?
33. What is the purpose of a power supply filter?
34. What is the purpose of a coupling capacitor?
35. What is the purpose of a bypass capacitor?

PROBLEMS

Basic Problems

Answers to odd-numbered problems are at the end of the book.

1. Convert each of the following values to μF:
   (a) 68 nF
   (b) 0.001 F
   (c) 4700 pF

2. Convert each of the following values to pF:
   (a) 68 nF
   (b) 0.047 μF
   (c) 0.0015 μF

3. Find the capacitance of a capacitor that stores 50 μC of charge with 10 V across the plates.

4. Find the capacitance of a capacitor that stores 5400 pC of charge with 20 V across the plates.

5. What is the voltage on a 10,000 pF capacitor that has a charge of 1 μC?

6. What is the voltage on a 220 μF capacitor that has a charge of 3.3 mC?

7. Find the charge on a 0.01 μF capacitor with 24 V across it.

8. Find the charge on a 470 μF capacitor with 60 V across it.

9. A capacitor is rolled from two rectangular sheets of foil that are 2.0 cm by 22 cm. The foils are separated by paper that is 0.5 mm thick. Determine the capacitance.

10. A ceramic capacitor is formed from square plates that are 1.6 cm on a side. The ceramic is 0.1 mm thick and has a dielectric constant of 1200. Determine the capacitance.

11. Calculate the total capacitance for each circuit in Figure 10–36.

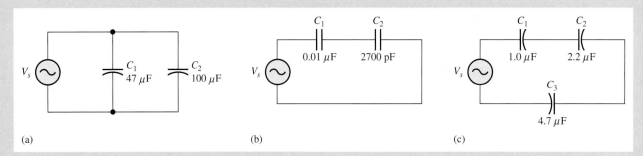

FIGURE 10–36

12. Calculate the total capacitance for each circuit in Figure 10–37.

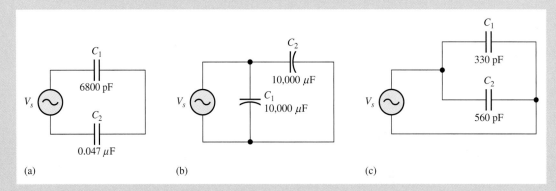

FIGURE 10–37

13. What is the total capacitance of four parallel 1000 μF capacitors?
14. What is the total capacitance of four series 1000 μF capacitors?
15. Refer to Figure 10–38. Initially, the capacitors are uncharged. What is the voltage across each capacitor after the switch is closed?

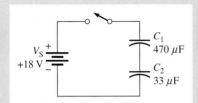

FIGURE 10–38

16. What is the RC time constant for the circuit in Figure 10–39?

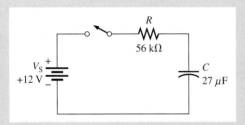

FIGURE 10–39

17. Refer to the circuit in Figure 10–39. What is the approximate voltage on the capacitor 3.0 s after the switch is closed?
18. Refer to the circuit in Figure 10–39. At what time after the switch is closed is the voltage across the capacitor equal to the voltage across the resistor?
19. Refer to the circuit in Figure 10–39. What is the current immediately after the switch closes?
20. Refer to the circuit in Figure 10–39. How long after the switch is closed has the current dropped to 0.1 mA?
21. Refer to the circuit in Figure 10–40. What is the approximate voltage on the capacitor 2.0 ms after the switch is closed?

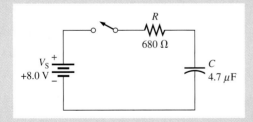

FIGURE 10–40

22. Refer to the circuit in Figure 10–40. At what time after the switch is closed is the voltage across the capacitor equal to the voltage across the resistor?
23. Refer to the circuit in Figure 10–40. How long after the switch is closed has the current dropped to 5.0 mA?

24. Find the capacitive reactance for a 10 μF capacitor for each of the following frequencies:
    (a) 10 Hz
    (b) 100 Hz
    (c) 1.0 kHz
    (d) 10 kHz

25. Find the capacitive reactance for a 0.1 μF capacitor for each of the following frequencies:
    (a) 10 Hz
    (b) 100 Hz
    (c) 1.0 kHz
    (d) 10 kHz

26. A capacitor has a reactance of 72 Ω at a frequency of 1.0 kHz. What is the capacitance?

27. At what frequency is the capacitive reactance of a 0.1 μF capacitor equal to 100 Ω?

28. For the circuit in Figure 10–41, find the capacitive reactance of C and the current in the circuit.

FIGURE 10–41

29. What is the total current in the circuit of Figure 10–41 if another 220 pF capacitor is placed in series with the first?

Basic-Plus Problems

30. Assume the capacitors are initially uncharged in the circuit of Figure 10–42. After the switch is closed, what is the charge on each capacitor?

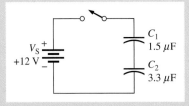

FIGURE 10–42

31. Assume a 300 pF capacitor has an air dielectric. If the capacitor is immersed in oil, what will be the capacitance?

32. (a) Find the dielectric constant (ε_r) of a capacitor that has a capacitance of 270 pF with a plate separation of 0.1 mm if the capacitor has square plates 3.5 cm on a side.
    (b) From the materials listed in Table 10–1, what material is the likely dielectric?

33. (a) A capacitor has a charge of 235 nC when the voltage across it is 50 V. If the capacitor is constructed from plates that are 24 cm by 4.5 cm that are separated by 0.1 mm, what is the dielectric constant?

    (b) From the materials listed in Table 10–1, what material is the likely dielectric?

34. What is the total capacitance in Figure 10–43?

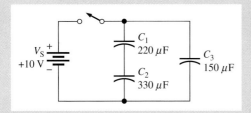

FIGURE 10–43

35. What is the voltage across each capacitor in Figure 10–43?
36. What is the time constant for the circuit in Figure 10–44?

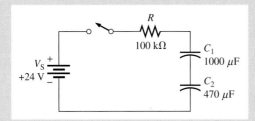

FIGURE 10–44

37. Refer to Figure 10–44. In one time constant after the switch has closed, what is the voltage across C_1?

38. Refer to Figure 10–45. Assume that the voltage across the capacitor and the resistor are equal. What is the frequency of the source?

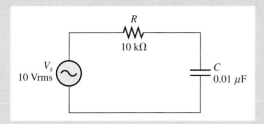

FIGURE 10–45

39. The cost difference between aluminum and tantalum capacitors was mentioned in Section 10–2. Use the Internet or other resources to find more about the cost difference and where you might use the higher-priced capacitor. Prepare a short report on your research.

Example Questions

ANSWERS

10–1: 100,000 nF

10–2: It is doubled.

10–3: 553 pF

10–4: 37,000 pF

10–5: 0.000 767 μF
10–6: 3.12 V
10–7: 89 μs
10–8: 8.9 V
10–9: 0.12 mA
10–10: 33.9 kHz
10–11: 4.28 mA

Review Questions

1. Capacitance is the ability to store a charge by means of an electrostatic field.
2. (a) 10^6 (b) 10^3 (c) 10^6
3. 270 μF
4. (a) increases (b) decreases
5. 6.37 nF
6. They have relatively high capacitance values in a small physical size and high working voltage ratings.
7. An electrolyte is a conductive material in which charge movement is by ions rather than electrons.
8. A very thin oxide layer that forms on a metallic plate.
9. Check that the polarity requirement is correct and that the working voltage rating is not exceeded.
10. Microfarads (μF) and picofarads (pF)
11. The total capacitance is the sum of the individual capacitors in parallel.
12. 50,000 μF
13. The total capacitance is the reciprocal of the sum of the reciprocals of the individual capacitors in series.
14. 2,000 μF
15. The smaller capacitor will have the larger voltage across it.
16. None
17. 155 ms
18. The universal time-constant curves are the transient response of an RC circuit, showing the charging and discharging current or voltage.
19. The series resistor
20. 37%
21. The capacitance and the frequency
22. The ohm
23. Resistance dissipates power; capacitive reactance does not.
24. $I = V/X_C$
25. Current leads the voltage by 90°.
26. The capacitor charges to the peak of the pulsating dc, and slowly discharges until the next pulse, keeping the output nearly constant.

27. They are available in the very large sizes required for power supply filtering.
28. To pass ac from one circuit to another while blocking dc
29. To pass unwanted transients on a dc power supply line to ground
30. The ability to store charge and increase or decrease the voltage in a controlled manner

Chapter Checkup

1. (d)	2. (a)	3. (a)	4. (c)	5. (c)
6. (b)	7. (d)	8. (a)	9. (c)	10. (d)
11. (d)	12. (a)	13. (a)	14. (b)	15. (b)
16. (c)				

11 CHAPTER

INDUCTORS

CHAPTER OUTLINE

11–1 Inductance
11–2 Types of Inductors
11–3 Series and Parallel Inductors
11–4 Inductors in DC Circuits
11–5 Inductors in AC Circuits
11–6 Applications of Inductors

INTRODUCTION

Like a capacitor, an inductor is also a fundamental passive component. Inductors are not as widely used as capacitors, but they are used in many circuits, especially at higher frequencies. Many properties of inductors have an analogous property with capacitors. Like capacitors, ideal inductors do not dissipate energy but can store it and return it later to the circuit. When a sine wave is applied to an inductor, the inductor produces an opposition to current called *inductive reactance*. This opposition shifts the phase of current and voltage in a circuit. In a capacitive circuit, current leads voltage; but in an inductive circuit, voltage leads current.

In this chapter, you will learn about inductors and their construction. You will learn the effects of connecting them in series and in parallel and learn how they respond in both dc and ac circuits. The chapter concludes with a survey of some important applications of inductors.

Study aids for this chapter are available at

http://www.prenhall.com/SOE

KEY OBJECTIVES

A section number is given for each objective. After completing this chapter, you should be able to

11-1 Define inductance, describe Lenz's law, and explain how inductance can be calculated for a coil

11-2 Compare the characteristics of various types of inductors

11-3 Calculate the total inductance of series and parallel inductors

11-4 Use the universal charging and discharging curves to predict the instantaneous current and voltage for an inductor in a dc circuit

11-5 Calculate the inductive reactance for an inductor in an ac circuit

11-6 List several important applications of inductors

COMPUTER SIMULATIONS DIRECTORY

The following figures have Multisim circuit files associated with them.

- Figure 11-16 Page 387
- Figure 11-22 Page 391

LABORATORY EXPERIMENTS DIRECTORY

The following exercises are for this chapter.

- **Experiment 20** Inductors
- **Experiment 21** Inductive Reactance

KEY TERMS

- Inductance
- Inductor
- Henry
- Inductive reactance
- Resonant circuit
- Positive feedback

CHAPTER 11

Most model trains have a dc motor that receives power through the track. When the train accelerates, a high current is in the rails. At a constant speed, the current is smaller and steady. If you suddenly turn the power supply down in order to stop the train, the current will continue for a short time even with no supply voltage. This current is created by the collapsing magnetic field of the coils in the motor. The motor coils are electrical inductors.

In electrical circuits, the shape of the current and voltage waveforms is due to these same magnetic effects. Moving electrical charges produce the magnetic field associated with the moving train, and the collapsing field produces current, similar to motor-generator action.

The current that continues after the voltage has been turned to zero has an analogy in physical systems with inertia. No physical mass (including our model train) can be stopped instantly. Inertia is the property of matter that tends to keep moving bodies in motion and keeps stationary bodies at rest. This is given in physics classes as Newton's third law of motion. With a body that has mass, its motion cannot change instantly; with the electrical inductor, current cannot change instantly.

11–1 INDUCTANCE

The magnetic field that surrounds a single wire is weak, but by forming the wire into a coil, the magnetic field is strengthened. Coils are often referred to as inductors because they exhibit the property of inductance.

In this section, you will learn the definition of inductance, describe Lenz's law, and explain how inductance can be calculated for a coil.

Inductance is the property of a conductor to oppose a change in current. When current starts in a conductor because of an applied voltage, a magnetic field expands outward from the conductor. This field stores energy. As the magnetic field expands, a voltage is induced in the conductor that tends to oppose the original voltage and thus a change in the original current. Once the current has reached a steady state, the magnetic field no longer expands, and no voltage is induced in the conductor.

Inductance is a property of all conductors, but it has only a tiny effect in straight conductors. It is much more pronounced in coils because of the enhanced magnetic field surrounding a coil. In fact, coils are often referred to as inductors because they are selected for their inductive properties. An **inductor** is a fundamental passive component in electronic circuits; it is designed to introduce a specific amount of inductance into a circuit for a given application. Lenz's law describes the property of inductance, which is given for coils as

When the current through a coil *changes*, an induced voltage is created across the coil that always opposes the change in current.

Figure 11–1 illustrates Lenz's law for an inductive circuit. In part (a), the current is constant and is limited by R_1. There is no induced voltage because the electromagnetic field is unchanging. In part (b), the switch suddenly is closed, placing R_2 in parallel with R_1, and thus reducing the total resistance. Naturally, the current tries to increase and the electromagnetic field begins to expand, but an induced voltage appears across the inductor that opposes this change in current for a brief time. Eventually, the current increases to a new higher value as shown in Figure 11–1(c), and the induced voltage drops to zero.

INDUCTORS

FIGURE 11–1

Demonstration of Lenz's law. When current tries to change due to a reduction in resistance, a voltage is induced across the coil that opposes the change in current.

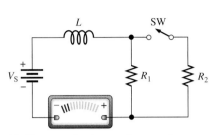
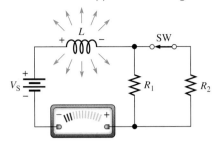
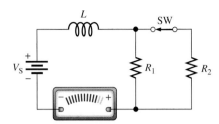

(a) The current has been on for some time. The total current is limited only by R_1.

(b) When SW closes, a voltage appears across the coil that opposes the *change* in current, so initially there is no change in the total current.

(c) After a certain time, the voltage across the coil decays and the total current has increased due to the lower resistance.

A similar effect occurs if the switch is now opened. A voltage will be induced across the coil that again opposes the change in current. This time the induced voltage is opposite to that shown in Figure 11–1(b). This is because the collapsing magnetic field returns energy to the circuit and induces a voltage that attempts to keep the current from changing from a higher to lower value.

The unit of inductance is the henry, named for Joseph Henry, who did pioneering work in electromagnetism. One **henry (H)** is the amount of inductance that induces one volt when the current changes at a uniform rate of one ampere per second. The henry is a large unit of inductance, so millihenries (mH) and microhenries (μH) are more commonly used.

Factors that Affect Inductance

Four parameters are important in establishing the inductance of a coil: permeability of the core material, number of turns of wire, core length, and cross-sectional area of the core.

As discussed in Chapter 7, an inductor is a coil of wire that surrounds a magnetic or nonmagnetic material called the core. Examples of nonmagnetic core materials are air, plastic, and glass. They have permeabilities close to a vacuum, which is $4\pi \times 10^{-7}$ Wb/At-m (or 1.26×10^{-6} Wb/At-m). Examples of magnetic materials are iron, nickel, steel, and cobalt. These have permeabilities that are hundreds or thousands of times greater than that of a vacuum. As you learned in Section 7–2, the permeability (μ) of the core material determines how easily a magnetic field can be established. The inductance is directly proportional to the permeability of the core material.

The number of turns of wire, the core length, and the cross-sectional area of the core are factors in setting the value of inductance of a coil. The inductance is inversely proportional to the length of the core and directly proportional to the cross-sectional area. In addition, the inductance is directly related to the number of turns squared. For coils wound such that the length is much longer than the diameter, this relationship is as follows:

$$L = \frac{N^2 \mu A}{l} \qquad (11\text{--}1)$$

where L is inductance in henries, N is number of turns of wire, μ is permeability in webers per ampere-turn-meter, A is cross-sectional area in meters squared, and l is coil length in meters.

HISTORICAL NOTE

Heinrich Lenz was born in Estonia (then Russia) and became a professor at the University of St. Petersburg. He carried out many experiments following Faraday's lead and formulated the principle of electromagnetism, which defines the polarity of induced voltage in a coil. The statement of this principle, Lenz's law, is named in his honor.

HISTORICAL NOTE

The unit of inductance, the henry, is named in honor of Joseph Henry (1797–1878). He began his career as a professor at a small school in Albany, New York, and later became the first director of the Smithsonian Institution. He was the first American after Franklin to undertake original scientific experiments. He superimposed coils of wire wrapped on an iron core and observed the effects of electromagnetic induction in 1830, a year before Faraday, but he did not publish his findings. Henry did obtain credit for the discovery of self-induction, however.

377

EXAMPLE 11–1

Problem
A student wraps 100 turns of wire on a pencil that is 7 mm in diameter as shown in Figure 11–2. The turns are 3.5 cm long. Determine the inductance of the coil that is formed. Because it is wound on a nonmagnetic material, assume the core length is the same as the coil length. The permeability of the pencil is approximately the same as for air or 1.26×10^{-6} Wb/At-m.

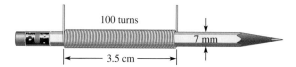

FIGURE 11–2

Solution
First, determine the area (in m^2).

$$A = \pi r^2 = \pi(0.0035 \text{ m})^2 = 38.5 \times 10^{-6} \text{ m}^2$$

Use Equation 11–1 to calculate the inductance.

$$L = \frac{N^2 \mu A}{l} = \frac{100^2(1.26 \times 10^{-6} \text{ Wb/At-m})(38.5 \times 10^{-6} \text{ m}^2)}{0.035 \text{ m}} = \mathbf{13.8 \ \mu H}$$

Question*
If the student wound the same coil on a steel bolt that is 3.5 cm long and 7 mm in diameter instead of a pencil, what is the inductance? The permeability (μ) of the steel is 5.5×10^{-4} Wb/At-m.

Winding Resistance

Most general-purpose inductors are wound with many turns of fine wire in order to produce a reasonable inductance in a small inductor. Fine wire has a higher resistance per unit length than heavy wire, so many small coils have a significant total resistance. This inherent resistance is called the dc resistance or the winding resistance (R_W). Although this resistance is distributed along the length of the wire, it effectively appears in series with the inductance of the coil, as shown in Figure 11–3. In many applications, the winding resistance may be small enough to be ignored and the coil considered as an ideal inductor (one with no resistance). In other cases, the resistance must be considered.

FIGURE 11–3
Winding resistance of an actual coil.

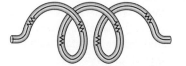

(a) The wire has resistance distributed along its length.

(b) Equivalent circuit

Answers are at the end of the chapter.

Winding Capacitance

When two conductors are placed side by side but not touching, there is always some capacitance between them. Thus, when many turns of insulated wire are together in a coil, a certain amount of stray capacitance, called winding capacitance (C_W), is a natural side effect. Usually, winding capacitance is not important except at high frequencies, where it may become quite important. The equivalent circuit for an inductor with both its winding resistance (R_W) and its winding capacitance (C_W) is shown in Figure 11–4. For this text, we will not be concerned further with winding capacitance.

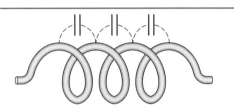
(a) Stray capacitance between each loop appears as a total parallel capacitance (C_W).

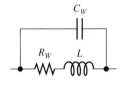
(b) Equivalent circuit

FIGURE 11–4

Winding capacitance of an actual coil.

Review Questions

Answers are at the end of the chapter.

1. What is inductance?
2. What is Lenz's law for coils?
3. What is the unit of inductance?
4. What are four factors that affect the inductance of a coil?
5. What is the equivalent circuit for a coil that has resistance and capacitance?

TYPES OF INDUCTORS 11–2

Like resistors, inductors are available in a number of standard values and forms. Generally, they are classified according to their core material.

In this section, you will learn the characteristics of various types of inductors.

Very small inductors generally are wound on a simple plastic or cardboard form that has no magnetic properties, so they behave the same as one with an air core. Larger coils take advantage of the higher permeability of iron or other core material to increase the magnetic flux and, therefore, the inductance. Variable inductors are constructed with a core that can be moved in and out of the coil with a screw-type adjustment, thus changing the inductance. The symbols for air-core, iron-core, variable, and ferrite-core inductors are shown in Figure 11–5.

(a) Air core   (b) Iron core   (c) Variable   (d) Ferrite core

FIGURE 11–5

Inductor symbols.

CHAPTER 11

Inductors are available in sizes with the same standard numeric values as resistors. Standard values of inductors include 1 µH, 1.2 µH, 1.5 µH, 1.8 µH, 2.2 µH, and so forth. Sometimes an inductor has the value stamped on it, and some use color bands like resistors. When color codes are used, they are read the same way as a resistor, except the value will be given in microhenries.

The appearance of inductors depends on their size and whether they are encapsulated or not. Encapsulated inductors have a coating that protects the fine wire and are usually color-coded, so they are sometimes confused for a resistor. Figure 11–6 shows some common types of inductors.

FIGURE 11–6
Common inductors.

(a) Encapsulated   (b) Wirewound, high current   (c) Torroid coil   (d) Variable

EXAMPLE 11–2

Problem
What is the inductance of an encapsulated coil that is marked with three red bands and a gold band?

Solution
The red bands all stand for the number 2. As with a resistor, the first two bands are the digits and the third (multiplier band) shows the number of zeros. The fourth (gold) band indicates 5% tolerance. Thus, the inductor is read as 2200 µH ± 5%. (This is equivalent to a 2.2 mH inductor.)

Question
What is the inductance if the third band is orange?

Testing inductors

Inductors are very reliable if they are operated within their current limit, but occasionally they fail. The most common failure is due to an open, which can easily be checked with an ohmmeter (the reading will be infinite if the inductor is open). Open windings are sometimes due to flexing the leads, a common problem in school labs due to handling. Shorted windings do not show up easily because a few shorted windings will only change the total resistance and total inductance a small amount. The best check is with a meter that can read inductance. An oscilloscope can indirectly test the inductance by measuring the time constant (described in Section 11–4).

Review Questions

6. What do solid lines alongside the schematic symbol of a coil indicate?
7. How is a variable inductor's value changed?

INDUCTORS

8. What is the color code for a 5% encapsulated 470 μH coil?
9. What is the value of an encapsulated inductor that is marked with three orange bands and a silver band?
10. How are inductors tested?

SERIES AND PARALLEL INDUCTORS 11–3

Inductors can be placed in series or parallel. In series, the total inductance increases; in parallel, the total inductance decreases.

In this section, you will learn to calculate the total inductance of series and parallel inductors.

Series Inductors

When inductors are connected in series, as in Figure 11–7, the total inductance is the sum of the individual inductors.

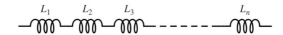

FIGURE 11–7

Series inductors.

Stated in equation form, the total inductance of n inductors in series is

$$L_T = L_1 + L_2 + L_3 + \cdots + L_n \tag{11-2}$$

This equation has the same form as Equation 5–1 for total resistance of resistors in series and Equation 10–5 for total capacitance of capacitors in parallel.

EXAMPLE 11–3

Problem
What is the total inductance of a 270 μH inductor that is in series with a 68 μH inductor?

Solution
The total inductance is the sum of the two inductors.

$$L_T = L_1 + L_2 = 270\ \mu\text{H} + 68\ \mu\text{H} = \mathbf{338\ \mu H}$$

Question
What is the total inductance if a third inductor of 100 μH is connected in series with the other two?

Parallel Inductors

When inductors are connected in parallel, as in Figure 11–8, the total inductance decreases. The formula for the total inductance of n inductors in parallel is

$$L_T = \cfrac{1}{\cfrac{1}{L_1} + \cfrac{1}{L_2} + \cfrac{1}{L_3} + \cdots + \cfrac{1}{L_n}} \tag{11-3}$$

CHAPTER 11

FIGURE 11–8
Parallel inductors.

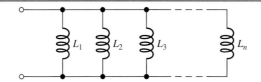

This general formula states that the total inductance of parallel inductors is equal to the reciprocal of the sum of the reciprocals of the individual inductors. This formula is similar to the formula developed for parallel resistors (Equation 5–8) and for series capacitors (Equation 10–10).

As in the case of two parallel resistors or two series capacitors, the formula for exactly two parallel inductors reduces to a product-over-sum form. Thus, for two inductors, the alternate form of Equation 11–3 is

$$L_T = \frac{L_1 L_2}{L_1 + L_2} \qquad (11\text{--}4)$$

EXAMPLE 11–4

Problem
What is the total inductance of a 270 μH inductor that is in parallel with a 68 μH inductor?

Solution
Apply Equation 11–4 for two inductors.

$$L_T = \frac{L_1 L_2}{L_1 + L_2} = \frac{(270\ \mu\text{H})(68\ \mu\text{H})}{270\ \mu\text{H} + 68\ \mu\text{H}} = \mathbf{54.3\ \mu H}$$

Notice that the total inductance is smaller than the smallest individual inductor.

Question
What is the total inductance if a third inductor of 100 μH is connected in parallel with the other two?

Review Questions

11. What is the rule for combining inductors in series?
12. What is the rule for combining several inductors in parallel?
13. When can you use the product-over-sum rule to calculate the total inductance?
14. What is the total inductance of three 4.7 mH inductors connected in series?
15. What is the total inductance of three 4.7 mH inductors connected in parallel?

11–4 INDUCTORS IN DC CIRCUITS

An inductor in a dc circuit will affect the current in the circuit when a change in conditions such as a switch closure occurs. The inductor responds to the change by developing a voltage across it that tends to oppose the change in current. Once the current is reestablished at a new level, the ideal inductor acts as a short to dc.

In this section, you will learn to use the universal charging and discharging curves to predict the instantaneous current and voltage for an inductor in a dc circuit.

Current in an Inductor

Recall that current in a conductor causes a magnetic field to surround the conductor. The field is intensified by forming the conductor into a coil. Energy is stored in this field whenever current is in the coil (inductor). When the current is a constant dc, the magnetic field surrounding it is constant. Ideally, there is no voltage across the inductor; however, in practice a small drop due to the winding resistance will be present. The inductance itself appears as a short to dc.

Lenz's law states that when a change occurs in a circuit that affects the amount of current in the inductor, an induced voltage appears across the inductor with a polarity that opposes the change in current.

Current cannot change instantaneously in an inductive circuit.

Interestingly, the shape of the response to a changing current in an inductive circuit has exactly the same shape as the voltage in a capacitive circuit, as described in Section 10–4. In fact, you can apply the same universal curves to the inductive case (as you will see), but you need to know when current is increasing or decreasing to apply the correct curve.

Keep in mind that an inductor opposes a change in current in a circuit. One way to think about the polarity of the voltage in a circuit with both resistance and inductance (*RL* circuit) is to consider Kirchhoff's voltage law, which must be satisfied at all times. Figure 11–9 illustrates how Kirchhoff's voltage law is applied to a basic series *RL* circuit. When the switch is open, as in Figure 11–9(a), a voltage equal to the source voltage appears across the switch, and the current is zero. The other components have no voltage across them and KVL is satisfied. Immediately after the switch is closed, as in Figure 11–9(b), the source voltage is induced across the coil that opposes the voltage source, and the current is still zero. KVL is still satisfied. As the magnetic field builds around the inductor, the voltage across the inductor (ideally) drops to zero and the voltage across the resistor rises until is equal to the source voltage, as illustrated in Figure 11–9(c). The current in the circuit rises to a value limited only by the circuit resistance.

SAFETY NOTE

You should avoid working on "live" circuits, even those powered by low voltages. An inductor can store energy when there is current in it. If the circuit is opened, the stored energy can build a very large voltage across the open, causing a spark. This voltage can be many times larger than the supply voltage and can cause a severe electrical shock if you are in contact with it.

FIGURE 11–9

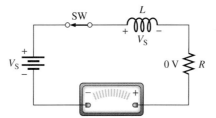
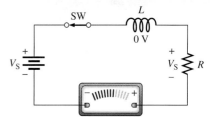

(a) Before the switch is closed, the source voltage appears across the switch.

(b) Immediately after closing the switch, the source voltage appears across the inductor.

(c) After the switch has been closed a long time, the voltage across the inductor decays to zero and the voltage across the resistor rises to the source voltage.

Current and Voltage Waveforms

The current and voltage waveforms for the *RL* circuit in Figure 11–9 follow the shapes illustrated in Figure 11–10. Immediately after the switch is closed, the voltage across the inductor is equal to the source voltage, shown as the initial voltage. It decays as illustrated in Figure 11–10(a). In accordance with KVL, as the voltage decays across the inductor, it rises across the resistor as shown in Figure 11–10(b). Because the current is proportional to the voltage across the resistor, as given by Ohm's law, the current in the circuit rises with the same shape as the voltage across the resistor.

CHAPTER 11

FIGURE 11–10
Shape of the voltage and current after a switch closure.

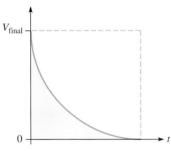
(a) After the switch closes, the voltage across the inductor decays.

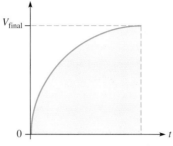
(b) At the same time, the voltage across the resistor rises.

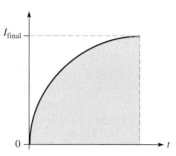
(c) The current has the same shape as the voltage across the resistor.

As in the case of *RC* circuits, the time required for the voltage or current to rise and fall in an *RL* circuit is determined by the values of *R* and *L*, but the shape of the curves are the same for all *RL* combinations.

If the current is suddenly interrupted by opening the switch in an *RL* circuit, the stored energy will attempt to maintain the current. Because the open switch represents an extremely high resistance, the voltage will build up significantly across the inductor and may arc across the gap of the open switch. The high voltage buildup is exploited in certain circuits (such as strobe lights and ignition systems) but can be a problem in others.

FIGURE 11–11
A series *RL* circuit driven by a square wave.

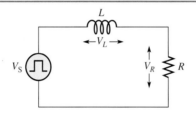

FIGURE 11–12
The response of the series *RL* circuit to a square wave. V_S is the source voltage, V_L is the voltage across the inductor, and V_R is the voltage across the resistor.

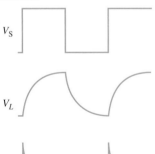

Instead of by opening and closing a switch, the series *RL* circuit can be driven by a square wave, as shown in Figure 11–11. The response is shown in Figure 11–12. The generator is set so that the square wave has a relatively long period compared to the circuit response. When the square wave first goes high, the response of the circuit is like that of the switch closure with a battery. The shape of the voltage across the inductor and resistor will be the same as that given for the switch closure case as previously discussed. However, when the pulse is low, the situation is completely different from the case of opening a switch. Instead of the extremely high resistance of the switch, which leads to an arc, the generator looks like a low resistance path. (The actual resistance is the Thevenin source resistance.) When the square wave drops to its low value, the voltage across the inductor is opposite to the case when the pulse was rising. Compare the waveforms to the waveforms for the *RC* circuit given in Figure 10–19.

A Practical Demonstration

To demonstrate inductive voltage buildup, let's look at the circuit shown in Figure 11–13(a). A neon bulb is connected across the inductor in a series *RL* circuit as shown. A large inductor (1 H or more) and a series resistor of 220 Ω will work nicely. (The primary of a small transformer will work for the inductor.) Two D cells in series supply the source voltage. When the switch is closed, the bulb does not light because the inductor builds a voltage of only 3 V to oppose the source and a neon bulb requires about 70 V to fire it. How-

INDUCTORS

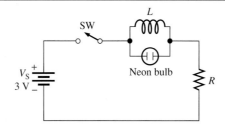

FIGURE 11–13

ever, when the switch is opened, the bulb fires for an instant because the voltage due to the collapsing magnetic field exceeds the required 70 V.

Another related circuit is shown in Figure 11–14. In this case, when the switch is closed, the bulb is in parallel with the source, which is too low a voltage to fire the bulb. When the switch is open, the inductive buildup will cause the bulb to light for an instant while the magnetic field of the inductor collapses.

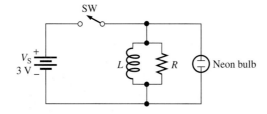

FIGURE 11–14

The *RL* Time Constant

When an inductor opposes the change in current in a series *RL* circuit, the current in the circuit cannot change instantaneously because of the opposing voltage that appears across the inductor. The rate of current change is determined by the time constant of the circuit.

The time constant for a series *RL* circuit is a time interval determined by the inductance divided by the resistance. It establishes the time required for current to become steady after a transient event such as a switch closure. Recall that the time constant in an *RC* circuit, which is the product of *R* and *C*, determines the time required for the voltage across the capacitor to become stable. As in the case of the *RC* circuit, the time constant is symbolized by the Greek letter τ. For the *RL* circuit, the formula for the time constant is

$$\tau = \frac{L}{R} \qquad (11\text{--}5)$$

where τ is time constant in seconds when the resistance is in ohms (Ω) and the inductance is in henries (H).

EXAMPLE 11–5

Problem
A 10 mH inductor is in series with a 2.2 kΩ resistor. What is the time constant?

Solution
Substituting into Equation 11–5,

$$\tau = \frac{L}{R} = \frac{10 \text{ mH}}{2.2 \text{ k}\Omega} = 4.55 \text{ }\mu\text{s}$$

Question
What is the time constant for a 68 mH inductor in series with a 330 Ω resistor?

CHAPTER 11

In an inductive circuit, the time constant is the amount of time that is required for the current in the circuit to change from an initial current to 63% of the final current. The final current is in the circuit when the voltage across the inductor has dropped to zero (ideally). Because the voltage on the resistor has the same shape as the current, you can measure the time constant by observing the time required for the voltage across the resistor to change from an initial voltage to 63% of the final voltage. This will give the same result as measuring the time for current to change.

For oscilloscope measurements, measuring the voltage across the resistor is the simplest way to measure the time constant. The key is to set the frequency low enough to see the entire waveform, from minimum to maximum on the resistor. The time constant can be found directly by measuring the time from starting voltage to 63% of the final voltage.

The rising and falling curves for inductive *RL* circuits follow the same exponential curves as those given in Section 10–4 for *RC* circuits. In the *RL* circuit, the rising exponential represents the current in the circuit (or voltage across the resistor) after a positive change occurs at the input; the falling exponential represents the current in the circuit (or voltage across the resistor) after a negative change on the input. Recall that in the *RC* circuit, the rising exponential represents the voltage across the capacitor after a positive change in the input and the falling exponential represents the voltage across the capacitor after a negative change in the input.

For reference, the universal curves are plotted again in Figure 11–15 and now are seen to represent the transient response of an *RC* or an *RL* circuit. Again, the *x* and *y* axes are not scaled to specific values but can be used to find voltage or current in a circuit at any instant of time, provided you know the time constant for the circuit and the shape of the response. The *y*-axis represents a percentage of the final value, either current or voltage. The *x*-axis represents the number of time constants that have elapsed after the transient event begins.

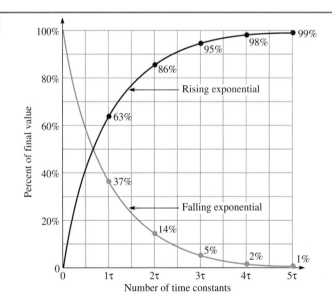

FIGURE 11–15

Universal time constant curves.

EXAMPLE 11–6

Problem

A 560 Ω resistor is in series with a 1.5 mH inductor and connected through a switch to a 10 V source, as shown in Figure 11–16. What is the approximate current in the circuit 1.5 μs after the switch is closed?

Solution

First, find the time constant.

$$\tau = \frac{L}{R} = \frac{1.5 \text{ mH}}{560 \text{ }\Omega} = 2.68 \text{ }\mu s$$

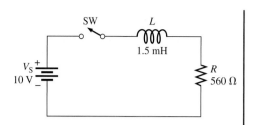

FIGURE 11–16

Next, find how many time constants are represented by 1.5 μs.

$$\text{Number of time constants} = \frac{1.5 \text{ }\mu s}{2.68 \text{ }\mu s} = 0.56$$

On the x-axis of the universal time-constant curve, find 0.56 time constants and note that the rising exponential is 43% of the final value (red curve). This reading of the curve is illustrated in Figure 11–17.

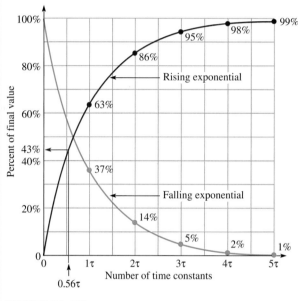

FIGURE 11–17

Determine the final value of current, I_f, by applying Ohm's law to the circuit. Ideally, the source voltage divided by the resistance gives the final current.

$$I_f = \frac{V_S}{R} = \frac{10 \text{ V}}{560 \text{ }\Omega} = 17.8 \text{ mA}$$

At 1.5 μs, the current is 43% of the final value. Thus,

$$I_{1.5 \text{ }\mu s} = 0.43 I_f = \textbf{7.7 mA}$$

Question

What is the current at a time of 2.5 μs after the switch is closed?

CHAPTER 11

COMPUTER SIMULATION

Open the Multisim file F11-16AC on the website. Using the oscilloscope, observe the waveform on the resistor. The dc voltage source has been replaced with a square wave in the exercise.

EXAMPLE 11–7

Problem
Find the time required for the current in Example 11–6 to reach 10 mA.

Solution
From Ohm's law, the final current is 17.8 mA. The percentage of the final current represented by 10 mA is

$$\% \text{ final current} = \frac{10 \text{ mA}}{17.8 \text{ mA}} \times 100\% = 56\%$$

This is the value read on the y-axis of the universal curve and corresponds to a time of approximately 0.82τ as read on the universal curve in Figure 11–18 (red curve).

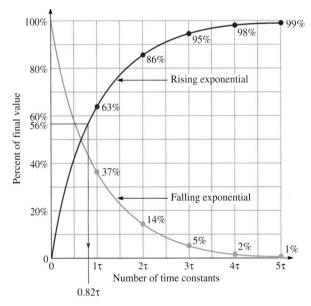

FIGURE 11–18

The time constant (from Example 11–6) is 2.68 μs. Therefore, the time until $I = 10$ mA is

$$t = 0.82\tau = 0.82(2.68 \text{ μs}) = \mathbf{2.2 \text{ μs}}$$

Question
What is the time required after the switch closure for the current to equal 12 mA?

388

Review Questions

16. In a series *RL* circuit, what happens if the current is suddenly interrupted by opening a switch?
17. What is the difference in the response of a series *RL* circuit driven with a square wave and one driven by a battery and switch that is opened and closed rapidly?
18. What happens in a series *RL* circuit driven by a square wave when the voltage drops from a high level to a low level?
19. What is the shape of the voltage across the inductor of a series *RL* circuit driven by a square wave? Assume the period of the square wave is long compared to the time constant.
20. What is the time constant when a 1 H inductor is in series with a 270 Ω resistor?

INDUCTORS IN AC CIRCUITS 11–5

An inductor in an ac circuit will have an opposition to the ac. This opposition is called inductive reactance and is determined, in part, by the frequency.

In this section, you will learn to calculate the inductive reactance for an inductor in an ac circuit.

Recall that in a dc circuit an inductor has an opposition to a change in current and the opposition is greater if the rate of change is greater. AC has a waveform that is constantly changing. Therefore, when an inductor is connected to an ac source, such as shown in Figure 11–19, the inductor responds to the attempted change in current by developing a voltage that opposes a change in current. The result is a phase shift between the voltage and current in the circuit.

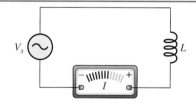

FIGURE 11–19

Current in an ac circuit is limited because of the inductor.

Inductive Reactance

The inductor in the circuit of Figure 11–19 does not have resistance (because it is ideal) and therefore does not dissipate any power from the source. Even so, it still limits the sinusoidal current (ac) that is in the circuit because it generates a voltage that opposes change. This opposition to ac is called **inductive reactance**. The symbol for inductive reactance is X_L, and its unit is the ohm (Ω), the same as capacitive reactance and resistance. As in the case of capacitors, the ideal inductor does not dissipate any energy; it only stores it in the magnetic field during a portion of the ac cycle and returns it to the source during another portion of the cycle.

How Frequency Affects Inductive Reactance

Figure 11–20 illustrates a function generator used as a source of constant amplitude ac voltage. If the frequency is increased, as in Figure 11–20(a), the current decreases. When the frequency of the source is decreased, as in Figure 11–20(b), the current increases.

CHAPTER 11

FIGURE 11–20

Current decreases when the frequency is raised; current increases when the frequency is lowered. Contrast this result for an inductive circuit with the capacitive circuit shown in Figure 10–25.

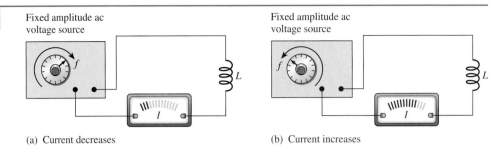

(a) Current decreases

(b) Current increases

When the frequency increases, its rate of change also increases, causing the opposing voltage across the coil to increase. This increase in the opposing voltage means the opposition to the current has *increased* with this increasing frequency. Conversely, when the frequency is lowered, the opposition will *decrease*. Therefore, the inductive reactance varies directly with frequency.

How Inductance Affects Inductive Reactance

Figure 11–21(a) shows that when a sinusoidal voltage with a fixed amplitude and fixed frequency is applied to an inductor, there will be a certain amount of alternating current in the circuit. If the total inductance in the circuit is doubled by adding a second identical inductor in series, as in Figure 11–21(b), half as much current will be in the circuit for the same frequency and applied voltage as before. Of course, this is equivalent to replacing the original inductor with one twice as large. Thus, more inductance in series implies less current or greater opposition to current. Inductive reactance is directly proportional to inductance.

FIGURE 11–21

A greater inductance decreases the total current in the circuit, which represents more opposition.

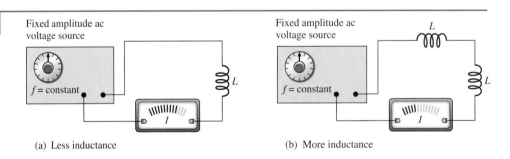

(a) Less inductance

(b) More inductance

Formula for Inductive Reactance

Inductive reactance is directly proportional to both the frequency and the inductance. The two ideas can be combined as a proportionality.

X_L is proportional to fL.

The constant of proportionality that relates X_L to fL is 2π. The 2π term, like that in the capacitive reactance equation, is based on radian frequency rather than hertz. The 2π term is introduced to convert f from radians/second to hertz. Therefore, the formula for inductive reactance (X_L) is

$$X_L = 2\pi fL \tag{11–6}$$

where X_L is inductive reactance in ohms, f is frequency in hertz, and L is inductance in henries.

INDUCTORS

EXAMPLE 11-8

Problem
Find the inductive reactance for a 4.7 mH inductor if the frequency is 10 kHz.

Solution
Apply Equation 11–6.

$$X_L = 2\pi f L = 2\pi(10 \text{ kHz})(4.7 \text{ mH}) = \mathbf{295 \ \Omega}$$

Question
What happens to the inductive reactance if the frequency is halved?

Ohm's Law in Inductive AC Circuits

The reactance of an inductor is analogous to the resistance of a resistor. As you saw in the last chapter, capacitive reactance is an opposition to current and is measured in ohms. Likewise, inductive reactance is an opposition to current, so it is also measured in ohms. Ohm's law applies to inductive circuits and can be stated as

$$I = \frac{V_L}{X_L} \quad (11\text{-}7)$$

where V_L is the voltage across the inductor.

You can apply Ohm's law in an inductive ac circuit, but keep in mind that you must express both the current and the voltage in a consistent way (rms, peak, peak-to-peak).

EXAMPLE 11-9

Problem
Find the current for the circuit in Figure 11–22. Assume the coil is ideal (no resistance).

Solution
First, find the inductive reactance.

$$X_L = 2\pi f L = 2\pi(15 \text{ kHz})(6.8 \text{ mH}) = 641 \ \Omega$$

Apply Ohm's law for an inductive circuit. The source voltage is applied directly to the inductor. Therefore,

$$I = \frac{V_L}{X_L} = \frac{10 \text{ V}}{641 \ \Omega} = \mathbf{15.6 \text{ mA}}$$

FIGURE 11-22

Question
What is the total current if a second 6.8 mH inductor is placed in parallel with the first?

COMPUTER SIMULATION

Open the Multisim file F11-22AC on the website. Verify the current in the circuit.

Inductive Phase Relationship

Recall that a sinusoidal voltage changes at its fastest rate at the zero crossing. At the peaks, the curve has a zero rate of change because it has just reached its maximum and is at the point of changing direction.

The induced voltage across an inductor is proportional to the rate of change of current in the inductor. Therefore, the maximum induced voltage occurs when the current is changing at its maximum rate (at the positive zero crossing). Likewise, the induced voltage is at its maximum negative value when the slope of the current is at its maximum negative value (at the negative zero crossing). At the peaks of the current waveform, the rate of change is zero, so the induced voltage across the inductor is zero. This phase relationship between the current in the inductor and the voltage across the inductor is illustrated in Figure 11–23. As you can see, the voltage peaks occur a quarter of a cycle before the current peaks. Thus, the voltage leads the current by 90° in an inductor.

FIGURE 11–23

The voltage is always leading the current in an inductor.

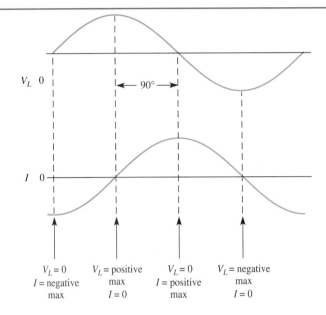

Review Questions

21. What two factors determine the inductive reactance of an inductor?
22. What is the measurement unit for inductive reactance?
23. Contrast inductive reactance with capacitive reactance. Which one increases when frequency increases?
24. What is Ohm's law as applied to an inductive circuit?
25. What is the phase relationship between current and voltage in an inductor for an ac circuit?

11–6 APPLICATIONS OF INDUCTORS

Inductors have more limited applications than capacitors. This is due to various factors including size, cost, and circuit requirements. Although inductors have fewer applications than capacitors, they are essential in radio frequency (rf) circuits.

In this section, you will learn some important applications of inductors.

Inductors at High Frequencies

Radio frequencies (rf) are generally considered to be those frequencies above the audio frequencies, starting with a low of about 10 kHz and extending to the microwave region (about 1 GHz). Within this range, physical components (resistors, capacitors, and inductors) are used in circuits but may behave differently at very high frequencies. At lower frequencies, ordinary components are often treated as ideal for calculations. At higher rf frequencies, the nonideal nature of components becomes more important. Interconnecting wires add inductance to any circuit and can be significant in the microwave region. For example, at microwave frequencies the leads of a capacitor look like an inductor. Likewise, the capacitance between the traces on a circuit board can be important. In the microwave region, these complicated interactions can affect circuits in unexpected ways. You need to be aware that circuit calculations are more approximate at high radio frequencies, particularly those involving inductors.

As you know, inductive reactance is dependent on the frequency, so inductance effects are more significant at rf than at audio frequencies. For lower frequencies, inductors are generally wound on cores. This has the advantage of giving higher inductance with fewer turns, which minimizes undesired magnetic coupling to other nearby components. At the very high radio frequencies, air core inductors are more common.

RF Choke

One application of inductors in rf circuits is to prevent radio frequencies in the rf section from getting into other parts of the system, such as the power supply or an audio frequency amplifier. An inductor is frequently used as a series filter to "choke off" any unwanted rf signals that may be picked up on a line (hence the name '*rf choke*'). It is common to see an rf choke in series with the power supply line for an rf section. At radio frequencies, the reactance of the rf choke becomes extremely large and essentially blocks the high-frequency current, while passing the dc from the power supply. A basic illustration of an inductor used as an rf choke is shown in Figure 11–24.

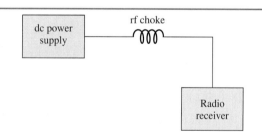

FIGURE 11–24

An rf choke used to block radio frequencies from the receiver to the power supply.

RF chokes are used in other applications where a large opposition to radio frequencies but low opposition to dc is desired. You may see rf chokes in high-frequency circuits in places you might expect a resistor at lower frequencies. You may also see them in digital circuits, where a fast transient event may cause interference with nearby circuits.

Resonant Circuits

One of the most important applications of inductors is for frequency selection in communications systems. In this application, an inductor is used in conjunction with a capacitor to select certain frequencies and block others. Frequency selective circuits are also known as **resonant circuits** (or *tuned circuits*). These circuits allow a narrow band of frequencies to be selected while all other frequencies are rejected. The tuner in your TV is based on this principle and permits you to select one channel out of the many that are available.

Frequency selectivity is based on the fact that the reactances of both capacitors and inductors depend on the frequency but in opposite ways. As frequency is increased, the reactance of inductors goes up, but the reactance of capacitors goes down. At a frequency where the

CHAPTER 11

reactance of a capacitor and an inductor are equal, the circuit has a unique response called *resonance*. Tuned *LC* circuits are covered in Sections 12–6 (series) and 13–5 (parallel).

Another application for tuned circuits is in oscillators, common in communication systems and in various instruments in the lab. Recall that an oscillator is a circuit that produces a periodic wave. Some oscillators use an *LC* circuit to produce a certain frequency. One widely used type, called the Hartley oscillator, uses a tapped coil to allow a small portion of the output to be returned to the input to reinforce the oscillations in a process called **positive feedback**. Reinforcement only occurs if the signal that is fed back is in phase with the input signal.

Filters

A filter is a frequency selective network. At radio frequencies, inductors are useful to help obtain the desired response of the network. At lower frequencies, inductors that can provide a reasonable inductive reactance tend to be large and expensive, so they are rarely used except in specialized applications. One such specialized application is in crossover networks for speaker systems. The network routes high frequencies to one set of speakers and low frequencies to another set.

Delay Lines

In color television, the color signal and black and white signal are processed separately. The two signals are brought back together for display at the picture tube, but any small difference in timing must be corrected first. Small inductors, called *delay line coils,* are used to delay the faster signal to synchronize it to the slower signal.

Delay lines are used in other systems as well. Another example is inside an analog oscilloscope, where you are likely to encounter a large coil. The purpose of this delay line is to start the trigger circuits, which move the beam across the CRT, before sending the signal to the display. After the signal starts the trigger circuits, the signal is delayed by passing it through the delay line.

Review Questions

26. What frequencies are considered to be rf?
27. When does circuit behavior tend to be nonideal?
28. What is an rf choke? What signals does it pass?
29. What two components are essential in a tuned circuit?
30. What is an oscillator?

CHAPTER REVIEW

Key Terms

Henry Unit of inductance; one henry is the amount of inductance that induces one volt when the current changes at a uniform rate of one ampere per second.

Inductance The property of a conductor to oppose a change in current. The effect is greatly reinforced in coils.

Inductive reactance The opposition of an inductor to ac; the unit is the ohm.

Inductor A fundamental passive component that consists of a coil of wire with specific inductive properties.

Positive feedback The process of returning a small portion of the output back to the input in a manner to reinforce the input signal. Reinforcement only occurs if the signal that is fed back is in phase with the input signal.

Resonant circuit A frequency selective *LC* circuit that allows a band of frequencies to be selected from all others.

Important Facts

- When the current through a coil changes, an induced voltage is created across the coil such that it always opposes the change in current.
- The amount of inductance of a coil depends on its diameter and length of its core as well as the number of turns and the material the coil is wound on.
- Actual inductors differ from ideal because they have resistance and capacitance associated with them.
- The total inductance of parallel inductors is found by the reciprocal of the sum of the reciprocals of the individual inductors.
- The total inductance of series inductors is the sum of the individual inductors.
- The time constant for an *RL* circuit is the ratio of the inductance to the resistance.
- The universal time constant curves can be applied to *RL* circuits to find the voltage or current at any instant of time after voltage is applied to the combination.
- The opposition to ac by an inductor is called inductive reactance. The symbol for inductive reactance is X_L, and its unit is the ohm (Ω).
- Inductive reactance varies directly with frequency.
- Alternating current in an inductor lags the voltage across the inductor by 90°.
- Inductors are used in tuned circuits, oscillators, filters, and delay lines.

Formulas

Inductance of a coil:

$$L = \frac{N^2 \mu A}{l} \tag{11-1}$$

Inductance of series inductors:

$$L_T = L_1 + L_2 + L_3 + \cdots + L_n \tag{11-2}$$

Inductance of parallel inductors:

$$L_T = \frac{1}{\dfrac{1}{L_1} + \dfrac{1}{L_2} + \dfrac{1}{L_3} + \cdots + \dfrac{1}{L_n}} \tag{11-3}$$

Inductance of two parallel inductors:

$$L_T = \frac{L_1 L_2}{L_1 + L_2} \tag{11-4}$$

Time constant for an *RL* circuit:

$$\tau = \frac{L}{R} \tag{11-5}$$

Inductive reactance:

$$X_L = 2\pi f L \qquad (11\text{–}6)$$

Ohm's law for in an inductive circuit:

$$I = \frac{V_L}{X_L} \qquad (11\text{–}7)$$

Chapter Checkup

Answers are at the end of the chapter.

1. The inductance of a coil is inversely proportional to its
   - (a) length
   - (b) diameter
   - (c) number of turns
   - (d) all of these answers

2. Inductance is measured in
   - (a) ohms
   - (b) farads
   - (c) henries
   - (d) webers

3. When current changes in an inductor, a voltage is induced that opposes the change. This statement is a summary of
   - (a) Ohm's law
   - (b) Kirchhoff's voltage law
   - (c) Faraday's law
   - (d) Lenz's law

4. An ohmmeter is generally used to find out if a coil is
   - (a) shorted
   - (b) open
   - (c) either (a) or (b)
   - (d) neither (a) nor (b)

5. The total inductance of two inductors in series is equal to the
   - (a) sum of the inductances
   - (b) product of the two inductances
   - (c) product over the sum of the inductances
   - (d) the sum of the reciprocals of the two inductances

6. The total inductance of two inductors in parallel is equal to the
   - (a) sum of the inductances
   - (b) product of the two inductances
   - (c) product over the sum of the inductances
   - (d) the sum of the reciprocals of the two inductances

7. When a dc voltage source is applied to an inductor, the current
   - (a) immediately goes to a value, then decays to zero
   - (b) is constant
   - (c) first is negative, then positive
   - (d) gradually builds to its final value

8. When a dc voltage source is applied to an inductor, the voltage across the inductor
   - (a) immediately goes to a value, then decays to zero
   - (b) is constant
   - (c) first is negative, then positive
   - (d) gradually builds to its final value

9. The time constant of an *RL* series circuit is directly proportional to the
   - (a) inductance
   - (b) resistance
   - (c) both (a) and (b)
   - (d) neither (a) nor (b)

10. Assume an *RL* series circuit is driven by a sinusoidal source. If the frequency is increased, the voltage across the resistor will

    (a) decrease (b) stay the same
    (c) increase

11. Assume a sinusoidal voltage is applied to an inductor. If the voltage is then increased, the reactance of the inductor will

    (a) decrease (b) stay the same
    (c) increase

12. In a series *RL* circuit, the voltage across the inductor

    (a) leads the current (b) is in phase with the current
    (c) lags the current

13. In a series *RL* circuit, the voltage across the resistor

    (a) leads the current (b) is in phase with the current
    (c) lags the current

14. A series inductor in a radio frequency line used as a filter to block high frequencies is also called a

    (a) choke (b) tuned circuit
    (c) high-pass filter (d) delay line

Questions

Answers to odd-numbered questions are at the end of the book.

1. What happens to the inductance of a coil if the number of turns is doubled but other factors remain unchanged?
2. What happens to the inductance of a coil if a magnetic core is used in place of an air core?
3. What causes stray capacitance in a coil and when is it a problem?
4. How is an iron core indicated on the schematic symbol of an inductor?
5. What is the symbol for a variable inductor?
6. Why does moving an iron core up or down in an inductor change the inductance?
7. What is the inductance of an encapsulated coil marked with yellow, violet, orange, gold bands?
8. Where is energy stored when there is current in an inductor?
9. What happens to the stored energy of an inductor if current suddenly drops?
10. When an *RL* circuit is driven by a square wave, the voltage across the inductor will alternate between positive and negative values. Explain why.
11. If the inductance of one inductor is *L*, what is the inductance of three equal-value inductors in series?
12. If the inductance of one inductor is *L*, what is the inductance of three equal-value inductors in parallel?
13. How would you use an oscilloscope to measure the time constant of a series *RL* circuit driven by a square wave generator?
14. When do you expect to see a falling exponential voltage across a resistor in an *RL* circuit driven by a square wave generator?
15. How many time constants are required for an exponentially rising waveform to go from 10% to 90% of the final value?

16. How does inductive reactance limit current in an ac circuit?
17. What is the relationship between the voltage induced across a coil and the current in the coil for an ac circuit?
18. What is an example of the use of a delay line?

PROBLEMS

Basic Problems

Answers to odd-numbered problems are at the end of the book.

1. Convert 270 μH to mH.
2. Convert 56 mH to μH.
3. An inductor has 400 turns of fine wire on a bakelite core that is 6 mm in diameter and 20 mm long. Assume bakelite has the same permeability as air. What is the inductance?
4. If the bakelite for the inductor in problem 3 is replaced with a core that has a permeability of 3.5×10^{-3} Wb/At-m, what is the new inductance?
5. What is the inductance of a 10 mm long inductor that has a cross-sectional area of 1.25×10^{-5} m^2 with 150 turns on a plastic core? (The permeability of plastic is the same as air.)
6. What is the total inductance if three 15 mH inductors are connected in series?
7. What is the total inductance if two 15 mH inductors are connected in parallel?
8. What is the total inductance of a 150 μH inductor that is in series with a 1.0 mH inductor?
9. What is the total inductance of a 150 μH inductor that is in parallel with a 1.0 mH inductor?
10. What is the time constant for a 47 mH inductor in series with a 470 Ω resistor?
11. (a) In Figure 11–25, what is the time constant?
    (b) What voltage is across the resistor 3 ms after the switch is closed?
    (c) What is the final current in the circuit after five time constants?

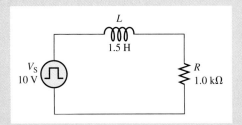

FIGURE 11–25

12. In a series *RL* circuit, determine the time it will take for current to build to full value after a switch is closed for each of the following combinations.
    (a) $R = 220 \, \Omega, L = 100$ mH
    (b) $R = 5.6$ k$\Omega, L = 470 \, \mu$H
    (c) $R = 6.8$ k$\Omega, L = 2.2$ mH
13. Find the inductive reactance for a 15 mH inductor if the frequency is 20 kHz.
14. Find the inductive reactance of a 33 μH inductor at 50 MHz.

15. Calculate the total inductance and the total reactance of the inductive circuit shown in Figure 11–26.

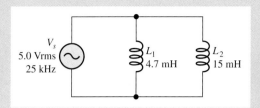

FIGURE 11–26

16. What is the total current and the current in each inductor in Figure 11–26?

Basic-Plus Problems

17. How many turns of wire are needed to form a 40 μH inductor, wound initially on a pencil? Assume the pencil has a diameter of 7 mm and the coil will be 25 mm long.

18. A low-carbon steel core that is 10 mm in diameter is wrapped with 250 turns of wire for a distance of 2 cm. Predict the time constant you should measure if the inductor is placed in series with a 1.0 kΩ resistor. Assume the permeability of the core is 1.0×10^{-3} Wb/At-m.

19. The total inductance of two parallel inductors is 333 μH. One of the inductors has an inductance of 560 μH. What is the inductance of the other inductor?

20. The time for current to change from 10% to 90% is measured for a series *RL* circuit and found to be 120 μs. The value of the resistor is 1.0 kΩ.

    (a) What is the time constant?

    (b) What is the inductance of the inductor?

21. For the circuit in Figure 11–27, calculate the frequency at which the voltage across the inductor is equal to the voltage across the resistor.

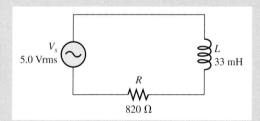

FIGURE 11–27

Example Questions

ANSWERS

11–1: 6.0 mH

11–2: 22,000 μH

11–3: 438 μH

11–4: 35.2 μH

11–5: 206 μs

11–6: 10.8 mA

11–7: 3.0 μs

11–8: The reactance is halved also; $X_L = 148\ \Omega$.

11–9: 31.2 mA

Review Questions

1. Inductance is the property of a conductor (or coil) to oppose a change in current.
2. When the current through a coil changes, an induced voltage is created across the coil that always opposes the change in current.
3. The henry
4. The core material, number of turns, cross-sectional area, and the length of the coil
5. The equivalent circuit consists of an inductor with a series resistor with a parallel capacitor across both. See Figure 11–4.
6. An iron core
7. Typically by moving a core material
8. Yellow, violet, brown, gold
9. 33 mH ± 10%
10. A simple test is to check the continuity with an ohmmeter. A test of the value is done with an inductance meter or indirectly with an oscilloscope.
11. The total inductance is the sum of the individual inductors.
12. The total inductance is the reciprocal of the sum of the reciprocals of each inductor.
13. For exactly two inductors
14. 14.1 mH
15. 1.57 mH
16. A large voltage appears across the coil (and switch) that opposes the change in current.
17. The square wave generator appears to be a low resistance path when the voltage goes to zero; the switch appears to be an extremely high resistance path.
18. The current drops exponentially toward zero at a rate determined by the time constant of the circuit.
19. The voltage across the inductor rises with the square wave, then decays toward zero as the magnetic field builds. When the square wave drops, the voltage across the inductor is equal and opposite to the square wave, then decays back to zero.
20. 3.70 ms
21. The inductance and the frequency
22. The ohm
23. Both inductive and capacitive reactances oppose current. Inductive reactance is larger as frequency is increased; capacitive reactance is smaller as frequency is increased.
24. $I = V_L/X_L$
25. Inductors shift the phase between voltage and current such that the current leads the voltage by 90°.
26. Generally they are between 10 kHz to 1 GHz.
27. At higher frequencies
28. An rf choke is an inductor designed to have high reactance at radio frequencies. It passes low frequencies and dc.

29. An inductor and a capacitor
30. An oscillator is a circuit that produces a periodic wave.

Chapter Checkup

1. (a)   2. (c)   3. (d)   4. (b)   5. (a)
6. (c)   7. (d)   8. (b)   9. (a)   10. (c)
11. (b) 12. (c) 13. (b) 14. (a)

SERIES AC CIRCUITS

INTRODUCTION

In this chapter, series ac circuits with resistance, capacitance, and inductance are discussed and series resonance is introduced. Phasor diagrams are introduced as a tool for visualizing the voltage relationships between R, L, and C. These diagrams require an understanding of some basic trigonometry—the mathematics of the right triangle. Right triangles are introduced and then applied to ac analysis of series R, L, and C circuits.

In a series circuit, current is the same in all components, so it is logical to consider how voltage shifts with respect to current. A capacitor always shifts the phase between voltage and current such that voltage lags the current. An inductor shifts the phase in the opposite direction; in this case, voltage leads the current. In a series circuit, when both components are present, the one with the larger reactance dominates because it will have the greater voltage across it.

When the reactance of the capacitor and the reactance of the inductor are equal, the condition is called resonance. Special frequency-selective circuits take advantage of the characteristics of resonant circuits. The chapter concludes with a discussion of these circuits.

CHAPTER OUTLINE

- 12–1 Representing Phasor Quantities
- 12–2 Impedance
- 12–3 Series *RC* Circuits
- 12–4 Series *RL* Circuits
- 12–5 Series *RLC* Circuits
- 12–6 Series Resonance

Study aids for this chapter are available at

http://www.prenhall.com/SOE

KEY OBJECTIVES

A section number is given for each objective. After completing the chapter, you should be able to

12–1 Explain how to describe a phasor using polar and rectangular notation

12–2 Develop the impedance triangle for a series *RL* or *RC* circuit

12–3 Calculate voltages in a series *RC* circuit, draw the phasor diagrams, and describe how frequency affects the phasors

12–4 Calculate voltages in a series *RL* circuit, draw the phasor diagrams, and describe how frequency affects the phasors

12–5 Analyze a series *RLC* circuit, showing the impedance and voltage phasor diagrams

12–6 Analyze series resonant circuits and describe series resonant filters

COMPUTER SIMULATIONS DIRECTORY

The following figures have Multisim circuit files associated with them.

- Figure 12–16 page 414
- Figure 12–18 page 416
- Figure 12–22 page 419
- Figure 12–34 page 429

LABORATORY EXPERIMENTS DIRECTORY

The following exercises are for this chapter.

- **Experiment 22** Series *RC* Circuits
- **Experiment 23** Series *RL* Circuits
- **Experiment 24** Series Resonance

KEY TERMS

- Imaginary number
- Complex number
- Rectangular notation
- Polar notation
- Impedance
- Resonant frequency
- Band-pass filter
- Band-stop filter
- Bandwidth
- Selectivity

CHAPTER 12

Solid objects have modes of vibration that depend upon the mechanical properties of the material and its geometry. A tuning fork is an example of a material that vibrates at a fixed frequency. The natural frequency of vibration is known as a resonance.

A spectacular example of mechanical resonance due to wind-induced vibrations is the failure of the Tacoma-Narrows bridge on November 7, 1940, shortly after it opened. The bridge's natural vibration frequency was enhanced by the wind, which induced unusual aerodynamic forces on the bridge. These forces have been the subject of much research since that fateful day. The wind-induced motion eventually caused the structure to collapse. The only fatality was a cocker spaniel.

Just as solid objects can have mechanical resonance, circuits can exhibit electrical resonance. Electrical resonance is an oscillating behavior due to the exchange of energy between a capacitor's electrical field and an inductor's magnetic field. This behavior is useful in frequency-selective circuits.

12–1 REPRESENTING PHASOR QUANTITIES

Most ac circuits have combinations of capacitors, inductors, and resistors. These circuits usually have voltages and currents that are out of phase with each other. The voltages or currents in ac circuits can be represented by phasor diagrams, which illustrate the magnitude and phase relationships.

In this section, you will learn how to describe a phasor using polar and rectangular notation.

Real Numbers

You are already familiar with numbers that make up the practical world of counting (integer numbers) as well as the infinite number of irrational numbers (such as π) that are used in science and math. Irrational numbers cannot be written as the ratio of integers. *Real numbers* can be integers or irrational numbers and can be represented on a number line. A typical number line is shown in Figure 12–1. It is drawn horizontally, with positive numbers located to the right and negative numbers located to the left. Real numbers describe the magnitude of a quantity, that is, its size. They can be positive or negative and can be whole (integer) numbers or decimals.

FIGURE 12–1

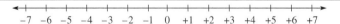

A number line showing integer real numbers. Real numbers include all numbers that can be shown on a number line.

The *j* Operator

In math, an *operator* is a command to do some mathematical operation. The mathematical symbols $+$, $-$, $\times$, $\div$, and $\sqrt{}$ are operators, which are shorthand instructions for performing mathematical operations.

The $-$ operator is normally associated with subtraction; however, it can be thought of as a rotational operator when it is appended to a number. Placing a minus ($-$) sign in front of a positive number on a number line is equivalent to rotating the number by 180° and results in the number being plotted on the negative side of the number line. As illustrated in Figure 12–2, the number 5, operated on by the $-$ operator, is rotated by 180° and becomes a negative number, -5. Likewise, operating on a negative number with the $-$ operator produces a 180° rotation back to a positive number.

SERIES AC CIRCUITS

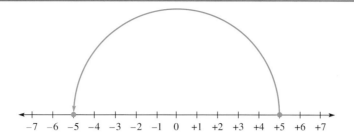

FIGURE 12–2

The − operator is equivalent to a 180° rotation.

Another operator is the *j* operator, which also is a rotational operator when appended to a number. In the case of the *j* operator, it rotates the number through a counterclockwise (CCW) rotation of 90°, rather than 180°. Figure 12–3 illustrates what happens when a positive real number is rotated by the *j* operator. In the figure, the real number 5 is operated on by *j* and is then shown along the *y*-axis as the number *j*5. The number is expressed with the *j* operator in front of the number.

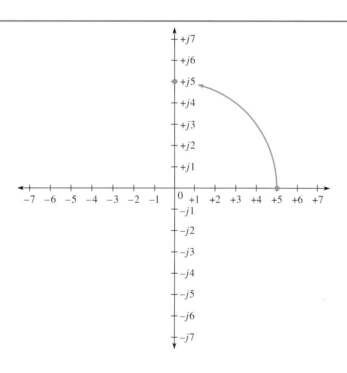

FIGURE 12–3

The *j* operator rotates a number by 90° counterclockwise.

Imaginary Numbers

As you have seen, when the − operator is appended to a number, the operation is equivalent to a 180° rotation, which is the same as two 90° rotations. Each 90° rotation must be equivalent to multiplying by $\sqrt{-1}$ because $\sqrt{-1}\sqrt{-1} = -1$. The $\sqrt{-1}$ intrigued mathematicians for many years because there is no number that, when multiplied by itself, will produce -1. Thus, the $\sqrt{-1}$ was said to be imaginary; and when it is appended to a number, the number is said to be an **imaginary number**. The word *imaginary* is an unfortunate choice for these numbers because it implies something that does not exist. Imaginary numbers are simply those that would be plotted along the *y*-axis in a rectangular coordinate system (frequently referred to as the Cartesian coordinate system). In electronics work, imaginary numbers are shown with the letter *j* appended to them (*j*33 and −*j*2.6, for example). The *j* operator is defined as follows:

$$j = \sqrt{-1}$$

405

CHAPTER 12

The property of the j operator can be observed by applying it successively to a number, as illustrated in Figure 12–4. In this case, the number 5 is multiplied by j and is written $j5$. A second multiplication by j produces $j^2 5$, which is equivalent to -5 as you have seen. Multiplying by j a third time produces $-j5$, which is a rotation to the negative y-axis. Finally, the fourth multiplication by j rotates the original number back to its original starting point because $j(-j5) = -j^2 5 = +5$. Imaginary numbers will always be plotted along the y-axis, and real numbers will be plotted along the x-axis.

FIGURE 12–4

Four successive multiplications of a number by j will rotate the number back to the starting point.

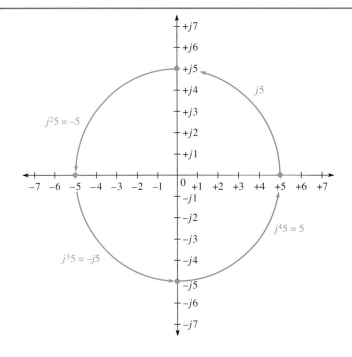

Complex Numbers

When a number is described by a real part (along the x-axis) and an imaginary part (along the y-axis), the number is said to be a **complex number**. In describing ac quantities, it is useful to show a "snapshot" of a rotating phasor as simply a complex number for a given instant of time.

Two common notations are used to describe complex numbers. Both notations provide a way to combine a real and an imaginary part. In **rectangular notation**, the distance along the x-axis is given first as the real term followed by the distance along the y-axis as the imaginary term. Figure 12–5 illustrates how complex numbers are expressed in rectangular notation.

Polar notation is the second way of expressing a complex number. With polar notation, the real and imaginary parts are combined into a single magnitude number followed by the angle. Figure 12–6 shows three examples of the same three complex numbers shown in Figure 12–5 but expressed using polar notation. Notice that the length of each line is based on the scale factor for the axes and the angle is measured from the positive x-axis.

Recall from Section 9–2 that a phasor is a rotating vector, which has magnitude and direction associated with it. Phasors are described by using complex numbers. For example, when the numbers shown in Figures 12–5 and 12–6 are applied to ac, they will be associated with a reactance, current, or voltage and have units assigned to them. In these cases, they are referred to as *phasors*.

406

SERIES AC CIRCUITS

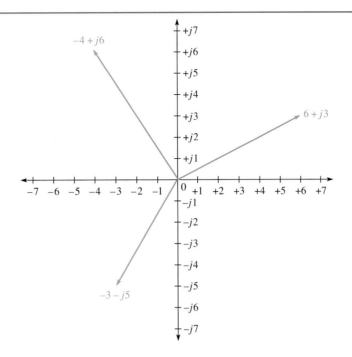

FIGURE 12-5

Complex numbers expressed using rectangular notation.

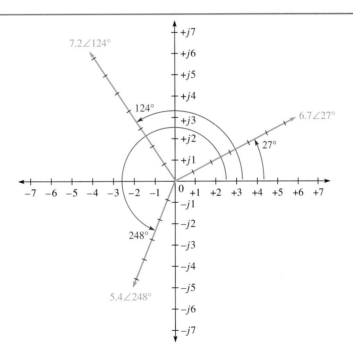

FIGURE 12-6

Complex numbers expressed using polar notation. These are the same complex numbers that are shown in Figure 12–5.

Right Triangles

Rectangular and polar notations for expressing complex numbers are related to a right triangle, which is a triangle with a 90° angle. By dropping a straight line from the end of any phasor to the *x*-axis, a right triangle is formed, as illustrated in Figure 12–7 for phasors in quadrant 1 and quadrant 4. The sides of the triangle are parallel to the *x*-axis and *y*-axis. The right triangle is a simplified way of showing the relationships in a complex number and will be used in developing ideas for ac series and ac parallel circuits. The vertical side can be defined based on the length along the *y*-axis (*j*), and the horizontal side can be defined based on the length along the *x*-axis.

CHAPTER 12

FIGURE 12–7

Phasors are directly related to right triangles.

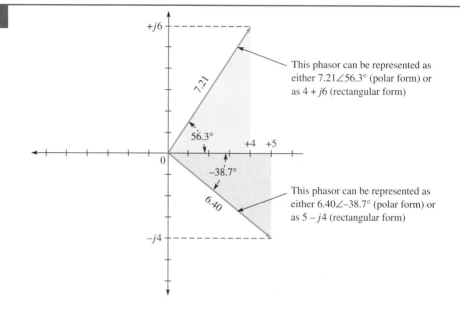

Figure 12–8 illustrates the common names associated with right triangles; the sides are labeled based on the reference angle, θ. The trigonometric functions of sine, cosine, and tangent (abbreviated sin, cos, and tan) are ratios of the sides as given in the following definitions:

$$\sin \theta = \frac{\text{opposite side}}{\text{hypotenuse}}$$

$$\cos \theta = \frac{\text{adjacent side}}{\text{hypotenuse}}$$

$$\tan \theta = \frac{\text{opposite side}}{\text{adjacent side}}$$

These definitions are basic to all right triangles, no matter what the size, and are useful for working with phasors. By applying these relationships, you can find an unknown angle or side for a given right triangle. For ac circuit calculations, the sides of the triangle will represent one of the quantities of interest, such as the voltage across a component.

FIGURE 12–8

Trigonometric relationships of a right triangle.

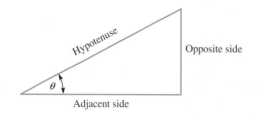

SERIES AC CIRCUITS

EXAMPLE 12–1

Problem
Using a trigonometric function, find the angle θ for each of the right triangles shown in Figure 12–9.

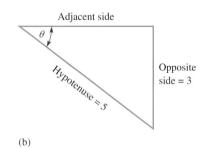

(a)   (b)

FIGURE 12–9

Solution
(a) You know the opposite side and the adjacent side. The ratio of the opposite side to the adjacent side define the tangent function.

$$\tan \theta = \frac{\text{opposite side}}{\text{adjacent side}} = \frac{4}{10} = 0.4$$

To isolate θ, put the tan on the right side and add the superscript -1.

$$\theta = \tan^{-1}(0.4) = \mathbf{21.8°}$$

The abbreviation $\tan^{-1}$ is read as "the angle whose tangent is".
For the TI-36X calculator, select the $\tan^{-1}$ function with second function key.

(b) You know the opposite side and the hypotenuse. The ratio of the opposite side to the hypotenuse define the sine function.

$$\sin \theta = \frac{\text{opposite side}}{\text{hypotenuse}} = \frac{3}{5} = 0.6$$
$$\theta = \sin^{-1}(0.6) = \mathbf{36.9°}$$

Question*
If you know the adjacent side and the hypotenuse, how can you find the angle that is formed?

An important property of right triangles is given by the famous *Pythagorean theorem,* which states

$$\text{hypotenuse}^2 = \text{adjacent side}^2 + \text{opposite side}^2$$

Given any two sides of a right triangle, you can calculate the third side with this theorem.

---

* Answers are at the end of the chapter.

CHAPTER 12

EXAMPLE 12-2

Problem
Using the Pythagorean theorem, calculate the unknown side of the right triangles shown in Figure 12–10.

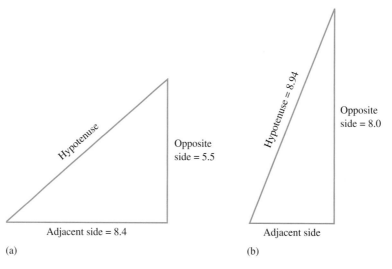

FIGURE 12–10

Solution

(a) The hypotenuse is the unknown side.

$$\text{hypotenuse}^2 = \text{adjacent side}^2 + \text{opposite side}^2$$

Taking the square root of both sides,

$$\text{hypoteneuse} = \sqrt{\text{adjacent side}^2 + \text{opposite side}^2} = \sqrt{8.4^2 + 5.5^2} = \mathbf{10.0}$$

(b) The adjacent side is the unknown. By rearranging the terms,

$$\text{adjacent side}^2 = \text{hypotenuse}^2 - \text{opposite side}^2$$

and taking the square root of both sides,

$$\text{adjacent side} = \sqrt{\text{hypoteneuse}^2 - \text{opposite side}^2} = \sqrt{8.94^2 - 8.0^2} = \mathbf{3.99}$$

Question
If the sides of a right triangle are each 2 units long, what is the length of the hypotenuse?

Review Questions

Answers are at the end of the chapter.

1. What does the *j* operator do?
2. On what axis are imaginary numbers plotted?
3. What are two methods for expressing complex numbers?
4. What is the definition of the tangent of an angle?
5. What is the Pythagorean theorem?

SERIES AC CIRCUITS

IMPEDANCE 12–2

Impedance is the combination of resistance, which opposes current because electrons give up energy by collisions, and reactance, in which energy is stored in a field and returned to the circuit later. You have seen how each of these effects, operating alone, limit current. Both of these effects can be combined with an impedance triangle.

In this section, you will learn how to develop an impedance triangle for a series RL or RC circuit.

Three fundamental passive components are resistors, capacitors, and inductors. Each of these components offers opposition to ac. With a resistive circuit, the opposition occurs when electrical energy is converted to heat. The opposition is due to this conversion process. In a purely resistive circuit, there is no phase shift between current and voltage.

In an ac circuit that has a source and a single capacitor, the opposition to current is the capacitive reactance, which is frequency dependent. Also, there is a phase shift between the current and voltage due to the capacitor in which the current leads the voltage by 90°.

Likewise, in an ac circuit that has a source and a single inductor, the opposition to current is the inductive reactance, which is also frequency dependent. The phase shift between the current and voltage is due to the inductor. In this case, the current lags the voltage by 90°.

Capacitive and inductive reactances oppose current by storing energy during part of the ac cycle and returning it in another part. Ideally, no net energy is dissipated in a capacitor or in an inductor, although in practice some energy is always dissipated in the internal resistance of the inductor. For our purposes, we will ignore these effects.

Most practical circuits consist of combinations of these three fundamental passive components plus other components. The total opposition to ac takes into account both resistive and reactive components and is a complex quantity called **impedance**. Total impedance includes both magnitude and phase (direction) and may be reported either in rectangular or polar notation. For many cases, only the magnitude of the impedance is required for a particular circuit calculation, so the angle is not included as part of the impedance that is reported, simplifying the calculations. The magnitude of impedance is abbreviated as Z.

Impedance Triangle

Recall that impedance represents the total opposition and includes both resistive and reactive effects. Because resistance does not shift phase but reactance does, it is helpful to visualize these phase shifts by plotting their values on the rectangular coordinate system as described in the Section 12–1. The resistance and reactance phasors are plotted depending on how they affect the phase angle between current and voltage. (This will be easier to visualize when the voltage phasors are introduced.) The basic rules for plotting resistive and reactance phasors for a series circuit are shown in Figure 12–11 and given as follows:

1. Resistance (R) is always plotted along the positive x-axis.
2. Capacitive reactance (X_C) is plotted along the negative y-axis ($-j$). In rectangular form, it is written as $-jX_C$, which shows that it is plotted along the negative y-axis.
3. Inductive reactance (X_L) is plotted along the positive y-axis ($+j$). In rectangular form, it is written as $+jX_L$, which shows that it is plotted along the positive y-axis.

The impedance is determined by finding the phasor sum of the resistance and reactance, using the rules for a right triangle. As an example of a resistive and reactive circuit, consider the basic series RC circuit shown in Figure 12–12(a). Assume that the frequency is set so that the reactance of the capacitor is equal to the resistance. Both R and X_C are treated as phasor quantities, as shown in the phasor diagram of Figure 12–12(b). Phasor R is drawn

CHAPTER 12

FIGURE 12-11

Directions for resistance and reactance phasors in a series circuit.

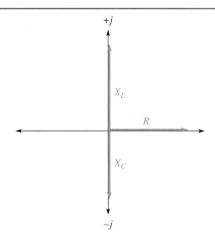

FIGURE 12-12

Development of the impedance triangle for a series RC circuit.

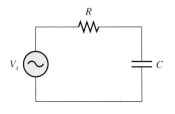

(a) A basic series RC circuit

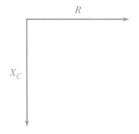

(b) Resistance and reactance phasors

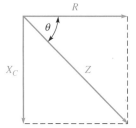

(c) Phasor sum of R and X_C

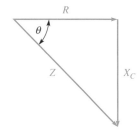

(d) The impedance triangle

along the x-axis, and phasor X_C is drawn along the −y-axis, which means the diagram is in quadrant 4 in the rectangular coordinate system. This relationship between R and X_C comes from the fact that the capacitor voltage in a circuit lags the current, and thus the resistor voltage, by 90°. Because Z is the phasor sum of R and X_C, it is represented as shown in Figure 12–12(c). A repositioning of the phasors, as shown in Figure 12–12(d), forms a right triangle. This is called the impedance triangle. The length of each phasor represents the magnitude in ohms. The angle θ (the Greek letter theta) is the phase angle and represents the phase difference of the circuit between the source voltage and current.

An equivalent triangle can be drawn for a series RL circuit. A basic series RL circuit is shown in Figure 12–13(a). For the RL circuit, the impedance triangle is drawn in quadrant 1 of the rectangular coordinate system because X_L is plotted along the positive y-axis, as indicated in Figure 12–13(b). The relationship between R and X_L comes from the fact that the

FIGURE 12-13

Development of the impedance triangle for a series RL circuit.

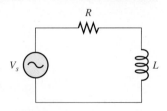

(a) A basic series RL circuit

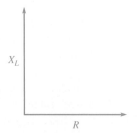

(b) Resistance and reactance phasors

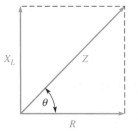

(c) Phasor sum of R and X_L

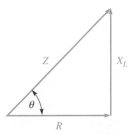

(d) The impedance triangle

inductor voltage leads the current, and thus the resistor voltage, by 90°. Again, the magnitude of the impedance, Z, is found by adding R and X_L as phasors, so it is represented as shown in Figure 12–13(c). A repositioning of the phasors, as shown in Figure 12–13(d), forms the impedance triangle.

The magnitude of the impedance phasor, Z, for both the series RC and the series RL circuits can be found by applying the Pythagorean theorem.

$$\text{hypotenuse}^2 = \text{adjacent side}^2 + \text{opposite side}^2$$

Substituting resistance for the adjacent side, reactance for the opposite side, and impedance for the hypotenuse results in the following relationship:

$$Z^2 = R^2 + X^2$$

Solving for Z,

$$Z = \sqrt{R^2 + X^2} \qquad (12\text{–}1)$$

where X represents X_C, X_L, or the phasor sum of X_C and X_L if both capacitance and inductance are present in the circuit.

The phase angle, θ, can be found by applying the tangent function from trigonometry.

$$\theta = \tan^{-1}\left(\frac{X}{R}\right) \qquad (12\text{–}2)$$

Again, X can be X_C, X_L, or the phasor sum of X_C and X_L. If X is capacitive, θ is plotted in quadrant 4; if X is inductive, θ is plotted in quadrant 1. The reason for these choices has to do with the natural phase shift that occurs with reactive components.

The following two examples illustrate how to find the impedance and the phase angle for basic series RC and RL circuits.

EXAMPLE 12–3

Problem

Calculate the magnitude of the impedance and the phase angle for the RC circuit shown in Figure 12–14. Draw the impedance phasor diagram and the impedance triangle.

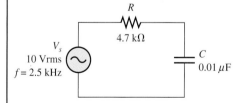

FIGURE 12–14

Solution

First, calculate the magnitude of the capacitive reactance at 2.5 kHz.

$$X_C = \frac{1}{2\pi f C} = \frac{1}{2\pi(2.5 \times 10^3 \text{ Hz})(0.01 \times 10^{-6} \text{ F})} = 6.37 \text{ k}\Omega$$

To show the direction is along the negative y-axis, $-j$ can be appended to this answer, giving $-j6.37 \text{ k}\Omega$.

Next, substitute the magnitude of X_C into Equation 12–1. The magnitude of the impedance is

$$Z = \sqrt{R^2 + X_C^2} = \sqrt{(4.7 \text{ k}\Omega)^2 + (6.37 \text{ k}\Omega)^2} = \mathbf{7.91 \text{ k}\Omega}$$

Notice that X_C is shown in this case for Equation 12–1 because there is only resistance and capacitive reactance in the circuit.

Calculate the phase angle by substituting values into Equation 12–2 with X_C used in place of X.

$$\theta = \tan^{-1}\left(\frac{X_C}{R}\right) = \tan^{-1}\left(\frac{6.37 \text{ k}\Omega}{4.7 \text{ k}\Omega}\right) = \tan^{-1}(1.36) = \mathbf{53.6°}$$

The impedance phasor diagram and impedance triangle are shown in Figure 12–15. The phase angle is measured between R and Z and is plotted in quadrant 4 because the circuit is capacitive.

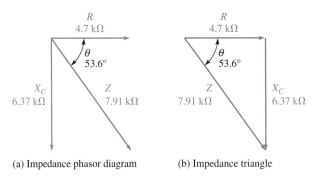

(a) Impedance phasor diagram   (b) Impedance triangle

FIGURE 12–15

Question
What happens to the total impedance if the frequency is reduced?

EXAMPLE 12–4

Problem
Calculate the magnitude of the impedance and the phase angle for the RL circuit shown in Figure 12–16. Show the impedance phasor diagram and the impedance triangle.

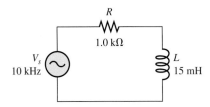

FIGURE 12–16

Solution
First, calculate the magnitude of the inductive reactance at 10 kHz.

$$X_L = 2\pi f L = 2\pi(10 \text{ kHz})(15 \text{ mH}) = 942 \text{ }\Omega$$

The $+j$ can be appended to this result to show the direction is along the positive y-axis: $+j942 \text{ }\Omega$.

SERIES AC CIRCUITS

Substituting the magnitude of X_L into Equation 12–1,

$$Z = \sqrt{R^2 + X_L^2} = \sqrt{(1.0\ \text{k}\Omega)^2 + (942\ \Omega)^2} = \mathbf{1.37\ k\Omega}$$

Calculate the phase angle by substituting values into Equation 12–2.

$$\theta = \tan^{-1}\left(\frac{X_L}{R}\right) = \tan^{-1}\left(\frac{942\ \Omega}{1.0\ \text{k}\Omega}\right) = \tan^{-1}(0.942) = \mathbf{43.3°}$$

The impedance triangle is plotted in quadrant 1 because X_L is along the positive y-axis. The impedance phasor diagram and the impedance triangle are shown in Figure 12–17.

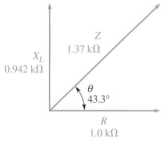
(a) Impedance phasor diagram

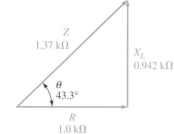
(b) Impedance triangle

FIGURE 12–17

Question
If the inductor is replaced with a smaller one, what happens to the phase angle?

COMPUTER SIMULATION

Open the Multisim file F12-16AC on the website. Using the oscilloscope, observe the waveform from the source and across the inductor. Verify that the inductor voltage is delayed from the source voltage.

Review Questions

6. What does each side of the impedance triangle represent?
7. If a resistance of 2 kΩ is added to a reactance of 2 kΩ, the result is not 4 kΩ. Why not?
8. For the rectangular coordinate system, on what axis is capacitive reactance plotted?
9. For the rectangular coordinate system, on what axis is inductive reactance plotted?
10. What is the phase angle of a series RL circuit in which the inductive reactance is equal to the resistance?

SERIES RC CIRCUITS 12–3

In any series circuit, the same current is in every component. This current develops a voltage across the resistive and reactive elements. In resistive components, the voltage is in phase with the current; but in reactive components, it is out of phase with the current.

In this section, you will learn how to calculate voltages in a series RC circuit, how to draw the phasor diagrams, and how frequency affects the phasors.

CHAPTER 12

SAFETY NOTE

Be aware of fire safety in the lab. Know the location of fire extinguishers. Electrical fires are never fought with water, which can conduct electricity. Use only a class-C extinguisher, which uses a nonconductive agent — typically CO_2. It is important to know what type of extinguisher you should use for any fire. The wrong choice can make the fire worse.

As you learned in the last section, impedance is the total opposition to ac. In any series circuit, impedance is the phasor sum of the resistance and reactance phasors. The current in the circuit can be calculated from the source voltage and impedance by applying Ohm's law.

$$I = \frac{V_s}{Z} \qquad (12\text{-}3)$$

Equation 12–3 is a general equation for any reactive circuit and is useful for finding the voltage across each component as long as you know the resistance or reactance of that component. As you know, in a series circuit, the current is the same in all components. This, of course, is true for any series circuit, no matter how many components are present. In the case of a series RC circuit, the voltage across the resistor and the voltage across the capacitor are simply

$$V_R = IR$$

and

$$V_C = IX_C$$

Likewise, the source voltage can be written as

$$V_s = IZ$$

From the three equations for voltage, you can see that current is the constant that relates the resistance, reactance, and impedance to the voltage. The impedance phasor diagram is directly related to the voltage phasor diagram. Given the impedance phasor diagram, multiply each phasor by the current, and you can draw the voltage phasor diagram, which will have the same shape.

EXAMPLE 12–5

Problem
Find the missing values on Table 12–1 for the circuit in Figure 12–18. Draw the impedance and voltage phasor diagrams.

Table 12–1

$R = 2.7\ k\Omega$	$I_R =$	$V_R =$
$X_C =$	$I_C =$	$V_C =$
$Z =$	$I_{tot} =$	$V_s = 10\ Vrms$

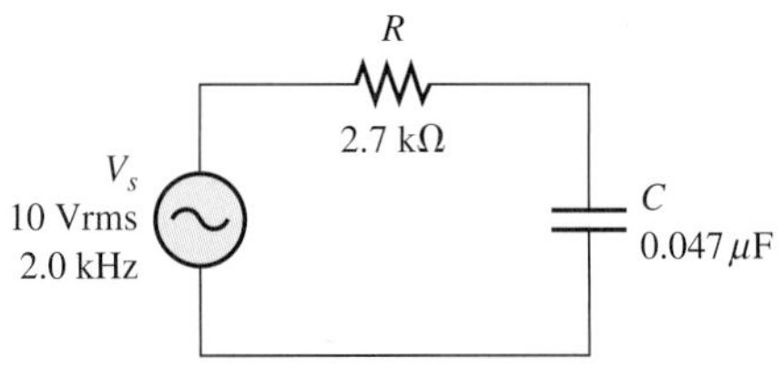

FIGURE 12–18

Solution
Step 1: Find the magnitude of the capacitive reactance at the source frequency of 2.0 kHz.

$$X_C = \frac{1}{2\pi f C} = \frac{1}{2\pi(2.0 \times 10^3\ Hz)(0.047 \times 10^{-6}\ F)} = 1.69\ k\Omega$$

The $-j$ can be appended to the answer to show the direction is along the negative y-axis as $-j1.69\ k\Omega$.

Step 2: Find the magnitude of the impedance and the phase angle.

$$Z = \sqrt{R^2 + X_C^2} = \sqrt{(2.7 \text{ k}\Omega)^2 + (1.69 \text{ k}\Omega)^2} = 3.19 \text{ k}\Omega$$

This result enables you to draw the impedance phasor diagram, as shown in Figure 12–19. The impedance is drawn as the diagonal on the phasor drawing.

The phase angle can be measured directly on the drawing or calculated as

$$\theta = \tan^{-1}\left(\frac{X_C}{R}\right) = \tan^{-1}\left(\frac{1.69 \text{ k}\Omega}{2.7 \text{ k}\Omega}\right) = 32°$$

The circuit is capacitive, so the angle is plotted in quadrant 4.

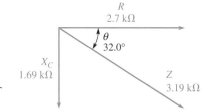

FIGURE 12–19
Impedance phasor diagram for the circuit in Figure 12–18.

Step 3: Use Ohm's law to determine the current. Of course, the current is the same throughout the circuit. The completed table is shown in Table 12–2.

$$I = \frac{V_s}{Z} = \frac{10 \text{ V}}{3.19 \text{ k}\Omega} = 3.13 \text{ mA}$$

Table 12-2

$R = 2.7$ kΩ	$I_R = 3.13$ mA	$V_R = 8.45$ Vrms
$X_C = 1.69$ kΩ	$I_C = 3.13$ mA	$V_C = 5.29$ Vrms
$Z = 3.19$ kΩ	$I_{tot} = 3.13$ mA	$V_s = 10$ Vrms

To find the voltage phasors, multiply each of the phasors in the impedance diagram by the current, as illustrated in Figure 12–20. As a check, you can verify that the voltage phasors satisfy the Pythagorean theorem (within round-off error).

$$10^2 = 5.29^2 + 8.45^2 \checkmark$$

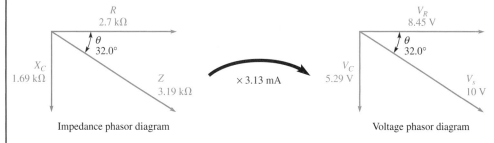

FIGURE 12–20
The voltage phasor diagram is related to the impedance phasor diagram.

Question
Will θ remain the same if V_s is increased? Explain your answer.

CHAPTER 12

COMPUTER SIMULATION

Open the Multisim file F12-18AC on the website. Verify the voltages and current in the circuit.

Variation of Impedance with Frequency in Series RC Circuits

As you know, capacitive reactance varies inversely with frequency. In a series circuit, the magnitude of the impedance is

$$Z = \sqrt{R^2 + X_C^2}$$

When X_C increases, the entire term under the square root sign increases and therefore the impedance increases; when X_C decreases, the impedance decreases. In a series RC circuit, impedance is inversely proportional to the frequency.

Figure 12–21 illustrates how the impedance phasors in a series RC circuit vary as the frequency increases or decreases. As the frequency increases, the capacitive reactance and impedance decrease. Think of each of the three triangles as a different impedance triangle for each frequency. Because capacitive reactance varies inversely with the frequency, the magnitude of the impedance and the phase angle also vary inversely with the frequency.

FIGURE 12–21

As frequency increases in an RC circuit, X_C decreases and Z decreases. Each frequency can be visualized as a different impedance triangle. The voltage phasors will have the same shape.

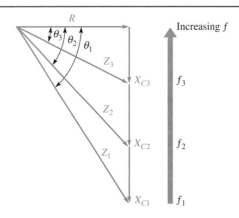

When the impedance of an RC circuit changes, the current in the circuit changes inversely. Thus, if the frequency is increased, the impedance decreases and the current increases (assuming a fixed source voltage). In the case of the voltage phasors, the voltage across the resistor and the voltage across the capacitor both change, but at any given frequency, the voltage phasor diagram always has the same shape as the impedance phasor diagram.

EXAMPLE 12–6

Problem

Draw the impedance and voltage phasor diagrams for the RC circuit shown in Figure 12–22 for 1.0 kHz and 10 kHz.

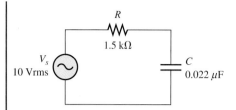

FIGURE 12–22

Solution
At 1.0 kHz:
First, find the magnitude of the reactance of the capacitor at the source frequency of 1.0 kHz.

$$X_C = \frac{1}{2\pi fC} = \frac{1}{2\pi(1.0 \times 10^3 \text{ Hz})(0.022 \times 10^{-6} \text{ F})} = 7.23 \text{ k}\Omega$$

Next, find the magnitude of the impedance and the phase angle at 1.0 kHz.

$$Z = \sqrt{R^2 + X_C^2} = \sqrt{(1.5 \text{ k}\Omega)^2 + (7.23 \text{ k}\Omega)^2} = 7.38 \text{ k}\Omega$$

Apply the tan function to calculate the phase angle.

$$\theta = \tan^{-1}\left(\frac{X_C}{R}\right) = \tan^{-1}\left(\frac{7.23 \text{ k}\Omega}{1.5 \text{ k}\Omega}\right) = 78.3°$$

The magnitude of the current is

$$I = \frac{V_s}{Z} = \frac{10 \text{ V}}{7.38 \text{ k}\Omega} = 1.35 \text{ mA}$$

Calculate the voltage phasors by multiplying each of the phasors in the impedance diagram by the current. Figure 12–23 illustrates the results for 1.0 kHz.

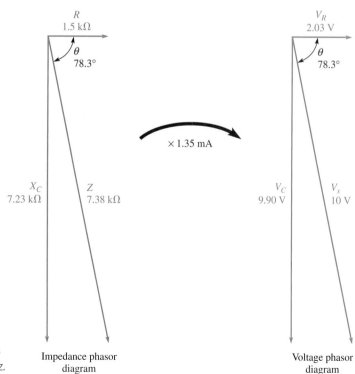

FIGURE 12–23
Impedance and voltage phasor diagrams for the circuit in Figure 12–22 at 1.0 kHz.

CHAPTER 12

At 10 kHz:

Use the same method as for 1.0 kHz. The reactance will be a factor of 10 smaller due to the frequency. Therefore, $X_C = 0.723$ kΩ and R, of course, remains the same at 1.5 kΩ. The magnitude of the impedance is

$$Z = \sqrt{R^2 + X_C^2} = \sqrt{(1.5 \text{ k}\Omega)^2 + (0.723 \text{ k}\Omega)^2} = 1.66 \text{ k}\Omega$$

The phase angle is

$$\theta = \tan^{-1}\left(\frac{X_C}{R}\right) = \tan^{-1}\left(\frac{0.723 \text{ k}\Omega}{1.5 \text{ k}\Omega}\right) = 25.7°$$

The angle is plotted in quadrant 4.

The magnitude of the current is

$$I = \frac{V_s}{Z} = \frac{10 \text{ V}}{1.66 \text{ k}\Omega} = 6.02 \text{ mA}$$

Multiply each phasor in the impedance phasor diagram by the current to calculate the voltage phasors. Figure 12–24 illustrates the results for 10 kHz. The impedance phasors are plotted with the same scale as those in Figure 12–23. Notice how the frequency change has changed the diagram. At 1 kHz, the capacitive reactance is the main contributor to the impedance; at 10 kHz, the resistance has a greater effect on the impedance than the capacitive reactance.

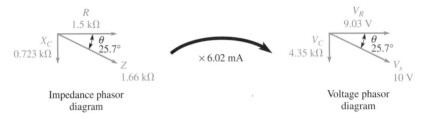

FIGURE 12–24
Impedance and voltage phasor diagrams for the circuit in Figure 12–22 at 10 kHz.

Question
Describe the impedance phasor diagram at a frequency of 4.82 kHz.

COMPUTER SIMULATION

Open the Multisim file F12-22AC on the website. Using the oscilloscope, observe the waveform from the source and across the capacitor. Verify the phase shift.

Review Questions

11. Why does the voltage phasor diagram always have the same shape as the impedance phasor diagram for a given frequency?

12. At a given frequency in a series RC circuit, if the voltage across the capacitor and the voltage across the resistor are equal, what can be said about the capacitive reactance and the resistance?

13. If the frequency is changed in a series RC circuit and the voltage across the capacitor increases as a result, is the new frequency higher or lower?

14. What happens to the magnitude of the impedance in an *RC* circuit if the frequency is increased?
15. What happens to the phase angle in a series *RC* circuit if the frequency is increased?

SERIES RL CIRCUITS 12–4

In an *RL* circuit, the phase shift is opposite to that of the series *RC* circuit, but the voltages across the components can be described in a similar manner. You will again see that the voltage phasor diagram is related to the impedance phasor diagram by the current in the circuit.

In this section, you will learn how to calculate voltages in a series *RL* circuit, how to draw the phasor diagrams, and how frequency affects the phasors.

Exept for the phase shift, a series *RL* circuit behaves like the series *RC* circuit. As in any series circuit, the impedance is the phasor sum of the resistance and reactance phasors. You can determine current by applying Ohm's law to the magnitude of the impedance.

$$I = \frac{V_s}{Z}$$

This current is the same in all components because it is a series circuit. As in the case of a series *RC* circuit, you can calculate the voltages in the series *RL* circuit by applying Ohm's law once you know the current. The voltages are

$$V_R = IR$$

and

$$V_L = IX_L$$

Likewise, the source voltage can be written as

$$V_s = IZ$$

As in the case of the series *RC* circuit, the voltage phasor diagram in a series *RL* circuit has the same shape as the impedance phasor diagram. Also, the voltage phasors are found by multiplying each phasor in the impedance diagram by the current in the circuit. In the case of the series *RL* circuit, both diagrams are drawn in quadrant 1.

EXAMPLE 12–7

Problem
Find the missing values on Table 12–3 for the circuit in Figure 12–25. Draw the impedance and voltage phasor diagrams.

Table 12–3

$R = 470\ \Omega$	$I_R =$	$V_R =$
$X_L =$	$I_L =$	$V_L =$
$Z =$	$I_{tot} =$	$V_s = 5.0$ Vrms

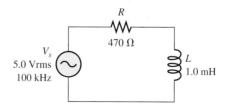

FIGURE 12–25

CHAPTER 12

Solution

The inductive reactance at 100 kHz is

$$X_L = 2\pi f L = 2\pi(100 \text{ kHz})(1.0 \text{ mH}) = 628 \text{ }\Omega$$

Substitute values into Equation 12–1.

$$Z = \sqrt{R^2 + X_L^2} = \sqrt{(470 \text{ }\Omega)^2 + (628 \text{ }\Omega)^2} = 785 \text{ }\Omega$$

The total current is

$$I = \frac{V_s}{Z} = \frac{5.0 \text{ V}}{785 \text{ }\Omega} = 6.37 \text{ mA}$$

Of course, this is the same current in the resistor and in the inductor. Applying Ohm's law,

$$V_R = IR = (6.37 \text{ mA})(470 \text{ }\Omega) = 3.00 \text{ V}$$
$$V_L = IX_L = (6.37 \text{ mA})(628 \text{ }\Omega) = 4.00 \text{ V}$$

Complete Table 12–4. The phasor diagrams are shown in Figure 12–26.

Table 12–4

$R = 470 \text{ }\Omega$	$I_R = 6.37$ mA	$V_R = 3.00$ Vrms
$X_L = 628 \text{ }\Omega$	$I_L = 6.37$ mA	$V_L = 4.00$ Vrms
$Z = 785 \text{ }\Omega$	$I_{tot} = 6.37$ mA	$V_s = 5.0$ Vrms

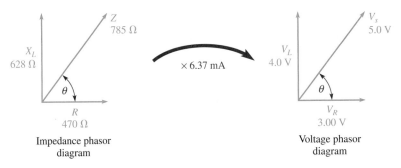

FIGURE 12–26

Question
What is the angle, θ, between the source voltage and current?

Variation of Impedance with Frequency in Series *RL* Circuits

The response of a series *RL* circuit to an increase in frequency is opposite to that of an *RC* circuit. Inductive reactance varies directly with frequency. Because the magnitude of the impedance is

$$Z = \sqrt{R^2 + X_L^2}$$

the impedance also varies directly with frequency.

Figure 12–27 illustrates how the impedance phasors in a series *RL* circuit vary as the frequency increases or decreases. As the frequency increases, the inductive reactance and impedance also increase. Think of each triangle as a different impedance triangle for each frequency. Because the inductive reactance varies directly with the frequency, the magnitude of the impedance and the phase angle also vary directly with the frequency.

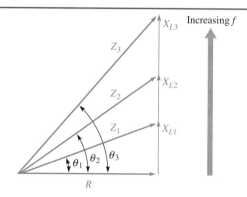

FIGURE 12–27

As frequency increases in an *RL* circuit, the inductive reactance and the total impedance both increase.

When the impedance of an *RL* circuit changes, the current in the circuit changes inversely. Thus if the frequency is increased, the impedance increases and the current decreases (assuming a fixed source voltage). In the case of the voltage phasors, the voltage across the resistor and the voltage across the inductor both change. As in the *RC* case, for any given frequency the voltage phasor diagram for an *RL* circuit always has the same shape as the impedance phasor diagram for that circuit.

EXAMPLE 12–8

Problem
Draw the impedance and voltage phasor diagrams for the *RL* circuit in the previous example if the frequency is raised to 200 kHz.

Solution
Calculate the magnitude of the reactance of the inductor and the impedance of the circuit.

$$X_L = 2\pi f L = 2\pi(200 \text{ kHz})(1.0 \text{ mH}) = 1.26 \text{ k}\Omega$$

$$Z = \sqrt{R^2 + X_L^2} = \sqrt{(0.47 \text{ k}\Omega)^2 + (1.26 \text{ k}\Omega)^2} = 1.34 \text{ k}\Omega$$

The phase angle is

$$\theta = \tan^{-1}\left(\frac{X_L}{R}\right) = \tan^{-1}\left(\frac{1.26 \text{ k}\Omega}{0.47 \text{ k}\Omega}\right) = 69.5°$$

Calculate the current.

$$I = \frac{V_s}{Z} = \frac{5.0 \text{ V}}{1.34 \text{ k}\Omega} = 3.72 \text{ mA}$$

Calculate the voltages.

$$V_R = IR = (3.72 \text{ mA})(0.47 \text{ k}\Omega) = 1.75 \text{ V}$$
$$V_L = IX_L = (3.72 \text{ mA})(1.26 \text{ k}\Omega) = 4.70 \text{ V}$$

Use the results to draw the impedance and voltage phasor diagrams, which are shown in Figure 12–28.

CHAPTER 12

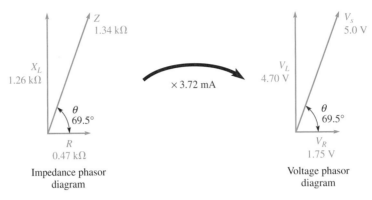

FIGURE 12-28

Question
If the only information you were given is the voltage phasor diagram and the current in the circuit, how could you find the impedance phasors?

Review Questions

16. In a certain series *RL* circuit, a 5 V source voltage is applied and 3 V is measured across the resistor. What is the voltage across the inductor?
17. For Question 16, what is the phase angle?
18. If the frequency is changed in a series *RL* circuit and the voltage across the inductor increases as a result, is the new frequency higher or lower?
19. What happens to the magnitude of the impedance in an *RL* circuit if the frequency is increased?
20. What happens to the phase angle in an *RL* circuit if the frequency is increased?

12-5 SERIES *RLC* CIRCUITS

A series *RLC* circuit has resistance, capacitance, and inductance. Before you can solve for current and voltage in a series *RLC* circuit, you must first combine the capacitive reactance and the inductive reactance into a single equivalent reactance. The resulting reactance can be combined with the resistance to find the impedance.

In this section, you will learn how to analyze a series *RLC* circuit, showing the impedance and voltage phasor diagrams.

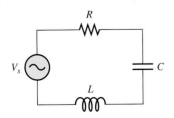

FIGURE 12-29

A series *RLC* circuit is illustrated in Figure 12-29. In the case of a series *RLC* circuit, each component contributes to the impedance; the impedance is the phasor sum of the individual reactances and resistances. After finding the impedance, it is useful to calculate current next. To find the current, apply Ohm's law using the total voltage divided by the impedance as you have seen in *RC* and *RL* circuits.

Impedance of Series *RLC* circuits

As you know, inductive reactance causes the current to lag the source voltage and capacitive reactance causes the current to lead the source voltage. Because of the opposing effects, when adding reactances, you must take into account the sign. The total reactance for a circuit with both inductive reactance and capacitive reactance in series is

$$X = |X_L - X_C| \qquad (12\text{--}4)$$

The vertical bars indicate the absolute value of the difference. With an absolute value, the sign of the difference is considered positive, no matter which reactance is greater. When $X_L > X_C$, the circuit as a whole will act like an inductive circuit, and the current will lag the source voltage. When $X_C > X_L$, the circuit as a whole will act like a capacitive circuit, and the current will lead the source voltage.

Although Equation 12–4 indicates that the total reactance is the absolute difference in the two reactances, the actual process is still addition. Inductive reactance is plotted on the positive *j* axis, and capacitive reactance is plotted on the negative *j* axis. The process is adding two numbers with opposite signs, so the result is the difference. The larger of the two reactances will determine on which axis to plot the result. The difference in the reactances is shown as an absolute value.

The magnitude of the impedance of a series *RLC* circuit is given by

$$Z = \sqrt{R^2 + |X_L - X_C|^2} \qquad (12\text{--}5)$$

The precedence of operation rules require that you first find the absolute difference in the reactances. After squaring this result, add it to the square of the resistance and take the square root of the total. Plot the result in quadrant 1 if $X_L > X_C$; otherwise plot it in quadrant 4.

EXAMPLE 12–9

Problem
Draw the impedance phasor diagrams for the *RLC* circuit shown in Figure 12–30.

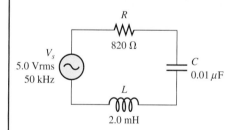

FIGURE 12–30

Solution
Step 1: Calculate the magnitude of the reactance of the inductor and the capacitor.

$$X_L = 2\pi f L = 2\pi(50 \text{ kHz})(2.0 \text{ mH}) = 628 \, \Omega$$

$$X_C = \frac{1}{2\pi f C} = \frac{1}{2\pi(50 \text{ kHz})(0.01 \, \mu\text{F})} = 318 \, \Omega$$

Step 2: With the known resistance, plot the reactances on a rectangular coordinate system plot, as shown in Figure 12–31(a), which shows all three phasors. Combining the two reactive phasors leads to the diagram in Figure 12–31(b). The inductive reactance is larger than the capacitive reactance, so the difference, $|X_L - X_C|$, is plotted on the positive *y*-axis.

Step 3: Combine the reactance and resistance phasors by applying Equation 12–1. The impedance is the phasor sum of R and $|X_L - X_C|$.

$$Z = \sqrt{R^2 + X^2} = \sqrt{820\ \Omega^2 + 310^2} = 877\ \Omega$$

Step 4: Calculate the phase angle using Equation 12–2.

$$\theta = \tan^{-1}\left(\frac{X}{R}\right) = \tan^{-1}\left(\frac{310\ \Omega}{820\ \Omega}\right) = 20.7°$$

The angle is in quadrant 1 because $X_L > X_C$.

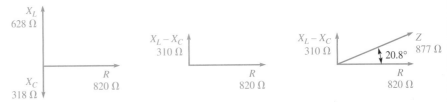

(a) Resistance and reactance phasors   (b) Sum of reactance phasors   (c) Impedance phasor is the sum of all three phasors

FIGURE 12–31

Question
What will happen to the impedance if the frequency is decreased to 40 kHz?

Voltage Phasor Diagram

As in the case of the series RC and RL circuits, the voltage phasor diagram in a series RLC circuit has the same shape as the impedance phasor diagram. Also, the voltage phasors are found by multiplying the impedance diagram phasors by the current in the circuit.

The magnitude of the source voltage of a series RLC circuit is given by the following formula:

$$V_s = \sqrt{V_R^2 + |V_L - V_C|^2} \qquad (12\text{–}6)$$

In a series circuit with both inductance and capacitance, the voltage across the inductor is always 180° out of phase with the voltage across the capacitor. As a result, the difference in V_L and V_C is always less than the largest individual voltage. In fact, the voltage across either or both components can exceed the source voltage. Usually the source voltage is known; in this case, Equation 12–6 can be used to solve for one of the other voltages.

EXAMPLE 12–10

Problem
Draw the voltage phasor diagrams for the RLC circuit shown in the last example. The source voltage is 5.0 Vrms.

Solution
Step 1: Use Ohm's law to calculate the current in the circuit.

$$I = \frac{V_s}{Z} = \frac{5.0\ \text{V}}{877\ \Omega} = 5.70\ \text{mA}$$

Step 2: Multiply the phasors shown in the impedance diagram in Figure 12–31(a) (Example 12–9) by this current (5.70 mA). Figure 12–32(a) shows the result.

$$V_R = IR = (5.70 \text{ mA})(0.82 \text{ k}\Omega) = 4.67 \text{ V}$$

$$V_L = IX_L = (5.70 \text{ mA})(0.628 \text{ k}\Omega) = 3.58 \text{ V}$$

$$V_C = IX_C = (5.70 \text{ mA})(0.318 \text{ k}\Omega) = 1.81 \text{ V}$$

Step 3: Add the voltage across the inductor to the voltage across the capacitor. The voltage across these two components is plotted along the y-axis as shown in Figure 12–32(b).

Step 4: Draw the phasor for V_s. This is the original 5.0 Vrms given in the problem, which can be checked by applying Equation 12–6. Thus,

$$V_s = \sqrt{V_R^2 + |V_L - V_C|^2} = \sqrt{4.67 \text{ V}^2 + |3.58 \text{ V} - 1.81 \text{ V}|^2} = 4.99 \text{ V}$$

The check is within round-off error. V_s is shown in Figure 12–32(c). Notice the similarity between the plots in Figure 12–31 and Figure 12–32.

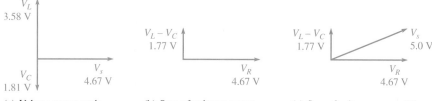

(a) Voltage across each component

(b) Sum of voltages across reactive components

(c) Sum of voltages across all three components produces V_s.

FIGURE 12–32

If you measure the voltage across any one component, you will see the results in Figure 12–32(a). If you measure the voltage across both the inductor and the capacitor, you will see the difference voltage plotted on the y-axis of Figure 12–32(b).

Question
What voltage is across both the capacitor and the resistor (excluding the inductor)?

Phase Shift in Series RLC Circuits

After combining the reactance phasors, you can simplify a series *RLC* circuit to a basic impedance or voltage right triangle. By applying basic trigonometry, you can write the phase shift in terms of either the impedance triangle or the voltage triangle. From the impedance triangle,

$$\theta = \tan^{-1}\left(\frac{|X_L - X_C|}{R}\right)$$

The voltage will lead the current if X_L is larger than X_C; otherwise it lags the current. An equivalent equation can be written in terms of voltage as

$$\theta = \tan^{-1}\left(\frac{|V_L - V_C|}{V_R}\right)$$

CHAPTER 12

Review Questions

21. In the equation $X = |X_L - X_C|$, what do the vertical bars mean?
22. In which quadrant is the impedance plotted when $X_C > X_L$ in a series RLC circuit?
23. If the frequency is increased in a series RLC circuit, will the circuit appear to be more inductive or more capacitive?
24. What is the impedance if a 1.0 kΩ resistor is in series with a capacitor with a reactance of 2.0 kΩ and an inductor with a reactance of 3.0 kΩ?
25. Does the source current lead or lag the source voltage for question 24?

12–6 SERIES RESONANCE

A series resonant circuit is an LC circuit operating at a frequency such that $X_L = X_C$. This frequency, called the resonant frequency, has special importance in certain applications such as band-pass and band-stop filters.

In this section, you will learn to analyze series resonant circuits and describe series resonant filters.

Condition for Resonance

All practical circuits have resistance, even if it is only the winding resistance of the inductor. When frequency is increased in an RLC circuit, the reactance of the inductor increases and the reactance of the capacitor decreases. Figure 12–33(a) shows a plot of the reactance of an inductor, the reactance of a capacitor, and the magnitude of the impedance as a function of frequency. Notice that there is one frequency where the magnitude of X_L and X_C are equal. This

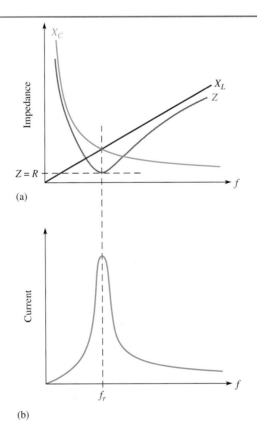

FIGURE 12–33

As frequency increases, the capacitive reactance decreases and the inductive reactance increases as shown in (a). At resonance, the reactances are equal and the circuit is resistive. At this frequency, current is a maximum as shown in (b).

is called the **resonant frequency**, f_r, or simply *resonance*. Because the capacitive reactance and the inductive reactance cancel each other at resonance, the circuit appears to be purely resistive at this one frequency, as indicated at the dip in the impedance curve. Below resonance, the capacitive reactance is dominant and contributes the most to the impedance. Above resonance, the inductive reactance is dominant and contributes the most to the impedance.

The canceling effect of the reactances at resonance gives rise to certain conditions. The impedance is minimum as indicated. This causes the current in the circuit to be maximum at resonance and limited only by the circuit resistance. (In the ideal case, with no resistance, the current would be infinite!) A plot of the current as a function of frequency is shown in Figure 12–33(b). When the frequency is zero (dc) there is no current due to the infinite reactance of the capacitor. The current rises to a maximum at resonance, then falls to a low value for all higher frequencies. Another consequence of the canceling effect is that the phase angle between the source current and voltage is zero at resonance.

The equation for finding the resonant frequency of an *LC* circuit can be developed by setting $X_C = X_L$ and solving for the frequency.

$$X_L = X_C$$
$$2\pi fL = \frac{1}{2\pi fC}$$
$$f^2 = \frac{1}{4\pi^2 LC}$$

$$f_r = \frac{1}{2\pi \sqrt{LC}} \qquad (12\text{--}7)$$

where f_r is the resonant frequency.

Equation 12–7 is based on ideal components; in actual circuits the resonant frequency is slightly different than calculated from Equation 12–7 due to resistance effects. Generally, these effects are small and can be ignored.

EXAMPLE 12–11

Problem
Calculate the resonant frequency for the series *RLC* circuit shown in Figure 12–34. Draw the impedance and voltage phasor diagrams for the circuit at resonance. The source voltage is 5.0 Vrms.

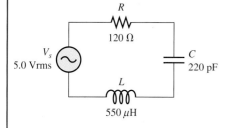

FIGURE 12–34

Solution
Calculate the resonant frequency from Equation 12–7.

$$f_r = \frac{1}{2\pi \sqrt{LC}} = \frac{1}{2\pi \sqrt{(550 \times 10^{-6})(220 \times 10^{-12})}} = \mathbf{458 \text{ kHz}}$$

For the TI-36X calculator, the key sequence is

At this frequency, calculate X_L.

$$X_L = 2\pi fL = 2\pi(458 \text{ kHz})(550 \text{ }\mu\text{H}) = 1.58 \text{ k}\Omega$$

X_C should have the same magnitude because the circuit is at resonance.

$$X_C = \frac{1}{2\pi fC} = \frac{1}{2\pi(458 \text{ kHz})(220 \text{ pF})} = 1.58 \text{ k}\Omega$$

The impedance phasor diagram is shown in Figure 12–35(a). When X_L is added to X_C, the reactances cancel, leaving only R as the impedance of the circuit at resonance.

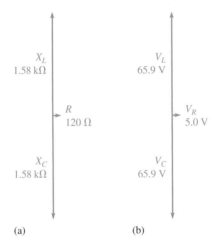

FIGURE 12–35

You need to know the current to determine the voltage phasors. Calculate the current by applying Ohm's law.

$$I = \frac{V_s}{R} = \frac{5.0 \text{ V}}{120 \text{ }\Omega} = 41.7 \text{ mA}$$

As in any series circuit, you can calculate the magnitude of the voltage phasors by multiplying the impedance phasors by the current.

$$V_L = IX_L = (41.7 \text{ mA})(1.58 \text{ k}\Omega) = 65.9 \text{ V}$$

The magnitude of the voltage across the capacitor is the same. The voltage phasors are as shown in Figure 12–35(b). You may be surprised to see that the voltages across the capacitor and inductor are actually larger than the source voltage. This is normal in a series resonant circuit. However, if you measure the voltage across both the inductor and the capacitor, it is zero.

Question
What happens to the resonant frequency if the inductor is twice as large and the capacitor is half as large?

SERIES AC CIRCUITS

COMPUTER SIMULATION

Open the Multisim file F12-34AC on the website. Using the DMM, observe the voltage across the capacitor and the inductor at resonance.

Series Resonant Filters

Series resonant circuits can be applied to making a frequency-selective filter for either passing or rejecting frequencies near the resonant frequency. Figure 12–36(a) shows how a series resonant circuit can be configured to pass frequencies around resonance and reject other frequencies. This type of filter is called a **band-pass filter**. Maximum current occurs at resonance; therefore, the voltage across the resistor is largest at this frequency. By taking the output across the resistor, the circuit will pass frequencies near resonance and reduce the amplitude of all other frequencies, as shown in Figure 12–36(b).

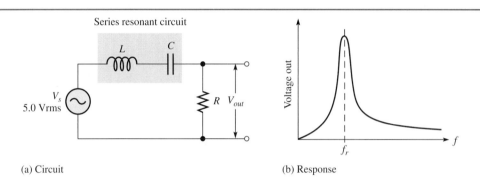

FIGURE 12–36

Series resonant band-pass filter.

(a) Circuit   (b) Response

A **band-stop filter** (or notch filter) is one that rejects the frequencies near resonance. A band-stop filter can be made by rearranging the components and taking the output across the resonant circuit instead of the resistor. Figure 12–37(a) shows the arrangement and Figure 12–37(b) shows the circuit response. This type of filter is useful for solving interference problems from a fixed-frequency source.

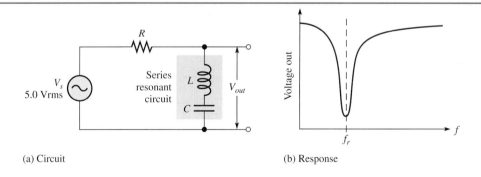

FIGURE 12–37

Series resonant band-stop filter. The voltage does not drop to zero at resonance because of winding resistance in the coil.

(a) Circuit   (b) Response

Bandwidth

The **bandwidth** of a band-pass filter is the range of frequencies for which the output voltage is 70.7% or more of the maximum value. For a resonant filter, the maximum value is the value at resonance. The frequencies at the 70.7% points are called the cutoff frequencies. For a band-stop filter, the bandwidth is between the two cutoff frequencies.

A narrow bandwidth is one in which the response curve is sharp and has cutoff frequencies that are close together. This is a measure of the selectivity of the circuit. **Selectivity**

CHAPTER 12

defines how well the *LC* resonant circuit passes certain frequencies and rejects others. The narrower the bandwidth, the greater the selectivity. Figure 12–38 illustrates the relationship between bandwidth and selectivity.

FIGURE 12–38

Comparative selectivity curves.

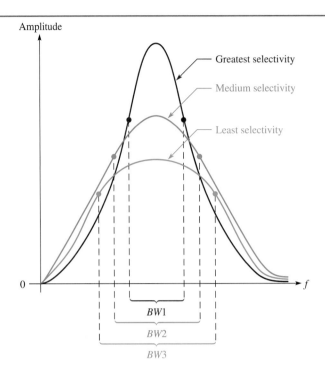

The bandwidth and selectivity are controlled by two principal factors. The first factor is the amount of resistance in the circuit; as resistance is larger, the bandwidth is wider and the selectivity is lower. In many resonant circuits, the resistance is primarily associated with the internal resistance of the coil, so it should be as small as possible for highest selectivity.

The question posed in Example 12–11 leads to the second factor that controls selectivity. There are many combinations of *L* and *C* that have the same resonant frequency because the frequency is determined by $\sqrt{LC}$. If *L* is made larger and *C* is made smaller by the same factor, the resonant frequency is unchanged. Although the resonant frequency is unchanged, the bandwidth and selectivity are changed. The bandwidth will be smaller if a larger inductance and a smaller capacitance are selected, provided the resistance is not changed.

Selectivity is also described by a term called the *quality factor* or *Q* of a resonant circuit. *Q* is a pure number (no units) that is directly proportional to the selectivity and inversely proportional to bandwidth. The relationship between *Q* and bandwidth is given by the formula:

$$Q = \frac{f_r}{BW} \qquad (12\text{–}8)$$

A larger bandwidth implies a lower selectivity and a lower *Q*. Likewise, a narrow bandwidth implies a highly selective circuit and high *Q*.

Equation 12–8 is quite accurate for cases where the *Q* of the circuit is 10 or more. In these cases the error is less than 1%.

EXAMPLE 12–12

Problem
Assume the bandwidth of the resonant circuit in Example 12–11 is 35 kHz. What is the *Q* of the circuit?

Solution
The resonant frequency is 458 kHz. Applying Equation 12–8,

$$Q = \frac{f_r}{BW} = \frac{458 \text{ kHz}}{35 \text{ kHz}} = \mathbf{13.1}$$

Question
What is the Q if the bandwidth is doubled for the same resonant frequency?

Review Questions

26. The impedance of a series resonant circuit is minimum at resonance. Why?
27. At resonance, is the current in a series resonant circuit high or low? Explain.
28. What is the resonant frequency of a series circuit with a 240 μH inductor in series with a 68 pF capacitor?
29. If a band-pass filter has a peak output of 10 V, what is the output at the cutoff frequencies?
30. What is meant by high selectivity?

Key Terms

CHAPTER REVIEW

Band-pass filter A circuit designed to pass frequencies around resonance and reject other frequencies.

Band-stop filter A circuit designed to reject frequencies around resonance and pass other frequencies.

Bandwidth The range of frequencies for which the output voltage is 70.7% or more of the maximum value.

Complex number A number that is described by both a real and an imaginary term (rectangular notation) or by showing a magnitude and direction (polar notation).

Imaginary number A number plotted along the positive or negative y-axis. Imaginary numbers in electronics are shown with the letter j appended to them.

Impedance The opposition to ac; it is a complex quantity with both resistive and reactive components. The magnitude of impedance is abbreviated Z.

Polar notation A method for describing a complex number by giving the magnitude first, followed by the angle with respect to the positive x-axis.

Rectangular notation A method for describing a complex number by giving the distance along the x-axis as the real part followed by the distance along the y-axis as the imaginary part.

Resonant frequency The frequency in an LC circuit where the magnitude of X_L and X_C are equal.

Selectivity A measure of how well a resonant circuit passes certain frequencies and rejects others.

Important Facts

- Real numbers are integers and irrational numbers that can be represented on a number line.
- The Pythagorean theorem is for a right triangle. It states

$$\text{hypotenuse}^2 = \text{adjacent side}^2 + \text{opposite side}^2$$

- In an impedance plot for a series circuit, resistance is plotted along the positive x-axis, capacitive reactance is plotted along the negative y-axis, and inductive reactance is plotted along the positive y-axis.
- The voltage across each component in a series circuit can be found by multiplying the resistance or reactance by the current.
- The voltage phasor diagram will have the same shape as the impedance phasor diagram because the phasors are related by the current in the circuit.
- In a series RLC circuit, when X_L and X_C are equal, they cancel. The frequency that this occurs is called the resonant frequency.
- At resonance, the impedance in a series circuit is at a minimum and the current is at a maximum.
- Series resonant circuits are used in band-pass filters to select a narrow band of frequencies and in band-stop filters to reject a narrow band of frequencies.

Formulas

Impedance of a series RL or RC circuit:

$$Z = \sqrt{R^2 + X^2} \tag{12-1}$$

Phase angle of a series RL or RC circuit:

$$\theta = \tan^{-1}\left(\frac{X}{R}\right) \tag{12-2}$$

Ohm's law in a reactive circuit:

$$I = \frac{V_s}{Z} \tag{12-3}$$

Total reactance in series:

$$X = |X_L - X_C| \tag{12-4}$$

Magnitude of the impedance in a series RLC circuit:

$$Z = \sqrt{R^2 + |X_L - X_C|^2} \tag{12-5}$$

Magnitude of the voltage in a series RLC circuit:

$$V_s = \sqrt{V_R^2 + |V_L - V_C|^2} \tag{12-6}$$

Resonant frequency:

$$f_r = \frac{1}{2\pi\sqrt{LC}} \tag{12-7}$$

Quality factor:

$$Q = \frac{f_r}{BW} \tag{12-8}$$

Chapter Checkup

Answers are at the end of the chapter.

1. An example of a complex number is
   - (a) −5.24
   - (b) $j3$
   - (c) $-j12.2$
   - (d) $2 + j3$

2. When the letter j is appended to a number, it means that the number is
   - (a) irrational
   - (b) imaginary
   - (c) real
   - (d) complex

3. With polar notation, a complex number is described by
   - (a) magnitude and angle
   - (b) altitude and azimuth
   - (c) length along x and y axes
   - (d) scientific notation

4. The opposite side of a right triangle divided by the adjacent side of the same triangle produces a ratio called the
   - (a) sine
   - (b) cosine
   - (c) tangent
   - (d) hypotenuse

5. In a purely resistive circuit, the phase shift between voltage and current is
   - (a) −90°
   - (b) 0°
   - (c) +45°
   - (d) +90°

6. The capacitive reactance (X_C) in a series circuit is plotted along the
   - (a) positive x-axis
   - (b) negative x-axis
   - (c) positive y-axis
   - (d) negative y-axis

7. If the frequency applied to an *RC* series circuit is increased, the total impedance will be
   - (a) smaller
   - (b) unchanged
   - (c) larger

8. If the voltage applied to an *RC* series circuit is increased, the total impedance will be
   - (a) smaller
   - (b) unchanged
   - (c) larger

9. The hypotenuse of the impedance triangle for a series *RL* circuit represents
   - (a) inductive reactance
   - (b) resistance
   - (c) capacitive reactance
   - (d) impedance

10. In a series *RLC* circuit, the current will lag the voltage if the
    - (a) capacitive reactance is larger than the resistance
    - (b) inductive reactance is larger than the resistance
    - (c) capacitive reactance is larger than the inductive reactance
    - (d) inductive reactance is larger than the capacitive reactance

11. In a series *RLC* circuit, the voltage phasor diagram has the same shape as the
    - (a) impedance phasor diagram
    - (b) current phasor diagram
    - (c) both of the above
    - (d) none of the above

12. In a series *RLC* circuit at resonance, the impedance is
    (a) the maximum value (b) the minimum value
    (c) 70.7% of maximum (d) 50% of maximum

13. If the frequency is raised in a series *RLC* circuit, the circuit will appear to be more
    (a) inductive (b) capacitive
    (c) resistive (d) none of these answers

14. If a variable inductor in a series resonant circuit is adjusted for more inductance, the resonant frequency will be
    (a) lower (b) the same
    (c) higher

15. The bandwidth of a band-pass filter is measured at the
    (a) peak level of the response
    (b) 33% level of the response
    (c) 50% level of the response
    (d) 70.7% level of the response

16. If a resonant circuit is highly selective, it has a
    (a) wide bandwidth (b) medium bandwidth
    (c) narrow bandwidth

Questions

Answers to odd-numbered questions are at the end of the book.

1. What numbers are considered to be irrational?
2. What is a mathematical operator?
3. What is the value of j^2?
4. What is the difference between a real and an imaginary number?
5. What is the difference between rectangular notation and polar notation?
6. Which trig function is expressed as the ratio of the adjacent side to the hypotenuse?
7. What is meant by $\tan^{-1}$?
8. What happens to electrical energy in an ideal capacitor or inductor in an ac circuit?
9. What is the phase angle in an *RC* circuit when $X_C = R$?
10. What is the phase angle in an *RL* circuit when $X_L = R$?
11. What happens to the phase angle when the frequency in a series *RC* circuit is increased?
12. What happens to the phase angle when a variable inductor in a series *RL* circuit is adjusted for more inductance?
13. Given the impedance phasor diagram for a series circuit, how do you convert it to a voltage phasor diagram?
14. How can you determine if a series *RLC* circuit is inductive or capacitive?
15. What is the condition for series resonance?
16. Why is current maximum at the resonant frequency for a series *RLC* circuit?
17. What is a band-stop filter?
18. How is the bandwidth defined for a band-stop filter?
19. What is selectivity?

20. What are two factors that affect the selectivity of a series resonant circuit?
21. What does a high Q imply for a resonant circuit?

Basic Problems

Answers to odd-numbered problems are at the end of the book.

1. Sketch the phasor on a rectangular coordinate system, represented by the complex number $2 - j4$.
2. Sketch the phasor on a rectangular coordinate system, represented by the complex number $3 + j1$.
3. Sketch the phasor on a rectangular coordinate system, represented by the complex number $3.5\angle 60°$.
4. Refer to the right triangle in Figure 12–39. Assume $A = 3.50$ and $B = 1.75$.
   (a) What is the length of C?
   (b) What is θ?
5. Refer to the right triangle in Figure 12–39. Assume $A = 8.0$ and $C = 9.0$. What is the length of B?

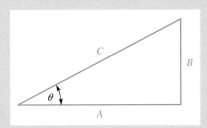

FIGURE 12–39

6. Refer to the right triangle in Figure 12–40. Assume $A = 2.5$ and $C = 5.6$.
   (a) What is the length of B?
   (b) What is θ?
7. Refer to the right triangle in Figure 12–40. Assume $C = 3.0$ and $\theta = 63°$. What is the length of A?
8. Calculate the magnitude of the impedance and the phase angle for the RL circuit shown in Figure 12–41. Sketch the impedance phasor diagram and the impedance triangle.

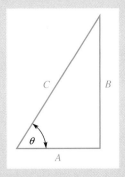   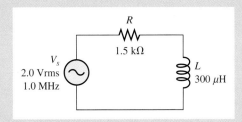

FIGURE 12–40   **FIGURE 12–41**

9. Calculate the magnitude of the impedance and the phase angle for the RC circuit shown in Figure 12–42. Sketch the impedance phasor diagram and the impedance triangle.

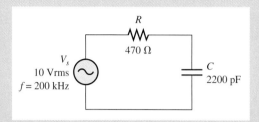

FIGURE 12–42

10. Calculate the current in the RL circuit shown in Figure 12–41.
11. Calculate the current in the RC circuit shown in Figure 12–42.
12. For the circuit in Figure 12–43, assume the frequency is set to 120 kHz. Complete Table 12–5. Draw the impedance and voltage phasor diagrams for the circuit.

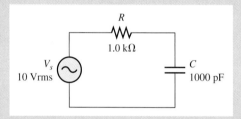

FIGURE 12–43

Table 12–5

$R = 1.0\ k\Omega$	$I_R =$	$V_R =$
$X_C =$	$I_C =$	$V_C =$
$Z =$	$I_{tot} =$	$V_s = 10$ Vrms

13. For the circuit in Figure 12–43, assume the frequency is changed to 200 kHz. Complete Table 12–6 for this frequency. Draw the impedance and voltage phasor diagrams for the circuit.

Table 12–6

$R = 1.0\ k\Omega$	$I_R =$	$V_R =$
$X_C =$	$I_C =$	$V_C =$
$Z =$	$I_{tot} =$	$V_s = 10$ Vrms

14. Refer to the circuit in Figure 12–43. To what frequency should the signal generator be set in order that the voltage across the capacitor is equal to the voltage across the resistor?

15. For the circuit in Figure 12–44, complete Table 12–7. Draw the impedance and voltage phasor diagrams for the circuit.

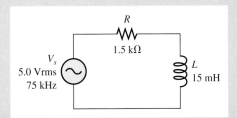

Table 12-7

R = 1.5 kΩ	I_R =	V_R =
X_L =	I_L =	V_L =
Z =	I_{tot} =	V_s = 5.0 Vrms

FIGURE 12-44

16. For the circuit in Figure 12-44, explain what will happen to the phase angle if the frequency is increased.

17. Complete Table 12-8 for the circuit in Figure 12-45.

Table 12-8

R = 270 Ω	I_R =	V_R =
X_C =	I_C =	V_C =
X_L =	I_L =	V_L =
Z =	I_{tot} =	V_s = 3.0 Vrms

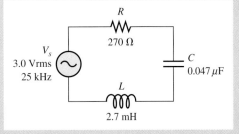

FIGURE 12-45

18. Draw the voltage phasors for the circuit in Figure 12-45. (You will need to show two phasor diagrams.)

19. The circuit in Figure 12-45 is inductive at the frequency shown. If the frequency is dropped to 10 kHz, will the circuit still look inductive? Justify your answer.

20. Calculate the resonant frequency for the series *RLC* circuit shown in Figure 12-46. Draw the impedance and voltage phasor diagrams for the circuit at resonance. The source voltage is 1.0 Vrms.

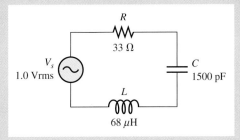

FIGURE 12-46

21. Calculate the resonant frequency of an *LC* circuit if a 0.068 μF capacitor is in series with a 180 μH inductor.

22. For the filter shown in Figure 12-47, determine the resonant frequency. Is this a band-pass or band-stop filter?

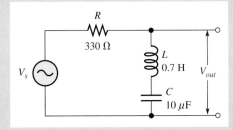

FIGURE 12-47

Basic-Plus Problems

23. In a series *RL* circuit, assume you set the source voltage for 5 V and measure 3 V across the resistor.

    (a) What is the voltage across the inductor?

    (b) What is the phase angle for the circuit?

24. An unknown capacitor is in series with a 1.0 kΩ resistor and connected to a 10 V source. The frequency is adjusted until the voltage across the resistor is equal to the voltage across the capacitor and is found to be 3.39 kHz.

    (a) What is the voltage across each component?

    (b) What is the capacitance?

25. A series *RLC* circuit is connected to a 10 V source. The voltage across the capacitor is twice the voltage across the inductor and is equal to the voltage across the resistor. What is the voltage across each component? (*Hint:* These are phasors!)

26. A series *RLC* circuit has 10 V across the resistor, 20 V across the capacitor, and 30 V across the inductor.

    (a) What is the source voltage?

    (b) What voltage would be measured across the resistor and the inductor together?

27. What value of inductance is required to resonant with a 220 pF capacitor to produce a resonant frequency of 1.5 MHz?

ANSWERS

Example Questions

12–1: Apply the cosine function.

12–2: 2.82

12–3: Impedance increases.

12–4: Phase angle decreases.

12–5: The phase angle remains the same. Although the voltage phasors are larger due to higher current, they all increase by the same amount.

12–6: At this frequency, $X_C = R = 1.5$ kΩ. The impedance phasor diagram will have equal length phasors for X_C and R. The total impedance is 2.12 kΩ. The angle between Z and R is 45°.

12–7: 53.2°

12–8: Divide each voltage phasor by the current.

12–9: X_L will be 502 Ω, X_C will be 398 Ω. Thus, the circuit appears less inductive than previously.

12–10: 5.0 V. $(5.0 \text{ V} = \sqrt{(1.81 \text{ V})^2 + (4.67 \text{ V})^2})$

12–11: The resonant frequency is unchanged.

12–12: Q is halved to 6.5.

Review Questions

1. It rotates a number by 90° counterclockwise.
2. The *y*-axis
3. Rectangular notation and polar notation

4. tan θ = opposite side / adjacent side
5. (hypotenuse)² = (adjacent side)² + (opposite side)²
6. The side parallel to the *x*-axis represents resistance; the side parallel to the *y*-axis represents reactance. The hypotenuse represents the impedance.
7. Resistance and reactance are plotted on different axes. The sum is done by applying the Pythagorean theorem and will be 2.83 kΩ.
8. The negative *y*-axis
9. The positive *x*-axis
10. +45°
11. The voltage phasors are found by multiplying the impedance diagram phasors by a constant, which is the current in the circuit.
12. They are equal.
13. Lower
14. Z decreases.
15. The phase angle θ decreases.
16. 4 V
17. +53.1°
18. Higher
19. Z increases.
20. θ increases.
21. The absolute (unsigned) value
22. Quadrant 4
23. More inductive
24. 1.41 kΩ
25. Lags
26. The reactances cancel at resonance, leaving only the resistance component.
27. Because impedance is a minimum, the current is maximum and limited only by the circuit resistance.
28. 1.24 MHz
29. 7.07 V
30. A high selectivity means the bandwidth is narrow and the response around resonance is steep.

Chapter Checkup

1. (d)	2. (b)	3. (a)	4. (c)	5. (b)
6. (d)	7. (a)	8. (b)	9. (d)	10. (d)
11. (a)	12. (b)	13. (a)	14. (a)	15. (d)
16. (c)				

13 CHAPTER

PARALLEL AC CIRCUITS

CHAPTER OUTLINE

13–1 Admittance
13–2 Parallel *RC* Circuits
13–3 Parallel *RL* Circuits
13–4 Parallel *RLC* Circuits
13–5 Parallel Resonance

INTRODUCTION

In this chapter, parallel ac circuits with resistance, capacitance, and inductance are discussed and parallel resonance is introduced. As in the last chapter with series ac circuits, parallel ac circuits are easy to visualize with phasor diagrams. For parallel circuits, the diagrams show admittance and current phasors instead of impedance and voltage phasors. *Admittance* may be a new term for you; it is the reciprocal of impedance. By understanding the basic circuits presented here, you will be able to solve more complicated problems. In many cases, a complicated circuit can be reduced to a two-component equivalent circuit.

In a parallel circuit, voltage is the same across all components, so it is logical to consider how current shifts with respect to voltage. The focus now is on the current, instead of on the voltage as it was in the series circuits. Recall that a capacitor always shifts the phase between current and voltage such that current leads the voltage. An inductor shifts the phase in the opposite direction; in this case, current lags the voltage. In a parallel circuit, when both reactive components are present, the one with the smaller reactance dominates because it will have the greater current through it.

As in series circuits, when the reactance of the capacitor and the reactance of the inductor are equal, the condition is called *resonance*. The chapter concludes with a discussion of parallel resonance.

Study aids for this chapter are available at

http://www.prenhall.com/SOE

KEY OBJECTIVES

A section number is given for each objective. After completing this chapter, you should be able to

13–1 Develop the admittance triangle for a parallel *RL* or *RC* circuit

13–2 Calculate the magnitude of currents in a parallel *RC* circuit, draw the admittance and current phasor diagrams, and describe how frequency affects the phasors

13–3 Calculate the magnitude of currents in a parallel *RL* circuit, draw the admittance and current phasor diagrams, and describe how frequency affects the phasors

13–4 Analyze a parallel *RLC* circuit, showing the admittance and current phasor diagrams

13–5 Analyze parallel resonant circuits and describe parallel resonant filters

COMPUTER SIMULATIONS DIRECTORY

The following figures have Multisim circuit files associated with them.

◆ Figure 13–8
  Page 450

◆ Figure 13–19
  Page 460

LABORATORY EXPERIMENTS DIRECTORY

The following exercises are for this chapter.

◆ **Experiment 25**
  Parallel *RC* Circuits

◆ **Experiment 26**
  Parallel *RL* Circuits

◆ **Experiment 27**
  Parallel Resonance

KEY TERMS

- Susceptance
- Admittance
- Tank circuit

ON THE JOB . . .

(Getty Images)

In many types of technical jobs, you will be required to write reports on projects. Reporting may be anything from simple logbook entries to formal technical reports. If you are involved in writing reports, find out the policy of the company. Some may require entries in ink and signed; others may have less formal requirements. If you are writing an explanation, keep in mind that an illustration of a problem may be a valuable addition to your report. If you are not a good speller, write out your report on a word processor and don't forget to use the spell checker!

CHAPTER 13

Many years ago, astronomers were amazed to discover that the temperature in the solar atmosphere rises rapidly as the height above the solar surface increases. The explanation of the heating source eluded them for years but is rooted in the concept of resonance.

In the center of the sun, the temperature is thought to be about 16,000,000°. At the surface the temperature is about 6000°. This tremendous temperature difference produces convection currents that carry with them magnetic field lines into the solar atmosphere (the corona). The magnetic field lines are shaken by the convection currents and vibrate like a spring. The vibration of one line affects nearby lines. The effect of multiple vibrating lines can cause the wave amplitude to grow, diminish, or even cancel. The vibrating waves carry energy into the corona.

Two mechanisms release the energy from the vibrating magnetic field lines. One is called *phase mixing*; the other is *resonant absorption*. Phase mixing occurs when different vibration frequencies mix. At certain points, the mixed frequencies can create high currents and release energy. Resonant absorption occurs when a number of waves of the same frequency are in step, creating regions where the amplitude of the wave grows rapidly as the waves add. This creates high currents, which can transfer energy to the corona, heating it to millions of degrees.

13-1 ADMITTANCE

The reciprocal of impedance is admittance. Admittance is the combination of conductance, which is the reciprocal of resistance, and susceptance, which is the reciprocal of reactance. These quantities are useful for analyzing parallel ac circuits containing resistive and reactive components.

In this section, you will learn to develop the admittance triangle for a parallel RL or RC circuit.

Conductance, Susceptance, and Admittance

Recall that conductance is the reciprocal of resistance. The measurement unit is the siemens, which is the reciprocal of the ohm.

$$G = \frac{1}{R} \tag{13-1}$$

In ac circuits, the opposition to current can be resistive or reactive depending on the component. For most circuits, we can treat all components as ideal; that is, resistive components are always resistors and reactive components can be either a pure capacitor or a pure inductor. The reciprocal of reactance is called **susceptance**, which is indicated with the letter B. Susceptance can be either capacitive or inductive. As in the case of reactances, a subscript is used to indicate if it is capacitive (B_C) or inductive (B_L). The capacitive susceptance is defined as

$$B_C = \frac{1}{X_C} \tag{13-2}$$

Likewise, the inductive susceptance is defined as

$$B_L = \frac{1}{X_L} \qquad (13\text{–}3)$$

For most practical circuits, the mS (millisiemens) and the μS (microsiemens) are more useful. For example, a 10 kΩ reactance has a conductance of 0.10 mS or 100 μS.

Admittance is the phasor sum of conductance and susceptance. Admittance can be thought of as the opposite to impedance; it indicates how readily current can be established. Mathematically, admittance is the reciprocal of impedance and is indicated with the letter Y. The measurement unit for admittance is also the siemens.

$$Y = \frac{1}{Z} \qquad (13\text{–}4)$$

The Admittance Triangle

In a parallel circuit, the reciprocals of resistance and reactance add to form the admittance, which is the reciprocal of impedance. Because phasors can be added with simple graphical methods, you can visualize circuit parameters from the standpoint of conductance, susceptance, and admittance. (Recall that in a series circuit, resistance and reactance phasors are added to form the total impedance.)

Because voltage is the same across all components that are in parallel, voltage is selected as the reference along the x-axis. The conductance and susceptance phasors are plotted on the rectangular coordinate system depending on how they affect the phase angle between current and voltage. Taking into account these phase angles, the basic rules for plotting conductance and susceptance phasors are shown in Figure 13–1 and given as follows:

1. Conductance (G) is always plotted along the positive x-axis.
2. Capacitive susceptance (B_C) is plotted along the positive y-axis ($+j$). (This is opposite to the series case for capacitive reactance.) In rectangular form, it is written $+jB_C$, which shows that it is plotted along the positive y-axis.
3. Inductive susceptance (B_L) is plotted along the negative y-axis ($-j$). In rectangular form, it is written $-jB_L$, which shows that it is plotted along the negative y-axis.

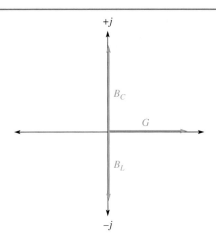

FIGURE 13–1

Directions for conductance and susceptance phasors in a parallel circuit.

Determine the admittance by calculating the phasor sum of the conductance and susceptance, using the rules for a right triangle. As an example, consider the basic parallel RC circuit shown in Figure 13–2(a). Assume that the frequency is set such that the conductance (G) and capacitive susceptance (B_C) are equal. Figure 13–2(b) shows how the phasors are plotted.

CHAPTER 13

FIGURE 13-2
Development of the admittance triangle for a parallel RC circuit.

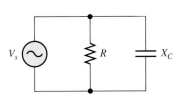
(a) A basic parallel RC circuit

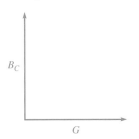
(b) Conductance and susceptance phasors

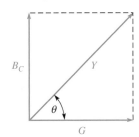
(c) Phasor sum of G and B_C

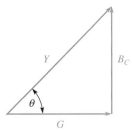
(d) The admittance triangle

Notice that the schematic shows resistance and reactance components (R and X_C) but the phasors that are plotted are the reciprocals of these quantities (G and B_C). The conductance phasor is plotted along the positive x-axis, and the capacitive susceptance phasor is plotted along the positive y-axis, which means the diagram is in quadrant 1 in the rectangular coordinate system. This relationship between G and B_C comes from the fact that the capacitor current in any circuit leads the voltage, and thus the resistor current, by 90°. Because Y is the phasor sum of G and B_C, it is represented as shown in Figure 13–2(c). A repositioning of the phasors, as shown in Figure 13–2(d), forms a right triangle. This is called the admittance triangle. The length of each phasor represents its magnitude in siemens, and the phase angle θ (the Greek letter theta) is the phase difference between the source voltage and current.

An equivalent triangle can be drawn for a parallel inductive circuit. Figure 13–3 illustrates a basic parallel RL circuit and the development of the admittance triangle. Notice that the schematic shows resistance and reactance, but the phasors that are plotted are the reciprocals of these quantities (G and B_L). The frequency is set such that the conductance (G) and inductive susceptance (B_L) are equal, as drawn in Figure 13–3(b). The conductance phasor is plotted along the positive x-axis, and the inductive susceptance phasor is plotted along the negative y-axis, which means the diagram is in quadrant 4 in the rectangular coordinate system. This relationship between G and B_L comes from the fact that the inductor current in any circuit lags the voltage, and thus the resistor current, by 90°. Because Y is the phasor sum of G and B_L, it is represented as shown in Figure 13–3(c). A repositioning of the phasors, as shown in Figure 13–3(d), forms the admittance triangle.

FIGURE 13-3
Development of the admittance triangle for a parallel RL circuit.

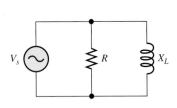
(a) A basic parallel RL circuit

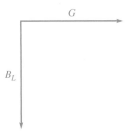
(b) Conductance and susceptance phasors

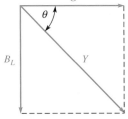
(c) Phasor sum of G and B_L

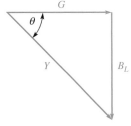
(d) The admittance triangle

As in any right triangle, the hypotenuse is equal to the square root of the sum of the squares of the sides, as given by the Pythagorean theorem. The sides of the right triangle are conductance and susceptance and the hypotenuse is admittance. For an RC parallel circuit, the magnitude of the admittance is

$$Y = \sqrt{G^2 + B_C^2} \qquad (13\text{-}5)$$

PARALLEL AC CIRCUITS

For a parallel RL circuit, the magnitude of the admittance is

$$Y = \sqrt{G^2 + B_L^2} \tag{13-6}$$

The phase angle, θ, can be written in terms of the capacitive or inductive susceptance and conductance phasors.

$$\theta = \tan^{-1}\left(\frac{B}{G}\right) \tag{13-7}$$

The following two examples show how to find the admittance for basic parallel RC and RL circuits.

EXAMPLE 13-1

Problem
Calculate the magnitude of the admittance and the phase angle for the RC circuit shown in Figure 13–4. Sketch the admittance phasor diagram and the admittance triangle.

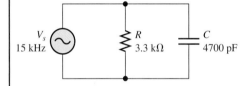

FIGURE 13-4

Solution
First, calculate the capacitive susceptance at 15 kHz. The capacitive susceptance is the reciprocal of the capacitive reactance. Therefore,

$$B_C = \frac{1}{X_C} = \frac{1}{\frac{1}{2\pi f C}} = 2\pi f C = 2\pi(15 \times 10^3 \text{ Hz})(4700 \times 10^{-12} \text{ F}) = 443 \text{ }\mu\text{S}$$

This is the magnitude of the capacitive susceptance. It may be shown with a $+j$ to indicate the direction is along the positive y-axis as $+j443$ μS.
Next, calculate the conductance,

$$G = \frac{1}{R} = \frac{1}{3.3 \text{ k}\Omega} = 303 \text{ }\mu\text{S}$$

and substitute into Equation 13–5, the equation for admittance in a parallel RC circuit.

$$Y = \sqrt{G^2 + B_C^2} = \sqrt{303 \text{ }\mu\text{S}^2 + 443 \text{ }\mu\text{S}^2} = \mathbf{537 \text{ }\mu\text{S}}$$

Determine the phase angle. Substitute the susceptance and conductance values into Equation 13–7.

$$\theta = \tan^{-1}\left(\frac{B}{G}\right) = \tan^{-1}\left(\frac{443 \text{ }\mu\text{S}}{303 \text{ }\mu\text{S}}\right) = \tan^{-1}(1.46) = \mathbf{55.6°}$$

CHAPTER 13

The admittance phasor diagram and admittance triangle are shown in Figure 13–5.

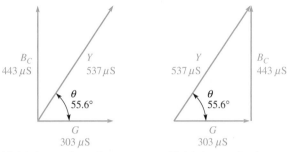

(a) Admittance phasor diagram (b) Admittance triangle

FIGURE 13–5

Question*
What happens to the admittance if the frequency is reduced?

EXAMPLE 13–2

Problem
Calculate the magnitude of the admittance and the phase angle for the *RL* circuit shown in Figure 13–6. Sketch the admittance phasor diagram and the admittance triangle.

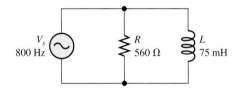

FIGURE 13–6

Solution
First, calculate the inductive susceptance at 800 Hz. The inductive susceptance is the reciprocal of the inductive reactance. Therefore,

$$B_L = \frac{1}{X_L} = \frac{1}{2\pi f L} = \frac{1}{2\pi(800\ \text{Hz})(0.075\ \text{H})} = 2.65\ \text{mS}$$

This is the magnitude of the inductive susceptance. It may be shown with $-j$ to indicate it is along the negative *y*-axis.
Calculate the conductance,

$$G = \frac{1}{R} = \frac{1}{560\ \Omega} = 1.78\ \text{mS}$$

and substitute into Equation 13–6.

$$Y = \sqrt{G^2 + B_L^2} = \sqrt{1.78\ \text{mS}^2 + 2.65\ \text{mS}^2} = \mathbf{3.20\ mS}$$

Calculate the phase angle. Substitute values into Equation 13–7.

$$\theta = \tan^{-1}\left(\frac{B}{G}\right) = \tan^{-1}\left(\frac{2.65\ \text{mS}}{1.78\ \text{mS}}\right) = \tan^{-1}(1.49) = \mathbf{56.1°}$$

---

*Answers are at the end of the chapter.

The admittance phasor diagram and admittance triangle are shown in Figure 13–7.

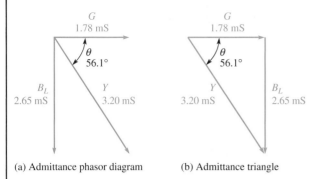

(a) Admittance phasor diagram   (b) Admittance triangle

FIGURE 13–7

Question
What happens to the admittance if the inductor is open?

Impedance in Two-Component Parallel Circuits

When a parallel circuit has two components (or can be reduced to two components), it is sometimes useful to consider the impedance, instead of the admittance. Recall that in parallel resistive circuits, the total opposition to current is *less* as more paths are added. With *RC* and *RL* circuits, the impedance is also less as more paths are added (but can actually be greater for *RLC* circuits, which are covered in Section 13–4).

For a two-component *RC* or *RL* circuit, the product-over-sum rule (developed in Section 5–5) can also be applied as long as you take into account the fact that the quantities are phasors. In this case, the magnitude of the impedance is

$$Z = \frac{RX}{\sqrt{R^2 + X^2}} \tag{13–8}$$

Notice that in the denominator, the sum is written as a phasor sum (square root of the sum of the squares) as discussed in Section 12–2 (and given in Equation 12–1). Also notice that X in Equation 13–8 stands for either the reactance of a capacitor, X_C, or the reactance of an inductor, X_L. The phase angle is

$$\theta = \tan^{-1}\left(\frac{R}{X}\right) \tag{13–9}$$

Equation 13–9 is equivalent to Equation 13–7 and produces the same result. The fractional part of the equation is the inverse of a series circuit. As in any circuit (series or parallel), current will always lead voltage in a capacitor and lag voltage in an inductor.

CHAPTER 13

EXAMPLE 13-3

Problem
Calculate the magnitude of the impedance and the phase angle for the parallel RC circuit shown in Figure 13-8. Using this result, calculate the total current.

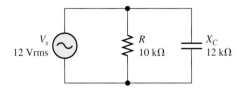

FIGURE 13-8

Solution
Substitute into Equation 13-8.

$$Z = \frac{RX}{\sqrt{R^2 + X^2}} = \frac{(10 \text{ k}\Omega)(12 \text{ k}\Omega)}{\sqrt{10 \text{ k}\Omega^2 + 12 \text{ k}\Omega^2}} = 7.68 \text{ k}\Omega$$

Calculate the phase angle by substituting into Equation 13-9.

$$\theta = \tan^{-1}\left(\frac{R}{X}\right) = \tan^{-1}\left(\frac{10 \text{ k}\Omega}{12 \text{ k}\Omega}\right) = 39.8°$$

Use Ohm's law to calculate the current.

$$I = \frac{V_s}{Z} = \frac{12 \text{ V}}{7.68 \text{ k}\Omega} = 1.56 \text{ mA}$$

Question
Does the current lead the voltage or does the voltage lead the current?

COMPUTER SIMULATION

Open the Multisim file F13-08AC on the website. Verify the current in each component and the total current.

Review Questions

Answers are at the end of the chapter.

1. What is susceptance?
2. What is admittance?
3. What is the measurement unit for conductance, susceptance, and admittance?
4. For what type of circuits (RC or RL) does the current lead the voltage?
5. When can the product-over-sum rule be applied?

13-2 PARALLEL RC CIRCUITS

Current phasors describe the relationships of the currents in the different branches of a parallel circuit and the total current. Because the voltage is the same across all components in parallel, it is the reference for a parallel circuit phasor diagram.

In this section, you will learn to calculate the currents in a parallel RC circuit, to draw the admittance and current phasor diagrams, and to describe how frequency affects the phasors.

In parallel circuits, the branch currents are independent of each other and depend only on the source voltage and the opposition to current of the component. In many circuits, the source voltage and opposition are known, and the current in any given branch is unknown. In this case, the branch current is calculated directly with Ohm's law. For example, if the current is needed in a resistor, Ohm's law can be written $I = V/R$; in an inductor or a capacitor, it is $I = V/X$. The total current from the source is $I = V/Z$.

Current Phasor Diagram

To calculate the magnitude of the branch currents or the total current in a parallel RC circuit, you can use the appropriate conductance, susceptance, or admittance quantity and multiply by the voltage. This is just another way of writing Ohm's law, but it is particularly convenient in parallel circuits. To find the current in a resistive branch, substitute conductance into Ohm's law.

$$I_R = \frac{V}{R} = \left(\frac{1}{R}\right)V$$

$$I_R = GV \tag{13-10}$$

To find the current in a capacitive branch, substitute susceptance into Ohm's law.

$$I_C = \frac{V}{X_C} = \left(\frac{1}{X_C}\right)V$$

$$I_C = B_C V \tag{13-11}$$

To find the total current, substitute admittance into Ohm's law.

$$I_{tot} = \frac{V}{Z} = \left(\frac{1}{Z}\right)V$$

$$I_{tot} = YV \tag{13-12}$$

Another way to calculate the total current from the source is to sum the branch currents. For a parallel RC circuit with just a resistor and a capacitor, the total current is the phasor sum of the currents.

$$I_{tot} = \sqrt{I_R^2 + I_C^2}$$

The currents described by Equations 13-10, 13-11, and 13-12 are phasor currents because they have magnitude and direction (phase). The equations are particularly useful for developing the relationship between admittance and current phasor diagrams. The current phasor diagram has the same shape and phase relationship as the admittance phasor diagram developed in Section 13-1.

CHAPTER 13

EXAMPLE 13–4

Problem
Draw the current phasor diagram for the RC circuit shown in Figure 13–9(a). This is the same circuit as shown in Example 13–1. For reference, the admittance phasor diagram is repeated as Figure 13–9(b). The source voltage is 12 V.

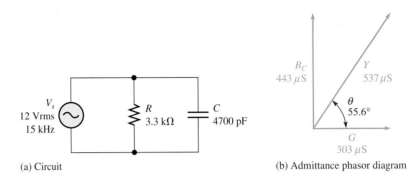

(a) Circuit   (b) Admittance phasor diagram

FIGURE 13–9

Solution
Calculate the currents by applying Equations 13–10, 13–11, and 13–12 as follows:

$$I_R = GV = (303\ \mu S)(12\ V) = 3.64\ mA$$
$$I_C = B_C V = (443\ \mu S)(12\ V) = 5.32\ mA$$
$$I_{tot} = YV = (537\ \mu S)(12\ V) = 6.44\ mA$$

The current phasors can be plotted along the same axis as the admittance phasors. They differ only by the voltage. To emphasize this relationship, the admittance and current phasors are plotted side-by-side in Figure 13–10 along with the operation that was performed to move from one to the other. Multiplying the admittance phasors by the circuit voltage gives the current phasors. The phase angle calculated in Example 13–1 is the same for both the admittance and current phasor diagrams, so it is not necessary to calculate it again.

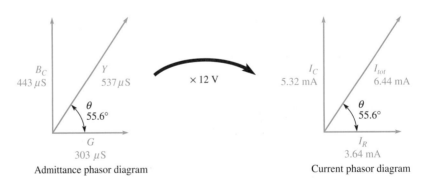

Admittance phasor diagram   Current phasor diagram

FIGURE 13–10

Question
Use the Pythagorean theorem to prove that the total current is correct (within round-off error).

PARALLEL AC CIRCUITS

EXAMPLE 13–5

Problem
Find the missing values on Table 13–1 for the circuit in Figure 13–11. Draw the admittance and current phasor diagrams.

Table 13–1

R = 6.8 kΩ	G =	I_R =	V_R =
X_C =	B_C =	I_C =	V_C =
Z =	Y =	I_{tot} =	V_s = 10 Vrms

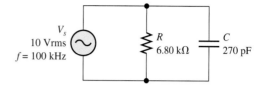

FIGURE 13–11

Solution
The reactance of the capacitor is

$$X_C = \frac{1}{2\pi f C} = \frac{1}{2\pi(100 \text{ kHz})(270 \text{ pF})} = 5.89 \text{ k}\Omega$$

Apply Equation 13–8 to find the magnitude of the impedance.

$$Z = \frac{(6.8 \text{ k}\Omega)(5.89 \text{ k}\Omega)}{\sqrt{(6.8 \text{ k}\Omega)^2 + (5.89 \text{ k}\Omega)^2}} = 4.45 \text{ k}\Omega$$

Calculate the conductance, susceptance, and admittance.

$$G = \frac{1}{R} = \frac{1}{6.8 \text{ k}\Omega} = 147 \text{ μS}$$

$$B_C = \frac{1}{X_C} = \frac{1}{5.89 \text{ k}\Omega} = 170 \text{ μS}$$

$$Y = \frac{1}{Z} = \frac{1}{4.45 \text{ k}\Omega} = 225 \text{ μS}$$

Determine the currents by multiplying the source voltage by the conductance, susceptance, and admittance.

$$I_R = GV = (147 \text{ μS})(10 \text{ V}) = 1.47 \text{ mA}$$
$$I_C = B_C V = (170 \text{ μS})(10 \text{ V}) = 1.70 \text{ mA}$$
$$I_{tot} = YV = (225 \text{ μS})(10 \text{ V}) = 2.25 \text{ mA}$$

As a check, the product of the total current and the impedance should equal the source voltage.

$$V_s = I_{tot} Z = (2.25 \text{ mA})(4.45 \text{ k}\Omega) = 10.0 \text{ V} \checkmark$$

Table 13–2 is the completed table. Using the values in Table 13–2, you can draw the admittance and current phasor diagrams. See Figure 13–12. The phase angle can be calculated by applying the tangent function to the current phasors.

Table 13–2

R = 6.8 kΩ	G = 147 μS	I_R = 1.47 mA	V_R = 10 Vrms
X_C = 5.89 kΩ	B_C = 170 μS	I_C = 1.70 mA	V_C = 10 Vrms
Z = 4.45 kΩ	Y = 225 μS	I_{tot} = 2.25 mA	V_s = 10 Vrms

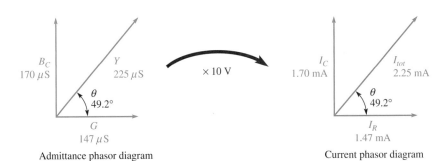

FIGURE 13–12

$$\theta = \tan^{-1}\left(\frac{I_C}{I_R}\right) = \tan^{-1}\left(\frac{1.70 \text{ mA}}{1.47 \text{ mA}}\right) = 49.2°$$

Question
What happens to the current phasor diagram if the resistor is replaced with a larger one?

Variation of Admittance with Frequency in Parallel *RC* Circuits

Capacitive susceptance varies directly with frequency; thus, as frequency is increased, there will be more current in the capacitor in a parallel circuit. The magnitude of the admittance of a parallel *RC* circuit is

$$Y = \sqrt{G^2 + B_C^2}$$

When B_C increases, the entire term under the square root sign increases and therefore the admittance increases. Conversely, when B_C decreases, the admittance decreases. In a parallel *RC* circuit, admittance is directly proportional to the frequency.

Figure 13–13 illustrates how the admittance phasors in a parallel *RC* circuit vary as the frequency increases or decreases. Think of each of the three triangles as a different admittance triangle for each frequency. The conductance is constant because it does not vary with frequency. Because capacitive susceptance varies directly with the frequency, the magnitude of the admittance and the phase angle also vary directly with the frequency.

FIGURE 13–13

As the frequency increases, the capacitive susceptance, B_C, increases and the admittance, Y, increases. Each value of frequency can be visualized as forming a different admittance triangle. The current phasors will have the same shape.

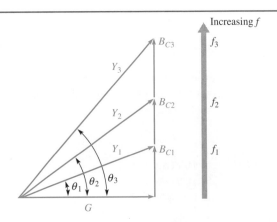

PARALLEL AC CIRCUITS

EXAMPLE 13-6

Problem
Draw the admittance and current phasor diagrams for the *RC* circuit in Example 13–5 if the frequency is increased to 200 kHz.

Solution
The frequency is doubled; therefore, the capacitive susceptance is doubled. The new value of B_C is 340 μS (rounded up from earlier result). The capacitive branch current will also be doubled to 3.40 mA.

The conductance is independent of the frequency, so it remains 147 μS and the current in the resistor remains at 1.47 mA.

Calculate the total current, I_{tot}, by applying the Pythagorean theorem to the branch currents.

$$I_{tot} = \sqrt{I_R^2 + I_C^2} = \sqrt{1.47 \text{ mA}^2 + 3.40 \text{ mA}^2} = 3.70 \text{ mA}$$

The phase angle can be calculated by applying the tangent function to the current phasors.

$$\theta = \tan^{-1}\left(\frac{I_C}{I_R}\right) = \tan^{-1}\left(\frac{3.40 \text{ mA}}{1.47 \text{ mA}}\right) = 66.6°$$

Figure 13–14 shows the admittance and current phasor diagrams. Notice the relationship between the admittance diagram and the current diagram.

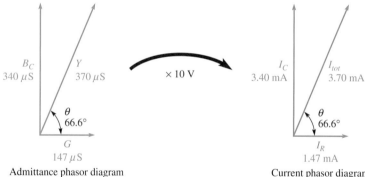

FIGURE 13–14

Question
What happened to the magnitude of the impedance when the frequency was increased?

Review Questions

6. What is the rule for combining branch currents in a parallel *RC* circuit?
7. What does the product of admittance and total current give in a parallel *RC* circuit?
8. What happens to the admittance in a parallel *RC* circuit if the capacitor is replaced with a larger one?
9. If the voltage is increased in a parallel *RC* circuit, what happens to the admittance phasors?
10. What happens to the phase angle if the frequency is decreased in a parallel *RC* circuit?

CHAPTER 13

13–3 PARALLEL RL CIRCUITS

The parallel RL circuit behaves similarly to the parallel RC circuit. The current in an inductor lags the voltage, so the current phasor for an inductor is drawn opposite to the current phasor for a capacitor.

In this section, you will learn how to calculate the magnitude of currents in a parallel RL circuit, draw the admittance and current phasor diagrams, and describe how frequency affects the phasors.

In parallel circuits, the current is usually the unknown and is calculated by applying Ohm's law. The total impedance when there are two components is given by the product-over-sum rule. With a resistor and an inductor, the product-over-sum rule is written

$$Z = \frac{RX_L}{\sqrt{R^2 + X_L^2}}$$

Ohm's law can be applied to the circuit as a whole as $I = V/Z$.

As in the case of the parallel RC circuit, the current phasor diagram in a parallel RL circuit has the same shape as the admittance phasor diagram. Also, the current phasors are found by multiplying the admittance phasors by the voltage in the circuit. To calculate the magnitude of a branch current, you can use the appropriate conductance, susceptance, or admittance quantity and multiply by the voltage.

To find the current in an inductive branch, Equation 13–11 is modified for the inductive susceptance.

$$I_L = \frac{V}{X_L} = \left(\frac{1}{X_L}\right)V$$

$$I_L = B_L V \qquad (13\text{–}13)$$

Because current lags the voltage in an inductor, the inductive susceptance and inductive current are plotted in the opposite direction to an RC circuit.

EXAMPLE 13–7

Problem
Find the missing values on Table 13–3 for the circuit in Figure 13–15. Draw the admittance and current phasor diagrams.

Table 13–3

$R = 470\ \Omega$	$G =$	$I_R =$	$V_R =$
$X_L =$	$B_L =$	$I_L =$	$V_L =$
$Z =$	$Y =$	$I_{tot} =$	$V_s = 5.0$ Vrms

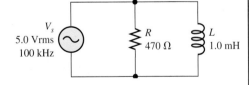

FIGURE 13–15

Solution
The voltage across each component is the same and is equal to the source voltage. Start on the right side of the table and enter 5.0 V for V_L and V_R.

Find the inductive reactance of L.

$$X_L = 2\pi fL = 2\pi(100 \text{ kHz})(1.0 \text{ mH}) = 628 \text{ }\Omega$$

Calculate the magnitude of the impedance by applying the product-over-sum rule (Equation 13–8).

$$Z = \frac{RX_L}{\sqrt{R^2 + X_L^2}} = \frac{(470 \text{ }\Omega)(628 \text{ }\Omega)}{\sqrt{470 \text{ }\Omega^2 + 628 \text{ }\Omega^2}} = 376 \text{ }\Omega$$

The conductance, susceptance, and admittance are

$$G = \frac{1}{R} = \frac{1}{470 \text{ }\Omega} = 2.13 \text{ mS}$$

$$B_L = \frac{1}{X_L} = \frac{1}{628 \text{ }\Omega} = 1.59 \text{ mS}$$

$$Y = \frac{1}{Z} = \frac{1}{376 \text{ }\Omega} = 2.66 \text{ mS}$$

Calculate the currents.

$$I_R = GV = (2.13 \text{ mS})(5.0 \text{ V}) = 10.64 \text{ mA}$$
$$I_L = B_LV = (1.59 \text{ mS})(5.0 \text{ V}) = 7.96 \text{ mA}$$
$$I_{tot} = YV = (2.66 \text{ mS})(5.0 \text{ V}) = 13.3 \text{ mA}$$

Complete Table 13–4.

Table 13–4

$R = 470 \text{ }\Omega$	$G = 2.13 \text{ mS}$	$I_R = 10.6 \text{ mA}$	$V_R = 5.0 \text{ Vrms}$
$X_L = 628 \text{ }\Omega$	$B_L = 1.59 \text{ mS}$	$I_L = 7.96 \text{ mA}$	$V_L = 5.0 \text{ Vrms}$
$Z = 376 \text{ }\Omega$	$Y = 2.66 \text{ mS}$	$I_{tot} = 13.3 \text{ mA}$	$V_s = 5.0 \text{ Vrms}$

The admittance and current phasor diagrams are plotted in Figure 13–16. The phase angle can be calculated by applying the tangent function to the current phasors.

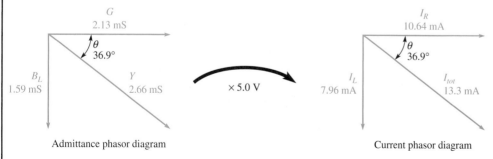

FIGURE 13–16

Question
Does Kirchhoff's current law apply to the circuit? Explain your answer.

CHAPTER 13

Variation of Admittance with Frequency in Parallel RL Circuits

As in the case of the parallel *RC* circuit, the admittance of a parallel *RL* circuit increases directly with the inductive susceptance. The magnitude of the admittance is

$$Y = \sqrt{G^2 + B_L^2}$$

Unlike the *RC* case, the inductive susceptance varies inversely with frequency; therefore, as frequency is increased, there will be less current in the inductor. In a parallel *RL* circuit, admittance is also inversely proportional to the frequency.

Figure 13–17 illustrates this concept. Because B_L varies inversely with the frequency, the magnitude of the admittance is smaller at higher frequencies. The phase angle is also smaller as frequency increases. Notice how this result differs from the capacitive case shown in Figure 13–13.

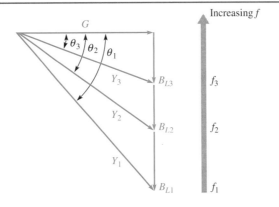

FIGURE 13–17

As the frequency increases, both the inductive susceptance, B_L, and admittance, *Y*, decrease. Each value of frequency can be visualized as forming a different admittance triangle. The current phasors will have the same shape.

Review Questions

11. What is the rule for combining conductance and inductive susceptance in a parallel *RL* circuit?
12. Why is the inductive susceptance drawn on the negative *y*-axis?
13. What happens to the admittance in a parallel *RL* circuit if the resistor is replaced with a larger one?
14. If the frequency is increased in a parallel *RL* circuit, what happens to the admittance phasors?
15. Given the current phasors, how could you convert them to admittance phasors?

13–4 PARALLEL *RLC* CIRCUITS

Before you can solve a parallel *RLC* circuit, you must first combine the capacitor and the inductor to form a single equivalent component. The equivalent component will appear to be an equivalent inductance or an equivalent capacitance. This is then combined with the resistance in the circuit.

In this section, you will learn to analyze a parallel *RLC* circuit, showing the admittance and current phasor diagrams.

Ohm's Law in Parallel RLC Circuits

A parallel *RLC* circuit is illustrated in Figure 13–18. As in any parallel circuit, the current in each component is independent of the other components and is dependent only on the applied voltage and the opposition to current of the component. To find the current in any one component, apply Ohm's law to the component. As you know, the current that is in reactive components will have a phase shift from the applied voltage. Inductive reactance causes the current to lag the source voltage, and capacitive reactance causes the current to lead the source voltage. The total current is the phasor sum of the individual branch currents.

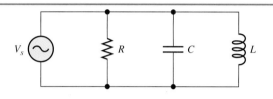

FIGURE 13–18

Admittance and Impedance of Parallel RLC Circuits

The admittance of a parallel ac circuit is the phasor sum of the individual conductances and susceptances in the branches. The magnitude of the admittance is the phasor sum of the conductance and susceptance phasors as given by

$$Y = \sqrt{G^2 + |B_C - B_L|^2} \qquad (13\text{–}14)$$

Compare this equation to Equation 12–5 for the impedance in a series *RLC* circuit. Both equations are solved in a similar manner. The capacitive and inductive susceptances cause the current to shift in opposite directions, therefore the difference term. Think of the total susceptance as the sum of a positive number and a negative number. The admittance is plotted in quadrant 1 if $B_C > B_L$; otherwise, it is in quadrant 4.

Having found the admittance, the easiest way to calculate the magnitude of the impedance is to take the reciprocal.

$$Z = \frac{1}{Y}$$

In resistive circuits, recall that the total resistance is always smaller than the smallest resistor. In parallel ac circuits with capacitance and inductance, the total impedance can be larger than the opposition from the smallest component because of phase shifts.

Current Relationships

In a parallel circuit, the current in the inductive branch and the current in the capacitive branch are always 180° out of phase with each other. As a result, the difference in I_C and I_L is always less than the largest individual current, and the total current in the inductive and capacitive branches is always less than the largest individual branch current. Of course, the current in the resistor is always shifted 90° with respect to the reactive currents. The equation for the total current in a parallel *RLC* circuit is

$$I_{tot} = \sqrt{I_R^2 + |I_C - I_L|^2} \qquad (13\text{–}15)$$

The next example illustrates the admittance and current phasor plots for a parallel *RLC* circuit.

EXAMPLE 13-8

Problem
Complete Table 13–5 for the circuit shown in Figure 13–19. Draw the admittance phasor diagram.

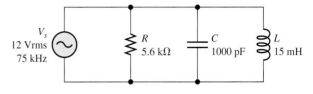

FIGURE 13–19

Table 13–5

Solution
Calculate the magnitude of the reactance of the capacitor and the inductor.

$$X_C = \frac{1}{2\pi fC} = \frac{1}{2\pi(75 \text{ kHz})(1000 \text{ pF})} = 2.12 \text{ k}\Omega$$

$$X_L = 2\pi fL = 2\pi(75 \text{ kHz})(15 \text{ mH}) = 7.07 \text{ k}\Omega$$

Calculate the conductance, capacitive susceptance, and inductive susceptance.

$$G = \frac{1}{R} = \frac{1}{5.6 \text{ k}\Omega} = 0.178 \text{ mS}$$

$$B_C = \frac{1}{X_C} = \frac{1}{2.12 \text{ k}\Omega} = 0.471 \text{ mS}$$

$$B_L = \frac{1}{X_L} = \frac{1}{7.07 \text{ k}\Omega} = 0.141 \text{ mS}$$

Use Equation 13–14 to calculate the magnitude of the admittance.

$$Y = \sqrt{G^2 + |B_C - B_L|^2} = \sqrt{0.178 \text{ mS}^2 + |0.471 \text{ mS} - 0.141 \text{ mS}|^2} = 0.375 \text{ mS}$$

The magnitude of the impedance is the reciprocal of the magnitude of the admittance.

$$Z = \frac{1}{Y} = \frac{1}{0.375 \text{ mS}} = 2.67 \text{ k}\Omega$$

Determine the currents by multiplying the conductance, susceptance, and admittance phasors by the applied voltage.

$$I_R = GV = (0.178 \text{ mS})(12 \text{ V}) = 2.14 \text{ mA}$$
$$I_C = B_C V = (0.471 \text{ mS})(12 \text{ V}) = 5.65 \text{ mA}$$
$$I_L = B_L V = (0.141 \text{ mS})(12 \text{ V}) = 1.69 \text{ mA}$$
$$I_{tot} = YV = (0.375 \text{ mS})(12 \text{ V}) = 4.50 \text{ mA}$$

Enter the results into Table 13–6.

Table 13–6

$R = 5.6 \text{ k}\Omega$	$G = 0.178 \text{ mS}$	$I_R = 2.14 \text{ mA}$	$V_R = 12 \text{ Vrms}$
$X_C = 2.12 \text{ k}\Omega$	$B_C = 0.471 \text{ mS}$	$I_C = 5.65 \text{ mA}$	$V_C = 12 \text{ Vrms}$
$X_L = 7.07 \text{ k}\Omega$	$B_L = 0.141 \text{ mS}$	$I_L = 1.69 \text{ mA}$	$V_L = 12 \text{ Vrms}$
$Z = 2.67 \text{ k}\Omega$	$Y = 0.375 \text{ mS}$	$I_{tot} = 4.50 \text{ mA}$	$V_s = 12 \text{ Vrms}$

The admittance phasor diagram is developed in Figure 13–20. Figure 13–20(a) shows the conductance and susceptance phasors only. Combining the susceptance phasors and then the conductance phasor produces the admittance phasor diagram in Figure 13–20(b).

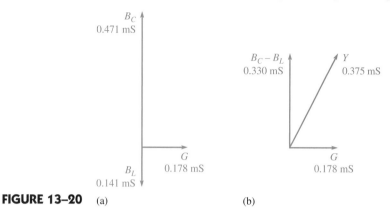

FIGURE 13–20 (a)   (b)

The current phasors are also shown in two plots. Figure 13–21(a) shows the branch currents. Combining the current in the capacitor and inductor and then combining the resistor current leads to the plot in Figure 13–21(b). Notice that the current phasor diagram has the same shape as the admittance phasor diagram because the current phasors are multiples of the conductance, susceptance, and admittance phasors.

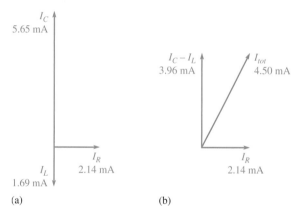

FIGURE 13–21 (a)   (b)

Question
Why is the total current less than the current in the capacitor?

COMPUTER SIMULATION

Open the Multisim file F13-19AC on the website. Verify the current in each component and the total current.

Phase Shift in Parallel RLC Circuits

After combining the susceptance phasors, you can describe a parallel *RLC* circuit with an admittance and current right triangle. By applying basic trigonometry, you can write the phase shift in terms of either the admittance triangle or the current triangle. From the admittance triangle,

$$\theta = \tan^{-1}\left(\frac{|B_C - B_L|}{G}\right)$$

CHAPTER 13

The current will lead the voltage if B_C is larger than B_L; otherwise it lags voltage. An equivalent equation can be written in terms of current as

$$\theta = \tan^{-1}\left(\frac{|I_C - I_L|}{I_R}\right)$$

Review Questions

16. Why are capacitive susceptance and inductive susceptance plotted in different directions?
17. What are two expressions for the phase shift in a parallel *RLC* circuit?
18. In a parallel *RLC* circuit, what determines if the current leads or lags the voltage?
19. What is the admittance of a parallel *RLC* circuit if a 10 kΩ resistor is in parallel with a capacitor with a reactance of 2.0 kΩ and an inductor with a reactance of 3.0 kΩ?
20. Does the source current lead or lag the source voltage in question 19?

13–5 PARALLEL RESONANCE

A parallel resonant circuit is an *LC* circuit operating at a frequency such that $X_L = X_C$. In parallel, this condition produces an entirely different response than what you saw in the series case.

In this section, you will learn how to analyze parallel resonant circuits and describe parallel resonant filters.

Condition for Resonance

When frequency is increased in an *LC* circuit, the reactance of the inductor increases and the reactance of the capacitor decreases. Just as in a series circuit, there is one frequency that the two reactances are equal. This is the resonant frequency, (f_r). The resonant frequency is found by the same equation (Equation 12–7) used to calculate the series resonant frequency, namely:

$$f_r = \frac{1}{2\pi\sqrt{LC}} \qquad (13\text{–}16)$$

In an ideal parallel resonant circuit, there is no resistance, only an ideal inductor (with no internal resistance) and an ideal capacitor, as shown in Figure 13–22(a). In this ideal

FIGURE 13–22 An ideal parallel *LC* circuit at resonance.

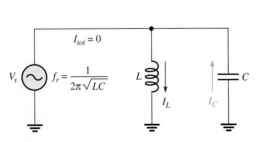
(a) Parallel circuit at resonance ($X_C = X_L$, $Z = \infty$)

(b) Current phasors

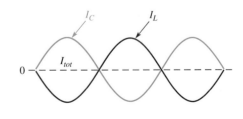
(c) Current waveforms

case, the currents are equal and opposite as illustrated in Figure 13–22(b) and 13–22(c). The total current from the source is zero, as shown, but there is current in the capacitor and the inductor. The current in the capacitor and inductor is rather like a pendulum — current goes from one to the other in a constant oscillation. The currents in the ideal circuit are equal and opposite in phase and sum to zero!

A parallel resonant circuit is often called a **tank circuit**. The word *tank* refers to the storage idea in parallel resonance. Energy is stored first in the electric field and then in the magnetic field. The stored energy is transferred back and forth on alternate half cycles of the sine wave. If an ideal circuit existed, the oscillation, once started, could continue forever; but the inevitable resistance dissipates energy, causing the oscillations to die out. Although the ideal tank circuit does not exist, tank circuits are widely used in oscillators, circuits that produce repetitive waveforms.

Currents in a Parallel Resonant Circuit

In practical circuits, resistance is present, so there will always be some current from the source at all frequencies. The resistance eventually causes oscillations to decay. Practical oscillators use an amplifier to compensate for these resistive losses. Resistance also produces a very small effect on the resonant frequency, which in most cases can be ignored.

Figure 13–23 illustrates what happens in an *RLC* circuit as the frequency is varied. At dc (0 Hz), there will be a current (not shown) in the inductor limited only by its internal resistance. As the frequency is raised, current in the inductor drops as the reactance increases and current in the capacitor increases. At resonance, the two currents are equal and have opposite phases, so they cancel; the source current, I_s, is determined only by the circuit resistance (including winding resistance of the inductor). At frequencies above resonance, the current in the capacitor is highest and dominates.

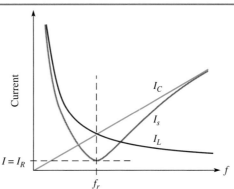

FIGURE 13–23

Current in a parallel *RLC* circuit. At resonance, the total current is a minimum.

At resonance the current in a parallel *RLC* circuit is a minimum; recall that this is the opposite of the current in a series *RLC* circuit. It follows that if the current is minimum, the impedance must be at its maximum value. As in the series case, the circuit appears to be purely resistive at the resonant frequency. Below resonance, the inductive susceptance is high (low reactance) so the circuit looks inductive. The source current will lead the source voltage, as you would expect with an inductor. At resonance, the inductive susceptance and capacitive susceptance cancel so the circuit looks resistive and the source current and voltage are in phase ($\theta = 0$). Above resonance, the capacitive susceptance is high (low reactance) so the circuit looks capacitive. The source current will lag the source voltage.

CHAPTER 13

EXAMPLE 13-9

Problem
Calculate the resonant frequency for the parallel *RLC* circuit shown in Figure 13-24. Draw the admittance and current phasor diagrams for the circuit at resonance. The source voltage is 3.0 Vrms.

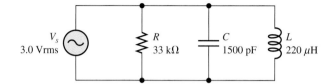

FIGURE 13-24

Solution
Calculate the resonant frequency from Equation 13-16.

$$f_r = \frac{1}{2\pi\sqrt{LC}} = \frac{1}{2\pi\sqrt{(220 \times 10^{-6})(1500 \times 10^{-12})}} = 277 \text{ kHz}$$

At this frequency, calculate X_L.

$$X_L = 2\pi fL = 2\pi(277 \text{ kHz})(220 \text{ }\mu\text{H}) = 383 \text{ }\Omega$$

The inductive susceptance is

$$B_L = \frac{1}{X_L} = 2.61 \text{ mS}$$

X_C and B_C have the same magnitude because the circuit is at resonance.

$$B_C = 2\pi fC = 2\pi(277 \text{ kHz})(1500 \text{ pF}) = 2.61 \text{ mS}$$

The conductance of *R* is

$$G = \frac{1}{R} = \frac{1}{33 \text{ k}\Omega} = 0.030 \text{ mS}$$

The admittance phasor diagram is shown in Figure 13-25(a). When B_L is added to B_C, the susceptances cancel, leaving only *G* as the admittance of the circuit at resonance.

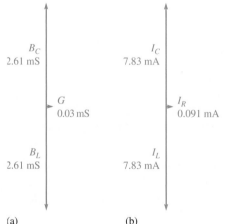

FIGURE 13-25 (a) (b)

PARALLEL AC CIRCUITS

Calculate the current by multiplying the admittance phasors by the voltage.

$$I_R = GV = (0.030 \text{ mS})(3.0 \text{ V}) = 0.91 \text{ mA}$$
$$I_C = B_CV = (2.61 \text{ mS})(3.0 \text{ V}) = 7.83 \text{ mA}$$
$$I_L = B_LV = (2.61 \text{ mS})(3.0 \text{ V}) = 7.83 \text{ mA}$$

The current phasors are plotted in Figure 13–25(b).

Question
What happens to the current phasors if the frequency is above resonance?

Parallel Resonant Filters

Like series resonant circuits, parallel resonant circuits can also be used to form frequency-selective filters. They can be configured as band-pass filters or band-stop filters, depending on where the output is taken.

Figure 13–26 shows a basic band-pass filter circuit and response. The output is taken across the tank circuit (shown in the colored box) in this application. At low frequencies, the tank circuit has low impedance; and therefore most of the source voltage is dropped across the resistor. At resonance, the impedance of the tank circuit is a maximum, so most of the source voltage drops across it. Because the output is across the tank circuit, it too has a peak at resonance. At high frequencies, the tank circuit has low impedance and voltage drops across the resistor again.

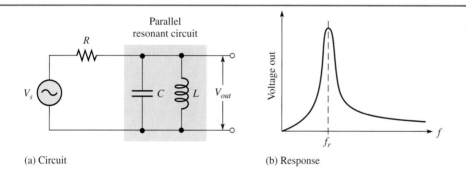

FIGURE 13–26

(a) Circuit (b) Response

Reversing the resistor and the tank circuit will form a band-stop filter. Figure 13–27 shows the circuit and its response. The output is now across the resistor, so the response is reversed from the band-pass filter, with a minimum output at the resonant frequency.

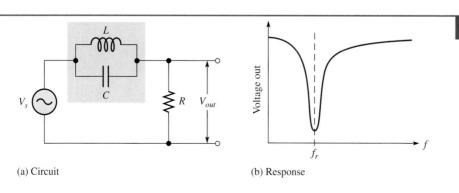

FIGURE 13–27

(a) Circuit (b) Response

CHAPTER 13

Bandwidth

As in the case of a series filter, the bandwidth of a parallel filter is measured at the points where the output voltage is 70.7% of the maximum value. The specific values of L and C and the circuit resistance control the bandwidth and selectivity (see Section 12–6). If L is larger, and C is smaller by the same factor, the resonant frequency is not changed, but the bandwidth will be larger. The selectivity is specified by the quality factor, Q. If Q is high, the circuit is highly selective and has a narrow bandwidth; if Q is low, the selectivity is lower and the bandwidth is wider. The equation $Q = f_r/BW$, introduced in Section 12–6 for series circuits, is valid in parallel circuits also.

The resistance can have a significant effect on the shape of the response for both band-pass and band-stop filters. One effect of resistance that may be overlooked is the effect of connecting the filter circuit to a load resistor. The load resistor becomes part of the filter circuit and affects its response. This idea is illustrated for a band-pass filter in Figure 13–28. The load resistor causes the response to be wider and not as peaked. A simple "fix" is to replace the load resistor with an amplifier that has a very high input resistance.

FIGURE 13–28

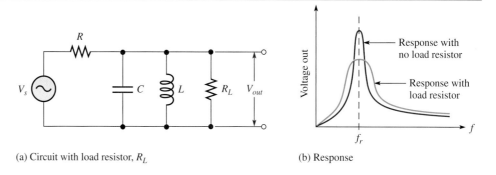

(a) Circuit with load resistor, R_L

(b) Response

Review Questions

21. How does the impedance differ between series and parallel resonance at the resonant frequency?
22. At resonance, what is the phase shift between current and voltage in a parallel resonant circuit?
23. At frequencies above resonance, does current lead or lag the voltage?
24. How does a parallel resonant band-pass filter differ from a parallel resonant band-stop filter?
25. What factors affect the selectivity of a resonant filter?

CHAPTER REVIEW

Key Terms

Admittance The phasor sum of conductance and susceptance; it indicates how readily current can be established.

Susceptance The reciprocal of reactance, indicated with the letter B. Susceptance can be either capacitive or inductive.

Tank circuit A parallel resonant circuit.

Important Facts

- Susceptance is the reciprocal of reactance and is indicated with the letter B. Susceptance can be either capacitive or inductive.
- Admittance is the reciprocal of impedance and is indicated with the letter Y. It is also the phasor sum of conductance and susceptance.
- The unit for measuring conductance, susceptance, and admittance is the siemens.
- Parallel ac circuits can be described in terms of admittance phasors, which include conductance and susceptance.
- In an admittance phasor diagram for a parallel ac circuit, conductance is plotted along the positive x-axis, capacitive susceptance is plotted along the positive y-axis and inductive susceptance is plotted along the negative y-axis.
- The current in each branch can be found by multiplying the conductance or susceptance of the branch by the voltage.
- The current phasor diagram for a given parallel ac circuit has the same shape as the admittance phasor diagram because the phasors are related by the voltage in the circuit.
- Frequency is directly proportional to capacitive susceptance and inversely proportional to inductive susceptance.
- The phase angle in an RLC parallel circuit can be calculated by applying trigonometry rules for a right triangle to the admittance or current phasor diagram.
- At resonance, the impedance in a parallel circuit is a maximum and the current is a minimum.
- Like series resonant circuits, parallel resonant circuits are also used in band-pass and band-stop filters.

Formulas

Conductance:

$$G = \frac{1}{R} \qquad (13\text{–}1)$$

Capacitive susceptance:

$$B_C = \frac{1}{X_C} \qquad (13\text{–}2)$$

Inductive susceptance:

$$B_L = \frac{1}{X_L} \qquad (13\text{–}3)$$

Admittance as the reciprocal of impedance:

$$Y = \frac{1}{Z} \qquad (13\text{–}4)$$

Admittance as the phasor sum of conductance and capacitive susceptance:

$$Y = \sqrt{G^2 + B_C^2} \qquad (13\text{–}5)$$

Admittance as the phasor sum of conductance and inductive susceptance:

$$Y = \sqrt{G^2 + B_L^2} \qquad (13\text{–}6)$$

Phase angle of a parallel RC or RL circuit:

$$\theta = \tan^{-1}\left(\frac{B}{G}\right) \quad (13\text{--}7)$$

Magnitude of impedance for a parallel two-component RC or RL circuit:

$$Z = \frac{RX}{\sqrt{R^2 + X^2}} \quad (13\text{--}8)$$

Alternate formula for the phase angle of a parallel RC or RL circuit:

$$\theta = \tan^{-1}\left(\frac{R}{X}\right) \quad (13\text{--}9)$$

Current in a resistive branch:

$$I_R = GV \quad (13\text{--}10)$$

Current in a capacitive reactive branch:

$$I_C = B_C V \quad (13\text{--}11)$$

Total current in a parallel ac circuit:

$$I_{tot} = YV \quad (13\text{--}12)$$

Current in an inductive reactive branch:

$$I_L = B_L V \quad (13\text{--}13)$$

Magnitude of the admittance in a parallel RLC circuit:

$$Y = \sqrt{G^2 + |B_C - B_L|^2} \quad (13\text{--}14)$$

Total current in a parallel RLC circuit:

$$I_{tot} = \sqrt{I_R^2 + |I_C - I_L|^2} \quad (13\text{--}15)$$

Resonant frequency:

$$f_r = \frac{1}{2\pi\sqrt{LC}} \quad (13\text{--}16)$$

Chapter Checkup

Answers are at the end of the chapter.

1. The reciprocal of impedance is
   (a) susceptance
   (b) admittance
   (c) conductance
   (d) reactance

2. The capacitive susceptance (B_C) in the phasor plot for a parallel circuit is plotted along the
   (a) positive x-axis
   (b) negative x-axis
   (c) positive y-axis
   (d) negative y-axis

3. Admittance is the phasor sum of
   (a) resistance and reactance
   (b) impedance and reactance
   (c) impedance and resistance
   (d) conductance and susceptance

4. When the frequency in a parallel RC circuit increases, the conductance will
   (a) decrease (b) stay the same
   (c) increase

5. When the frequency in a parallel RC circuit increases, the capacitive susceptance will
   (a) decrease (b) stay the same
   (c) increase

6. In a parallel RC circuit, if the value of C is increased, the current will
   (a) decrease (b) stay the same
   (c) increase

7. In a parallel RLC circuit, if B_C is larger than B_L, the source current will
   (a) lead the source voltage
   (b) be in phase with the source voltage
   (c) lag the source voltage

8. In a certain parallel RC circuit, the resistance is 3.0 kΩ, and the capacitive reactance is 4.0 kΩ. The total impedance will be
   (a) 1.71 kΩ (b) 2.4 kΩ
   (c) 5.0 kΩ (d) 7.0 kΩ

9. In a certain parallel RL circuit, the conductance is 1.0 mS and the inductive susceptance is 2.0 mS. The total admittance will be
   (a) 0.67 mS (b) 2.24 mS
   (c) 2.5 mS (d) 3.0 mS

10. In a certain parallel RLC circuit, the source current and the source voltage are in phase. The frequency is
    (a) below resonance (b) at resonance
    (c) above resonance (d) any of these answers

11. To tune a parallel resonant circuit to a lower frequency with a variable inductor, the inductor should be adjusted for
    (a) more inductance (b) less inductance

12. A parallel resonant band-pass filter has the output taken across the
    (a) tank circuit (b) resistor
    (c) none of the above

13. In a parallel resonant band-stop filter, assume that L is increased and C is decreased by the same amount. This will affect the
    (a) resonant frequency (b) selectivity
    (c) both of the above (d) none of the above

14. In a parallel resonant filter, raising the source voltage will cause the bandwidth to
    (a) increase (b) stay the same
    (c) decrease

15. In a parallel resonant filter, using a higher value for a load resistor will cause the bandwidth to
    (a) increase (b) stay the same
    (c) decrease

Questions

Answers to odd-numbered questions are at the end of the book.

1. Under what conditions can the product-over-sum rule be applied to an ac parallel circuit?
2. What is the reciprocal of resistance, reactance, and impedance?
3. What is the reciprocal of 1.0 mS?
4. What happens to capacitive susceptance if frequency is increased?
5. What happens to conductance if frequency is increased?
6. How do you convert an admittance phasor drawing to a current phasor drawing?
7. On which axis is inductive susceptance plotted?
8. For a parallel RC circuit, what happens to the admittance triangle if the frequency is increased?
9. Why are capacitive susceptance and inductive susceptance plotted in opposite directions?
10. In a parallel RLC circuit, under what conditions will the phase angle of current lag the voltage?
11. How is the admittance phasor diagram affected if a variable inductor in a parallel RL circuit is adjusted to a larger value?
12. In the equation $Y = \sqrt{G^2 + |B_C - B_L|^2}$, what do the vertical bars mean?
13. What is an ideal inductor?
14. What happens to the impedance as the frequency of a parallel resonant circuit is moved from below resonance to above resonance?
15. What happens to the admittance as the frequency of a parallel resonant circuit is moved from below resonance to above resonance?
16. How does a parallel resonant band-pass filter differ from the series resonant band-pass filter discussed in Chapter 12?
17. In what way does a load resistor affect the shape of the response of a parallel resonant filter?

PROBLEMS

Basic Problems

Answers to odd-numbered problems are at the end of the book.

1. What is the reactance of an inductor if the inductive susceptance is 100 μS?
2. Calculate the magnitude of the admittance and the phase angle for the RC circuit shown in Figure 13–29. Sketch the admittance phasor diagram.

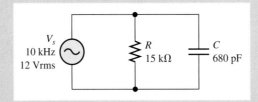

FIGURE 13–29

3. Assume the frequency in Problem 2 is changed to 15 kHz. Sketch the admittance phasor diagram.

4. Calculate the inductive susceptance for a 1.50 H inductor operating at a frequency of 60 Hz.

5. Calculate the impedance if a 33 kΩ resistor is in parallel with a capacitor that has a reactance of 25 kΩ.

6. Calculate the phase angle for the *RC* circuit in Problem 5.

7. A 200 μH inductor is in parallel with a 680 Ω resistor. The frequency is 400 kHz. What is the impedance?

8. A 470 pF capacitor is in parallel with a 5.6 kΩ resistor. The frequency is 50 kHz. What is the impedance of the circuit?

9. Calculate the phase angle if a parallel *RL* circuit has a 10 kΩ resistor in parallel with an inductor with a reactance of 2.5 kΩ.

10. Find the missing values on Table 13–7 for the circuit in Figure 13–30. Draw the admittance and current phasor diagrams.

Table 13–7

$R = 2.2$ kΩ	$G =$	$I_R =$	$V_R =$
$X_C =$	$B_C =$	$I_C =$	$V_C =$
$Z =$	$Y =$	$I_{tot} =$	$V_s = 10$ Vrms

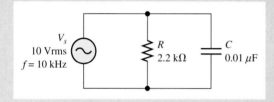

FIGURE 13–30

11. Find the missing values on Table 13–8 for the circuit in Figure 13–31. Draw the admittance and current phasor diagrams.

Table 13–8

$R = 8.2$ kΩ	$G =$	$I_R =$	$V_R =$
$X_L =$	$B_L =$	$I_L =$	$V_L =$
$Z =$	$Y =$	$I_{tot} =$	$V_s = 8.0$ V$_{rms}$

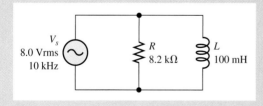

FIGURE 13–31

12. Complete Table 13–9 for the circuit shown in Figure 13–32. Draw the admittance phasor diagrams.

Table 13–9

$R = 3.3\ \text{k}\Omega$	$G =$	$I_R =$	$V_R =$
$X_C =$	$B_C =$	$I_C =$	$V_C =$
$X_L =$	$B_L =$	$I_L =$	$V_L =$
$Z =$	$Y =$	$I_{tot} =$	$V_s = 10\ \text{Vrms}$

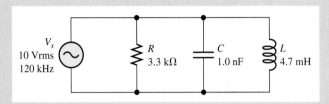

FIGURE 13–32

13. Find the admittance and the impedance of the circuit in Figure 13–33.

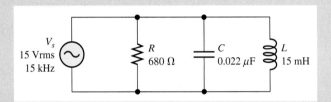

FIGURE 13–33

14. Calculate the current in each branch of the circuit in Figure 13–33 and the total current.

15. Calculate the resonant frequency of the circuit shown in Figure 13–34.

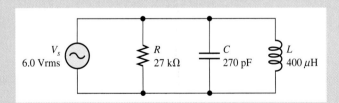

FIGURE 13–34

16. At resonance, what is the current from the source in Figure 13–34?

17. A parallel LC circuit has a 470 pF capacitor. The circuit resonates at 156 kHz. What is the value of the inductor?

Basic-Plus Problems

18. The parallel RC circuit in Figure 13-35 has a total current of 16.0 mA from the source.

    (a) What is the value of X_C?

    (b) What is the value of C?

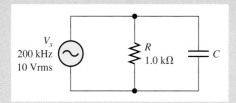

FIGURE 13-35

19. What is the frequency of a source if a 1.0 mH inductor has the same magnitude of current as a 3.3 kΩ resistor that is in parallel with it?

20. An RLC circuit is connected to a 10 kHz source and the current in the capacitor is four times that of the current in the inductor. To what frequency must the source be set for resonance?

21. Assume you replace an inductor in a resonant circuit with one that is 10% smaller than the original one. If the resonant frequency was 10 kHz before the change, what is the new resonant frequency?

22. Assume you need a band-stop filter for 19 kHz. You have available a 10 mH inductor, a 10 kΩ resistor, and a selection of capacitors. Draw the schematic for the filter, showing the capacitance you select.

Example Questions

ANSWERS

13-1: The frequency is smaller.

13-2: The admittance will be equal to the conductance of the resistor (1.78 mS).

13-3: The current leads the voltage.

13-4: $I = \sqrt{3.64 \text{ mA}^2 + 5.32 \text{ mA}^2} = 6.44$ mA

13-5: G will be smaller; B_C is unchanged. The result is that Y is smaller and the phase angle is larger.

13-6: I_C will be higher; therefore, the impedance is smaller.

13-7: KCL applies but the currents are phasors. The sum of the branch currents is equal to the source current.

13-8: The reactive current is the sum of the capacitive and inductive currents, which are out of phase. This causes the reactive current to be smaller than the capacitive current. Even adding the current in the resistor does not produce a current higher than the one in the capacitor alone.

13-9: The current in the capacitor will increase and the current in the inductor will decrease. The current in the resistor is unaffected.

Review Questions

1. Susceptance is the reciprocal of reactance; it can be either capacitive or inductive.

2. Admittance is the reciprocal of impedance.

3. The siemens.
4. *RC*
5. Only to two components in parallel; with ac you use phasor arithmetic.
6. The current in the resistor and the current in the capacitor are added as the square root of the sum of the squares.
7. The voltage
8. The admittance will be larger.
9. They are unaffected.
10. The phase shift is smaller.
11. The conductance and the inductive susceptance are added as the square root of the sum of the squares.
12. Because the current lags the voltage in an inductor. The negative *y*-axis implies a lagging current with current phasors and the admittance phasors have the same shape as the current phasors.
13. The admittance will be smaller.
14. The conductance is unaffected but the susceptance will be smaller. Hence, the admittance and phase angle will also be smaller.
15. Divide each current by the voltage.
16. The capacitive susceptance and inductive susceptance shift current by 90° but in opposite directions. The admittance plot indicates this difference by plotting them in different directions.
17. In terms of admittance:

$$\theta = \tan^{-1}\left(\frac{|B_C - B_L|}{G}\right)$$

In terms of current:

$$\theta = \tan^{-1}\left(\frac{|I_C - I_L|}{I_R}\right)$$

18. Current will lead the voltage if $B_C > B_L$; otherwise it will lag.
19. 0.194 mS
20. Leads
21. Impedance is a minimum in a series resonant circuit; it is maximum in a parallel resonant circuit.
22. 0°
23. Above resonance the current leads the voltage.
24. The output is taken across the tank circuit in a band-pass filter and across the resistor in a band-stop filter.
25. Selectivity is controlled by the resistance of the circuit and the specific values of *L* and *C*.

Chapter Checkup

1. (b)	2. (c)	3. (d)	4. (b)	5. (c)
6. (c)	7. (a)	8. (b)	9. (b)	10. (b)
11. (a)	12. (a)	13. (b)	14. (b)	15. (c)

APPENDIX A

Definitions of the Fundamental Units

Fundamental Unit	Definition
Meter (length)	The meter is the length of the path traveled by light in a vacuum during a time interval of 1/(299,792,458) second (1983).
Kilogram (mass)	The mass of the international prototype kilogram (1889).
Second (time)	The second is the duration of 9,192,631,770 periods of the radiation corresponding to the transition between two hyperfine levels of the ground state of the cesium 133 atom (1967).
Ampere (electric current)	The ampere is that constant current which, if maintained in two straight parallel conductors of infinite length, of negligible circular cross section, and placed 1 meter apart in a vacuum, would produce between these conductors a force equal to 2×10^{-1} newton per meter of length (1948).
Kelvin (temperature)	The kelvin unit of thermodynamic temperature is the fraction 1/273.16 of the thermodynamic temperature of the triple point of water (1967).
Candela (luminous intensity)	The candela is the luminous intensity, in a given direction, of a source that emits monochromatic radiation of frequency 540×10^{12} hertz and that has a radiant intensity in that direction of 1/683 watt per steradian (1979).
Mole (amount of substance)	The mole is the amount of substance of a system that contains as many elementary entities as there are atoms in 0.012 kilogram of carbon 12 (1971).

APPENDIX B

Table of Standard Resistor Values

Resistance Tolerance (±%)

0.1%				0.1%				0.1%				0.1%				0.1%				0.1%			
0.25%	1%	2%	10%	0.25%	1%	2%	10%	0.25%	1%	2%	10%	0.25%	1%	2%	10%	0.25%	1%	2%	10%	0.25%	1%	2%	10%
0.5%		5%		0.5%		5%		0.5%		5%		0.5%		5%		0.5%		5%		0.5%		5%	
10.0	10.0	10	10	14.7	14.7	—	—	21.5	21.5	—	—	31.6	31.6	—	—	46.4	46.4	—	—	68.1	68.1	68	68
10.1	—	—	—	14.9	—	—	—	21.8	—	—	—	32.0	—	—	—	47.0	—	47	47	69.0	—	—	—
10.2	10.2	—	—	15.0	15.0	15	15	22.1	22.1	22	22	32.4	32.4	—	—	47.5	47.5	—	—	69.8	69.8	—	—
10.4	—	—	—	15.2	—	—	—	22.3	—	—	—	32.8	—	—	—	48.1	—	—	—	70.6	—	—	—
10.5	10.5	—	—	15.4	15.4	—	—	22.6	22.6	—	—	33.2	33.2	33	33	48.7	48.7	—	—	71.5	71.5	—	—
10.6	—	—	—	15.6	—	—	—	22.9	—	—	—	33.6	—	—	—	49.3	—	—	—	72.3	—	—	—
10.7	10.7	—	—	15.8	15.8	—	—	23.2	23.2	—	—	34.0	34.0	—	—	49.9	49.9	—	—	73.2	73.2	—	—
10.9	—	—	—	16.0	—	16	—	23.4	—	—	—	34.4	—	—	—	50.5	—	—	—	74.1	—	—	—
11.0	11.0	11	—	16.2	16.2	—	—	23.7	23.7	—	—	34.8	34.8	—	—	51.1	51.1	51	—	75.0	75.0	75	—
11.1	—	—	—	16.4	—	—	—	24.0	—	24	—	35.2	—	—	—	51.7	—	—	—	75.9	—	—	—
11.3	11.3	—	—	16.5	16.5	—	—	24.3	24.3	—	—	35.7	35.7	—	—	52.3	52.3	—	—	76.8	76.8	—	—
11.4	—	—	—	16.7	—	—	—	24.6	—	—	—	36.1	—	36	—	53.0	—	—	—	77.7	—	—	—
11.5	11.5	—	—	16.9	16.9	—	—	24.9	24.9	—	—	36.5	36.5	—	—	53.6	53.6	—	—	78.7	78.7	—	—
11.7	—	—	—	17.2	—	—	—	25.2	—	—	—	37.0	—	—	—	54.2	—	—	—	79.6	—	—	—
11.8	11.8	—	—	17.4	17.4	—	—	25.5	25.5	—	—	37.4	37.4	—	—	54.9	54.9	—	—	80.6	80.6	—	—
12.0	—	12	12	17.6	—	—	—	25.8	—	—	—	37.9	—	—	—	56.2	—	—	—	81.6	—	—	—
12.1	12.1	—	—	17.8	17.8	—	—	26.1	26.1	—	—	38.3	38.3	—	—	56.6	56.6	56	56	82.5	82.5	82	82
12.3	—	—	—	18.0	—	18	18	26.4	—	—	—	38.8	—	—	—	56.9	—	—	—	83.5	—	—	—
12.4	12.4	—	—	18.2	18.2	—	—	26.7	26.7	—	—	39.2	39.2	39	39	57.6	57.6	—	—	84.5	84.5	—	—
12.6	—	—	—	18.4	—	—	—	27.1	—	27	27	39.7	—	—	—	58.3	—	—	—	85.6	—	—	—
12.7	12.7	—	—	18.7	18.7	—	—	27.4	27.4	—	—	40.2	40.2	—	—	59.0	59.0	—	—	86.6	86.6	—	—
12.9	—	—	—	18.9	—	—	—	27.7	—	—	—	40.7	—	—	—	59.7	—	—	—	87.6	—	—	—
13.0	13.0	13	—	19.1	19.1	—	—	28.0	28.0	—	—	41.2	41.2	—	—	60.4	60.4	—	—	88.7	88.7	—	—
13.2	—	—	—	19.3	—	—	—	28.4	—	—	—	41.7	—	—	—	61.2	—	—	—	89.8	—	—	—
13.3	13.3	—	—	19.6	19.6	—	—	28.7	28.7	—	—	42.2	42.2	—	—	61.9	61.9	62	—	90.9	90.9	91	—
13.5	—	—	—	19.8	—	—	—	29.1	—	—	—	42.7	—	—	—	62.6	—	—	—	92.0	—	—	—
13.7	13.7	—	—	20.0	20.0	20	—	29.4	29.4	—	—	43.2	43.2	43	—	63.4	63.4	—	—	93.1	93.1	—	—
13.8	—	—	—	20.3	—	—	—	29.8	—	—	—	43.7	—	—	—	64.2	—	—	—	94.2	—	—	—
14.0	14.0	—	—	20.5	20.5	—	—	30.1	30.1	30	—	44.2	44.2	—	—	64.9	64.9	—	—	95.3	95.3	—	—
14.2	—	—	—	20.8	—	—	—	30.5	—	—	—	44.8	—	—	—	65.7	—	—	—	96.5	—	—	—
14.3	14.3	—	—	21.0	21.0	—	—	30.9	30.9	—	—	45.3	45.3	—	—	66.5	66.5	—	—	97.6	97.6	—	—
14.5	—	—	—	21.3	—	—	—	31.2	—	—	—	45.9	—	—	—	67.3	—	—	—	98.8	—	—	—

NOTE: These values are generally available in multiples of 0.1, 1, 10, 100, 1 k, and 1 M.

APPENDIX C

Lab Manual Figures

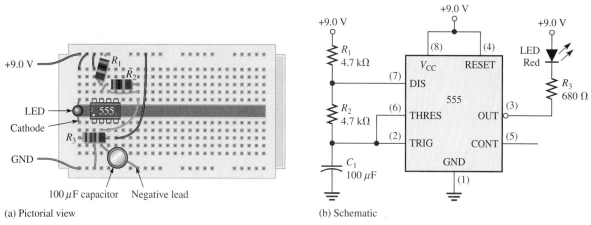

(a) Pictorial view

(b) Schematic

Figure 2–5 Light blinker.

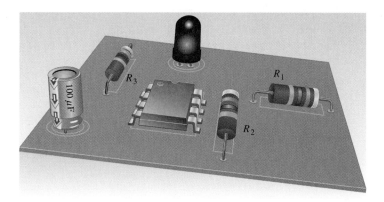

Figure 2–6 3-D view of the light blinker (Picture adapted from the computer program Ultiboard).

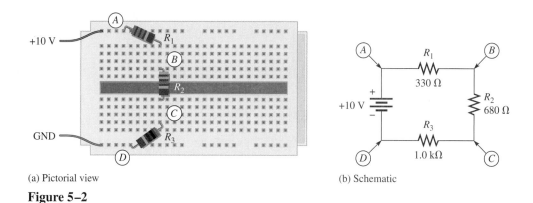

(a) Pictorial view

(b) Schematic

Figure 5–2

477

LAB MANUAL FIGURES

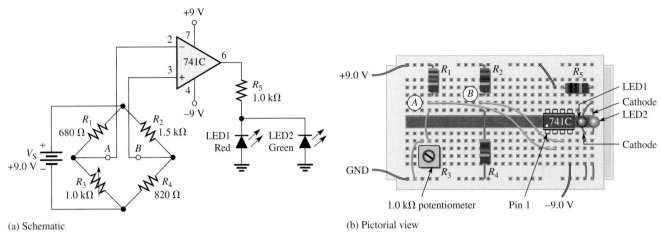

(a) Schematic (b) Pictorial view

Figure 12–4

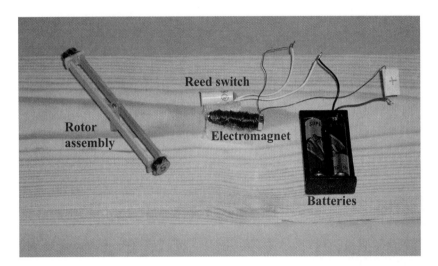

Figure 15–5

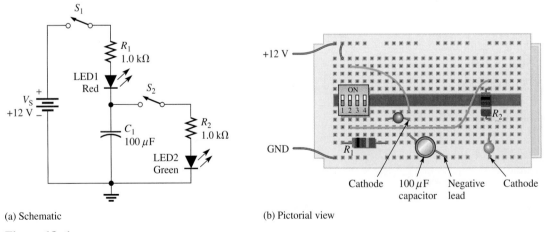

(a) Schematic (b) Pictorial view

Figure 18–1

ANSWERS TO ODD-NUMBERED QUESTIONS

Chapter 1

1. Small size and low power dissipation
3. Electronic engineers are generally involved in the design of new products, including testing, manufacturing, and installation of electrical and electronic equipment.
5. The National Fire Protection Association (NFPA)
7. None
9. Potential energy has the ability to do work because of position or configuration. Kinetic energy has the ability to do work because of motion.
11. Chemical potential energy
13. Only a very tiny fraction of the total number of electrons in a substance can be transferred by friction. Only these electrons are involved in electrostatic experiments.
15. Coulomb's law
17. The atomic number is equal to the number of protons in the nucleus.
19. Metals have one, two, or three outer shell electrons that are loosely bound to the atom. Nonmetals do not have free electrons.

Chapter 2

1. 50×10^{-6}
3. 10^6
5. 10^9
7. Any unit in the SI system that is not one of the seven fundamental or two supplementary units
9. A working standard is an instrument or device that is used for routine calibration and certification of test equipment or component.
11. Long-term stability, accuracy, and insensitivity to environmental conditions
13. No. Accuracy requires that a measurement is very close to an accepted value. If a measured value is imprecise, the various measurements used to determine precision are scattered and cannot be trusted for accuracy.
15. A binomial is an expression with two terms.
17. Starting with quadrant 1 on the top right side, they are numbered counterclockwise.
19. It is not necessary to have a line on a graph pass through every data point (except in calibration curves). The line should be drawn as a smooth curve to show the trend of the data.

Chapter 3

1. Metals are held by metallic bonds, in which electrons hold the positive ions of the metal together. The electrons are free to move about.
3. In a mercury vapor lamp, the charge carriers are negative electrons and positive mercury ions.
5. A diode
7. A plasma is a state of matter in which all of the electrons are stripped from the atoms. Plasmas behave so differently from ordinary gases that they have their own classification.

9. Electrons leave the negative terminal of a battery, travel through a spring, a metal strip, a switch, and a metal reflector to the bulb. They return from the bulb to the positive side of the battery.
11. Hertz
13. RMS means root-mean-square.
15. The "charging" process is actually causing a reversible chemical reaction to store chemical energy, not charge.
17. 20 hours
19. It produces less heat.
21. A superconductor is a specialized material that is capable of carrying current with no resistance but must be kept extremely cold.
23. A rheostat can be connected as a potentiometer only if it has three terminals. The common connection between one side and the variable terminal needs to be opened to allow connections to all three terminals.
25. Linear taper means that the resistance is proportional to the angular position of the control; nonlinear taper means that the resistance is not proportional to the angular position.
27. The tolerance is 2%.
29. No
31. American Wire Gauge
33. A meter that selects the optimum range automatically for displaying the reading
35. A multimeter is connected for measuring current by selecting the current function and moving the leads to the current jacks. The meter is connected in series with the circuit, never in parallel.
37. Normally it is a nonlinear scale with increasing values from right to left.

Chapter 4

1. It increases.
3. 4.95 V
5. 1.39 mA
7. 16.4 V
9. 11 MΩ
11. 0.1 S
13. 46.9 Ω
15. It is a unit of energy that is equivalent to a power level of 1 kilowatt maintained for one hour.
17. The horsepower is a unit of power equivalent to 746 W. It is frequently used to specify motor power.
19. 18 cents
21. 21.8 A
23. 3.16 A
25. It increases.

ANSWERS TO ODD-NUMBERED QUESTIONS

Chapter 5

1. 1.67 MΩ
3. 1.83 MΩ
5. 28.2 kΩ
7. 270 Ω
9. 0.75 mA in each
11. 2.0 V
13. 18 V
15. The voltage drop across any given resistor in a series circuit is equal to the ratio of that resistor to the total resistance, multiplied by source voltage.
17. The end terminals are connected between V_S and ground. The output is connected between the center terminal and ground.
19. 5.83 mS
21. 405 kΩ
23. The sum of the currents entering a node is equal to the sum of the currents leaving the node.
25. 300 mA

Chapter 6

1. Add R_3 and R_4 because they are in series. Combine this result in with parallel R_2, then add the resistance of R_1. As a formula, $R_T = [(R_3 + R_4) \parallel R_2] + R_1$.
3. They will sum to zero.
5. They will sum to zero.
7. Current in the equivalent resistance and the series resistor are identical.
9. None are in series
11. R_2 is directly across the source, so the current in it can be found directly by applying Ohm's law. $I_2 = \dfrac{V_S}{R_2}$
13. A voltage source and a series resistor
15. Yes
17. Because it is in parallel with the voltage source and cannot affect the output
19. The divider in (a) is "stiff" for a 10 kΩ load, because the load is at least 10X larger than the divider resistors and will have little effect on the output voltage. The others are not.
21. The meter has the least loading affect on the circuit with the smallest equivalent resistance — (a).
23. The sum is an algebraic sum and therefore the sign is necessary to obtaining a correct result.
25. 1. Remove the load resistor.
    2. Calculate the voltage between A and ground and B and ground. These voltages are the Thevenin voltages for the left and right side of the bridge.

ANSWERS TO ODD-NUMBERED QUESTIONS

3. Calculate the resistance between *A* and ground and *B* and ground with the voltage source replaced with a short. These resistances are the Thevenin resistances for the left and right side of the bridge.

4. Draw two facing Thevenin circuits, one for each side of the bridge.

27. 110 Vac. There is no current in the circuit, therefore the other bulbs have no voltage drops. Kirchhoff's voltage law shows that the input voltage must appear across the open.

Chapter 7

1. Moving charge

3. The ampere-turn/meter

5. *Permeability* is a measure of the flux density (B) that will occur for a given magnetic field intensity (H) and depends on the material. *Relative permeability* is a ratio of the permeability of a given material to that of a vacuum.

7. Both are oppositions. Reluctance is the opposition to establishing magnetic flux; resistance is the opposition to current.

9. Residual magnetism occurs in ferromagnetic materials because of hysteresis; a magnetic field remains in the material after the magnetic field intensity (H) has been removed.

11. Current

13. An air gap adds significant reluctance to a magnetic path; a smaller air gap has a smaller reluctance, making it easier to establish a magnetic field.

15. An electrically activated valve that is used to control the flow of air, water, steam, refrigerants, and other fluids

17. A large relay designed for high voltages and/or current

19. Switches can be shown in series with a load or other switches on a "rung" of the ladder diagram. They are never in parallel with a load or relay coil.

21. A line diagram

Chapter 8

1. Yes. If the conductor is moved parallel to magnetic field lines, no voltage is induced.

3. The induced voltage increases.

5. The rate of change of the magnetic flux with respect to the coil and the number of turns of wire in the coil.

7. The field is intensified between the conductors and they are pushed apart.

9. In a stationary field generator, the magnetic field is stationary and the conductor rotates. In a rotating field generator, the magnetic field rotates and the conductor is stationary.

11. The stator

13. A wound rotor produces higher flux density and can have more power than a permanent magnet.

15. An air gap is for clearance for the rotor to move.

17. Across from each other (180°)

19. From the changing electrical ac waveform

21. A connection method that uses three coils connected in series like a triangle

23. A prony brake is a test instrument that consists of a drum and braking arrangement for testing motors. It measures motor torque.

25. A shunt-wound motor
27. An induction motor
29. If the rotor could turn at the synchronous speed, it would not cut any field lines and the torque would drop to zero.
31. The bars of the cage are the moving conductors of the rotor.

Chapter 9

1. All other periodic waves can be constructed from a set of sine waves of the appropriate amplitude and frequency.
3. 90°
5. Rotating electrical machines (alternators) and electronic oscillators
7. Each is the reciprocal of the other.
9. A phasor is represented by an arrow; the magnitude is represented by its length and the direction is represented by where it is pointing.
11. A phase shift only has meaning if it is compared to another waveform. The reference waveform is used as the waveform for comparison.
13. No. If the frequencies are different, there is a constantly changing phase between the waves.
15. A pulse is a very rapid transition from one voltage to another and then, after an interval of time, a very rapid transition back to the original level.
17. For a pulse, the amplitude is its height measured from the baseline.
19. A sawtooth waveform
21. Radio waves with frequencies higher than about 1 GHz.
23. The dc offset control adds or subtracts a dc component to a waveform.
25. Analog and digital
27. Whenever you want to synchronize the display with the ac line voltage, such as looking for ripple on a power supply.
29. A small variable capacitor is adjusted on the probe itself or in a small box connected to the oscilloscope.
31. Five automated measurements are voltage, peak-to-peak voltage, average voltage, frequency, and period.

Chapter 10

1. A dielectric is an insulating material. In a capacitor it is the material between two conducting surfaces.
3. Micro (μ), which means 10^{-6} and pico (p), which means 10^{-12}
5. 100,000 pF
7. Farads/meter (F/m)
9. A spark goes across the dielectric. This can be a temporary failure if the dielectric is oil, or a permanent failure if the dielectric is a solid material.
11. Ceramic
13. Aluminum and tantalum
15. They are polarized, have generally low breakdown voltages, have high leakage current, and typically have 20% tolerances.
17. 2200 pF

19. 180 μF
21. 3579 pF
23. 103 ms
25. Five
27. Capacitive reactance
29. It increases
31. At the zero crossing points
33. To smooth variations in the dc
35. Provide a low reactance path around a resistor without affecting the dc

Chapter 11

1. It is quadrupled.
3. Stray capacitance is between the turns due to conductors separated by an insulator. It is normally a problem only at very high frequencies.
5.
7. 47,000 μH $\pm$5%
9. The magnetic field collapses, generating a voltage.
11. 3L
13. Set the frequency low enough to see the entire waveform and view the waveform across the resistor. Measure the time from starting voltage to 63% of the final voltage.
15. 2.2
17. The voltage leads the current by 90°.

Chapter 12

1. Numbers that cannot be written as the ratio of integers
3. -1
5. Rectangular notation describes a complex number as the sum of a real part and an imaginary part. Polar notation describes a complex number in terms of a length (magnitude) and an angle (direction).
7. "The angle whose tangent is"
9. $-45°$
11. It decreases.
13. Multiply by the current in the circuit.
15. $X_L = X_C$
17. A circuit designed to block a group of frequencies
19. Selectivity defines how well a resonant circuit passes certain frequencies and rejects others. The narrower the bandwidth, the greater the selectivity.
21. High Q implies high selectivity and narrow bandwidth.

Chapter 13

1. It is applied only to two components and the fact that they are phasors must be taken into account (except for the special case of two resistors).
3. 1.0 kΩ
5. Nothing; it is unchanged.
7. The –j axis
9. Because they shift the phase of current in opposite directions
11. The susceptance is decreased; therefore, the admittance will decrease and the phase angle will be smaller.
13. It is an inductor with no resistance or stray capacitance. Although an ideal inductor does not exist, it is a useful approximation for many circuits.
15. The admittance falls to a minimum at resonance, then rises after resonance.
17. It causes the band-pass to be wider and not as peaked.

ANSWERS TO ODD-NUMBERED PROBLEMS

Chapter 1

1. 66,000 N-m
3. 2.97 N

Chapter 2

1. (a) 7.3×10^4 (b) 2.2×10^{-4}
   (c) -9.2×10^7 (d) -5.1×10^{-3}
   (e) 4.7×10^5
3. (a) 160,000 (b) 0.0056
   (c) $-800,000$ (d) -0.071
   (e) 0.42
5. (a) 220 kΩ (b) 47 μF
   (c) 5.0 kW (d) 55 ns
   (e) 1.5 MHz (f) 10 mH
7. (a) 2200 kΩ (b) 0.100 μs
   (c) 10 MW (d) 0.220 kΩ
   (e) 4000 μV (f) 0.000 100 μs
9. (a) 50,5̲00 (b) 22̲0
   (c) 465̲0 (d) 11.0
   (e) 1.00
11. $l = RA/\rho$
13. $C = L/Z_0^2$
15. See Figure ANS–1.

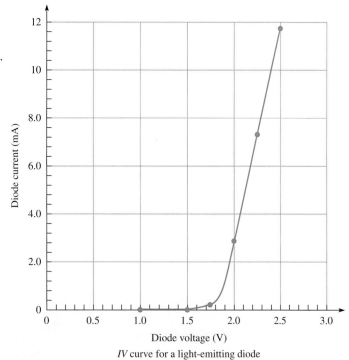

IV curve for a light-emitting diode

FIGURE ANS–1

17. (a) 2.25×10^3 (b) 2.41×10^6
    (c) 16.5×10^3 (d) 5.35×10^{-3}
19. (a) $2.67 \text{ M}\Omega$ (b) 20 nF
    (c) 670 MHz (d) 10.65 nF
21. $a = \sqrt{c^2 - b^2}$
23. $N = \sqrt{\dfrac{lL}{\mu A}}$
25. $C_1 = \dfrac{1}{\dfrac{1}{C_T} - \dfrac{1}{C_2}}$
27. See Figure ANS–2.

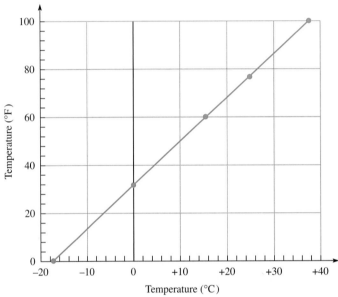

FIGURE ANS–2

Chapter 3

1. 20 A
3. 2 s
5. 20 h
7. 15 W
9. 375 W
11. (a) $3.6 \text{ k}\Omega \pm 5\%$ (b) $680 \text{ k}\Omega \pm 10\%$
    (c) $2.2 \text{ }\Omega \pm 5\%$ (d) $10 \text{ M}\Omega \pm 10\%$
13. (a) green, violet, blue, red, brown
    (b) white, brown, black, black, red
    (c) red, blue, brown, gold, brown
    (d) violet, green, black, brown, red

15. 475 ft
17. 0.276 Ω
19. 0.180 Ω-mm²/m

Chapter 4

1. (a) 37.0 mA (b) 2.2 A
   (c) 0.244 mA (d) 1.30 μA
3. (a) 6.0 V (b) 15.0 V
   (c) 110 V (d) 9.92 V
5. (a) 1.0 MΩ (b) 270 kΩ
   (c) 10 Ω (d) 3.91 kΩ
7. (a) 100 mS (b) 370 μS
   (c) 100 nS
9. (a) The current will drop to 1/3 of its original value.
   (b) The current will be three times its original value.
   (c) The current will be 1/2 its original value.
   (d) The current will be twice its original value.
11. (a) 20.0 V (b) 30.2 V
13. $I_{min} = 81.3$ mA; $I_{max} = 89.8$ mA
15. 0.833 A
17. (a) 1.1 kW (b) 107 mW
    (c) 3.7 W (d) 504 mW
19. 545 mA
21. 2.00 kWh
23. $ 3.60
25. $ 7.31
27. See Figure ANS–3.

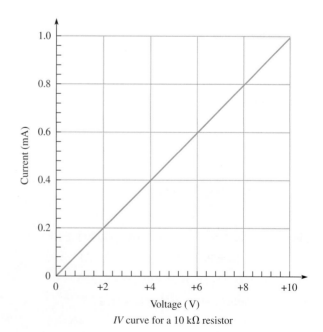

FIGURE ANS–3

29. 12.8 V
31. For $V_{GS} = 0$ V, $r_{ds} = 400\ \Omega$;
    For $V_{GS} = -1.5$ V, $r_{ds} = 800\ \Omega$;
    For $V_{GS} = -2.5$ V, $r_{ds} = 1.67\ k\Omega$;
    For $V_{GS} = -3$ V, $r_{ds} = 10\ k\Omega$

Chapter 5

1. 297 kΩ
3. 27 kΩ
5. 21.8 kΩ
7. 0.459 mA
9. 3.33 kΩ
11. 6.4 V
13. 2.0 V
15. 359 Ω
17. 1.08 kΩ
19. 77.6 Ω
21. 6.4 mA
23. $R_1 = 8.29$ kΩ; $I_1 = 2.89$ mA
25. 30 Ω
27. See Figure ANS–4.

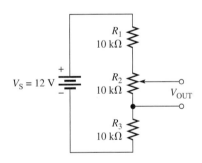

FIGURE ANS–4

29. $I_3 = 7.1$ mA
31. 11.8 mA
33. For each lamp, $I = 0.87$ A.
    $I_T = 2.6$ A

ANSWERS TO ODD-NUMBERED PROBLEMS

Chapter 6

1. See Table ANS–1.

   Table ANS–1

$I_1 = 1.0$ mA	$R_1 = 3.3$ kΩ	$V_1 = 3.3$ V
$I_2 = 0.67$ mA	$R_2 = 10$ kΩ	$V_2 = 6.7$ V
$I_3 = 0.33$ mA	$R_3 = 10$ kΩ	$V_3 = 3.3$ V
$I_4 = 0.33$ mA	$R_4 = 10$ kΩ	$V_4 = 3.3$ V
$I_T = 1.0$ mA	$R_T = 9.97$ kΩ	$V_S = 10.0$ V

3. $R_L = 825$ Ω

5. See Table ANS–2.

   Table ANS–2

$I_1 = 7.35$ mA	$R_1 = 180$ Ω	$V_1 = 1.32$ V
$I_2 = 2.59$ mA	$R_2 = 510$ Ω	$V_2 = 1.32$ V
$I_3 = 9.94$ mA	$R_3 = 270$ Ω	$V_3 = 2.68$ V
$I_4 = 9.94$ mA	$R_4 = 100$ Ω	$V_4 = 0.99$ V
$I_T = 9.94$ mA	$R_T = 503$ Ω	$V_S = 5.0$ V

7. $V_L = 11.8$ V
9. 8.31 V
11. (a) 5.50 V (b) 116 Ω (c) 116 Ω (d) 65.3 mW
13. 47.5 mA
15. $V_{AB} = 0.61$ V
17. $R_3 = 1.186$ kΩ
19. R_4 open
21. $R_2 = 400$ Ω
23. $R_L = 176$ Ω (minimum value)
25. $I_2 = -0.477$ mA
27. $I_S = 2.82$ mA

Chapter 7

1. 7.5 mWb
3. 390 At
5. 1.0×10^7 At/Wb
7. 1875 At/m
9. (a) 80 At (b) 64 t
11. 44.8×10^6 At/Wb
13. 7.0×10^{-4} Wb/At-m
15. 20.0 μWb
17. 1.56 A

ANSWERS TO ODD-NUMBERED PROBLEMS

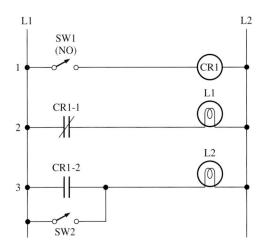

FIGURE ANS–5

19. See Figure ANS–5.

21. If CR1 is not energized and down limit switch is closed, momentarily pressing SW3 (CLOSE) causes CR2 to energize, latching contacts CR2-2. If down limit switch opens, power is removed from CR2.

23. See Figure ANS–6.

25. Answers will vary.

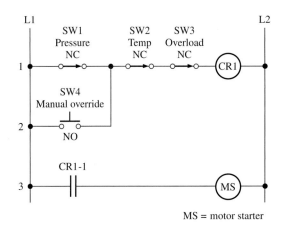

FIGURE ANS–6

Chapter 8

1. 135 mV
3. 0.184 T
5. 0.633 T
7. 0.096 N
9. 6000 rpm
11. 0.36 hp
13. (a) Positive (with respect to other end) (b) The induced force will oppose the motion; it is downward.
15. Four poles. Each pair of poles produces 1 cycle. When rectified, each cycle produces 2 pulses.

491

ANSWERS TO ODD-NUMBERED PROBLEMS

Chapter 9

1. (a) 0.52 rad (b) 1.57 rad
   (c) 2.62 rad (d) 4.36 rad
   (e) 5.67 rad
3. (a) 45° (b) 120°
   (c) 86° (d) 143°
   (e) 360°
5. 42 Vpp
7. 200 W
9. 4.0 kHz
11. (a) 0.333 ms (b) 3.0 kHz
13. 12.0 Vrms
15. (a) 4.23 V (b) 9.66 V
    (c) −7.07 V (d) −5.00 V
17. (a) 35 V (b) 22°
    (c) lagging
19. 6.0 Vpp
21. (a) 120 mVrms (b) 16.7 kHz
23. (a) 25 μs (b) 6.25 μs
25. 0.5 ms/div

Chapter 10

1. (a) 0.068 μF (b) 1000 μF
   (c) 0.0047 μF
3. 5.0 μF
5. 100 V
7. 0.24 μC
9. 195 pF
11. (a) 147 μF (b) 2.12 nF
    (c) 600 nF
13. 4000 μF
15. $V_{C1} = 1.18$ V; $V_{C2} = 16.8$ V
17. 10.3 V
19. 0.214 mA
21. 3.7 V
23. 2.7 ms
25. (a) 159 kΩ (b) 15.9 kΩ
    (c) 1.59 kΩ (d) 159 Ω
27. 15.9 kHz
29. 0.207 mA

ANSWERS TO ODD-NUMBERED PROBLEMS

31. $C_{oil} = 1200$ pF
33. (a) $\varepsilon_r = 4.92$ (b) mica
35. $V_{C1} = 6.0$ V; $V_{C2} = 4.0$ V; $V_{C3} = 10$ V
37. 4.83 V
39. Answers will vary.

Chapter 11

1. 0.270 mH
3. 285 μH
5. 35.4 μH
7. 7.5 mH
9. 130 μH
11. (a) $\tau = 1.5$ ms (b) $V_R = 8.6$ V
    (c) $I = 10.0$ mA
13. 1.88 kΩ
15. $L_{tot} = 3.58$ mH $X_{L(tot)} = 562$ Ω
17. 144 turns
19. 821 μH
21. 3.95 kHz

Chapter 12

1. See Figure ANS–7.

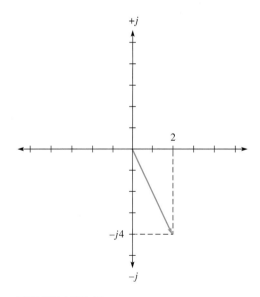

FIGURE ANS–7

493

ANSWERS TO ODD-NUMBERED PROBLEMS

3. See Figure ANS–8.

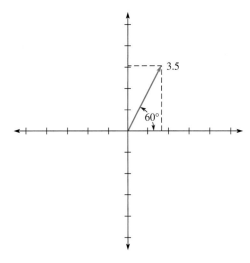

FIGURE ANS–8

5. 4.12
7. 1.36
9. See Figure ANS–9.

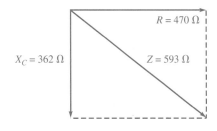 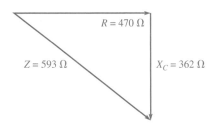

FIGURE ANS–9

11. 16.9 mA
13. See Table ANS–3 and Figure ANS–10.

Table ANS–3

$R = 1.0\ \text{k}\Omega$	$I_R = 7.82\ \text{mA}$	$V_R = 7.82\ \text{Vrms}$
$X_C = 796\ \Omega$	$I_C = 7.82\ \text{mA}$	$V_C = 6.23\ \text{Vrms}$
$Z = 1.28\ \text{k}\Omega$	$I_{tot} = 7.82\ \text{mA}$	$V_s = 10\ \text{Vrms}$

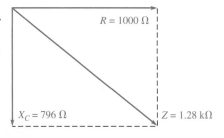

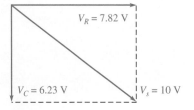

FIGURE ANS–10

ANSWERS TO ODD-NUMBERED PROBLEMS

15. See Table ANS–4 and Figure ANS–11.

Table ANS–4

$R = 1.5$ kΩ	$I_R = 0.692$ mA	$V_R = 1.04$ Vrms
$X_L = 7.07$ kΩ	$I_L = 0.692$ mA	$V_L = 4.89$ Vrms
$Z = 7.23$ kΩ	$I_{tot} = 0.692$ mA	$V_s = 5.0$ Vrms

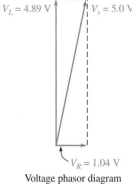

FIGURE ANS–11

17. See Table ANS–5.

Table ANS–5

$R = 270$ Ω	$I_R = 7.59$ mA	$V_R = 2.05$ Vrms
$X_C = 135$ Ω	$I_C = 7.59$ mA	$V_C = 1.03$ Vrms
$X_L = 424$ Ω	$I_L = 7.59$ mA	$V_L = 3.22$ Vrms
$Z = 396$ Ω	$I_{tot} = 7.59$ mA	$V_s = 3.0$ Vrms

19. $X_L = 170$ Ω; $X_C = 338$ Ω; Because $X_C > X_L$, the circuit will no longer look inductive.
21. 45.5 kHz
23. (a) 4.0 V   (b) 53.1°
25. $V_L = 4.47$ V; $V_C = 8.94$ V; $V_R = 8.94$ V
27. 51 μH

Chapter 13

1. 10 kΩ
3. $Y = 92.5$ μS; $\theta = 43.9°$; See Figure ANS–12.

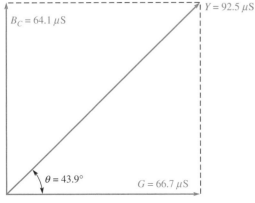

Admittance phasor diagram

FIGURE ANS–12

ANSWERS TO ODD-NUMBERED PROBLEMS

5. 19.9 kΩ
7. 404 Ω
9. 76°
11. See Table ANS–6 and Figure ANS–13.

Table ANS–6

$R = 8.2$ kΩ	$G = 122$ μS	$I_R = 0.976$ mA	$V_R = 8.0$ Vrms
$X_L = 6.28$ kΩ	$B_L = 159$ μS	$I_L = 1.27$ mA	$V_L = 8.0$ Vrms
$Z = 4.99$ kΩ	$Y = 200$ μS	$I_{tot} = 1.60$ mA	$V_s = 8.0$ Vrms

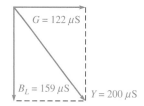 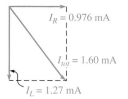

Admittance phasor diagram   Current phasor diagram

FIGURE ANS–13

13. $Y = 2.01$ mS; $Z = 498$ Ω
15. 484 kHz
17. 2.2 mH
19. 525 kHz
21. 10.54 kHz

GLOSSARY

Abscissa The *x*-axis (horizontal axis) in the Cartesian coordinate system.

Accuracy The difference between the measured value and accepted or true value of a measurement.

AC resistance The resistance at a point on the *IV* curve for a device if a small *change* in voltage is divided by a corresponding *change* in current. AC resistance is designated with the letter *r*.

Admittance The phasor sum of conductance and susceptance; it indicates how readily current can be established.

Alternating current (ac) Current that changes direction or "alternates" back and forth between positive and negative values.

Alternation A half-cycle of a sinusoidal wave.

Alternator An ac generator.

American Wire Gauge (AWG) A specification of wire size; the larger the gauge number, the smaller the wire.

Ampere The unit for current; the fundamental electrical unit. One ampere is one coulomb per second.

Ampere-hours (Ah) A measure of the capacity of a battery to deliver current over time.

Amplitude The peak value of a sinusoidal wave; it is measured from the center to its maximum. For a pulse, it is its height measured from the baseline.

Armature That part of an alternator in which a voltage is induced due to the relative motion between a conductor and a magnetic field.

Armature coil A conductor wound on the armature of a motor or generator for the purpose of providing a magnetic field.

Armature reaction The distortion of the magnetic field from the field coils due to an additional field created by the armature current.

Atomic number The number of protons in the nucleus of an atom; the atomic number identifies the specific element.

Attenuator A calibrated resistive divider that reduces a waveform by a known amount.

Autoranging A feature on a DMM in which the meter automatically selects the optimum range for displaying the reading.

Average value For a sinusoidal wave, the average of all points assuming all values are positive. It is 0.637 times the peak value.

Back emf (electromotive force) A voltage developed across a spinning armature that opposes the original applied voltage.

Balanced With respect to a Wheatstone bridge, the condition where the output voltage is zero.

Band-pass filter A circuit designed to pass frequencies around resonance and reject other frequencies.

Band-stop filter A circuit designed to reject frequencies around resonance and pass other frequencies.

Bandwidth The range of frequencies for which the output voltage is 70.7% or more of the maximum value.

Battery A dc voltage source that converts chemical energy to electrical energy.

Binomial An expression that contains two terms.

Branch One current path in a parallel circuit.

Brushes Conductive contacts usually made from carbon or carbon-graphite that make contact with the moving slip rings and provide an electrical path from a source to the rotor of a generator or motor.

Capacitance The ability to store a charge by means of an electrostatic field.

Capacitive reactance Opposition to ac by a capacitor.

Capacitor A fundamental component that has two conductive plates separated by an insulating region called the dielectric. A capacitor has the ability to store an electric charge.

Cell The basic unit of a battery. A single cell will have a certain fixed voltage when new.

Closed shell An energy level for electrons within the atom that contains the maximum permissible number of electrons.

Coefficient A number written in front of a term.

Commutator A segmented ring that is used in dc generators and motors to reverse the polarity of the rotor.

Complex number A number that is described by both a real and an imaginary term (rectangular notation) or by showing a magnitude and direction (polar notation).

Compound A substance in which two or more elements combine in a definite proportion.

Conductance The reciprocal of resistance, it is a measure of the ability of a material to conduct current. The unit for conductance is the siemens, S.

Conduction band A band of energies in the energy level of an atom just above the highest normally occupied level (the valence band).

Conduction electron An electron that is not bound to any particular atom.

Conductor A material that allows the free movement of charge; it can be a solid, liquid, or gas.

Constant A quantity that never changes.

Conventional current The movement of charge. The assumed direction is from the positive terminal to the negative terminal of the source.

GLOSSARY

Current The rate charge moves past a point.

Current divider A parallel circuit in which the current entering a node is divided into different branches.

DC offset A control on a function generator that adds or subtracts a dc component to the output signal.

DC resistance The resistance of a device that is found by dividing the voltage at some point by the corresponding current. DC resistance is designated with the letter R.

Dependent variable A variable that is influenced by changes in another variable.

Derived unit A unit that uses two or more fundamental or supplementary units in its definition.

Dielectric The insulating material that is between the conductive plates of a capacitor.

Dielectric constant Another name for relative permittivity.

Digital multimeter (DMM) An instrument that can measure ac or dc voltage, ac and dc current, and resistance and shows the result as a number in a display.

Diode A device that passes current in only one direction.

Direct current (dc) Current that is uniform in one direction.

Duty cycle The ratio of the pulse width to the period for a pulse waveform. It is often expressed as a percentage.

Efficiency The ratio of the output power, P_{OUT}, to the input power, P_{IN} of a device.

Electrolyte (1) A material that forms ions in water solution, thus becoming a good conductor. (2) A conductive material in which charge movement is by ions. The electrolyte in combination with a plastic film forms the negative plate of an electrolytic capacitor.

Electromagnet A temporary magnet formed by passing a current through a coil which is wrapped around a magnetic core.

Electron A negatively charged subatomic particle that accounts for the flow of charge in solid conductors. In the Bohr model, electrons orbit the nucleus at discrete distances.

Electron flow The movement of electrons in a conductor. The assumed direction is from the negative terminal to the positive terminal of the source.

Energy The ability or capacity for doing work.

Engineering notation A method of writing a large or small number as an ordinary number (between 1 and 1000) times a power of ten. The exponent can only take on values that are evenly divisible by three.

Equation A mathematical statement that sets one expression or quantity equal to another.

Equivalent circuit A circuit that has characteristics that are electrically the same as another circuit and is generally simpler than the original circuit.

Error The difference between the true or best-accepted value of some quantity and the measured value.

Exciter An external source that supplies dc power for the field coils of an electrical device (such as an alternator).

Expression One or more terms.

Fall time The time required for a pulse to go from 90% to 10% of its amplitude.

Farad The unit of capacitance; one farad is the capacitance when a charge of one coulomb is stored with a potential difference of one volt across the plates.

Faraday's law A law stating that the voltage that is induced across a wire that is moving relative to a magnetic field is determined by the strength of the field, length of the conductor, speed of the conductor, and the angle it makes with respect to the magnetic field. For a coil, Faraday's law gives the voltage as the product of the number of turns and the rate of change of magnetic flux.

Field coil A winding on a motor or generator for the purpose of providing a magnetic field; it is wound on a field pole.

Filter A low-pass network that smooths out variations in dc.

Fixed point A reference to numbers that are expressed with a decimal point and a number of zeros as needed.

Floating point A number written with the decimal point not in a fixed location. Floating point numbers are frequently written using scientific notation.

Flux Imaginary lines used to describe the magnitude and shape of a magnetic field.

Flux density A value that is proportional to the number of lines of flux that are perpendicular to a given area.

Frequency The number of complete cycles that occur in one second; it is measured in hertz.

Function generator A versatile instrument for producing a variety of low-frequency waveforms.

Fundamental unit One of the seven measurement units (along with two supplementary units) that normally can be defined independently of all other units. For the SI system, there are seven fundamental units. (An exception in the SI system is the fundamental electrical unit, the ampere, which uses another fundamental unit, the second, in its definition.)

Generator A device that converts mechanical work to electricity.

Heat sink A metal surface designed to conduct and radiate heat.

GLOSSARY

Henry (H) The unit of inductance; one henry is the amount of inductance that induces one volt when the current changes at a uniform rate of one ampere per second.

Hertz (Hz) The measurement unit for frequency; one hertz is one cycle per second.

Horsepower (hp) A commonly used unit for measuring the power of motors. One hp is equal to 746 W.

Hysteresis The effect in a magnetic material that occurs when the flux density lags behind the application of the magnetic field intensity.

Imaginary number A number plotted along the positive or negative y axis. Imaginary numbers in electronics are shown with the letter j appended to them.

Impedance Opposition to ac; it is a complex quantity with both resistive and reactive components. The magnitude of impedance is abbreviated Z.

Independent variable A variable that is controlled or changed to observe the effect on another variable.

Inductance The property of a conductor to oppose a change in current. The effect is greatly reinforced in coils.

Induction motor An ac motor that achieves excitation to the rotor by transformer action.

Inductive reactance The opposition of an inductor to ac; the unit is the ohm.

Inductor A fundamental passive component that consists of a coil of wire with specific inductive properties.

Instantaneous current The current in a circuit measured at an instant in time.

Insulator A material that prevents the free movement of charge; a nonconductor.

Integrated circuit A combination of circuit elements connected together as a single entity on a small piece of semiconductive material.

Ion A charged particle that is formed when one or more electrons have been added or removed from a neutral atom or molecule.

Ionization potential The energy to remove an electron from an atom.

Joule The unit for energy or work; it is equal to one newton-meter.

Kilowatt-hour (kWh) An amount of energy equivalent to a power level of 1 kilowatt maintained for one hour. It is equal to 3.6 million joules (J).

Kinetic energy The energy due to motion of matter.

Kirchhoff's current law A circuit law stating that the total current into a node is equal to the total current out of the node.

Kirchhoff's voltage law A circuit law stating that the sum of all the voltage drops around a single closed path in a circuit is equal to the total source voltage in that closed path. An equivalent statement for Kirchhoff's voltage law is *The algebraic sum of the voltages around a single closed path in a circuit equals zero.*

Ladder diagram A drawing that shows the logic associated with relay control circuits. Two vertical legs are connected to a voltage source with relay coils or other loads connected to the "rungs" of the ladder.

Lag A sine wave that is shifted along the x-axis later than a reference wave.

Lead A sine wave that is shifted along the x-axis earlier than a reference wave.

Light-emitting diode (LED) A diode constructed from GaAs that emits light when current passes through it.

Load A component that uses the power provided by the rest of the circuit.

Magnetic field intensity The magnetomotive force per unit length.

Magnetic domains A group of millions of atoms that have naturally aligned their magnetic fields within certain types of magnetic materials.

Magnetization curve A B-H curve for a particular material showing the flux density as a function of the magnetic field intensity.

Magnetomotive force The cause of magnetic flux. It is the product of the number of turns of a conductor and the current in that conductor.

Manual ranging A method used on a multiple scale meter in which the user must select an appropriate range for the measurement.

Metallic bonding A chemical bonding where a "sea" of negatively charged electrons hold the positive ions of a metal together in fixed positions.

Mixture A nonchemical combination of two or more substances that are combined in any proportion.

Monomial An expression that contains only a single term.

Neutron An uncharged subatomic particle found in the nucleus of an atom.

Newton-meter The metric unit for work; it is equal to one joule.

Node A connection point in a circuit that has a common path for two or more components.

N-type impurity A material added to a silicon crystal that has five electrons in its outer shell and provides an "extra" electron in the crystal.

Nucleus The central part of an atom that holds positively charged protons and uncharged neutrons.

GLOSSARY

Ohm The unit of resistance. One ohm is one volt per ampere.

Ohm's law A fundamental law of electricity that states that the current in a circuit is proportional to the applied voltage and inversely proportional to the resistance.

Ordinate The y-axis (vertical axis) in the Cartesian coordinate system.

Origin The intersection of the x-axis and y-axis.

Oscillator An electronic circuit that produces a repetitive waveform.

Oscilloscope A versatile instrument that shows a graph of the voltage in a circuit as a function of time.

Oxidation-reduction A chemical reaction in which one or more electrons are transferred from one reactant to another.

Parallel circuit A circuit that has two or more current paths connected to a common voltage source.

Peak-to-peak value The magnitude of a sine wave measured from the maximum negative level to the maximum positive level.

Peak value The magnitude of a sinusoidal wave measured from the center to its maximum positive level.

Period The time required for 1 cycle (or one repetition) of a wave.

Permeability A parameter that defines the ease with which a magnetic field can be established in a given material.

Phase An angular measurement that specifies the position of a wave relative to a reference.

Phase shift The angular difference between two waves of the same frequency.

Phasor A rotating vector.

Polar notation A method for describing a complex number by giving the magnitude first, followed by the angle with respect to the positive x-axis.

Pole (1) For a switch, the separate line the switch can control. (2) For a magnet, the surface on which magnetic field lines enter or leave.

Polynomial An expression with more than two terms.

Positive feedback The process of returning a small portion of the output back to the input in a manner to reinforce the input signal. Reinforcement only occurs if the signal that is fed back is in phase with the input signal.

Potential energy The capacity to do work because of the position or configuration of a substance.

Potentiometer A three terminal variable resistor with two fixed terminals; a third terminal is connected to a wiper between the fixed terminals.

P-type impurity A material added to a silicon crystal that has three electrons in its outer shell and provides a missing bonding position or "hole" in the crystal.

Power The rate that energy is transferred or converted to heat or other form. Power is measured in watts (W).

Power supply Generally refers to a device that changes the alternating utility voltage into a constant (dc) voltage. AC power supplies convert the utility voltage to an ac output.

Precision A measure of the repeatability (or consistency) of a series of data points.

Pretrigger capture The acquisition of data before a trigger event.

Primary winding The input winding on a transformer.

Prime mover The energy source, such as a diesel engine, that spins the rotor of a generator.

Programmable logic controller A specialized computer that is designed for industrial control applications.

Prony brake A device designed to measure motor torque.

Proton A subatomic particle with a positive charge that is part of the nucleus.

Pulse A waveform that consists of a very rapid transition (leading edge) from one voltage or current level (baseline) to an amplitude level, and then, after an interval of time, a very rapid transition (trailing edge) back to the original baseline level.

Pulse width The time between the point on the rising edge where the value is 50% of amplitude to the point on the falling edge where the value is 50% of amplitude.

Radian An angular measurement based on a circle. One radian is the angle formed when the arc of a circle is equal to its radius; it is equal to 57.3°.

Rectangular notation A method for describing a complex number by giving the distance along the x-axis as the real part followed by the distance along the y-axis as the imaginary part.

Rectifier A circuit that converts ac into pulsating dc.

Rectifying The process of converting ac to dc.

Relative permeability For a given material, the ratio of absolute permeability (μ) to the permeability of a vacuum (μ_0).

Relative permittivity A pure number that compares the effectiveness of a dielectric to that of air.

Relay An electrically controlled switch.

Reluctance The opposition to the establishment of a magnetic field in a material.

Resistance The opposition to current.

Resistivity A constant for each type of material that is related to its ability to conduct electricity. Metric units are the ohm-millimeter squared/meter ($\Omega\text{-mm}^2/\text{m}$) or the

GLOSSARY

ohm-meter (Ω-m). The English-system unit for wires is the ohm-circular mil per foot (Ω-CM/ft).

Resistor A component that is designed to have a certain amount of resistance between its leads or terminals.

Resolution A measure of the ability to discern differences in a measurement. For meters, it is often given as the smallest difference in voltage, current, or resistance that the meter can display.

Resonant circuit A series or parallel *LC* circuit that allows a narrow band of frequencies to be selected from all others.

Resonant frequency The frequency in an *LC* circuit where the magnitude of X_L and X_C are equal.

Rest energy The equivalent energy of matter because it has mass.

Retentivity The ability of a material, once magnetized, to maintain a magnetized state without the presence of a magnetizing force.

Rheostat A two-terminal variable resistor, with one fixed terminal and one terminal connected to a variable wiper.

Rise time The time required for a pulse to go from 10% to 90% of its amplitude.

RMS current A measure of ac; it is the amount of current that delivers the same power to a load as dc of the same magnitude. For a sinusoidal wave it is 0.707 times the peak current.

RMS voltage A measure of ac voltage; it is the amount of voltage that delivers the same power to a load as a dc voltage of the same magnitude. For a sinusoidal wave it is 0.707 times the peak voltage.

Rotor The moving part of a motor or generator.

Round off The process of dropping one or more digits to the right of the last significant digit on a number.

Saturated With respect to a magnetic core, the level where the flux density tends to remain constant even as the magnetomotive force increases.

Scale The value of each division along the *x*-axis or *y*-axis.

Schematic A drawing that uses standard symbols to show the logical connection of the various components in a circuit.

Scientific notation A method of writing a large or small number as an ordinary number between 1 and 10 times a power of ten.

Secondary winding The output winding on a transformer.

Selectivity A measure of how well a resonant circuit passes certain frequencies and rejects others.

Semiconductor A material that has four electrons in the outer shell and whose conductive properties can be controlled by the addition of certain impurities.

Series aiding When two or more batteries are connected in a manner to increase the voltage.

Series circuit A circuit with only one current path.

Series opposing A method of connecting batteries to have less total voltage. (Normally this is not a useful connection.)

Shell A permitted energy level for electrons within the atom.

Significant digits The digits known to be correct in a measured number.

Sinusoidal The cyclic pattern of alternating current, which has the same shape as the trigonometric sine function in mathematics.

Sinusoidal wave An electrical waveform with the same shape as the trigonometric sine function in mathematics.

Slip The difference between the synchronous speed of the stator and the rotor speed in an induction motor.

Slip ring A solid circular ring connected to the rotating coil in a generator or motor that is part of the electrical path for the rotor.

Slope A measure of how steep the graph of *x* and *y* is. It is found by dividing a change in the *y*-variable by a corresponding change in the *x*-variable.

Slope-intercept form A method of writing a linear equation; $y = mx + b$ where *x* and *y* stand for variables and *m* stands for the slope.

Solenoid A device designed to convert an electrical signal into a mechanical motion.

Solenoid valve An electrically controlled valve for control of air, water, steam, oils, refrigerants, and other fluids.

Specific gravity A dimensionless number that compares the density of a substance to that of water.

Squirrel cage An aluminum frame within the rotor of an induction motor that forms the electrical conductors for the circulating rotor current.

Stator The stationary part of a motor or generator.

Substrate The supporting material on which an integrated circuit is fabricated.

Superconductor A material that, in extremely cold temperatures, is capable of carrying current with no resistance.

Susceptance The reciprocal of reactance, indicated with the letter *B*. Susceptance can be either capacitive or inductive.

Superposition method A way to determine currents and voltages in a linear circuit that has multiple sources by taking one source at a time and algebraically summing the results.

Supplementary unit One of two basic angle measurements that are part of the SI system for measurement.

Switch A control element that can open or close one or more lines of a circuit.

GLOSSARY

Symmetry A control on a function generator that enables the adjustment of the rise and fall time of sine and sawtooth waves and the duty cycle of pulses; may be called a duty cycle control.

Synchronous motor A motor in which the rotor moves at the same rate as the rotating magnetic field of the stator.

Tank circuit A parallel LC combination operating at the resonant frequency.

Taper Refers to how the angular position of a variable resistor is related to the resistance between two terminals. A linear taper means the resistance is proportional to the angular position.

Term A group of variables and/or constants that are not separated with plus or minus signs.

Terminal equivalency Two different circuits that respond identically from the viewpoint of the two output terminals.

Thevenin's theorem A theorem that states that any two-terminal, resistive circuit can be replaced with a voltage source (V_{TH}) and a series resistance (R_{TH}) when viewed from two output terminals.

Three-phase An ac system in which three wires carry voltages that are out of phase with the other by one-third of a cycle (120°); generally a fourth conductor is used for neutral or ground.

Throw The number of contacts (or connected positions) on a switch.

Time constant A fixed time interval that determines the time response of an RC circuit or an RL circuit.

Transducer A device that converts energy from one form to another.

Transformer A device formed by two or more windings that are magnetically coupled to each other and provide a transfer of power electromagnetically from one winding to the other.

Transistor A three-terminal, solid-state device fabricated from silicon or germanium that has the ability to amplify a signal or act as an electronic switch.

Trimmer capacitor A small adjustable capacitor for very fine adjustments of capacitance in a circuit.

Turns ratio (for electronic power transformers) The number of turns in the secondary winding divided by the number of turns in the primary winding.

Universal curves A graph of the charging and discharging curves for RC and RL circuits that are plotted as a percentage of the final value of current or voltage versus the number of time constants that have elapsed.

Valence band A continuous band of energies caused by a large number of atoms in a metallic crystal, which blurs the discrete energy levels of the valence electrons.

Valence electrons Electrons in the outermost-filled shell of an atom.

Variable A quantity that changes.

Volt The unit of voltage: One volt is one joule per coulomb.

Voltage Work required (or energy expended) to move a unit of positive charge from a negative point to a more positive point.

Voltage-divider A circuit consisting of series resistors across which one or more output voltages are taken.

Volt-Ohm-Milliammeter (VOM) A portable analog multimeter that can measure voltage, current, or resistance. It uses a needle against a scale to show the reading.

Watt The unit of power. One watt is equal to one joule per second.

Watt's law A law that states the relationships of power to current, voltage and resistance.

Wheatstone bridge A circuit that is used for precise resistance measurements. It consists of four resistive arms, a dc voltage source, and a detector.

Work The product of a force and a distance; the force applied must be measured in the same direction as the distance.

Working standard An instrument or device that is used for routine calibration and certification of test equipment or a component.

Working voltage The maximum sustained dc voltage a capacitor can withstand.

INDEX

Abscissa, 44
Absolute value, 425
Accuracy, 34–45
Admittance, 444–449, 459
Admittance triangle, 445
Air gap, 235
Algebra, 38
Alternating current, 65–66, 292
Alternation, 65
Alternator, 223, 267–269
    frequency, 271
    magnetic structure, 270
    three-phase, 274
American Wire Gauge, 78
Ampere, 33
Ampere, Andre, 33
Ampere-hour, 69
Amplitude, 308, 313
Angular measurement, 293
Armature, 269
    reaction, 271
Atom, 12
    conduction band, 15
    electrons, 14–15, 226, 228
    nucleus 13, 226
    protons, 13
    valence band, 15
Atomic number, 13
Attenuator, 313
Audio frequencies, 393
Average value, 297

Back emf, 276
Bandwidth, 431–432, 466
Bardeen, John, 4
Baseline, 308
Battery, 67
    ampere-hour ratings, 68
    lead -acid, 68
B-H curve, 229–234
Binomial, 39
Biot-Savart's law, 264
Bohr, Neils, 13
Branch, 148
Brattain, Walter, 4
Brushes, 269

Calculator, 27, 29
Capacitance, 336
    factors affecting, 339
    winding, 379
Capacitive reactance, 356–358, 411, 418
Capacitor, 336–363
    ac coupling, 361
    applications, 360–363
    breakdown, 340
    dc blocking, 361
    ceramic, 342
    charging, 349–351
    dielectric, 336–340
    discharging, 350–351
    electrolytic, 343–344
    labeling, 345
    mica, 341
    parallel, 345, 348
    phase relationship, 359
    plastic film, 342
    power line decoupling, 362
    series, 346–349
    trimmer capacitors, 345
    variable, 344
    voltage divider, 348
    working voltage, 340
Careers, 5
Cartesian coordinate system, 44, 405
Cell, 68
Circular mil, 79
Clamp-on ammeter, 83
Closed shell, 14
Coefficient, 38
Color code, 75–78
Commutator, 272
Complex numbers, 302, 406
Compound, 15
Conductance, 102, 150, 444
Conduction band, 15
Conduction electron, 15
Conductor, 15, 58
Constant, 38
Conventional current, 62
Coulomb, 11
Coulomb, Charles A., 10
Coulomb's law, 11, 24
Covalent bond, 16
Current, 33, 96–99
    divider, 159
    instantaneous, 65
    rms, 66

dc offset, 313
DeForest, Lee, 4
Degree, 293, 301
Delay line, 394
Delta connection, 275
Democritus, 12
Dependent variable, 45
De Prony, Gaspard, 277
Descartes, Rene, 44
Diamagnetic, 228
Dielectric, 336–340
Dielectric constant, 340
Digital multimeter, 81–85
    accuracy, 85
    autoranging, 81
    continuity testing, 85
    current measurement, 83
    manual ranging, 81
    resistance measurement, 84
    resolution, 85
Diode, 61

Direct current, 63–66
Distribution rule, 40
Duty cycle, 309

Edison, Thomas, 4
Effective value, 297
Efficiency, 71–72
Electrodes, 58
Electrolyte, 59, 343
Electromagnet, 266
Electromagnetic spectrum, 292
Electron flow, 62
Electronic power supply, 71
Energy, 8, 108
    chemical, 67–68
    kinetic, 9
    potential, 9
    rest, 10
Engineering notation, 27
Equation, 39–44
Equivalent circuits, 174–185
Error, 35
Exponential curves, 353
Expression, 38

Factoring, 40
Fall time, 309
Falling edge, 308
Farad, 337
Faraday, Michael, 260, 337
Faraday's law, 260–263
Ferrites, 228
Ferromagnetic, 228
Feynman, Richard, 258
Field coil, 269–270
Filter, 360, 394
    band-pass, 431, 465
    band-stop, 431, 465
    parallel resonant, 465
    series resonant, 431
Fire hazard, 6
Fixed point, 24
Flashlight, 64
Fleming, Thomas, 4, 259
Floating-point, 27
Fluorescent lamp, 59
Flux, 223
Flux density, 223–224, 235
Franklin, Benjamin, 62
Frequency, 65, 298
Function generator, 312–315

Generator, 70, 223, 267
Grad, 293
Graphing, 44–47
Ground, 33

Heat sink, 117
Henry, 377
Henry, Joseph, 377

503

INDEX

Hertz, 65
Hertz, Heinrich, 65
Horsepower, 113
Huygens, Christian, 292
Hysteresis, 231–232

Ideal current source, 196
Ideal voltage source, 196
Imaginary numbers, 405–406
Impedance, 411, 449, 459
Impedance triangle, 412
Independent variable, 45
Inductance, 376
    factors affecting, 377
Inductive reactance, 389–390, 411
Inductor, 376–394
    ac response, 389
    parallel, 381
    series, 381
    symbols, 379
Instantaneous current, 65
Instantaneous voltage equation, 299
Insulator, 16, 59
Integrated circuit, 4
Internal resistance, 188
Interpoles, 272
Inverse square law, 24
Ion, 15, 58
Ionization potential, 14
IV characteristic, 102

j operator, 404–405
Joule, 10, 108
Joule, James Prescott, 8

Kilby, Jack, 4
Kilowatt-hour, 109
Kirchhoff, Gustav, 139
Kirchhoff's current law, 156–159, 185–187
Kirchhoff's voltage law, 139–142, 185–187

Ladder diagram, 244
Lagging wave, 301–307
Laser, 96
Lawrence Livermore National Lab, 96
Leading wave, 301–307
Left-hand rule, 264–265
Lenz, Heinrich, 377
Lenz's law, 376
Light-emitting diode, 61, 64
Load, 63
Loading, 35, 174, 192–196

Magnetic domains, 228
Magnetic field, 222–227, 263–266
Magnetic field intensity, 227
Magnetic materials, 228
Magnetic properties, 228

Magnetization curve, 234
Magnetomotive force, 226
Maxwell, James Clerk, 292
Measurement units, 32
Metallic bonding, 15, 59
Metric prefixes, 28–31
Microwave frequencies, 393
Millikan, Robert, 11
Mixture, 15
Monomial, 39
Motors, 275–282
    ac, 279–282
    dc, 275–278
    compound, 279
    induction, 280–281
    series dc, 278
    shunt dc, 278–279
    synchronous, 279, 282
    universal, 280

Nanotechnology, 258
National Electrical Code, 6, 267
National Fire Protection Association, 6
National Institute for Standards and Technology, 35
Neutron, 13
Newton, 7
Newton, Issac, 292
Newton-meter, 7
Node, 148
Noyce, Robert, 4
N-type impurity, 61

Occupational Safety and Health Administration, 6
Oersted, Hans Christian, 222
Ohm, 34
Ohm, Georg, 34, 96–97
Ohm's law, 34, 96–108, 136–138, 149–150, 152–156, 358, 391, 416, 451, 456, 459
Ohm's law for magnetic circuits, 233
Ordinate, 34
Origin, 34
Oscillator, 295, 312, 394, 463
Oscilloscope, 315–325
    analog oscilloscope, 315–322
    automated measurements, 324
    block diagram, 316–317
    controls, 318
    digital oscilloscope, 323–325
    graticule, 316
    measurements, 320
    probes, 319–320
    triggering, 324
Oxidation-reduction reaction, 68

Parallel circuit, 147–160
    equivalent parallel circuit, 177
Paramagnetic, 228

Peak-to-peak value, 296
Peak value, 296
Period, 65, 298
Periodic, 309
Permeability, 229, 377
Permittivity, 340
Phase angle, 412–415, 446–449
Phase shift, 301, 359–360, 392, 454, 458, 461
    measurement, 323
Phasor, 300–307, 406
Plasma, 58
Polar notation, 406
Polynomial, 39
Positive feedback, 394
Potentiometer, 73, 146–147
Power, 34, 71–72, 277
    measuring, 113, 277
Power supply filtering, 360
Precision, 35
Prime mover, 269
Product-over-sum rule, 41
Programmable logic controller, 243–246
Prony brake, 277
Proton, 13
P-type impurity, 61
Pulse, 308
Pulse width, 309
Pythagorean theorem, 409, 446

Q, 432, 466
Quark, 58

Radian, 293, 301
Radio frequencies, 393
RC circuit
    frequency effect, 418
    parallel, 451–455
    series, 415
    square-wave response, 352
    time constant, 352
Real numbers, 404
Rectangular notation, 406
Rectifier, 360
Rectifying, 272
Relative permeability, 230
Relative permittivity, 340
Relay, 242
    solid state, 243
Reluctance, 230
Resistance, 34, 73, 96–99
    ac, 119
    dc, 117–118
    parallel resistors, 149–152
    series resistors, 133–134
    winding, 378
    wire, 78
Resistivity, 79

INDEX

Resistor, 73
    carbon-composition, 73
    color-codes, 75–78
    fixed, 73
    metal-film, 74
    network, 73
    potentiometer, 73–74, 146–147
    power dissipation, 111, 114–115
    precision, 74
    rheostat, 73
    specifications, 75
    surface-mount, 73–74
    thermistor, 75
    variable, 73–75
    wirewound, 74
Resonant circuits, 393, 428–432, 462–466
Resonant frequency, 429, 462
Retentivity, 232
rf choke, 393
Rheostat, 73
Right-hand rule, 259–260
Right triangle, 407–410
Rise-time, 308
Rising-edge, 308
RL circuit
    frequency effect, 422–423, 458
    parallel, 456–458
    series, 421
    square-wave response, 384
    time constant, 385
RLC circuit, 424–427
rms, 66, 82, 296
Rotor, 269
Rounding off, 37

Safety, 5–7
    appliance, 108
    battery, 69
    capacitor, 343, 350
    class-C extinguisher, 416
    electrolytic capacitors, 343
    emergency cut-off switch, 205
    eye protection, 6
    fire, 416
    inductors, 383
    jewelry, 296
    lab, 205
    liquids, 157
    magnets, 229
    multimeter, 85
    National Electrical Code, 267
    resistor, 74
    transformer, 239
    wire size, 114
Salient poles, 282
Saturation, 229, 233

Sawtooth waveform, 310–311
Scale, 46
Schematic, 64
Schumann, Robert, 316
Scientific notation, 24–27
Selectivity, 431–432, 466
Semiconductor, 16, 59
Series aiding, 64
Series circuit, 132–147
Series opposing, 64
Series-parallel combination circuits 174–207
    equivalent parallel circuit, 177–180
    equivalent series circuit, 174–177
Shackelford, Joel, 316
Shells, 14
Shockley, William, 4
SI system, 32–33,
    derived units, 32, 224
    fundamental units, 32
    supplementary units, 32
Siemens, 102
Signal sources, 312
Significant digits, 35
Sine function, 260, 292
Sinusoidal wave, 65, 295, 313
    equation, 299
    parameters, 296–299
Slip, 281
Slip ring, 269
Slope, 39, 102
Slope-intercept form, 39, 46
Solenoid, 240
Solenoid valve, 240–241
Solid-state relay, 243
Speaker, 241
Specific gravity, 69
Square wave, 310, 313
Squirrel-cage, 281
Standards, 34
Static electricity, 10
Stator, 269
Subscript notation, 135
Substrate, 4
Superconductor, 73
Superposition method, 196–200
Surveying, 132
Susceptance, 444, 454, 458
Sweep, 314
Switch, 64, 241
    closed, 64
    open, 64
    pole, 241
    throw, 241
Symmetry control, 314

Tank circuit, 463
Taper, 75
Term, 38
Terminal equivalency, 191
Tesla, 224
Tesla, Nikola, 224
Thermistor, 75
Thevenin, Leon, 191
Thevenin's theorem, 188–191
Thomson, J.J., 11
Three-phase, 274, 280
Time constant, 352–353, 385–386
Torque, 8, 277
Transducer, 202
Transformer, 237
    high-frequency, 239–240
    isolation, 240
    power, 239
    primary winding, 237
    safety, 237
    secondary winding, 237
    step-down, 238
    step-up, 238
    turns ratio, 238
Transistor, 4, 61
Triangular waveform, 310–311, 313
Troubleshooting, 205–207

Underwriter's Laboratories, 6
Universal curves, 353

Vacuum tube, 4
Valence band, 15
Valence electrons, 14
Van de Graaff generator, 11
Variable, 38
Vector, 302, 406
Volt, 33
Volta, Alessandro, 33
Voltage, 33, 96–99
    divider 142–147, 192–196, 201
    drop, 135
    rms, 66
VOM, 85–87

Watt, 34
Watt, James, 34, 111
Watt's law 112
Weber, 223
Weber, Wilhelm, Eduard, 223
Wheatstone bridge, 200–204
Wire resistance, 78–80, 230
Work, 7, 108
Working standard, 34
wye connection, 275

TK
7816
.B788

2005